Impact of Digitalization on Communication Dynamics

Albérico Travassos Rosário
GOVCOPP, IADE, Universidade Europeia, Portugal

Anna Carolina Boechat
Universidade Europeia, Portugal

IGI Global
Scientific Publishing
Publishing Tomorrow's Research Today

Vice President of Editorial	Melissa Wagner
Managing Editor of Acquisitions	Mikaela Felty
Managing Editor of Book Development	Jocelynn Hessler
Production Manager	Mike Brehm
Cover Design	Phillip Shickler

Published in the United States of America by
IGI Global Scientific Publishing
701 East Chocolate Avenue
Hershey, PA, 17033, USA
Tel: 717-533-8845
Fax: 717-533-8661
E-mail: cust@igi-global.com
Website: https://www.igi-global.com

Library of Congress Cataloging-in-Publication Data

CIP Data Pending
ISBN:979-8-3693-3579-6
eISBN:979-8-3693-3580-2

British Cataloguing in Publication Data
A Cataloguing in Publication record for this book is available from the British Library.

Table of Contents

Detailed Table of Contents

The combination of Artificial Intelligence (AI) with Augmented Reality/Virtual Reality (AR/VR) is causing a dramatic shift in communication technology. This talk examines the fascinating intersections between these domains, emphasising how AI fosters communication and how advancements in AR/VR technology broaden the scope of connection. Section 1: AI Redefining Communication analysis at how artificial intelligence (AI) is changing text, audio, and video—the traditional communication channels—to provide personalised content delivery, sentiment analysis, and real-time language translation. Section 2: AR/VR Transforming Communication: Analysis at how AR and VR technology can submerge individuals into communal virtual environments, resulting in more meaningful interactions and enhanced communication experiences. then about the difficulties and moral implications of these new technologies, putting a focus on inclusive involvement and responsible development.

Extended Reality (XR) is changing how we interact with digital information and communicate with each other by transforming how people experience and interact with their surroundings. It can be used to create immersive experiences like virtual tours or to promote companies, products, and brands. Although the industry and scientific community have improved recently, XR must overcome some issues to be fully deployed and understood in multiple areas. For example, it is essential to understand user experience guidelines to create realistic XR applications. Also, those applications require a lot of processing power, which can result in performance issues. Furthermore, integrating artificial intelligence opens a world of possibilities, presenting new research trends to explore. This chapter examines the definition of XR to demonstrate what it is possible to do with it, the emerging challenges, and how XR technologies can be applied to the communication field.

This chapter delves into the reciprocal relationship between communication technology and innovation. It traces the historical evolution of communication technology, from the telegraph to the digital age, analyzing how each technological breakthrough transformed communication practices, societal structures, and human interactions. It assesses the multifaceted impact of communication technology across sectors like business, education, healthcare, and governance, illustrating its role in driving economic growth, facilitating collaboration, and empowering individuals. Highlighting the catalytic effect of technological advancements on innovation, the chapter explores the impact of communication infrastructure investments on economic development, job creation, and growth. Furthermore, it explores emerging trends such as 5G, AI, and IoT, projecting their transformative potential across various industries.

This abstract looks at how enhanced connection and immersive interactivity may be used by 5G technology to change virtual event experiences. The essay examines how 5G networks may provide more network capacity, very low latency, and quicker data rates. These developments make it possible to utilize virtual reality apps, communicate in real time, and transmit content smoothly. Moreover, it underscores the impact of 5G on several sectors such as business, education, and entertainment, showcasing its ability to create virtual worlds that faithfully replicate the live event experience.

Major disasters and public health crises require effective communication strategies to disseminate vital information and save lives. During the recent COVID-19 pandemic, use of digital communication to communicate information became critical as governments enacted measures to protect public health, including COVID-19 curtailment strategies intended to save lives. To increase response effectiveness to critically important messaging, it is imperative to understand how to communicate with different communities to best achieve cooperation and compliance. Studies have identified culture as being a strong determinant of how people reacted to COVID-19 messaging. The following chapter describes a survey of 184 Hispanic and Latino Americans and non-Hispanic and Latino Americans to determine their attitudes towards COVID-19 mitigation protocol compliance and effective messaging strategies. The study found that significant differences between Hispanic and Latino and non- Hispanic and Latino culture and preferences for specific messaging strategies during the pandemic.

Crisis communication in today's digital age, shaped by social media and Volunteered Geographic Information (VGI), has revolutionized response strategies. Social media-based VGI can enable rapid dissemination of information and foster two-way communication, empowering communities to report incidents, seek help, and share experiences in real-time. This transparency builds trust between authorities and the public, dispels rumors, and ensures adherence to safety measures. VGI can amplify the voices of those affected, humanizing crises and mobilizing support for relief efforts. However, using VGI requires careful verification and ethical considerations like privacy protection and combating misinformation. Techniques such as cross-referencing with credible sources and utilizing geolocation data enhance reliability. Collaborative platforms can streamline information exchange, facilitating efficient crisis response and volunteer coordination. Actionable updates with visual aids can improve public engagement.

In recent years, several social movements have emerged on the international scene, some of which have a global reach, in part due to the digital environment, and in particular to social media. Digital activism describes different forms of activism that use digital technology and it has gained importance, as it enables users, activists and institutions to share content that has the potential to reach a very significant number of people. It is against this backdrop that this research was developed. Our main objective is to contribute to the understanding of how digital activism works in Portugal, in particular how activist organizations use digital media to communicate with their audiences. We try to understand whether these communication channels really represent an added value to publicize the mission, causes, campaigns and work carried out by the institutions, and if they can effectively mobilize the members of the public, so that they participate in activist actions proposed and organized by the institutions.

Over the past few decades, the manufacturing industry has become a hotbed of research on the rapidly developing trend of smart technology. Numerous cutting-edge technologies, including the internet of things, industrial internet of things (IOT), cyberphysical production systems, artificial intelligence, block chains, and systematic data analysis, were included in smart technology (ST). In order to close the gap between humans and machines, we give an overview of various smart technologies that can be used to regulate the sophisticated characteristics of computer and sensor technologies. The entire literature review on ST to improve machinery equipment production and maintenance, as well as prospective threats to ST, is presented in this research article. Additionally, the survey report comparing the adoption rate currently to that of the previous five years is included in this article in order to reduce the amount of human labor required in the industrial industry through the use of various sensors, cost-effective methodologies, and real-time outcomes.

This chapter delves into China's strategic Digital Silk Road initiative and its impact on Africa's digital transformation. It looks at the impact that Chinese IT giants have had on the African digital environment and how the continent may learn from China's experience. This chapter acts as a guide through the opportunities and problems presented by topics such as data security and digital sovereignty. In-depth case studies of China's digital projects in Africa show the scope and potential of these efforts. Its goal is to help African stakeholders reap the most benefits from and limit any negative consequences of this relationship, so fostering long-term, equitable progress in the digital sphere for all.

The "eatertainment" phenomenon has become widespread as one of the latest trends for many restaurant brands. The main goal of the present work is to examine the relationship between the visual presentation of food and its influence on the multidimensional notion of brand experience. The research was exploratory and three focus group discussions were held. Findings revealed that food presentation enhances various dimensions of brand experience, although its impact varies: it significantly influences sensory brand experiences, mainly the visual and taste perceptions; it is less prominent in affective brand experiences, where other factors play a more significant role; it also shapes behavioral brand experiences, encouraging photography, online sharing, and culinary exploration; and its impact on intellectual brand experiences varies, with some seeing it inspiring and others merely aesthetic. However, plating is just one aspect of a holistic brand experience, necessitating a multifaceted strategy for contemporary diners.

Companies and businesses have responsibilities towards their customers and the society at large. Such responsibilities should be revealed in their business actions and marketing communications done by them. The objective of the study is to understand the various aspects of socially responsible marketing communications. The methodology adopted is a conceptual analysis of the various aspects of marketing communications done by companies. Companies should understand their roles in socially responsible marketing communications viz. advertising, sales promotion, and personal selling. They should realize their responsibilities towards promotion of social responsibility and should adopt initiatives towards socially responsible marketing communications. Such initiatives will instill a sense of trust and belief in the minds of customers and the society at large. This will in turn allow companies to fulfill their social responsibilities and to achieve business excellence in the long run.

Companies and businesses have roles in marketing communications and target marketing done by them. They should ensure that such communications are socially responsible. The objective of the study is to understand the various aspects of socially responsible marketing communications and target marketing, and ethical and legal aspects and concerns related to such issues. The methodology adopted is a conceptual analysis of the various aspects and issues. Enlightened companies adopt initiatives for socially responsible marketing communications and target marketing. The discussions will allow academicians and practitioners to understand the importance of social responsibility in marketing communications and target marketing and the various issues related to such communications. Policy makers may investigate about the ethical and legal concerns related to the issues and may implement rules and regulations which will ensure the triple bottom line – people, profit, and planet.

This chapter analyses current technologies and the challenges involved in extracting and classifying articles and news headlines from historical journals, as well as converting images to text format. The work to develop a tool focused on digitising historical journals was carried out by a multidisciplinary team of experts in media studies, artificial intelligence, image processing, and cultural heritage preservation. The data used derives from two historic Portuguese journals, Diário de Notícias and Jornal de Notícias, which were created in the mid-19th century. This project is based on a mixture of heuristics, computer vision, pattern recognition, and other artificial intelligence and machine learning techniques. The main challenges included the variability in the design of historical journals, preserving the quality of images over time, and continuously improving image processing and OCR techniques to adapt to different styles and periods of newspapers.

Preface

The rapid evolution of digital technologies has fundamentally transformed the way we communicate, both at an individual level and across global networks. In the past few decades, digitalization has emerged as a powerful force reshaping communication dynamics, enabling new forms of interaction, collaboration, and information exchange. This book, *Impact of Digitalization on Communication Dynamics*, offers an in-depth exploration of this transformation, presenting a collection of diverse perspectives and research findings that examine the multifaceted impact of digitalization on communication in various contexts.

At the heart of this transformation is the integration of Artificial Intelligence (AI) with communication technologies, which has redefined traditional modes of interaction. AI has penetrated text, audio, and video channels, bringing innovations like personalized content delivery, sentiment analysis, and real-time language translation. With AI's ability to process vast amounts of data instantaneously, communication has become more adaptive, responsive, and tailored to individual preferences. Augmented Reality (AR) and Virtual Reality (VR) technologies complement these advances, creating immersive environments that offer novel ways to connect, collaborate, and share experiences. This convergence of AI, AR, and VR heralds a new era of digital communication, yet it also raises significant ethical questions, particularly around data privacy, inclusivity, and the potential for technology to reinforce societal biases.

Extended Reality (XR), encompassing AR, VR, and Mixed Reality (MR), is further pushing the boundaries of how we experience digital communication. XR technologies have introduced unprecedented levels of interactivity, allowing users to engage with digital content in ways previously unimaginable. However, the development of XR faces significant challenges, such as the need for substantial processing power, the creation of user-friendly experiences, and the integration of AI capabilities. These technologies hold immense potential for transforming industries ranging from education and healthcare to entertainment and marketing, but realizing this potential requires overcoming technical and design obstacles that continue to evolve.

The historical progression of communication technology is another key theme of this book. From the invention of the telegraph to the rise of the internet and the advent of 5G networks, each technological breakthrough has marked a paradigm shift in the way we communicate. The introduction of 5G, with its enhanced network capacity and ultra-low latency, is poised to revolutionize real-time communication and facilitate the seamless integration of Internet of Things (IoT) devices. This will enable new applications in smart cities, autonomous vehicles, and remote healthcare, illustrating the profound impact of digitalization on various sectors of the economy.

The impact of digitalization on education and professional development cannot be overstated. The rise of digital learning platforms and the integration of AI-driven tools in instructional design have redefined the educational experience. Distance education, once seen as a secondary alternative, has now become a primary mode of learning for millions worldwide. The use of interactive technologies and digital simulations has enhanced learner engagement and provided new opportunities for personalized education. However, these developments also bring challenges, such as ensuring equitable access, maintaining academic integrity, and addressing the complexities of digital content delivery.

Public health crises, such as the COVID-19 pandemic, have highlighted the critical importance of effective digital communication. During the pandemic, digital platforms became essential for disseminating information, guiding public behavior, and coordinating response efforts. The variation in cultural responses to digital messaging underscored the need for tailored communication strategies that resonate with diverse audiences. This book delves into the complexities of digital communication during crises, examining how different communities perceive and respond to health-related messaging and the role of digital literacy in shaping public understanding.

Digital activism has emerged as a powerful tool for social change in the digital age. The rise of social media platforms has given voice to marginalized communities and facilitated global movements, amplifying messages of social justice, climate action, and human rights. The case of digital activism in Portugal, explored in this book, highlights the strategies used by activists to mobilize public support and engage with audiences in meaningful ways. While digital platforms offer unprecedented opportunities for advocacy, they also pose risks related to misinformation, echo chambers, and the potential for surveillance and censorship.

The commercial landscape has also been significantly altered by digitalization. Businesses now rely on digital communication strategies to enhance brand visibility, engage customers, and build trust. Trends such as "eatertainment" in the hospitality industry showcase how visual presentation and digital experiences are being leveraged to create memorable brand encounters. By integrating digital storytelling and interactive content, companies can foster deeper connections with consumers, driving brand loyalty and competitive differentiation.

On the global stage, initiatives like China's Digital Silk Road demonstrate the strategic use of digital infrastructure to foster international development. By investing in digital projects across Africa, China aims to extend its influence while simultaneously promoting digital inclusion. However, these efforts raise important questions about data sovereignty, cybersecurity, and the equitable distribution of technological benefits. This book examines the implications of such initiatives, offering insights into the opportunities and challenges they present for developing nations.

The democratization of information, facilitated by digital platforms, has transformed academic and educational resources. The shift from traditional textbooks to online resources has expanded access to knowledge, yet it also highlights issues of digital equity and the digital divide. Institutions and educators must navigate these challenges, leveraging digital tools to enhance learning while ensuring that no student is left behind. This book explores the dynamics of digital resource availability, emphasizing the need for inclusive approaches to education in a digital era.

In conclusion, *Impact of Digitalization on Communication Dynamics* offers a comprehensive examination of the ways digital technologies are reshaping the landscape of human communication. The diverse collection of research presented in this book reflects the complex, multifaceted nature of digital transformation. As we continue to embrace new digital tools and platforms, it is essential to approach these changes with a critical eye, recognizing both the opportunities they offer and the ethical dilemmas they pose. By understanding the evolving dynamics of digital communication, we can work towards a more connected, equitable, and informed future.

This book invites readers from all fields—academics, practitioners, policymakers, and the general public—to engage with these discussions, offering a rich tapestry of insights and analyses that illuminate the profound impact of digitalization on our world. Together, we can navigate the challenges and harness the potential of digitalization to enhance communication, foster innovation, and build stronger, more resilient communities.

Albérico Travassos Rosário
GOVCOPP, IADE, Universidade Europeia, Portugal

Anna Carolina Boechat
Universidade Europeia, Portugal

Chapter 1
AI in Communication Technology and Innovation of AR/VR in Communication

R. Ramya
https://orcid.org/0000-0002-8071-9343
Sathyabama Institute of Science and Technology, India

S. Ramamoorthy
SRM Institute of Science and Technology, India

ABSTRACT

The combination of Artificial Intelligence (AI) with Augmented Reality/Virtual Reality (AR/VR) is causing a dramatic shift in communication technology. This talk examines the fascinating intersections between these domains, emphasising how AI fosters communication and how advancements in AR/VR technology broaden the scope of connection. Section 1: AI Redefining Communication analysis at how artificial intelligence (AI) is changing text, audio, and video—the traditional communication channels—to provide personalised content delivery, sentiment analysis, and real-time language translation. Section 2: AR/VR Transforming Communication: Analysis at how AR and VR technology can submerge individuals into communal virtual environments, resulting in more meaningful interactions and enhanced communication experiences. then about the difficulties and moral implications of these new technologies, putting a focus on inclusive involvement and responsible development.

DOI: 10.4018/979-8-3693-3579-6.ch001

1. INTRODUCTION

AI's Redefining Touch on Communication: From Channels to Solutions

AI is transforming communication across text, audio, and video, impacting content delivery, sentiment analysis, and even language translation. (Xiaoxia, 2022) Here's a breakdown of its influence and potential:

1. Personalized Content Delivery:

Text: AI analyzes user data (e.g., browsing history, search queries) to suggest relevant articles, emails, or chat messages. Think of personalized news feeds or targeted marketing campaigns.

Audio: Music streaming services use AI to recommend songs based on listening habits, while podcasts can dynamically insert ads tailored to individual preferences.

Video: AI-powered platforms curate video recommendations on YouTube or suggest relevant shows on streaming services based on watch history.

2. Sentiment Analysis:

Text: Social media platforms and customer service tools use AI to gauge user sentiment from written text, helping businesses understand customer satisfaction and adjust their approach accordingly.

Audio: AI analyzes speech patterns and tone to detect emotions in calls or voice messages, potentially improving customer service interactions or even flagging potential issues.

Video: Facial recognition and expression analysis can reveal emotional reactions to video content, offering valuable insights for marketing or understanding audience engagement.

3. Real-time Language Translation:

Text: Translation services like Google Translate use AI to translate text in real-time, breaking down language barriers for communication and information access.

Audio: AI-powered devices like simultaneous interpretation headsets translate spoken conversations in real-time, facilitating multilingual meetings and conferences.

Video: Live captioning and subtitles generated by AI can make video content accessible to a wider audience, including deaf and hard-of-hearing individuals.

Beyond Channels: AI Solutions for Enhanced Communication

User Experience (UX) Enhancement: AI chatbots can answer user questions, solve problems, and guide them through complex tasks, improving website and app navigation.
Collaboration Tools: AI-powered platforms like Slack use sentiment analysis and topic modeling to summarize conversations, suggest relevant information, and streamline teamwork.
Accessibility Solutions: AI-powered text-to-speech and speech-to-text tools enable communication for individuals with disabilities, promoting inclusivity and participation.
However, AI-powered communication also raises concerns:
Bias and Fairness: AI algorithms trained on biased data can perpetuate existing inequalities in communication, requiring careful development and monitoring.
Privacy and Security: Processing personal data for personalized content or sentiment analysis raises privacy concerns, necessitating strong data protection measures.
Job Displacement: Automation through AI-powered communication tools could potentially displace some jobs, requiring workforce adaptation and reskilling initiatives.

Overall, AI is revolutionizing communication, offering exciting possibilities for personalized experiences, seamless collaboration, and improved accessibility. However, responsible development and ethical considerations are crucial to ensure AI benefits everyone and fosters a more inclusive and equitable communication landscape.

The combination of Artificial Intelligence (AI) with Augmented Reality/Virtual Reality (AR/VR) is causing a dramatic shift in communication technology. This talk explains how artificial intelligence is changing traditional communication channels. Sentiment analysis, real-time translation, and personalized delivery Artificial Intelligence is driving a metamorphosis of traditional communication channels such as text, audio, and video (AI). Let's examine the three main ways that AI is affecting these channels:

1. Tailored Content Distribution:

Algorithmic curation: To comprehend preferences and anticipate content that appeals to each individual, AI systems examine user data, including search histories, browsing patterns, and social interactions. Imagine receiving news feeds that are customised to your individual interests, receiving music recommendations based on your listening habits, or receiving movie recommendations based on your taste in movies.

AI-powered dynamic content adaptation: content may be instantly personalised by changing language, style, and images according on user demographics, geography, or even emotional state. When a visitor visits an e-commerce website on a laptop or a mobile device, for instance, the website may highlight distinct product aspects.

Micro-targeting: AI enables extremely focused marketing and advertising initiatives. Imagine getting advertisements for goods you might actually consider purchasing or for instructional materials that are pertinent to your present learning objectives.

2. Evaluation of Sentiment:

Recognising emotions: AI assesses emotional sentiment by analysing text, audio, and video data. AI chatbots can be used in customer care to recognise disgruntled clients and correctly escalate their problems. Similar to this, social media companies can spot possible hostility or bullying and take appropriate action by using sentiment analysis.

Market research and brand monitoring: AI can examine how the general public feels about particular brands, goods, or advertising campaigns, giving companies insightful information. This makes it possible to manage crises and modify marketing plans in real time during surges in unfavourable sentiment.

assistance for mental health: AI-powered programmes are able to identify indicators of depression, anxiety, and other mental health issues by analysing speech patterns and written language. These programmes may then provide users with resources for assistance or put them in touch with experts.

3. Instantaneous Language Interpretation:

Linguistic barriers disappear: More and more advanced AI-powered translation solutions provide accurate, near-instantaneous translation between languages. This makes it possible for people, businesses, and tourists to communicate globally.

Custom translation styles: AI is capable of modifying translations to suit particular audiences and situations. For instance, translating social media posts may benefit from a more informal style, yet translating corporate documents may call for a more professional tone.

Multimodal translation: AI is going beyond text translation by combining picture analysis and speech recognition to translate written material contained in images as well as spoken discussions, hence reducing barriers to communication.

Possible Difficulties and Things to Think About

Even while AI has many interesting potential applications, there may be certain drawbacks to be aware of. Data privacy and bias: Because algorithms rely on data, it is possible for such data to contain biases that produce unfair or discriminating results. Sensible AI development must address data protection and bias mitigation. Digital divide and accessibility: Not everyone has equal access to digital literacy or technology. Ensuring that AI-powered communication tools are inclusive and available to everyone is imperative. Overuse of technology Even if AI can improve communication, genuine human connection is still crucial. Technology use and encouraging real human connection need to coexist in harmony.

In conclusion, a new era of individualised, perceptive, and barrier-free communication is being ushered in by AI. We can open the door to a future where communication breaks down barriers, promotes understanding, and gives people and communities around the world more power by wisely utilising its potential. examines the fascinating intersections between these domains, emphasising how AI fosters communication and how advancements in AR/VR technology broaden the scope of connection.

AI-Powered Communication Solutions: Improving Accessibility, Cooperation, and User Experiences

Artificial intelligence (AI) is radically changing the communication environment by providing cutting-edge solutions that improve accessibility for a larger range of audiences, facilitate seamless collaboration, and improve user experiences. Here are a few noteworthy developments:

Tailored Communication: Chatbots and virtual assistants: AI-driven chatbots may customise user journeys, respond to frequently asked questions, and offer immediate customer service. Envision chatbots providing personalised responses based on user preferences and suggesting pertinent content.

Content adaptation: AI adapts content delivery to the interests, demographics, and immediate needs of the user. Imagine news feeds displaying personalised articles or learning platforms that dynamically change the difficulty of articles.

Sentiment analysis: AI can identify user emotions from both text and audio, allowing for more sympathetic dialogue. While learning platforms can modify content according to student engagement, customer service bots are able to recognise frustration and escalate problems.

Improved Cooperation

Real-time language translation: Artificial intelligence (AI) eliminates language barriers to enable smooth teamwork between heterogeneous groups. Imagine remote work settings with real-time translation support, or international gatherings where everyone understands one other quickly.

Speech-to-text and note-taking: Artificial intelligence (AI) records meetings and presentations, facilitating the easy retrieval and searching of information. This enables varied teams with various learning preferences and capacities to actively engage.

Tools for project management: AI organises work, sets up meetings, and foresees obstacles to make project management easier. This facilitates effective teamwork and simplifies procedures.

Enhanced Usability

Text-to-speech and voice navigation: AI turns text into audio so that persons with visual impairments can access information. Imagine having AI assistants narrate documents and webpages so that voice commands may be used for navigating.

Recognition and production of sign language: AI bridges the divide and promotes inclusivity by facilitating communication between people who use sign language and those who do not.

Adaptive interfaces: AI modifies keyboard shortcuts, colour contrast, and text size to suit user preferences and accommodate a range of abilities and limitations.

Extra Advantages

Enhanced productivity: Artificial Intelligence streamlines monotonous jobs, providing more time for advanced dialogue and creativity.

Decreased expenses: AI-driven solutions have the ability to optimise communication procedures, which may result in a reduction of business operating expenses.

Audiences that are engaged: More engaged users create stronger connections and better user experiences. This is achieved through personalised communication and easily available options.

Obstacles and Things to Think About

Data privacy and bias: Preventing data misuse and algorithmic bias requires ethical AI development.

Digital divide: Access to technology and digital literacy are not shared by all. Inclusion must be considered when designing solutions.

Human connection: Real human engagement might be hampered by an over-reliance on technology. Maintaining personal communication while utilising AI-powered solutions requires a delicate balance.

AI-powered communication tools have the ability to completely change the way people communicate, work together, and access information. We can make communication more accessible, interesting, and inclusive for all by adopting new advances responsibly and resolving any issues that may arise.

AI Algorithms and Applications for Efficient Communication

Table 1 shows the details of AI Algorithms and Applications for Efficient Communication

Table 1. AI Algorithms and Applications for Efficient Communication

AI Algorithms	Advantages	Applications
Natural Language Processing (NLP)	Analyzes and understands human language	Chatbots for customer service and virtual assistants
Recommender Systems:	Recommends relevant content based on user preferences	Personalized news feeds and social media content
Machine Learning (ML):	Learns from data to improve predictions and decision-making	Spam filtering and email security
Deep Learning:	Learns complex patterns from large datasets	Automatic speech recognition and transcription
Computer Vision:	Analyzes and interprets visual information	Facial recognition for security and accessibility

- These are just a few examples, and many other algorithms and techniques are used for various applications.
- The specific algorithms used depend on the desired outcome and the type of communication data being processed.
- The effectiveness of AI-powered communication depends heavily on the quality and quantity of data used to train the algorithms.

- Ethical considerations are crucial when developing and deploying AI-powered communication solutions, such as avoiding bias and ensuring user privacy.

Overall, AI algorithms offer a powerful tool for improving communication efficiency and effectiveness across various channels and applications. However, responsible development and ethical considerations are essential to ensure these technologies benefit everyone and contribute to a positive communication landscape.

2. AR/VR: EXPANDING OUR COMMUNICATION HORIZONS

2.1. Introduction to AR/VR Innovations

Innovation in AR and VR is exploding with fascinating possibilities, obfuscating the distinction between the actual and virtual worlds. (Akbari, 2022) Here's a brief look:

1. Deeper Immersion: As haptic technology and spatial mapping advance, experiences become more interactive and realistic. Imagine being able to navigate naturally across virtual spaces and experiencing the texture of virtual objects.
2. AI-Powered Collaboration: AI is improving the tools for AR and VR collaboration. Imagine surgeons operating remotely using augmented reality overlays guided by AI-powered analysis, or architects working together to jointly manipulate 3D construction models in real-time.
3. The Revolution in Education: Picture yourself studying historical events through realistic simulations or performing virtual organ dissections in biology class. AI-powered learning systems combined with AR/VR are making education a more dynamic and interesting experience.
4. Advances in Healthcare: AR overlays enable surgeons do intricate surgeries, and VR treatment helps patients overcome phobias. In order to improve treatment outcomes, AI is further personalising VR experiences and analysing patient data.
5. Entertainment Evolution: Picture going to concerts where you're in the thick of things or engaging in virtual reality gaming where everything seems real. AI in conjunction with AR/VR and dynamic storytelling is expanding the possibilities for entertainment.

Obstacles Still Exist: It is imperative to address ethical issues pertaining to bias, manipulation, and data privacy. To minimise the negative effects of these technologies and close the digital divide, equitable access and responsible development are essential.

Innovation in AR and VR is developing quickly and has the potential to change many facets of our life. We may use it to create a more interconnected and rewarding future by carefully handling ethical and accessibility issues.

2.2. AR/VR in Communication

Imagine entering a collaborative virtual environment where communication is not restricted by spatial boundaries. That's exactly what AR and VR technology accomplishes, building immersive worlds that excitingly change communication.

Virtual Community Experiences:

Meaningful encounters: Despite geographic limitations, AR and VR enable in-person connections between people. Shared virtual spaces provide for deeper connection by allowing for the expression of emotions and nonverbal clues that are sometimes lost in text or video calls.

Improved communication experiences: Picture yourself working on a project in a virtual office while collaborating on a real-time 3D model manipulation, or learning a language by interacting with virtual characters in a simulated setting. Barriers are broken down via AR and VR, resulting in more dynamic and engaging communication.

Modern Applications:

Collaboration over long distances is made easy with AR/VR platforms that include AI. While doctors can remotely counsel patients using augmented reality overlays for medical checks, architects can collaborate to visualise building plans in a virtual space.

Education: Picture taking a virtual field trip to explore the Great Wall of China or dissecting a virtual frog in biology class. Immersion and interactive learning environments are produced when AR and VR are combined with AI-powered learning platforms.

Healthcare: AR overlays advise surgeons with intricate surgeries, and VR treatment helps patients conquer phobias. AI is able to customise VR experiences and analyse patient data to improve therapy results.

Entertainment: Interactive games blur the boundaries between reality and fantasy, while AR/VR concerts put you right in the thick of things. AI is able to customise gameplay and even produce dynamic narratives that change based on decisions made by the player.

Challenges and Ethical Repercussions:

Inclusionary participation: The lack of accessibility to AR/VR technologies may lead to a digital divide. To guarantee that everyone can participate, inclusive design and accessibility concerns are essential.

Responsible development: Data privacy, potential biases, and manipulation are ethical challenges raised by AI-powered AR/VR systems. To safeguard users and guarantee that these technologies are utilised for good, ethical standards and responsible development are crucial.

AI-powered AR and VR has the potential to drastically change learning, collaboration, and communication. To guarantee that new technologies benefit everyone and help create a more connected and fair world, ethical considerations and inclusive design are essential.

3. EXAMINING IN MORE DETAIL: HOW AR AND VR FOSTER VALUABLE EXCHANGES IN COMMUNITY VIRTUAL ENVIRONMENTS

By immersing people in shared virtual spaces, augmented reality (AR) and virtual reality (VR) have the potential to completely transform communication. These technologies present special opportunities for deeper connections and improved communication experiences. Let's examine this analysis in more detail:

Both Submersion and Joint Presence: Breaking down geographical barriers: Envision being virtually present in the same space while attending a meeting or presentation with coworkers or students who are physically situated anywhere in the world. This creates a feeling of unity and shared presence that is not location-specific.

Enhanced involvement: By appealing to the senses—sight, hearing, and in certain situations, touch—the immersive qualities of AR and VR encourage deeper engagement. Higher levels of participation, improved information retention, and more targeted conversations can result from this.

Emotional connection: Effective communication relies heavily on nonverbal clues like body language and facial emotions. These indications can be replicated by AR/VR avatars, promoting empathy and emotional connection in virtual environments.

Novel Modes of Communication

Interactive environments: Picture working together in real time on a 3D design project in a virtual setting, dynamically exchanging ideas and manipulating things. This opens up new channels of communication and goes beyond standard video conferencing.

Gamification and storytelling: By combining interactive storytelling components and game dynamics into AR/VR experiences, communication can become more memorable and captivating. To improve your negotiating skills, consider studying history through role-playing games or immersive simulations.

Multimodal communication: AR/VR can incorporate facial expressions, spatial motions, and even haptic feedback in addition to text and voice, making for a more complex and nuanced communication experience.

Improved Inclusivity and Accessibility

Overcoming physical restrictions: Virtual reality (VR) enables people with mobility limitations to explore new places and engage in activities they might not be able to physically due to its ability to generate simulated situations that are inaccessible in the actual world.

Sensory customisation: AR/VR experiences can be tailored to each user's requirements and preferences by changing components such as sounds, graphics, and interaction styles to suit a range of skill levels and impairments.

Language obstacles: Envision incorporating real-time translation into AR/VR settings to facilitate smooth communication and cooperation across varied linguistic communities. This promotes inclusivity and global understanding.

Obstacles and Things to Think About

Technical restrictions: The accessibility, affordability, and sensory fidelity of current AR/VR technology are constrained. It is imperative to address these issues in order to promote inclusion and greater adoption.

Security and privacy issues: To safeguard user information, data acquired in virtual environments requires strong security mechanisms and well-defined privacy regulations.

Possibility of social isolation: Although AR and VR technologies improve communication, an excessive dependence on virtual interactions may result in social isolation. Achieving equilibrium between virtual and physical connections is essential.

Although AR and VR technologies are still in their infancy, they have the potential to greatly improve communication by fostering more meaningful connections and improving communication experiences through the creation of immersive, shared virtual spaces. We can set the stage for a time when technology fosters global understanding and deeper connections by tackling problems responsibly and creating inclusive solutions.

4. CHALLENGES OF USING AI AND AR/VR IN COMMUNICATION REDEFINITION

Difficulties in Redefining Communication with AI, AR, and VR: Even if AI, AR, and VR have the potential to completely redefine communication, a number of issues must be resolved to enable ethical and inclusive development:

Moral Issues

Bias and discrimination: AI systems that have been trained on biassed data may reinforce preexisting communication disparities and produce discriminating results. It is imperative to employ mitigation techniques such as human monitoring and various data sets.

Data security and privacy: Processing personal information for AI-powered communication presents privacy issues. Transparency, user permission, and strong data protection protocols are crucial.

disinformation and manipulation: Deepfakes and disinformation can be disseminated via AI-generated material. To counter this, fact-checking resources, user education, and moral standards are required.

Inclusivity and Accessibility

Digital divide: There may be a new one as not everyone has access to AI and AR/VR technology. Initiatives for digital literacy, accessibility features, and cost-effective solutions are crucial.

Physical restrictions: People with specific physical restrictions may not be a good fit for AR/VR technology. Alternative interfaces and inclusive design concepts are essential.

Neurodiversity concerns: AR/VR experiences can be overpowering or unsettling for some people with neurodiverse problems. Customisation choices and sensitive design are essential.

Impact on Social and Psychological Domains

Addiction and isolation: Immersion in virtual environments for extended periods of time can cause addiction and a disconnection from social relationships in real life. Promoting real-world interaction and balanced usage are crucial.

Mental health issues: Anxiety, phobias, and other mental health problems can be brought on by AR and VR encounters. It is essential to provide user warnings, responsible content development, and mental health support.

Identity and embodiment: These concepts become more ambiguous as boundaries between the actual world and virtual environments become less distinct. To negotiate these difficult concerns, ethical discourse and responsible development are necessary.

Technical Difficulties

Technical restrictions: The user experience and immersion may be hampered by current AR/VR technology constraints such as latency, resolution, and field of view. It is necessary to conduct further research and development.

Interoperability and standards: A user's experience may be shattered if several AR/VR platforms are incompatible with one another. For broader adoption, interoperability and open standards are essential.

Cybersecurity risks: Cyberattacks may be possible with AR/VR experiences. To secure users and data, strong security measures and user education are crucial.

Although AI, AR, and VR have the potential to revolutionise communication, it is crucial to carefully weigh these issues. We can make sure that these technologies benefit everyone and contribute to a more connected and equitable future by giving ethical, inclusive, and responsible development first priority.

How to Overcome the Challenges of Using AI and AR/VR in Communication Redefinition

Redefining communication through AI and AR/VR presents a number of obstacles that must be overcome. These challenges include social, technical, ethical, and accessibility issues. Here are a few possible fixes:

Challenges of Ethics

Discrimination and bias: Variable data sets In order to train AI systems and prevent bias from being perpetuated, actively gather and use data from a variety of populations.

Human oversight: To detect and lessen potential biases in AI outputs, use human review procedures.

Accountability and transparency: Establish explicit accountability procedures together with open and transparent AI development and deployment processes.

Data Security and Privacy

robust data security protocols Put strong data security measures in place, such as anonymization and encryption.

User consent: Obtain the user's express, informed consent before collecting and using their data.

Minimise the amount of data you collect and utilise, only what is required to fulfil the intended goal.

False Information and Manipulation

Tools for fact-checking: Provide and make use of technologies to recognise and mark content generated by AI that may be harmful.

User education: Teach users how to assess information critically and spot possible deception.

Establish and uphold moral standards for the creation and application of artificial intelligence in communication.

Inclusivity and Accessibility

Budget-friendly solutions: Make research and development investments in low-cost AR/VR technology and content. Include features that make content more accessible to individuals with impairments, such as closed captioning, voice control, and alternative input methods. Initiatives for digital literacy: In order to close the digital divide and encourage access to technology, offer support and training initiatives.

Physical Restrictions

Principles of inclusive design: To ensure accessibility for everybody, take into account physical restrictions during the design and development process. Different interfaces: For people who might not be able to utilise conventional AR/VR devices, offer alternate interfaces.

Concerns about neurodiversity:Design that is sensitive: Create AR and VR experiences while being mindful of possible triggers and overpowering stimuli. Options for customisation: Let consumers choose the settings and level of rigour of AR/VR experiences.

Mental health support: Offer information and assistance to users who might be negatively impacted by AR or VR in terms of their mental health.

Impact on Social and Psychological Domains

Addiction and seclusion: Promote moderate use: Encourage people to use technology responsibly and to participate in offline activities. Provide time management tools: Integrate time management features into AR and VR platforms to assist users in staying out of overly immersed situations.

Concerns About Mental Health

User alerts: Issue explicit cautions regarding possible hazards and activate alerts for delicate material.

Conscientious generation of content: Provide recommendations for producing morally and responsibly themed AR and VR entertainment that supports mental wellness. Support for mental health: Include tools and avenues of assistance for people who are struggling with mental health issues.

Self-Awareness and Embodiment

Talks about ethics: Encourage candid conversations on how identity and embodiment are affected by AR and VR. Development that is responsible: Create and apply AR/VR technology in a way that honours user autonomy and agency.

Technical Difficulties

Technical restrictions: Research and development should be continued in order to enhance network, software, and hardware capabilities. Open collaboration: To exchange knowledge and hasten development, promote cooperation between users, developers, and academics.

Mutual Compatibility and Norms

Create open standards: To guarantee interoperability, open standards for AR/VR platforms and content should be established and promoted. Cross-platform compatibility: Motivate content creators to provide something that works on several platforms.

Cybersecurity Threats

Strong security protocols: To secure user data and systems, put strong security protocols and encryption solutions into place. Frequent security updates: To fix new vulnerabilities, update devices and software on a frequent basis. User education: Inform users about the best practices for cybersecurity so they can stay safe online.

Recall that conquering these obstacles is a continual process that calls for constant cooperation and adjustment. We can make sure that AI and AR/VR contribute to a more moral, welcoming, and constructive communication environment for all by aggressively tackling these concerns.

5. FUTURE RESEARCH DIRECTIONS: AI IN COMMUNICATION TECHNOLOGY AND AR/VR INNOVATION

Future Research Directions: AI in Communication Technology and AR/VR Innovation

The convergence of AI and communication technology, particularly in AR/VR, holds immense potential for reshaping communication paradigms. Here are some exciting future research directions:

1. Enhanced Natural Language Processing (NLP) for Seamless Communication:
   - Develop AI-powered conversational agents understanding complex emotions, context, and nuances in human communication.
   - Research advanced NLP techniques for real-time translation and interpretation in AR/VR environments, eliminating language barriers.

Explore AI-driven sentiment analysis to personalize communication based on individual emotions and preferences.

2. Multisensory Storytelling and Immersive Experiences:

   Research how AI can dynamically adapt AR/VR content based on user preferences, creating personalized and engaging narratives.

   Explore the use of AI for generating realistic and interactive virtual characters, fostering deeper emotional connections in VR communication.

   Investigate AI-powered haptic feedback systems for AR/VR, replicating real-world sensations for richer immersive experiences.

3. AI-Assisted Collaboration and Teamwork in AR/VR:

   Develop AI-powered meeting platforms within AR/VR, utilizing spatial audio and avatars for enhanced collaboration and team dynamics.

   Research AI systems that analyze group interactions in AR/VR, providing real-time feedback and suggestions for improved teamwork.

   Explore the use of AI for generating shared virtual workspaces with dynamic content and information visualizations for efficient collaboration.

4. Ethical Considerations and Societal Impact:

   Research the potential biases and ethical implications of using AI in communication technologies and AR/VR experiences.

   Develop frameworks for ensuring user privacy and data security in AR/VR environments powered by AI.

   Investigate the socio-economic impact of AI-powered communication, addressing potential inequalities and ensuring equitable access to technologies.

5. Accessibility and Inclusivity:

   Develop AI-powered interfaces for AR/VR that cater to diverse abilities and needs, ensuring accessibility for all users.

   Research AI-driven tools for language translation and sign language interpretation within AR/VR, fostering inclusive communication.

   Explore the use of AI for adapting AR/VR content to individual cognitive styles and learning preferences, promoting inclusivity in educational and training contexts.

6. Integration with Brain-Computer Interfaces (BCI):

   Research how AI can interpret brain signals and translate them into intuitive controls for AR/VR experiences, creating more natural interactions.

   Explore the use of AI for real-time emotional recognition and feedback in AR/VR communication, fostering deeper understanding and empathy.

   Investigate the ethical implications and potential risks of merging AI, BCI, and communication technologies, ensuring responsible development and use.

These are just a few examples, and the possibilities are constantly evolving. The future of AI in communication technology and AR/VR innovation is bright, but it's crucial to consider both the exciting opportunities.

6. CONCLUSION

This study highlights the potential of AI and AR/VR to transform communication by examining their interactions. In conclusion, the integration of Artificial Intelligence (AI) with Augmented Reality/Virtual Reality (AR/VR) is revolutionizing communication technology. AI is reshaping traditional communication channels by providing personalized content delivery, sentiment analysis, and real-time language translation. AR/VR technology is creating immersive virtual environments that enhance communication experiences. However, it is crucial to consider the ethical implications and ensure inclusive involvement and responsible development in the advancement of these technologies.

REFERENCES

Abdullah, M. (2023). Analyzing augmented reality (AR) and virtual reality (VR) recent development in education. *Social Sciences & Humanities Open*, 8(1), 100532. Advance online publication. DOI: 10.1016/j.ssaho.2023.100532

Ahmed, I., Jeon, G., & Piccialli, F. (2021). From artificial intelligence to explainable artificial intelligence in industry 4.0: A survey on what, how, and where. *IEEE Transactions on Industrial Informatics*, 18(8), 5031–5042. DOI: 10.1109/TII.2022.3146552

M, Akbari., N, Ha., & S, Kok. (2022). A systematic review of AR/VR in operations and supply chain management: maturity, current trends and future directions. Journal of Global Operations and Strategic Sourcing.

Al, Ansi., A, M., & Al-Ansi, A. (2020). Future of education post covid-19 pandemic: Reviewing changes in learning environments and latest trends. *Solid State Technology*, 63(6), 201584–201600.

Al-Ansi, A. (2022). Investigating characteristics of learning environments during the COVID-19 pandemic: A systematic review. *Canadian Journal of Learning and Technology*, 48(1). Advance online publication. DOI: 10.21432/cjlt28051

Al-Ansi, A. M., & Fatmawati, I. (2023). Integration of ICT in higher education during covid-19 pandemic: A case study. *International Journal of Learning and Change*, 15(2).

Al-Ansi, A. M., Garad, A., & Al-Ansi, A. (2021). ICT-based learning during Covid-19 outbreak: Advantages, opportunities and challenges. *Gagasan Pendidikan Indonesia*, 2(1), 10–26. DOI: 10.30870/gpi.v2i1.10176

Al-Azawi, R. (2018). Embedding augmented and virtual reality in educational learning method: Present and future. In In2018 9th international conference on Information and communication systems (ICICS) (pp. 218–222). IEEE.

Alalwan, N., Cheng, L., Al-Samarraie, H., Yousef, R., Alzahrani, A. I., & Sarsam, S. M. (2020). Challenges and prospects of virtual reality and augmented reality utilization among primary school teachers: A developing country perspective. *Studies in Educational Evaluation*, 66, 100876. DOI: 10.1016/j.stueduc.2020.100876

Ali, W. (2020). Online and remote learning in higher education institutes: A necessity in light of COVID-19 pandemic. *Higher Education Studies*, 10(3), 16–25. DOI: 10.5539/hes.v10n3p16

Alizadehsalehi, S., Hadavi, A., & Huang, J. C. (2021). Assessment of AEC students' performance using BIM-into-VR. *Applied Sciences (Basel, Switzerland)*, 11(7), 3225. DOI: 10.3390/app11073225

Ayoub, A., & Pulijala, Y. (2019). The application of virtual reality and augmented reality in Oral & Maxillofacial Surgery. *BMC Oral Health*, 19(1), 1–8. DOI: 10.1186/s12903-019-0937-8 PMID: 31703708

Ban, H., & Ning, J. (2021). *Online English teaching based on artificial intelligence internet technology embedded system* (Vol. 2021). Mobile Information Systems.

Beck, D. (2019). Augmented and virtual reality in education: Immersive learning research. *Journal of Educational Computing Research*, 57(7), 1619–1625. DOI: 10.1177/0735633119854035

Bernacki, M. L., Greene, J. A., & Crompton, H. (2020). Mobile technology, learning, and achievement: Advances in understanding and measuring the role of mobile technology in education. *Contemporary Educational Psychology*, 60, 101827. DOI: 10.1016/j.cedpsych.2019.101827

Bin, Y., & Mandal, D. (2019). English teaching practice based on artificial intelligence technology. *Journal of Intelligent & Fuzzy Systems*, 37(3), 3381–3391. DOI: 10.3233/JIFS-179141

Biswas, P., Orero, P., Swaminathan, M., Krishnaswamy, K., & Robinson, P. (2021). Adaptive accessible AR/VR systems. In InExtended abstracts of the 2021 CHI Conference on human Factors in computing systems (pp. 1–7).

Boyles, B. (2017). Virtual reality and augmented reality in education. West Point, Ny: Center For Teaching Excellence, United States Military Academy.

Chen, C. Y. (2022). Immersive virtual reality to train preservice teachers in managing students' challenging behaviours: A pilot study. *British Journal of Educational Technology*, 53(4), 998–1024. DOI: 10.1111/bjet.13181

Chen, P., Liu, X., Cheng, W., & Huang, R. (2017). A review of using augmented reality in education from 2011 to 2016. Innovations in smart learning, 13–18.

Chen, Y. L., Hsu, C. C., & Lin, C. Y. (2021). Engaging language learners through the use of virtual reality and a robot. In Society for Information Technology & Teacher Education International Conference Association for the Advancement of Computing in Education (AACE)., pp. 906–909.

Cheng, K. H., & Tsai, C. C. (2019). A case study of immersive virtual field trips in an elementary classroom: Students' learning experience and teacher-student interaction behaviors. *Computers & Education*, 140, 103600. DOI: 10.1016/j.compedu.2019.103600

Childs, E., Mohammad, F., Stevens, L., Burbelo, H., Awoke, A., Rewkowski, N., & Manocha, D. (2021). An overview of enhancing distance learning through augmented and virtual reality technologies. arXiv preprint arXiv:2101.11000.

Cieri, R. L., Turner, M. L., Carney, R. M., Falkingham, P. L., Kirk, A. M., Wang, T., Jensen, B., Novotny, J., Tveite, J., Gatesy, S. M., Laidlaw, D. H., Kaplan, H., Moorman, A. F. M., Howell, M., Engel, B., Cruz, C., Smith, A., Gerichs, W., Lian, Y., & Farmer, C. G. (2021). Virtual and augmented reality: New tools for visualizing, analyzing, and communicating complex morphology. *Journal of Morphology*, 282(12), 1785–1800. DOI: 10.1002/jmor.21421 PMID: 34689352

Cortez, P., Moro, S., Rita, P., King, D., & Hall, J. (2018). Insights from a text mining survey on expert systems research from2000 to 2016. *Expert Systems: International Journal of Knowledge Engineering and Neural Networks*, 35(3), e12280. DOI: 10.1111/exsy.12280

Criollo-C, S., Altamirano-Suarez, E., Jaramillo-Villacís, L., Vidal-Pacheco, K., Guerrero-Arias, A., & Luján-Mora, S. (2022). Sustainable teaching and learning through a mobile application: A case study. *Sustainability (Basel)*, 14(11), 6663. DOI: 10.3390/su14116663

Criollo-C, S., Guerrero-Arias, A., Jaramillo-Alcázar, Á., & Luján-Mora, S. (2021). Mobile learning technologies for education: Benefits and pending issues. *Applied Sciences (Basel, Switzerland)*, 11(9), 4111. DOI: 10.3390/app11094111

de Oliveira, M. E., & Corr^ea, C. G. (2020). Virtual reality and augmented reality applications in agriculture: A literature review. In 2020 22nd symposium on virtual and augmented reality (svr) (pp. 1–9). IEEE. DOI: 10.1109/SVR51698.2020.00017

Divekar, R. R., Drozdal, J., & Chabot, S. (2021). Foreign language acquisition via artificial intelligence and extended reality: Design and evaluation. *Computer Assisted Language Learning*, , 1–29.

Du, R., Turner, E., Dzitsiuk, M., Prasso, L., Duarte, I., Dourgarian, J., . . . Kim, D. (2020). DepthLab: Real-time 3D interaction with depth maps for mobile augmented reality. In InProceedings of the 33rd annual ACM Symposium on user interface Software and technology (pp. 829–843).

Eckert, M., Volmerg, J. S., & Friedrich, C. M. (2019). Augmented reality in medicine: Systematic and bibliographic review. *JMIR mHealth and uHealth*, 7(4), e10967. DOI: 10.2196/10967 PMID: 31025950

Elmqaddem, N. (2019). Augmented reality and virtual reality in education. Myth or reality? *International Journal of Emerging Technologies in Learning*, 14(3), 234. DOI: 10.3991/ijet.v14i03.9289

Fu, Y., Hu, Y., & Sundstedt, V. (2022). A systematic literature review of virtual, augmented, and mixed reality game applications in healthcare. [HEALTH]. *ACM Transactions on Computing for Healthcare*, 3(2), 1–27. DOI: 10.1145/3472303

Gandedkar, N. H., Wong, M. T., & Darendeliler, M. A. (2021). Role of virtual reality (VR), augmented reality (AR) and artificial intelligence (AI) in tertiary education and research of orthodontics: An insight. [WB Saunders]. *Seminars in Orthodontics*, 27(2), 69–77. DOI: 10.1053/j.sodo.2021.05.003

Garad, A., Al-Ansi, A. M., & Qamari, I. N. (2021). The role of e-learning infrastructure and cognitive competence in distance learning effectiveness during the covid-19 pandemic. *Journal Cakrawala Pendidikan*, 40(1), 81–91. DOI: 10.21831/cp.v40i1.33474

Gargrish, S., Mantri, A., & Kaur, D. P. (2020). Augmented reality-based learning environment to enhance teaching-learning experience in geometry education. *Procedia Computer Science*, 172, 1039–1046. DOI: 10.1016/j.procs.2020.05.152

Garzon, ´. J., Pavon, ´. J., & Baldiris, S. (2019). Systematic review and meta-analysis of augmented reality in educational settings. *Virtual Reality (Waltham Cross)*, 23(4), 447–459. DOI: 10.1007/s10055-019-00379-9

Guan, C., Mou, J., & Jiang, Z. (2020). Artificial intelligence innovation in education: A twenty-year data-driven historical analysis. *International Journal of Innovation Studies*, 4(4), 134–147. DOI: 10.1016/j.ijis.2020.09.001

Gudoniene, D., & Rutkauskiene, D. (2019). Virtual and augmented reality in education. *Baltic Journal of Modern Computing*, 7(2), 293–300. DOI: 10.22364/bjmc.2019.7.2.07

Guo, X., Guo, Y., & Liu, Y. (2021). The development of extended reality in education: Inspiration from the research literature. *Sustainability (Basel)*, 13(24), 13776. DOI: 10.3390/su132413776

Hantono, B. S., Nugroho, L. E., & Santosa, P. I. (2018). Meta-review of augmented reality in education. In In2018 10th international conference on information technology and electrical engineering (ICITEE) (pp. 312–315). IEEE. DOI: 10.1109/ICITEED.2018.8534888

Huang, J., Saleh, S., & Liu, Y. (2021). A review on artificial intelligence in education. *Academic Journal of Interdisciplinary Studies*, 10(3), 206–206. DOI: 10.36941/ajis-2021-0077

Huang, K. T., Ball, C., Francis, J., Ratan, R., Boumis, J., & Fordham, J. (2019). Augmented versus virtual reality in education: An exploratory study examining science knowledge retention when using augmented reality/virtual reality mobile applications. *Cyberpsychology, Behavior, and Social Networking*, 22(2), 105–110. DOI: 10.1089/cyber.2018.0150 PMID: 30657334

Huang, S. (2019). Augmented reality and virtual reality: The power of AR and VR for business: M. Claudia tom dieck, timothy jung. Switzerland: Springer, 2019, 335 pp.119.99, ISBN 978-3-030-06245-3.

Imottesjo, H., & Kain, J. H. (2022). The Urban CoCreation Lab—an integrated platform for remote and simultaneous collaborative urban planning and design through web-based desktop 3D modeling, head-mounted virtual reality and mobile augmented reality: Prototyping a minimum viable product and developing specifications for a minimum marketable product. *Applied Sciences (Basel, Switzerland)*, 12(2), 797. DOI: 10.3390/app12020797

Javornik, A. (2016). Augmented reality: Research agenda for studying the impact of its media characteristics on consumer behaviour. *Journal of Retailing and Consumer Services*, 30, 252–261. DOI: 10.1016/j.jretconser.2016.02.004

Johnson, W. L., & Valente, A. (2009). Tactical language and culture training systems: Using AI to teach foreign languages and cultures. *AI Magazine*, 30(2), 72–72. DOI: 10.1609/aimag.v30i2.2240

Johnson, W. L., Vilhjálmsson, H. H., & Marsella, S. (2005). Serious games for language learning: how much game, how much AI? In AIED, vol. 125, no. 1, pp. 306–313.

Kaminska, ´. D., Sapinski, ´. T., Wiak, S., Tikk, T., Haamer, R. E., Avots, E., Helmi, A., Ozcinar, C., & Anbarjafari, G. (2019). Virtual reality and its applications in education: Survey. *Information (Basel)*, 10(10), 318. DOI: 10.3390/info10100318

Ke, F., Pachman, M., & Dai, Z. (2020). Investigating educational affordances of virtual reality for simulation-based teaching Wireless Communications and Mobile Computing training with graduate teaching assistants. *Journal of Computing in Higher Education*, 32(3), 607–627. DOI: 10.1007/s12528-020-09249-9

Kim, C., Yoon, H. C., Kim, D. H., & Do, Y. R. (2018). Spectroscopic influence of virtual reality and augmented reality display devices on the human nonvisual characteristics and melatonin suppression response. *IEEE Photonics Journal*, 10(4), 1–11. DOI: 10.1109/JPHOT.2021.3107852

Kim, N. Y., Cha, Y., & Kim, H. S. (2019). Future English learning: Chatbots and artificial intelligence. *Multimedia-Assisted Language Learning*, 22(3), 32–53.

Klimova, B., & Polakova, P. (2020). Students' perceptions of an EFL vocabulary learning mobile application. *Education Sciences*, 10(2), 37. DOI: 10.3390/educsci10020037

Kose, ¨. H., & Güner-Yildiz, N. (2021). Augmented reality (AR) as a learning material in special needs education. *Education and Information Technologies*, 26(2), 1921–1936. DOI: 10.1007/s10639-020-10326-w

Krüger, J. M., Buchholz, A., & Bodemer, D. (2019). Augmented reality in education: Three unique characteristics from a user's perspective. In Proc. 27th Int. Conf. on Comput. in Educ, 412–422.

Kumar, P. P., Thallapalli, R., Akshay, R., Sai, K. S., Sai, K. S., & Srujan, G. S. (2022). Stateof-the-Art: Implementation of augmented reality and virtual reality with the integration of 5G in the classroom. InAIP Conference Proceedings, 2418, 20069 (AIP Publishing LLC).

Lai, N. Y. G., Wong, K. H., Yu, L. J., & Kang, H. S. (2020). Virtual reality (VR) in engineering education and training: A bibliometric analysis. In InProceedings of the 2nd world symposium on software engineering (pp. 161–165).

Lavrentieva, O., Arkhypov, I., Kuchma, O., & Uchitel, A. (2020). Use of simulators together with virtual and augmented reality in the system of welders' vocational training: Past, present, and future.

Lewis, R. B. (1999). SIMSTAT with wordstat: A comprehensive statistical package with a content analysis module. *Field Methods*, 11(2), 166–179. DOI: 10.1177/1525822X9901100207

Li, X., Yi, W., Chi, H. L., Wang, X., & Chan, A. P. (2018). A critical review of virtual and augmented reality (VR/AR) applications in construction safety. *Automation in Construction*, 86, 150–162. DOI: 10.1016/j.autcon.2017.11.003

Luo, M., & Cheng, L. (2020). Exploration of interactive foreign language teaching mode based on artificial intelligence. *International Conference on Computer Vision, Image and Deep Learning (CVIDL)*, pp. 285–290. DOI: 10.1109/CVIDL51233.2020.00-84

Ma, L. (2021). An immersive context teaching method for college English based on artificial intelligence and machine learning in virtual reality technology. Mobile Information Systems, vol. 2021, Article ID 2637439, 7 pages. DOI: 10.1155/2021/2637439

Maunder, R. E. (2018). Students' peer relationships and their contribution to university adjustment: The need to belong in the university community. *Journal of Further and Higher Education*, 42(6), 756–768. DOI: 10.1080/0309877X.2017.1311996

Mystakidis, S., Fragkaki, M., & Filippousis, G. (2021). Ready teacher one: Virtual and augmented reality online professional development for K-12 school teachers. *Computers*, 10(10), 134. DOI: 10.3390/computers10100134

Nguyen, V. T., & Dang, T. (2017). Setting up virtual reality and augmented reality learning environment in unity. In In 2017 IEEE international symposium on Mixed and augmented reality (ISMAR-Adjunct) (pp. 315–320). IEEE. DOI: 10.1109/ISMAR-Adjunct.2017.97

Nguyen, V. T., Jung, K., & Dang, T. (2019). Creating virtual reality and augmented reality development in classroom: Is it a hype? In In 2019 IEEE international conference on artificial Intelligence and virtual reality (AIVR) (pp. 212–2125). IEEE.

Oberdorfer, ¨. S., Birnstiel, S., Latoschik, M. E., & Grafe, S. (2021). Mutual benefits: Interdisciplinary education of pre-service teachers and hci students in vr/ar learning environment design. [Frontiers Media SA]. *Frontiers in Education*, 6, 693012. DOI: 10.3389/feduc.2021.693012

Olbina, S., & Glick, S. (2022). Using integrated hands-on and virtual reality (VR) or augmented reality (AR) approaches in construction management education. *International Journal of Construction Education and Research*, , 1–20.

Pan, Z., Sun, Y., Yao, Z. W., & Li, M. (2021). Application of virtual reality in English teaching in 2021. *3rd World Symposium on Artificial Intelligence (WSAI)*, IEEE, pp. 64–71. DOI: 10.1109/WSAI51899.2021.9486322

Patel, S., Panchotiya, B., & Ribadiya, S. A. (2020). Survey: Virtual, augmented and mixed reality in education. *IJERT*, 9, 1067–1072.

Permanasari, A. E., Nugroho, H. E., Ardhanarko, H. B. S., Fauziati, S., Utami, I. K., & Sholihin, M. (2022). Development of VR and AR learning media for enterprise business ethic scenarios. In In2022 international conference on information technology Systems and innovation (ICITSI) (pp. 225–230). IEEE. DOI: 10.1109/ICITSI56531.2022.9970892

Phakamach, P., Senarith, P., & Wachirawongpaisarn, S. (2022). The Metaverse in education: The future of immersive teaching & learning. *RICE Journal of Creative Entrepreneurship and Management*, 3(2), 75–88.

Poláková, P., & Klímová, B.Pol'akov´a. (2019). Mobile technology and Generation Z in the English language classroom—A preliminary study. *Education Sciences*, 9(3), 203. DOI: 10.3390/educsci9030203

Qian, J. (2022). Research on artificial intelligence technology of virtual reality teaching method in digital media art creation. *Journal of Internet Technology*, 23(1), 125–132.

Raja, M., & Lakshmi Priya, G. G. (2022). Using virtual reality and augmented reality with ICT tools for enhancing quality in the changing academic environment in COVID-19 pandemic: An empirical study. In *InTechnologies, artificial Intelligence and the Future of learning post-COVID-19* (pp. 467–482). Springer. DOI: 10.1007/978-3-030-93921-2_26

Ramya, R. (2024). *Analysis and Applications Finding of Wireless Sensors and IoT Devices With Artificial Intelligence/Machine Learning. AIoT and Smart Sensing Technologies for Smart Devices.* IGI Global., DOI: 10.4018/979-8-3693-0786-1.ch005

Ramya, R., & Ramamoorthy, S. (2022). Analysis of machine learning algorithms for efficient cloud and edge computing in the loT, Challenges and Risks Involved in Deploying 6G and NextGen Networks, pp. 72–90.

Ramya, R., & Ramamoorthy, S. (2022). Survey on Edge Intelligence in IoT-Based Computing Platform. *Survey on Edge Intelligence in IoT-Based Computing Platform, Lecture Notes in Networks and Systems, Springer*, 356, 549–556. DOI: 10.1007/978-981-16-7952-0_52

Ramya, R., & Ramamoorthy, S. "Analysis of machine learning algorithms for efficient cloud and edge computing in the loT", Challenges and Risks Involved in Deploying 6G and NextGen Networks, pp. 72–90, 2022.

Ramya, R., & Ramamoorthy, S. (2022). Survey on Edge Intelligence in IoT-Based Computing Platform. *Lecture Notes in Networks and Systems, Springer*, 356, 549–561. DOI: 10.1007/978-981-16-7952-0_52

Rau, P. L. P., Zheng, J., Guo, Z., & Li, J. (2018). Speed reading on virtual reality and augmented reality. *Computers & Education*, 125, 240–245. DOI: 10.1016/j.compedu.2018.06.016

Remolar, I., Rebollo, C., & Fern'andez-Moyano, J. A. (2021). Learning history using virtual and augmented reality. *Computers*, 10(11), 146. DOI: 10.3390/computers10110146

Shorey, S., Ang, E., Yap, J., Ng, E. D., Lau, S. T., & Chui, C. K. (2019). A virtual counseling application using artificial intelligence for communication skills training in nursing education: Development study. *Journal of Medical Internet Research*, 21(10), e14658. DOI: 10.2196/14658 PMID: 31663857

Xiaohong, W., & Yanzheng, W. (2021). The application of artificial intelligence in modern foreign language learning. *4th International Conference on Big Data and Education*, pp. 34–37. DOI: 10.1145/3451400.3451406

Xiaoxia, L. (2022). Practice of Artificial Intelligence and Virtual Reality Technology in College English Dialogue Scene Simulation. Wireless Communications and Mobile Computing, vol. 2022, Article ID 4922675, 9 pages. https://doi.org/DOI: 10.1155/2022/4922675

Chapter 2
Exploring the Frontiers of Communication With Extended Reality (XR):
Applications, Challenges, and Future Directions

João A. Dias
https://orcid.org/0000-0001-8036-4709
UNIDCOM, IADE, Universidade Europeia, Portugal

Miguel Boavida
https://orcid.org/0000-0002-5895-6828
IADE, Universidade Europeia, Portugal

Jacinto Estima
https://orcid.org/0000-0001-8837-4637
Department of Informatics Engineering, CISUC, University Coimbra, Portugal

ABSTRACT

Extended Reality (XR) is changing how we interact with digital information and communicate with each other by transforming how people experience and interact with their surroundings. It can be used to create immersive experiences like virtual tours or to promote companies, products, and brands. Although the industry and scientific community have improved recently, XR must overcome some issues to be fully deployed and understood in multiple areas. For example, it is essential to understand user experience guidelines to create realistic XR applications. Also, those applications require a lot of processing power, which can result in performance issues. Furthermore, integrating artificial intelligence opens a world of possibilities,

DOI: 10.4018/979-8-3693-3579-6.ch002

presenting new research trends to explore. This chapter examines the definition of XR to demonstrate what it is possible to do with it, the emerging challenges, and how XR technologies can be applied to the communication field.

INTRODUCTION

Extended Reality (XR) is revolutionizing our communication and interaction with the digital world, offering transformative and engaging experiences across various industries. These experiences enable us to reshape our perception of reality by overlaying digital images of the natural world or constructing entirely virtual environments. XR also opens innovative ways to experience and explore the world, facilitating the sharing and comprehending of information in previously unimaginable ways (Rauschnabel et al., 2022). The concept of Extended Reality (XR) encompasses both Virtual Reality (VR) and Augmented Reality (AR). It represents a spectrum of environments that merge physical and virtual elements to create immersive user experiences.

VR fully immerses users in a digitally simulated environment, isolating them from the physical world and allowing interaction within a completely virtual space. In "Defining Virtual Reality: Dimensions Determining Telepresence," (Steuer, 1992) explores the concept of telepresence in VR and its implications for media consumption and interpersonal communication.

Conversely, Augmented Reality (AR) has become a game changer in media, technology, and entertainment as AR apps blend elements into reality smoothly, allowing users to engage with them and opening a world of possibilities. AR overlays digital information onto the real world, enhancing one's perception of reality through smartphones, tablets, or AR glasses. This technology can revolutionize how narratives are crafted and consumed and offer new avenues for research in communication behaviors and processes. By adding this layer of interactivity, AR contributes to redefining and revolutionizing the user experience and how people are engaged (Grubert et al., 2017).

While distinct, these technologies share the goal of enhancing sensory perception and user interaction. XR thus capitalizes on the strengths of both VR and AR, offering versatile applications that range from entertainment and education to complex industrial and communication tasks, ultimately broadening the scope of digital interaction and connectivity.

This chapter is organized as follows: it starts with "Extended Reality Fundamentals," a section that explores the concept of extended reality and the essential elements needed to develop XR solutions. The following section, "XR Applied to Communication," provides a detailed review of how these technologies can be applied in different communication areas, such as content production for the media, education and literacy for society, and media training for public figures. The following chapters address XR solutions' main identified challenges and limitations and point to future directions and XR trends and communication. Finally, some final remarks about the transformative potential of XR in shaping modern communication paradigms.

EXTENDED REALITY (XR) FUNDAMENTALS

XR technologies have the potential to revolutionize different domains, from education and healthcare to entertainment and industrial applications. One of the defining features of XR technologies is their ability to update the stimulus based on the user's eye, head, or body movements (Kasowski et al., 2023). This dynamic interactivity allows for highly immersive and engaging experiences, blurring the lines between the physical and digital realms. XR has grown in recent years, with applications in educational, biomedical, and commercial settings, capitalizing on its potential to teach new skills and enhance learning (Santos et al., 2023). This growing interest in XR can be attributed to the increased affordability and accessibility of the necessary hardware and software. Furthermore, the advancements in head-mounted displays, computing power, display technologies, and sensor integration, which facilitate audio and visual immersion, have played a crucial role in adopting XR technologies within academia, leading to the development of increasingly sophisticated and user-friendly XR systems. This field has sparked a surge of creativity and exploration as educators and practitioners from various disciplines seek to integrate these technologies into their practices.

The concept of XR has its roots in the early 1950s, with the development of Virtual Reality (VR) and Augmented Reality (AR) technologies. Since then, these technologies have evolved and expanded, leading to the emergence of Mixed Reality (MR) and the broader concept of XR. Everything started between 1950s and 1970s when the first AR glasses were created to improve the user's world perception (Rheingold, 1992). The first attempts to create VR glasses and controllers emerged with the boom in video game consoles. However, they were not as well received as they would have liked since the types of solutions designed required processing power that was not available at the time. Thus, XR technologies lay dormant until mid-2015, with the tech giants focusing on developing devices capable of processing experiences with the quality needed to be accepted by the industry and its end users (Burdea & Coiffet, 2024). Although XR can be divided into the three above-presented areas, all of them share the same key components to enable it: i) a camera to capture the surrounding environment, ii) a processor to understand the environment and simulate virtual objects in real-time, and iii) supporting hardware, like headsets, google, or smartphones. In this way, the evolution of XR technologies has been closely linked to the evolution of the hardware components needed to use them.

Virtual Reality allows users to be immersed in a virtual environment without being able to visualize and interact with the real world. In this way, users can only immerse themselves and interact through VR glasses and devices (e.g., controllers and sensors) that communicate with them. The user's movements are captured using motion detection technologies, which, together with sensors that enhance the different senses of the human being, create truly immersive experiences that transport users to multiple places and environments. In other words, a VR environment is a computer-generated simulation in which users may experience a realistic experience by being immersed in a digital world. This approach can potentially innovate and revolutionize how people interact with information, communicate, and even consume media and digital content (Muñoz-Saavedra et al., 2020). The level of immersion created by VR applications allows different industries to consider its deployment. For example, VR may be used in entertainment, such as games or the film industry. By creating immersive and engaging environments, VR can transport users to new worlds, allowing them to be part of narratives and become active participants instead of only observers. Another example of its applicability is using VR for educational and training purposes. Health professionals can use VR for medical training, rehabilitation, and patient education by simulating real-world scenarios without compromising or damaging the users (Bansal et al., 2022). From a communication perspective, there is a large specter of opportunities that are starting to be adopted or at least explored. As a starting point, VR can foster more engaging and interactive forms of remote collaboration, enabling individuals to meet and work together in shared virtual spaces, regardless of physical location (Han et al., 2024). However, there are also exciting opportunities for creating new media content and training. The use of VR to enhance content perception by adopting a transmedia approach that enables the viewer to explore the base content autonomously is discussed in the next chapter with a range of examples from the news to documentaries or fiction. The use of VR as a communication tool that can contribute to society by creating awareness or improving literacy in areas like education, health, or inclusion is also discussed. Finally, examples of media training and using VR to prepare public speeches or political debates are discussed.

For developing AR experiences, it is essential to consider the following components: a device with a display (e.g., a smartphone or a tablet), a tracking system to determine the user's location and orientation (e.g., a camera), and software capable of rendering the virtual content and precisely aligning it with the natural world (Dieter Schmalstieg & Hollerer, 2016; Lee et al., 2023). AR applications can be deployed in a variety of scenarios. For example, they can be used for education and training, where users may interact with virtual simulations to enhance learning and understanding. In tourism, AR may add additional information to visitors when visiting a museum or a garden (Bangay & McKenzie, 2022). AR has changed how people experience and interact with content in media and communication. For example, this technology allows users to apply filters to their images and videos on social platforms, creating more engaging and immersive communication experiences. This type of content opens new opportunities for creative expression and creation (Al-Amri et al., 2023). The use of AR technology in communication is, for the moment, different from the examples associated with VR. AR technology has become available to almost everyone with a smartphone, who can see the content captured by the phone's camera with additional layers of information on the display. As a starting point, most applications will require the user to be on location and eventually point the device to something that will be identified and trigger additional information. Good application examples are associated with museums or people visiting the physical place. Nevertheless, AR can also be used in brand activation scenarios where the audience is invited to recreate their (virtual) presence on a big concert stage, in a movie setting, or in a famous event. A simple but effective example of AR, which could be translated to other communication areas, is the application Marine Traffic, developed by Kpler and used by ocean sailors to share their position and some information related to the sailing conditions. Marine Traffic has an AR feature that makes it possible to point to a nearby boat and see information about the vessel and its sailing conditions in an augmented way.

Finally, Mixed Reality (MR) combines the physical and digital worlds, almost seamlessly integrating virtual objects into the natural environment. This unique feature creates a sense of presence and interaction beyond what VR or AR can achieve independently. This approach is already being applied in fields like remote assistance and training (Fidalgo et al., 2023) or collaborative learning (Pan et al., 2021). It is worth noticing that the name mixed reality emerged in technological areas, the first to develop and use devices that could change from a closed immersive experience to a situation where the user could see through the device. Nowadays, this type of device is starting to become more common and accessible. Still, the name mixed reality tends to be less used, as the name XR emerged as a more flexible term that can be adapted to future developments in immersive technologies. Mixed Reality refers explicitly to the blending of physical and virtual worlds, something that is now seen as a subset of XR. The industry has widely adopted XR as it provides clarity and inclusiveness. XR emphasizes the user experience across various applications, from fully immersive VR experiences to interactive AR applications. This holistic view is essential for industries like education, training, gaming, and healthcare, which may utilize multiple forms of immersive technologies.

Implementing XR technologies requires careful hardware, software, and user experience design analysis. Hardware such as headsets, haptic devices, and motion controllers play a crucial role in creating immersive experiences, while software development focuses on creating realistic environments and interactive elements. In turn, user experience design ensures that XR applications are intuitive and easy to use, allowing users to navigate virtual environments and interact with digital objects. As XR technologies evolve, their potential to transform sectors such as education, entertainment, and communication is becoming increasingly evident. The benefits of using XR go far beyond creating user-oriented experiences. They make it possible to offer better training and development tools since it is possible to create realistic training scenarios in safe environments without real risks. Areas such as healthcare, manufacturing, and aviation, where mistakes can be costly and take lives, can benefit greatly from these technologies, allowing workers to develop their skills through experience and resulting in fewer mistakes and more confidence in their work. Furthermore, in areas such as education, these technologies allow students, for example, to explore complex scientific concepts or access historical sites through more effective and memorable learning experiences (Geroimenko, V., 2020; Elliott et al., 2018). In the area of health and well-being, it is possible to improve the quality of healthcare and the patient experience. For example, AR can help healthcare professionals overcome some limitations in physiotherapy and rehabilitation sessions. As far as communication is concerned, XR is expected to have a significant impact, offering new ways of interacting, collaborating, sharing information, and changing the customer's experience, enabling them to make better decisions and feel more satisfied with their choices (Flavián et al., 2019). The following section details how XR can be applied to the communication field, providing examples and case studies that have already been implemented.

XR APPLIED TO COMMUNICATION

When considering the use of XR technologies in communication sciences, it is essential to envision its use as a way for the user or viewer to go deeper into the main content delivered by traditional media, making his experience more interactive and personalized. Over the last few years, with the boom of social networks, people have gotten used to having content delivered to them with increasing personalization based on the information systems that can collect about user preferences and behavior. That is one of the main difficulties some traditional media face when trying to retain their viewers or readers who do not want to spend time seeing content irrelevant to them.

Theoretically, this may be seen as a learning approach where users actively construct their knowledge through their experience in a social process mediated through, as in constructivism, being part of a social process mediated through a culture's symbols and language (Merriam & Bierema, 2005). In the context of constructivism and experiential learning, it is essential that learners can enter real-world situations and "authentic" environments where learning is situational (Gerry et al., 2006).

This section presents and discusses the use of XR in communication, considering three significant areas: content perception, society, and media training. For content perception, XR can be used in the news to further analyze parts of the news content in detail, like a map or a 3D representation of a specific building or infrastructure or used as an extended media format that enables viewers to have a personalized experience about a movie or a musical concert. Several areas can be identified and explored when discussing the use of these technologies in this context. XR can enhance news and current affairs, documentaries and educational programs, sports, drama and fiction series, recreation and music programs, commercials, and advertising. For society, XR opens a new world of possibilities concerning content interaction in education (learning and training), business, health (therapy and literacy), accessibility and inclusion, or cultural preservation and ecology. Furthermore, XR can also be used, and it is already being used as a training tool for public speaking and preparing meetings or public debates.

XR FOR CONTENT PERCEPTION

When discussing the use of XR in communication, the first examples people will imagine are related to broadcast media and the different flavors of TV, radio, and online platforms. However, newspapers and magazines can also benefit from its use. This section will discuss practical examples of how using XR can extend traditional content.

News and Current Affairs

It is easy to imagine the use of XR in the news and how it allows the viewer to go deeper into the content presented, allowing the user to observe, for example, the details of a map or a 3D representation of a specific building or infrastructure. There is already a tendency for viewers to start browsing additional information on the Internet. This way, XR allows them to explore the content, keep connected with the same content provider, and experience the same look and feel of the original experience.

The use of VR/XR in the news industry started around 2016, with the BBC and the Guardian, for example, beginning to create VR content. The BBC project "We Wait" (Steed et al., 2018), which recreated the experience of a Syrian refugee family crossing the sea, is one of those early examples. However, VR was growing slower than predicted, and years later, the BBC VR hub closed. According to the data consultancy firm Statista (Statista - the Statistics Portal, 2024), VR is finally growing, starting with 5.41 million headsets sold in 2019 to 8.52 million in 2022, and it is expected to rise to 14.31 million in 2024. However, according to the "Journalism, Media and Technology Trends and Predictions 2023 report" published by the Reuters Institute for the Study of Journalism, despite the sustained growth of VR in many areas, its use in journalism is not keeping pace with this growth (Newman, 2023). Some examples can be used to increase the interest for this technology in this field. One example of VR content available today is the one provided by ABC News (News, 2024). However, most VR experiences today are based on 360-degree videos, which require more profound levels of interaction. Those are still missing and will only be available with 3D content created from the original footage that makes "visiting" the virtual location a more immersive experience.

One key aspect of using XR as complementary media is the underlying level of personalization it enables. The spatial context of the subject can be analyzed and observed by the user at their own pace and with the level of detail they want. Not only spatial representations like maps (e.g., of conflict areas), data that will evolve (e.g., poles or elections), or the architectural details of a building or a historical scene can gain with this extended version. The prior examples suggest that a TV program is the leading media in which a news report or documentary can interrupt the normal flow of the news piece to observe the content in detail. However, this approach can also be applied to newspapers or books, such as hyperlinks connecting related information. News consumption is increasingly personalized, and different users will want to observe various aspects of the piece in varying levels of detail. That is why this type of content suits the use of XR as a complementary source of information very well. On the other hand, the time needed for content production is limited, making its adoption difficult.

Documentaries

This type of media content typically uses 3D modeling to showcase architectural elements that are no longer available or significantly deteriorated. This 3D reconstruction of the environment (Mudička & Kapica, 2022) is the ideal place to create an interactive experience where the user can escape from the normal flow of the program and explore the area at his own will. In some cases, like documentaries targeted to a younger audience, the XR experience may gain some gamification elements. In addition to the natural exploration of the environment, the user should discover or identify elements that contain additional elements or clues that make the viewer/user focused on the content. Finding treasures or solving enigmas may be a more exciting approach to (not only) younger audiences, introducing gamification elements that trigger the competitive nature of the viewer/user. This approach is an exciting way to provide a transmedia flavor to content, keeping the original media self-contained but opening space for alternative narratives in different devices. XR allows viewers to look at the scenery or the action from the perspective of one of the main characters. This content is probably more prepared to be explored in XR since the production teams already use part of the technology needed.

Sports

Sports shows are a particular type of content, considering a hybrid approach of live and curated content, where virtual content is superimposed on top of the live stream. Sports content typically falls into three main groups: live coverage of events or matches, event analysis, and educational content that explains the sport's rules or technical and tactical aspects.

The educational content will benefit the most from using XR since it allows users to look at a specific situation, not only from the outside but through the eyes of the participant. It is important to note that participants are not only the athletes but also referees, umpires, and other officials and staff involved in the sport. In team sports, the tactical analysis may move from the traditional bird's eye view to the perspective of each player (or official), and the technical analysis enables the users to explore the details of the movement from multiple perspectives or angles. Explaining the rules of the sport and how those rules will influence the tactical behavior of the athlete, or the team is crucial for making the viewer more knowledgeable about a sport, causing him to watch with more enthusiasm and for extended periods. This situation is particularly true for popular sports, where the details, such as most Olympic sports, could be more known. Event analysis is like educational content. Still, once again, there are time constraints associated with this type of content needing to be available soon after the event (Yun-chao et al., 2023). The advances in AI will eventually make it possible to have reliable 3D models automatically generated from multicamera footage. Nevertheless, nowadays, educational content in sports is one of the most exciting areas where XR could be explored and developed.

Drama and Fictional Series

The use of XR in movies or TV series is an exciting way to address the multiple layers of interest the user may have, like some details of the plot, the characters, or the setting, broadly seen as the time and place in which the story is told (Times, 2016). As a starting point, the viewer sees or at least starts to see the content in its typical or traditional form. However, depending on the type of movie or series, some people will be curious about the historical aspects of the moment and place where the story takes place. This curiosity can be about the political background or aspects of a specific building or landscape the characters visit. Another type of viewer may be interested in the production details of a film or series and be curious about the set or where and how a specific scene was shot.

XR is an exciting way to deliver this extended layer of information to the viewer, who probably will have an additional reason to revisit the film or episode just to explore these new layers of content. Providing this opportunity to move and observe locations appeals to many viewers who are real fans of a particular production, but it is also an opportunity to make the viewer, now in a "user" role, interact or solve some challenges related to or unrelated to the story. It is interesting to note that depending on the type of project, most of this content may have already been developed in preproduction, which is an excellent way to deliver it to the audience. In some cases, it can also allow some sponsors to place promotional material or assume a more gamified perspective, giving the viewer/user some challenges when interacting with the content.

However, working with fictional content also allows XR to explore a transmedia narrative perspective, where the user/viewer can unfold additional elements related to a specific character or a side story. This exploration opens many possibilities for developing more interesting side stories to give relevance to secondary characters by telling their story or testing possible solutions for the project to evolve, such as a new movie or a new season (Ge & Hsiao, 2020).

Recreational and Music Programs

When considering recreational content like talk shows, game shows, reality shows, and music concerts, it is vital to assume that target audiences are significantly different and willing to interact in various ways and with varying intensity levels, creating the need to address them differently. A talk show is typically designed around the presenter, a well-known personality who conducts the show, addresses several topics, and interviews several guests. Some blocks are pre-recorded and may cover events or aspects associated with the life of one of the guests. These pre-recorded pieces can be augmented similarly to the one discussed in the documentaries section.

However, talk shows also enable people to participate virtually in a show that usually has a live audience. The co-existence of those physical and virtual live audiences is an exciting way of creating interaction by giving the viewer/user a specific role in the show, applauding or even making some basic moves. Having some images of the virtual live audience broadcasted is also a teasing element that promotes the participation or at least the presence of those who value being, even virtually, on TV (Hamilton et al., 2017).

Usually revolving around a quiz or physical challenge, the so-called game shows may benefit from using XR differently. The core element of the program is the competition, so the gamification elements involved may be translated into a virtual scenario, making the user play something identical to the virtual twin of the original program. The competition can be done while the live show is running or asynchronous, like a game that people can play online and may produce some ranking that presents the best virtual player in the competition. Extending the sense of playing the game at the individual level is natural since most viewers are already in the "participant's shoes," experiencing the show from a first-person perspective. Having XR almost as a game companion of the show (available 24/7) or as part of the live show as an extended way for the user to not only see the program but participate virtually may be very interesting for product placement and sponsors or may be monetized in a similar way that is done nowadays through phone calls or equivalent.

Reality shows are a form of informal social ethology in which people are focused on understanding the behaviors of a community in a closed environment surrounded by cameras and microphones. Scripted or not, the program evolves around the idea that people like to observe other people's behavior without being noticed. The confined space and the competition are enough to fuel tensions that can emerge naturally or artificially. A confined space, like a house, allows one to explore the house and look at the details of the compartments. Most of these programs assume that communication with the outside world is forbidden, so creating a digital twin where fans could interact with the participants does not make sense. Nevertheless, the gamification of the user experience, like daily challenges where the users need to find specific objects related to the ongoing stories that take place in the reality show, may be an exciting motivation for people to keep exploring and, at the same time, to get some form of analytics from the user/viewer. It is important to note that these daily challenges are an excellent place for product placement or to create interaction points with the sponsors.

Music concerts and other art performances need to be addressed as live events, where the viewer can be at the locations or see the show remotely (on television), or pre-recorded events, where most of the solutions are identical to those presented in documentaries. Using XR in live performances is mainly about the opportunity to see the show from a different perspective, and probably the most interesting perspective is the artist's perspective. Having the chance to see the show from the eyes of each band member, has regular or premium content, will undoubtedly be something that some fans will adore. On the other hand, being present in a concert as a virtual companion could be interesting for those who could not be physically present but could enjoy the perspective of their close ones. The video walls used in large venues are already trying to overcome the problem of the difficulty of feeling the live concert experience and, at the same time, being able to see the artists. In a different perspective, in January 2024, Meta, in partnership with The Diamond Bros, created an immersive VR concert experience that allowed users to see Doja Cat's concert live without leaving the comfort of their homes (Experience Doja Cat's Concert in VR, 2024).

Commercials and Advertising

Different examples of advertisement, product placement, and interaction with the audience using XR were already given. Still, globally, the main challenge of advertisement is to keep the viewers/users focused without distracting them or even disconnecting them from the experience. Part of the solution is to involve the user in the experience and make him a participant rather than a passive viewer. The viewer's participation is oriented toward a reward that can be delivered as nonpublic content, a gift, or a prize.

XR may be a natural extension of the typical advertisement content. Still, it can provide some form of gamification that needs to be rewarded and may have the added value of collecting information about user behavior. Interestingly, not all XR applications require XR googles; the Displaced by The New York Times (Silverstein, 2015) or 6x9 by The Guardian (6×9: A Virtual Experience of Solitary Confinement, 2016) are excellent examples of this approach being used in public flat screens or, more precisely, video walls.

XR FOR THE SOCIETY

In the previous section, the focus was on the use of XR in creating extended layers of information that can enhance several types of content used in broadcast media, traditional radio and television, or new online digital channels or infrastructures. In this section, we shift to a more conceptual level and discuss how these XR technologies can improve literacy and provide new forms of communication that can benefit specific areas of society. Education and training, business, health (therapy and literacy) accessibility and inclusion, cultural preservation, and ecology are the examples chosen to showcase XR's role in this type of communication.

Education and Training

The use of XR in education and training is one of the areas where this technology is much more established, with several real-life examples functioning and being used worldwide. Because of the costs involved, one of the first areas to use XR to its full extent was technical training, particularly in areas where the price of the actual equipment or the danger involved justified the investment. Aeronautics and aerospace were used to buy and develop simulators for training pilots and staff long before XR was an option (Pausch et al., 1992). So, in this area, the cost of an XR infrastructure was minimal compared to the price of the simulators. The availability and competitiveness of the XR equipment led to other industries using XR to train their personnel. One example is the automotive industry, which started using XR in the preliminary training of operators working in their assembly lines. Nowadays, the use of XR in assembly lines or maintenance training comes mainly in two flavors: VR-based approaches that enable the trainee to get acquainted with the equipment and the technical procedures involved, and AR-based approaches that use the actual equipment as the background where additional layers of information are presented (Fraga-Lamas et al., 2018).

The benefits of using XR in the industry apply directly to technological education in areas where universities do not have the necessary equipment. However, the advantages of using XR in education are broader than just the technical simulations in technological courses. Almost all areas can benefit from more reliable simulations where the user can experience a more profound sense of immersion. From Psychology to Architecture, Astronomy, or Life sciences, being able to explore and analyze experiments that simulate actual events provides a unique way to promote decision-making. It places the student in a unique position of learning by doing, or at least by deciding the actions to perform in real-time, making it a perfect tool for training for complex surgeries (Moglia et al., 2016) or military training. XR is also an exciting tool for use at lower education levels. It helps younger generations develop new learning strategies with a higher level of interactivity, and where some gamification approaches can make the content more appealing. Playing with situations and understanding the strategy needed to overcome the problems is the most effective way of learning something, and it is no different from playing games. "Through the Solar System" (Zhang & Wang, 2021) is an exciting example of the use of XR in science education.

According to (Scavarelli et al., 2021), the future of VR/AR in education will involve the use of a platform, not unlike current LMS/CMS systems used within educational institutions, built with more significant consideration of accessibility and the interplay between the virtual and physical, social and individual, in mind. However, the greatest challenge will be to use this technology not as a replacement for the physical classroom but to enable activities and access facilities that are impossible in physical settings (Arya et al., 2011).

Business

The previous section addressed the use of XR in business as a training tool, but its use in different business areas is growing as the equipment, know-how, and Internet speed evolve. In some business areas, XR is used as a planning tool, such as urban design and visualization. In contrast, tools like Arkio (Arkio, 2024) allow planners to visualize and design urban spaces interactively. XR may be used to conduct public consultations or to discuss urban planning within the community collaboratively. XR is beneficial when the virtual representation of a physical space is needed, and that is why it is being increasingly used in the actual state industry to showcase apartments or buildings that do not exist physically yet.

For the same reason, XR will also be crucial in Tourism and Hospitality as a tool for travelers to visualize accommodations and attractions and explore destinations, museums, and historical sites. Retail can also benefit from using XR, increasing the quality of product visualization, particularly when shopping online. The use of XR in the entertainment industry was, by its nature and target audience, one of the first areas of adoption, with the creation of new types of interaction in video games and physical spaces prepared for VR gaming experiences (VR Paintball, 2024). XR technology's effectiveness and success are undeniable, making its use in the military industry unavoidable due to the amount of money available to create and develop new war machines. The defense industry's armed forces widely use XR.

Besides its use in particular business-oriented applications, XR may shape the future of offices as a tool people use daily to work and interact with others. Several aspects must be improved to replace traditional computers and displays with XR devices: comfort, usability, and internet velocity. Nowadays, an increasing number of users are already using multiple displays and additional devices, such as tablets and phones, in the ecosystem of their workplace. Replacing these displays with their virtual counterparts seems promising and may present several benefits. Support for user collaboration may make this even more effective, with the same infrastructure being used as a personal workstation or a virtual place for meetings and presentations. The bottleneck, once again, is comfort and internet velocity.

Healthcare and Medicine

The use of XR in health goes from literacy and education to clinical or therapeutic use. XR in health can be seen as an effective way to increase health literacy by providing insightful information about health, medicine, and the signs people should look for to promote their health and well-being (Li et al., 2023). This approach works with kids or younger people, helping them to understand some biological principles of life and for adults and seniors to understand the mechanisms underlying certain diseases. Improving health and quality of life can also be done with virtual coaching that helps people exercise their bodies and minds. From physical training to cognitive stimulation, several solutions have already used VR immersion to create a proper environment for physical training and meditation or to develop some brain work (Supernatural, 2024).

However, the role of XR is also associated with its use as a professional tool. In medical training, XR applications provide a safe environment to practice medical procedures, teach medical concepts, and serve as a therapeutic tool. The Immersive Medical Tension Relief System (IMEDTRS, 2023), created by Inklusion Entertainment, a Portuguese company, is a practical example of the use of XR as a tool. It presents a solution to be used by patients in a medical appointment to reduce stress levels, for example, when performing some dental care procedures.

Accessibility and Inclusion

Extended reality encompasses many possibilities, although most examples use visual images and the Oculus. However, the solutions must be designed with accessibility features that make them usable to diverse user groups, with or without different impairments. XR may be presented as an assistive technology that provides audio descriptions and haptic feedback for visually impaired users. It can also generate real-time captioning and visual alerts to assist those with earring impairments. XR may allow the creation of immersive learning environments for students with special needs or provide training for ADL (Activities of daily living) or IADL (Instrumental Activities of Daily Living) for people with health conditions or impairments.

However, XR's role in promoting inclusion and the participation of those unable to participate may be even more subtle. Sharing an improved experience with those who are not able to join us in a particular physical space, like visiting a museum, going to a concert, or visiting a foreign country, is also a way to improve inclusion if those who can do it physically could virtually and in real-time participate in our experiences.

Cultural Preservation and Ecology

Above all, cultural preservation and ecology require people to understand their relevance, what they represent, and how they can be preserved. It involves everyone's participation, but it's even more relevant that the younger generations step forward to defend it. Generating awareness about cultural preservation, ecology, or environmental protection is the first step to ensuring its defense.

Communication plays a crucial role in educating without being boring. XR solutions may help present these topics to younger generations with more interactivity and innovative gamification approaches. Playing with the problems and "seeing" the results can make the content more appealing and exciting. Older people may also enjoy this approach, as it allows them to see this content at a more relaxed pace (like a documentary) and allows them to discover some aspects and details that are eventually presented too quickly in traditional formats. It also creates an opportunity to capture special groups like people with health conditions or impairments and people with different cultural backgrounds.

There are some interesting examples using these technologies in the context of the exhibits (Alexander et al., n.d.; Lacoche et al., 2017; Snibbe & Raffle, 2009; Sylaiou et al., 2010), using reality-based interactions (Jacob et al., 2008), simulating the visiting of the museum (Li et al., 2018), and using hybrid narratives across virtual and physical contexts (Kilteni et al., 2012).

XR FOR MEDIA TRAINING

The role of XR in enhancing interpersonal communication is vast and covers a large and diverse number of situations. As a starting point, it is valid to assume that XR may create a more immersive environment for people who are geographically apart to communicate with a level of proximity and intimacy that will not compete with the actual physical presence but will be far better than the tools we use today, like audio or video calls (Markowitz & Bailenson, 2019). The idea of using XR as the supporting technology for a meeting is the natural evolution of actual platforms where people interact using audio, video, and screen sharing using multiple displays simultaneously. XR may present a more natural way for the participants to interact, making them share the same "virtual" environment and look "virtually" in the same direction during a presentation. Using multiple displays in a meeting with just one camera in a fixed position sometimes creates that odd feeling that the participant is not paying attention or talking to the other participants. Extended Reality, creating a spatial coherence between all the participants, will help create a more interesting (virtual) meeting environment.

However, XR and VR are also exciting ways to train and test the behavior of communicators or people who need to be prepared for public speaking or other forms of social interaction. For example, XR may enhance a speaker's preparation to address a public audience in the same way that it can desensitize a participant in a debate, using a realistic animated representation of the contenders. XR may also test and evaluate nonverbal communication, mainly body language since the oculus covers the face. Training or preparation for public speaking is an area where XR is already being used (Bachmann et al., 2023), and a comparison between its use and the traditional methods is still limited since most of the studies focus on overcoming the anxiety of the presenter (Palmas et al., 2021). This means that the use of XR for public presentation training must be addressed at different levels, mainly as a teaching tool for those who are learning to talk in public, as a practical tool for those who need to overcome anxiety or who are not comfortable at speaking in front of a large audience and finally as a tool for professional presenters that want to test new ideas or use this type of solution as a warm-up procedure before the event.

XR applications may recreate the environment where the presentation takes place and prepare the presenter for the vision of the audience he will have when on stage. This can be used to teach solutions and evaluate how the presenter "looks" at the audience while speaking. However, the solution can be more sophisticated and generate noise or disturbing factors that could interfere with the quality of the presentation. Because the head and hand motions can be recorded, it's an exciting tool not only for immediate feedback but also for observing the evolution of the presenter. As an example, since the virtual environment contains the relative position of the projection screen and the audience, it makes it possible to measure the amount of time the presenter is not looking at the audience or if it is staring at a specific location, common mistakes by people that are not comfortable with speaking in public (VirtualSpeech - Soft Skills Training with VR, 2019).

But XR can also be used to prepare meetings and debates in a more personalized way and focused on a more direct confrontation. Commercial meetings and political discussions are distinct in how XR can be used. Still, both share the benefits of the team members being acquainted with the space and the relative position of the participants. Meetings sometimes have a script to support the narrative and interventions of the different participants, and programmed avatars or actors playing the role of the other intervenients can be very effective in creating a more natural flow. In political debates, XR can be used as a simulation tool, which can be used after the traditional analytics performed on the footage available about the opponent. The analysis's initial step typically involves identifying behavior patterns usually associated with facial expressions or other non-verbal forms. This initial labeling identifies the signs that indicate the eminence of an attack, or that "damage" has been inflicted. These patterns can be used in planned simulation trials, using an actor as an opponent or pre-programming an avatar. On the same path, the virtual opponent may desensitize the participant to questions or subjects that make him uncomfortable. Besides its role as a simulation tool, XR is also beneficial for evaluation, identification of behavior patterns that need to be improved, and an overall assessment of the evolution of the performance because, as mentioned, XR equipment keeps track of head movement and hand positioning. Collecting this type of measurement means that an XR-based solution makes it possible to identify the head positioning (where they are looking) and hand positioning (what they are doing with them).

What XR technology offers to this exciting area of communication makes it possible to state that this will be one of the areas with the most significant potential for expansion. VR-based solutions for public speaking, like Public Speach and Ovation (VR Speaking, 2022), are already available for different platforms. The secrecy evolving backstage of political work means that no solution is officially presented as being used for that purpose.

OVERVIEW

When discussing the potential for using XR in the communication sciences, the number of relevant areas and contexts in which its use will likely emerge soon stands out. This section identifies three main areas with different objectives and working methodologies: XR for content perception, XR for Society, and XR for media training.

XR for Content Perception explores how extended reality can be integrated into content production to personalize the user experience and make the content richer by allowing the user to immerse themselves in the main content at their own pace and according to their interests. In a non-exhaustive way, six types of content were identified, which, due to their characteristics, constitute an excellent opportunity to redesign or extend the user experience using complementary interactive XR content. These six types of content, with several overlaps, represent most of the content produced for the media and are identified as (i) News and Current Affairs, (ii) Documentaries, (iii) Sports, (iv) Drama and Fictional Series, (v) Recreational and Music Programmes, and (vi) Commercials and Advertising. For all of them, practical examples of using XR as complementary content were given, and practical aspects of its implementation were discussed.

XR for Society explores ways XR can improve literacy, develop new forms of communication that benefit society, and generate awareness about relevant topics. This area presents an alternative way to address some of society's significant challenges, such as education, health, and ecology. The five domains identified as the ones where XR can provide a more profound impact are (i) Education and Training, (ii) Business, (iii) Healthcare and medicine, (iv) Accessibility and Inclusion, and (v) Cultural Preservation and Ecology. As with XR for Content Perception, each context is discussed using practical examples and describing situations where XR can enhance perception and raise awareness about its problems.

Finally, XR for Media Training, as the name implies, discusses using XR to train and test communicators or people who need to be prepared for public speaking or other forms of social interaction, like debates or interviews. The use of XR-based tools for this purpose ranges from teaching how to present to constraining nonverbal aspects of communication or as a meeting or debate simulator. Globally, these three main areas reflect the potential that XR is expected to have or is already starting to have in communication sciences.

Challenges and Limitations of XR Applications

Despite all the advances and future perspectives, XR technology still needs to overcome several challenges and limitations to allow interactive applications to grow in importance and begin to be widely used (Ratcliffe et al., 2021). This section will cover the most challenging limitations, ranging from privacy and technical issues to barriers to adopting this type of application and the costs associated with its development.

Nowadays, data privacy is one of the significant ethical issues technological applications faces. In XR, massive data, including sensitive personal information (e.g., location, movement, or facial expressions), is used to create and deploy interactive experiences. Using this information may lead to privacy violations that bad-intentioned individuals can exploit (Krauß et al., 2023). Another known issue is the automated treatment systems and algorithms performing on top of data. In the context of XR, this type of approach can lead to unwanted effects, such as behavior and perception manipulation. For example, these algorithms can target advertising in a particular way, raising questions about manipulation and consent. XR also raises concerns about information protection, such as its collection, storage, and use. The collected information, including biometric data or information regarding the user's surroundings, can be susceptible. Preventing data exposure is essential to ensure its confidentiality. However, this can be particularly problematic since users may share sensitive information without realizing it due to the immersive nature of XR technology.

One of the biggest obstacles to deploying XR applications is related to hardware and software limitations. Hardware devices such as servers, network devices, or storage may limit the performance and scalability of XR applications, forcing developers to implement efficient resource utilization strategies. Since XR hardware technologies will evolve rapidly, regular upgrades are required to maintain relevance and compatibility. Besides hardware limitations, it is also essential to note that developing applications that will be deployed for multiple XR platforms represents an additional challenge since each platform has its API (Application Program Interface), SDK (Software Development Kit), and requirements. An API defines how an application communicates with the platform, while an SDK is a collection of tools (software) to develop applications for that specific platform. This means that developing for different platforms with different APIs and SDKs represents recreating the same application several times with different programming code. This limitation makes XR integration in the current content management systems and other marketing and communication tools more complex and time-consuming. Another critical obstacle is the level of optimization required. The software must be optimized to guarantee a fluid and immersive experience on devices that still have limited capabilities and resources.

Creating immersive and high-standard XR content requires 3D modeling, different design elements, animation, and specialized programming. The lack of quality in the content produced will affect adoption, as users expect increasingly immersive and realistic experiences. The technical complexity needed increases development costs, especially if the team has yet to gain experience with XR. Those solutions rely on specific devices, such as VR/AR glasses, compatible smartphones, or headsets. The equipment can still be expensive for end users, and companies wishing to adopt XR solutions must consider the costs of acquiring and maintaining these devices. Also, integrating XR solutions with existing IT systems (such as databases, servers, and applications) can take time and effort, which leads to increasing costs. As always, adding to the initial development costs, it's also important to consider additional costs associated with updates, bug fixes, and ongoing improvements that exist in every project but are more prevalent when dealing with new emerging technologies. Globally, companies must evaluate how XR solutions fit into their current infrastructures and consider these costs over the long term. The adoption of XR solutions depends on public acceptance. If users don't see value or usefulness, adoption can be slow. Educating the public about the benefits of XR solutions is essential.

As mentioned before, the main challenge associated with adopting XR in the different communication areas is adapting the traditional production pipeline to accommodate the planning of all the elements needed to support XR so that the additional content can be available on time. It's essential to notice that these changes also impact the structure of the production team.

A further and more holistic challenge is the emergence of multidisciplinary teams that look at final media products in a unified way. In these teams, these new types of media content are seen and identified as transmedia elements that should be produced almost simultaneously.

Future Directions and XR TRENDS in Communication

The future of XR is mainly constrained by the internet infrastructure, which needs to increase in velocity and throughput for the connected devices to be able to share the increasing amount of information that a virtual environment requires. This will be achieved in a couple of years, with latency (understood as the perceived delay) as one of the most problematic aspects that needs to be solved.

Besides the technical limitations, XR will be increasingly used in virtual events and remote meetings, with participants emulating in a 3D space what they do in Microsoft Teams, Google Meet, Zoom, or other online meeting platforms. As stated before, a virtual environment creates a more coherent interaction space. It provides spatial orientation to the participants, giving them a sense of relative positioning that cannot be achieved with today's platforms. Using XR also creates the opportunity to design the location where the meeting takes place, and tools like XR event planner from Accenture allow the virtual reconstruction of the physical locations, facilitating remote viewing and event configuration.

XR may improve productivity for those who spend a lot of time in online meetings, reducing videoconference fatigue and minimizing distractions like message notifications or emails.

Collaborative XR environments will trigger several opportunities discussed in Section 2, creating opportunities to use those immersive spaces to develop interactive narratives in VR or AR and reinventing the transmedia approach by creating opportunities to experience the narrative from different perspectives. Films, documentaries, and even books can benefit from the general adoption of XR devices, providing a new dimension to storytelling.

XR allows natural forms of interaction, with the viewers being able to interact and having the opportunity to actively participate in discovering new story elements or hidden aspects of the plot. This will help create emotional bonds with the audience and retain the viewers for more extended periods. And in the future, XR environments will offer richer, immersive experiences involving multiple senses. Sounds, sights, and tactile sensations can be incorporated, making the experience more meaningful and memorable.

CONCLUSION

Extended Reality in Communication is a meeting point of multiple areas and represents a massive opportunity for innovation. It also poses many problems articulating different techniques, languages, and processes. However, the potential of XR in communication is enormous and may constitute a special communication channel for the new generations and the communities associated with video games and emerging technologies, targets that are less connected with traditional media. On the other hand, there are still numerous difficulties in using XR solutions, firstly because of the equipment's price and discomfort and secondly because the competitive advantage of using them has yet to be evident to everyone. XR solutions require exciting and specific content to motivate their adoption, together with robust technological support. This chapter has tried to highlight the opportunities arising from technology, creativity, and the development of new communication solutions.

The growth of XR solutions will probably be associated with what we may call 'adoption for ease,' i.e., the effort related to adopting the technology must be offset by the added facilities created by using it. This way, it is possible to identify some areas of communication where this adoption will be naturally faster. The areas that are already using 3D models and visual representations of scenarios, like the news and documentaries, are probably the first adopters. Those products will also allow increasing levels of personalization to the user. For example, the details of a map or a 3D representation of a historic building or infrastructure may be viewed and explored at their own pace and with the level of detail desired. With the boom of social networks, people have gotten used to increasing content personalization and go to the Internet for additional information while seeing this type of content. This type of content may have some level of gamification associated with it, providing an opportunity for social interaction and attracting a younger audience. XR may also be used in movies and TV series, enabling the audience to explore the scenario or to better understand the setting, especially if there are some historical elements involved. However, from a fictional perspective, XR allows a deeper exploration of a transmedia narrative, where the user/viewer can unfold additional elements related to a specific character or a side story.

The use of XR in education and training is one of the areas where this technology is much more established, especially in areas where the price of the actual equipment or the danger involved justifies the investment. Urban planning and real estate are also starting to use XR to visualize spaces that still need to exist physically. For the same reason, XR will also be crucial in Tourism and Hospitality as a tool for travelers to visualize accommodations and attractions and explore destinations, museums, and historical sites. Generating awareness about cultural preservation, ecology, and environmental protection, which require society to understand their relevance, can benefit from using XR to make the message more appealing and effective to the younger generations. As the supporting technology for a meeting, XR is the natural evolution of platforms where people interact using audio, video, and screen sharing using multiple displays simultaneously. Extended Reality allows the creation of a spatial coherence between all the participants that doesn't exist today. It will be (already is) a tool for training people to be prepared for public speaking and to prepare the participants in meetings and debates where a more direct confrontation is expected.

Despite all the advances already made by XR technologies, some limitations still prevent them from proliferating further. Accessibility and inclusion must be considered, as there is a significant gap in understanding the needs of people with disabilities in immersive environments. Using traditional development standards and best practices does not solve the challenge of XR technologies in this field. Recent studies highlight the importance of considering these concerns from the outset of developing XR experiences rather than relying on third-party solutions (Mott et al., 2019). Still, on this subject, there is a need to study the effect of XR technologies on users' mental health. Although their use is seen as beneficial, no studies still show their long-term effect on the human brain. In this way, the future of XR technologies should involve creating ubiquitous and continuous experiences in our lives that empower individuals and promote an inclusive digital environment (Kenwright, 2023).

In conclusion, we rarely have the privilege of being part of a revolution, and even more rarely do we can be aware that we are in the middle of it. A revolution capable of changing the way we live in society. The latest examples might be the emergence of the internet or, more recently, the appearance of smartphones. So, looking over all the existing difficulties and constraints, it's important to think about the pleasure of being part of the change, which, as in previous cases, takes us to a reality that we can hardly imagine today. And the revolution has already begun!

REFERENCES

6. 9: A virtual experience of solitary confinement. (n.d.). The Guardian. https://www.theguardian.com/world/ng-interactive/2016/apr/27/6x9-a-virtual-experience-of-solitary-confinement

Al-Amri, S., Hamid, S., Noor, N. F. M., & Gani, A. (2023). A framework for designing interactive mobile training course content using augmented reality. *Multimedia Tools and Applications*, 82(20), 30491–30541. Advance online publication. DOI: 10.1007/s11042-023-14561-4 PMID: 36855614

Alexander, J., Barton, J., & Goeser, C. (n.d.). Transforming the art museum experience: gallery one. Museumsandtheweb.Com. Arya, A., Hartwick, P., Graham, S., & Nowlan, N. (n.d.). Virtual Space as a Learning Environment: Two Case Studies. http://www.avaya.com

Arkio. (2024). Arkio. Arkio. https://www.arkio.is/

Bachmann, M., Subramaniam, A., Born, J., & Weibel, D. (2023). Virtual reality public speaking training: Effectiveness and user technology acceptance. *Frontiers in Virtual Reality*, 4, 1242544. Advance online publication. DOI: 10.3389/frvir.2023.1242544

Bangay, S., & McKenzie, S. (2022). A Classification Schema for Designing Augmented Reality Experiences. *Informing Science and IT Education Conference*. DOI: 10.28945/4942

Bansal, G., Rajgopal, K., Chamola, V., Xiong, Z., & Niyato, D. (2022). Healthcare in Metaverse: A Survey on Current Metaverse Applications in Healthcare. *IEEE Access : Practical Innovations, Open Solutions*, 10, 1–1. DOI: 10.1109/ACCESS.2022.3219845

Burdea, G. C., & Coiffet, P. (2024). *Virtual Reality Technology* (3rd ed.). John Wiley & Sons Inc. DOI: 10.1002/9781119512608

Elliott, C., Rose, M., & van Arnhem, J.-P. (2018). *Augmented and virtual reality in libraries*. Rowman and Littlefield.

Experience Doja Cat's Concert in VR. (2024, January 9). Meta. https://about.fb.com/news/2024/01/doja-cat-vr-8concert

Fidalgo, C. G., Yan, Y., Cho, H., Sousa, M., Lindlbauer, D., & Jorge, J. (2023). A Survey on ijuRemote Assistance and Training in Mixed Reality Environments. 29(5), 2291–2303. .DOI: 10.1109/TVCG.2023.3247081

Flavián, C., Ibáñez-Sánchez, S., & Orús, C. (2019, July 1). The impact of virtual, augmented, and mixed reality technologies on the customer experience. *Journal of Business Research*, 100, 547–560. DOI: 10.1016/j.jbusres.2018.10.050

Fraga-Lamas, P., Fernandez-Carames, T. M., Blanco-Novoa, O., & Vilar-Montesinos, M. A. (2018). A Review on Industrial Augmented Reality Systems for the Industry 4.0 Shipyard. *IEEE Access : Practical Innovations, Open Solutions*, 6, 13358–13375. DOI: 10.1109/ACCESS.2018.2808326

Ge, R., & Hsiao, T.-C. (2020, May 1). A Summary of Virtual Reality, Augmented Reality and Mixed Reality Technologies in Film and Television Creative Industries. IEEE Xplore. .DOI: 10.1109/ECBIOS50299.2020.9203607

Geroimenko, V. (2020). *Augmented Reality in Education. Springer Series on Cultural Computing*. Springer International Publishing. DOI: 10.1007/978-3-030-42156-4

Gerry, S., Koschmann, T., & Suthers, D. (2006). Computer-Supported Collaborative Learning. In *Cambridge handbook of the learning science* (pp. 409–426). CambridgeUniversity Press.

Grubert, J., Langlotz, T., Zollmann, S., & Regenbrecht, H. (2017). Towards Pervasive Augmented Reality: Context-Awareness in Augmented Reality. *IEEE Transactions on Visualization and Computer Graphics*, 23(6), 1706–1724. DOI: 10.1109/TVCG.2016.2543720 PMID: 27008668

Hamilton, T., II. E. C., Hamilton, T., Welsh, C., II, E. C., & Welsh, C. (2017, February 20). Glitched: A VR Talk Show. IMDb. https://www.imdb.com/title/tt10318060/

Han, Y., Miao, Y., Sha, H., & Liu, Y. (2024). Exploring Influences of Appearance and Voice Realism of Virtual Humans on Communication Effectiveness in Social Virtual Reality. .DOI: 10.1109/VRW62533.2024.00157

IMEDTRS. (2023). IMEDTRS. https://imedtrs.com/

Jacob, R. J. K., Girouard, A., Hirshfield, L. M., Horn, M. S., Shaer, O., Solovey, E. T., & Zigelbaum, J. (2008). Reality-based interaction. *Proceedings of the SIGCHI Conference on Human Factors in Computing Systems*, 201–210. DOI: 10.1145/1357054.1357089

Kasowski, J., Johnson, B. A., Neydavood, R., Akkaraju, A., & Beyeler, M. (2023). A systematic review of extended reality (XR) for understanding and augmenting vision loss. *Journal of Vision (Charlottesville, Va.)*, 23(5), 5. DOI: 10.1167/jov.23.5.5 PMID: 37140911

Kenwright, B. (2023). Impact of XR on Mental Health: Are we Playing with Fire? Cornell University. https://doi.org//arxiv.2304.01648.DOI: 10.48550

Kiesberg, P. (2015). A structured approach to defense simulation train- ing. *ERCIM News*, •••, 28–29.

Kilteni, K., Groten, R., & Slater, M. (2012). The Sense of Embodiment in virtual reality. In Presence: Teleoperators and Virtual Environments (Vol. 21, Issue 4, pp. 373–387). MIT Press Journals. DOI: 10.1162/PRES_a_00124

Krauß, V., Berkholz, J., Recki, L., & Boden, A. (2023). Beyond Well-Intentioned: An HCI Students' Ethical Assessment of Their Own XR Designs. .DOI: 10.1109/ISMAR59233.2023.00020

Lacoche, J., Pallamin, N., Boggini, T., & Royan, J. (2017). Collaborators awareness for user cohabitation in co-located collaborative virtual environments. *Proceedings of the 23rd ACM Symposium on Virtual Reality Software and Technology*, 1–9. DOI: 10.1145/3139131.3139142

Lee, B., Sedlmair, M., & Schmalstieg, D. (2023). Design Patterns for Situated Visualization in Augmented Reality. *IEEE Transactions on Visualization and Computer Graphics*, •••, 1–12. DOI: 10.1109/TVCG.2023.3327398 PMID: 37883275

Li, B. J., Peña, J., & Jung, Y. (2023). Editorial: VR/AR and wellbeing: The use of immersive technologies in promoting health outcomes. *Frontiers in Virtual Reality*, 3, 1119919. Advance online publication. DOI: 10.3389/frvir.2022.1119919

Li, Y., Ch'ng, E., Cai, S., & See, S. (2018). Multiuser Interaction with Hybrid VR and AR for Cultural Heritage Objects. 2018 3rd Digital Heritage International Congress (DigitalHERITAGE) Held Jointly with 2018 24th International Conference on Virtual Systems & Multimedia (VSMM 2018), 1–8. DOI: 10.1109/DigitalHeritage.2018.8810126

Markowitz, D., & Bailenson, J. (2019). *Virtual Reality and Communication*. Communication., DOI: 10.1093/obo/9780199756841-0222

Merriam, S., & Bierema, L. (2005). *Adult Learning: Linking Theory and Practice*. Willey.

Moglia, A., Ferrari, V., Morelli, L., Ferrari, M., Mosca, F., & Cuschieri, A. (2016). A Systematic Review of Virtual Reality Simulators for Robot-assisted Surgery. *European Urology*, 69(6), 1065–1080. DOI: 10.1016/j.eururo.2015.09.021 PMID: 26433570

Mott, M., Cutrell, E., Gonzalez Franco, M., Holz, C., Ofek, E., Stoakley, R., & Ringel Morris, M. (2019). Accessible by Design: An Opportunity for Virtual Reality. 2019 IEEE International Symposium on Mixed and Augmented Reality Adjunct (ISMAR-Adjunct). DOI: 10.1109/ISMAR-Adjunct.2019.00122

Mudička, Š., & Kapica, R. (2022). Digital Heritage, the Possibilities of Information Visualisation through Extended Reality Tools. *Heritage*, 6(1), 112–131. DOI: 10.3390/heritage6010006

Muñoz-Saavedra, L., Miró-Amarante, L., & Domínguez-Morales, M. (2020). Augmented and Virtual Reality Evolution and Future Tendency. *Applied Sciences (Basel, Switzerland)*, 10(1), 322. DOI: 10.3390/app10010322

Newman, N. (2023). Journalism, media, and technology trends and predictions 2023. In ora.ox.ac.uk. Reuters Insitute for the Study of Journalism. https://ora.ox.ac.uk/objects/uuid:38c0393f-3634-459a-a505-c53f2eb637a3

News, A. B. C. (2024). Virtual Reality. ABC News. https://abcnews.go.com/VR

Paintball, V. R. (2024). Fun in Prague. Retrieved July 20, 2024, from https://funinprague.eu/en/vr-paintball

Palmas, F., Reinelt, R., Cichor, J. E., Plecher, D. A., & Klinker, G. (2021, March 1). Virtual Reality Public Speaking Training: Experimental Evaluation of Direct Feedback Technology Acceptance. IEEE Xplore. .DOI: 10.1109/VR50410.2021.00070

Pan, X., Zheng, M., Xu, X., & Campbell, A. G. (2021). Knowing Your Student: Targeted Teaching Decision Support Through Asymmetric Mixed Reality Collaborative Learning. *IEEE Access : Practical Innovations, Open Solutions*, 9, 164742–164751. DOI: 10.1109/ACCESS.2021.3134589

Pausch, R., Crea, T., & Conway, M. (1992). A Literature Survey for Virtual Environments: Military Flight Simulator Visual Systems and Simulator Sickness. *Presence (Cambridge, Mass.)*, 1(3), 344–363. DOI: 10.1162/pres.1992.1.3.344

Ratcliffe, J., Soave, F., Bryan-Kinns, N., Tokarchuk, L., & Farkhatdinov, I. (2021). Extended Reality (XR) Remote Research: a Survey of Drawbacks and Opportunities. *Proceedings of the 2021 CHI Conference on Human Factors in Computing Systems*. DOI: 10.1145/3411764.3445170

Rauschnabel, P. A., Babin, B. J., tom Dieck, M. C., Krey, N., & Jung, T. (2022). What is augmented reality marketing? Its definition, complexity, and future. *Journal of Business Research*, 142(1), 1140–1150. DOI: 10.1016/j.jbusres.2021.12.084

Rheingold, H. (1992).. . *Virtual Reality (Waltham Cross)*.

Santos, T. T., Piuvezam, G., Medeiros, G. C. B. S., & Mata, Á. N. de S. (2023). Extended reality as a health education strategy of adolescents at school: Protocol for systematic review and meta-analysis. *BMJ Open*, 13(7), e072438. DOI: 10.1136/bmjopen-2023-072438 PMID: 37407033

Scavarelli, A., Arya, · Ali, Robert, ·, & Teather, J. (2021). Virtual reality and augmented reality in social learning spaces: a literature review. 25, 257–277. .DOI: 10.1007/s10055-020-00444-8

Dieter Schmalstieg, & Hollerer, T. (2016). Augmented Reality: Principles and Practices. Addison-Wesley Professional.

Snibbe, S. S., & Raffle, H. S. (2009). Social immersive media: Pursuing best practices for multi-user interactive camera/projector exhibits. Conference on Human Factors in Computing Systems - Proceedings, 1447–1456. DOI: 10.1145/1518701.1518920

Speaking, V. R. LLC (2022). Ovationvr.com. https://www.ovationvr.com/

Statista - The Statistics Portal. (n.d.). Statista. Retrieved July 20, 2024, from https://www.statista.com/search/?q=VR+headsets+sold&p=1

Steed, A., Pan, Y., Watson, Z., & Slater, M. (2018). "We Wait"—The Impact of Character Responsiveness and Self Embodiment on Presence and Interest in an Immersive News Experience. *Frontiers in Robotics and AI*, 5, 112. Advance online publication. DOI: 10.3389/frobt.2018.00112 PMID: 33500991

Steuer, J. (1992). Defining Virtual Reality: Dimensions Determining Telepresence. *Journal of Communication*, 42(4), 73–93. DOI: 10.1111/j.1460-2466.1992.tb00812.x

Supernatural. (n.d.). Supernatural. https://www.getsupernatural.com/

Sylaiou, S., Mania, K., Karoulis, A., & White, M. (2010). Exploring the relationship between presence and enjoyment in a virtual museum. *International Journal of Human-Computer Studies*, 68(5), 243–253. DOI: 10.1016/j.ijhcs.2009.11.002

Times, T. N. Y. (2016, May 19). Seeking Pluto's Frigid Heart. The New York Times. https://www.nytimes.com/interactive/2016/05/19/science/space/seeking-plutos-frigid-heart-nytvr.html

VirtualSpeech - Soft Skills Training with VR. (2019). Virtualspeech.com. https://virtualspeech.com/

Yun-chao, M., Ren, M., & Li, X. (2023). Application of virtual simulation technology in sports decision training: A systematic review. *Frontiers in Psychology*, 14, 1164117. Advance online publication. DOI: 10.3389/fpsyg.2023.1164117 PMID: 37275736

Zhang, Y., & Wang, J. (2021). "Through the Solar System" - XR science education system based on multiple monitors. 2021 IEEE Conference on Virtual Reality and 3D User Interfaces Abstracts and Workshops (VRW). DOI: 10.1109/VRW52623.2021.00096

Chapter 3 Communication Technology and Innovation: Trends, Opportunities, and Challenges

Anushka Islam Mim
https://orcid.org/0009-0000-2999-7236
Islamic University of Technology, Bangladesh

Md. Abul Kalam Azad
Islamic University of Technology, Bangladesh

ABSTRACT

This chapter delves into the reciprocal relationship between communication technology and innovation. It traces the historical evolution of communication technology, from the telegraph to the digital age, analyzing how each technological breakthrough transformed communication practices, societal structures, and human interactions. It assesses the multifaceted impact of communication technology across sectors like business, education, healthcare, and governance, illustrating its role in driving economic growth, facilitating collaboration, and empowering individuals. Highlighting the catalytic effect of technological advancements on innovation, the chapter explores the impact of communication infrastructure investments on economic development, job creation, and growth. Furthermore, it explores emerging trends such as 5G, AI, and IoT, projecting their transformative potential across various industries.

DOI: 10.4018/979-8-3693-3579-6.ch003

1.INTRODUCTION

Today, technology has seamlessly integrated into our daily communication patterns, molding the ways in which individuals, businesses, and societies exchange information. The rapid evolution of communication technology has catalyzed transformations in interaction modalities. It has also acted as a catalyst for novel innovations across diverse sectors. This chapter delves into the dynamic landscape of communication technology and its pervasive impact on our daily lives (Binkley et al., 2012; Järvelä & Renninger, 2014).

1.1 What is Communication Technology?

Consider a situation where your friend resides abroad, and you wish to converse with them. Even a few years ago, this would have been an unbelievable act. However, today, it is not only possible to engage in verbal communication but also to visually connect through video calls. This remarkable capability is made possible by communication technology. It has redefined the very concept of communication, rendering the entire process wholly dependent on technology. In essence, communication technology refers to the application of technology in facilitating communication. Briefly stated, communication technology encompasses the tools, systems, and methods employed to transmit, receive, and exchange information. This encompasses a wide array of technologies, including telephones, email, social media, video conferencing, and many others. The overarching goal is to enable efficient and effective communication between individuals or groups, often transcending vast distances and spanning various platforms. It encompasses a diverse range of tools and systems meticulously designed to facilitate the exchange of data, messages, and media between individuals or groups.

The history of communication technology is complicatedly woven throughout human history, evolving from basic forms of communication to the sophisticated systems of the present day. Early humans relied on non-verbal forms of communication, including gestures, body language, facial expressions, drums, and smoke signals for long-distance communication. Ancient civilizations developed writing systems, such as cuneiform in Mesopotamia and hieroglyphs in Egypt. The invention of the printing press revolutionized information dissemination, rendering books and printed materials more accessible. The development of the telegraph, enabling long-distance communication through electrical signals, marked a significant milestone. The telephone, invented by Alexander Graham Bell, further transformed personal and business communication. The beginning of broadcast communication, with the introduction of radio and later television, brought news and entertainment to a broad audience. The rise of computers and the internet, particularly the last, has had a profound impact on global communication, connecting people across the world in real-time. The development of mobile phones marked a significant shift, allowing individuals to communicate wirelessly on the go. Smartphones further revolutionized communication by integrating various functions into a single device. The 21st century indicated the rise of social media platforms and the concept of Web 2.0, emphasizing user-generated content and interactive communication, which has transformed how people share information and connect online. Ongoing innovations in artificial intelligence, virtual reality, and other emerging technologies continue to shape the landscape of communication, introducing new possibilities and challenges.

2. INNOVATION

Today, we find ourselves immersed in the era of modern technology, where every day ushers in novel inventions that challenge and reshape the ways in which tasks were previously accomplished. Electric vehicles, smartphones, online streaming services, 3D printing, blockchain technology, and artificial intelligence (AI) are but a few examples of innovation. Innovation is the process of creating something entirely new or significantly improving upon existing ideas, products, services, processes, or methods to address a need, solve a problem, or capitalize on an opportunity. Ultimately, innovation boosts progress, fosters growth, and paves the way for advancements that mold the manner in which we live, work, and interact with the world.

2.1 Innovation in Communication Technology

Communication technologies have undergone a remarkable transformation over the years, driven by continuous innovation. From the pioneering inventions of the telegraph and telephone to the groundbreaking advent of the internet and smartphones, we have witnessed a relentless evolutionary journey. This ongoing innovation in communication technology continues to reshape the ways individuals and organizations interact, collaborate, and exchange information in an increasingly interconnected global landscape. Driving this innovation are several key factors, including advancements in semiconductor technology, which have enabled faster processors and smaller, more efficient devices. Moreover, the ever-growing demand from consumers and businesses for speedier, more reliable, and convenient communication fuels the innovation engine. Companies respond by developing products and services that provide to these needs, such as higher internet speeds, improved connectivity, and more intuitive user interfaces. The competitive landscape in the communication technology sector further accelerates innovation, as companies strive to gain an edge through faster product development cycles, enhanced features, and more affordable solutions for consumers. The increasing interconnectedness of the global economy presents opportunities for cross-border collaboration and innovation, allowing companies to tap into a diverse pool of talent, resources, and ideas from around the world to develop cutting-edge communication technologies and solutions.

Furthermore, companies actively gather and analyze data on how people utilize communication technologies, incorporating this valuable feedback into their product design and development processes. This user-centric approach enables the creation of solutions that more effectively meet the evolving needs and preferences of users. Emerging trends such as the Internet of Things (IoT), artificial intelligence (AI), virtual reality (VR), and 5G networking open up new frontiers for innovation in communication technology. Companies are actively exploring these transformative technologies to create novel communication experiences and capabilities that push the boundaries of what was previously possible. Collaboration between companies, research institutions, and government agencies can serve as a powerful catalyst for accelerating innovation by pooling resources, expertise, and knowledge. This synergistic approach fosters advancements in communication technology that improve connectivity, enhance user experiences, and fundamentally transform the ways in which people interact and communicate.

Innovative factors play a pivotal role in shaping the development and evolution of communication technology in several profound ways. They drive the rapid pace of advancement, leading to the continuous development of faster, more efficient, and reliable communication systems. Innovation gives rise to a diverse array of communication technology options, ranging from traditional landline telephones to modern solutions such as VoIP, social media platforms, and instant messaging apps, catering to the diverse needs and preferences of users.

Furthermore, innovations in communication technology continually improve connectivity, enabling seamless communication across vast distances through various channels, including text, voice, video, and data transmission. Innovative factors also focus on enhancing communicative capabilities and cost-efficiency, pushing the boundaries of what is possible in communication technology. This opens up new avenues for collaboration, creativity, and productivity through features like real-time collaboration, virtual meetings, and augmented reality communication. On a global scale, innovation in communication technology has a profound impact, breaking down geographical barriers, fostering cultural exchange, and facilitating economic growth and development through improved connectivity and collaboration. Ultimately, innovative factors drive the continuous improvement, expansion, and transformation of communication technology, shaping the ways in which people communicate, connect, and collaborate in today's interconnected world.

2.1. Artificial Intelligence

Suppose, you live alone and to remove your loneliness you are using chatbots to chat with. Chatbots are AI generated program. Artificial Intelligence (AI) refers to the simulation of human intelligence in machines, enabling them to perform tasks that typically require human intelligence. This includes learning, reasoning, problem-solving, perception, understanding natural language, and interacting with the environment. AI systems are designed to analyze data, make decisions, and adapt based on new information or experiences, often with the goal of achieving specific objectives or solving complex problems. AI encompasses a broad range of techniques and approaches. The ultimate aim of AI is to create systems that can autonomously replicate or even surpass human-level intelligence across various domains and applications.

Artificial Intelligence (AI) encompasses various subfields, each focusing on specific aspects of intelligent behavior and problem-solving. Some key subfields of AI include:

Machine Learning: Machine learning, a subset of AI that focus on developing algorithms and techniques that enable computers to learn from and make predictions or decisions based on data. This includes supervised learning, unsupervised learning, reinforcement learning, and deep learning.

Natural Language Processing (NLP): NLP involves the interaction between computers and humans using natural language. It includes tasks such as language translation, sentiment analysis, text summarization, and speech recognition.

Computer Vision: Computer vision enables computers to interpret and understand visual information from images or videos. This includes object detection, image classification, facial recognition, and image generation.

Robotics: Robotics combines AI with mechanical engineering to design, build, and program robots capable of performing tasks autonomously or with human assistance. Applications range from industrial automation to service robots in healthcare and household settings.

Expert Systems: Expert systems use AI techniques to emulate the decision-making abilities of human experts in specific domains. They rely on knowledge bases and inference engines to provide expert-level advice or solutions.

Knowledge Representation and Reasoning: This subfield focuses on how knowledge can be represented in a machine-readable format and how reasoning processes can manipulate this knowledge to reach conclusions or solve problems.

Reinforcement Learning: Reinforcement learning involves training AI agents to make sequential decisions by interacting with an environment and receiving feedback in the form of rewards or penalties. It is commonly used in autonomous systems and game playing.
Evolutionary Algorithms: Evolutionary algorithms are inspired by biological evolution and use techniques such as genetic algorithms and genetic programming to solve optimization and search problems.
AI Ethics and Fairness: With the increasing deployment of AI systems, there is a growing focus on addressing ethical considerations such as bias, fairness, transparency, and accountability in AI development and deployment.

These subfields of AI collectively contribute to the advancement and application of intelligent systems across various domains, from healthcare and finance to transportation and entertainment.

The purpose of Artificial Intelligence (AI) is multifaceted and encompasses various objectives and goals.

For instance, AI enables the automation of repetitive tasks, freeing up human resources to focus on more creative or strategic activities. This includes automating processes in industries such as manufacturing, customer service, finance, and healthcare.AI systems can analyze vast amounts of data, identify patterns, and make data-driven decisions or recommendations. This is particularly valuable in complex and dynamic environments where human decision-making may be limited by cognitive biases or information overload. AI technologies optimize processes and workflows, leading to increased efficiency and productivity, supply chains, resource allocation, scheduling, and logistics. AI also enables personalized experiences and recommendations by analyzing user data and preferences. Like, personalized product recommendations, content curation, and targeted marketing campaigns. AI systems excel at solving complex problems, ranging from scientific research and medical diagnosis to financial modeling and environmental analysis. It also fosters innovation by enabling the exploration of new ideas, designs, and solutions through techniques such as generative modeling and automated creativity. This includes applications in art, design, music, and literature. By advancing healthcare through personalized medicine, disease diagnosis, and treatment optimization, it enhances accessibility through assistive technologies for people with disabilities. The most crucial role AI plays is enhancing safety and security by analyzing data to detect anomalies, identify threats, and prevent risks in areas such as cybersecurity, surveillance, and disaster response. Nowadays, AI is also used to optimize energy consumption, resource utilization, and waste management, contributing to environmental sustainability. Ultimately, the purpose of AI is to augment human capabilities, enabling individuals and organizations to achieve their goals more effectively and efficiently. By complementing human intelligence with machine intelligence, AI has the potential to address complex challenges and unlock new opportunities for societal advancement.

AI also plays a crucial role in communication technology by enhancing various aspects such as: AI-powered chatbots and virtual assistants facilitate seamless interactions between users and systems, providing real-time assistance and automating responses to inquiries. AI-driven translation tools enable instant translation of text and speech, breaking down language barriers and fostering global communication. AI algorithms accurately transcribe spoken words into text, powering voice commands, dictation software, and voice-enabled interfaces, making communication more accessible and efficient. AI analyzes text and speech to gauge sentiment, allowing businesses to understand customer feedback, sentiment trends, and public opinion, enabling them to tailor communication strategies accordingly. It's algorithms analyze user data to deliver personalized communication experiences, such as targeted advertising, content recommendations, and customized messaging, improving user engagement and satisfaction. It also optimizes network traffic, predicts demand, and identifies potential issues to ensure reliable and efficient communication services, enhancing network performance and user experience.

Overall, AI enhances communication technology by enabling more efficient, personalized, and accessible communication experiences across various channels and platforms.

2.1.b Augmented Reality

Augmented reality (AR) is a cutting-edge technology that seamlessly blends digital content with the real-world environment, creating an immersive and interactive experience for users. Through various devices, such as smartphones, tablets, or specialized AR glasses, computer-generated elements like images, text, or animations are superimposed onto the user's perception of the physical surroundings, effectively merging the virtual and real worlds.

The key aspects of AR technology are multifaceted. Firstly, it allows for the overlaying of digital information onto the user's view of the tangible world, enhancing their perception of reality. Secondly, AR experiences often enable real-time interaction with virtual objects, fostering engagement and immersion. Thirdly, AR systems employ sensors and advanced algorithms to recognize and comprehend the user's surroundings, enabling accurate placement of virtual elements within that space. Additionally, AR applications can adapt to the user's environment, providing contextually relevant information based on factors such as location, object recognition, or user input.

The applications of AR span a diverse range of domains, including gaming, education, healthcare, retail, navigation, and design, offering practical solutions and enhancing experiences across various industries. However, AR experiences require compatible hardware, such as smartphones with AR capabilities or dedicated AR glasses, to view and interact with the virtual content.

Furthermore, AR often integrates with other cutting-edge technologies, such as computer vision, machine learning, and spatial mapping, to deliver seamless and immersive experiences. While AR holds immense potential, it also faces challenges, including ensuring accurate tracking, optimizing performance across different devices, addressing privacy concerns, and creating compelling content that enhances rather than distracts from the real world.

In the realm of communication technology, augmented reality has found numerous applications, transforming the way we interact and engage with others. Popular platforms like Snapchat, Facebook Messenger, and Instagram offer a wide range of AR filters, known as Lenses, that overlay virtual effects onto users' faces in real-time during video calls or when capturing photos and videos. These filters allow for creative self-expression and add an element of fun to digital communication. Google Duo and Microsoft Teams have also embraced AR technology, incorporating AR effects and filters into video calls, enabling users to add engaging virtual elements like masks, backgrounds, and animations to their conversations. Microsoft Teams' Together Mode takes it a step further by using AI segmentation technology to place participants in a shared virtual space, creating a more immersive and collaborative meeting experience.

While not strictly AR, platforms like Zoom have introduced virtual background features that allow users to replace their real backgrounds with virtual images or videos, enhancing the visual communication experience. Additionally, various messaging platforms and smartphones offer AR emojis, which use facial tracking and animation to create personalized 3D avatars that mimic users' expressions and gestures during video calls or messaging (Grubert, Langlotz, Zollmann, & Regenbrecht, 2016). These examples illustrate how augmented reality has transformed communication technology, adding immersive and interactive elements to conversations, making them more engaging, enjoyable, and expressive for users. As AR technology continues to evolve, its applications in communication and beyond will undoubtedly expand, offering new and innovative ways to interact with digital content and enhance our experiences in the real world (Billinghurst, Clark, & Lee, 2015).

2.1.c Virtual Reality

Virtual reality (VR) is a computer-generated simulation that transports users into a fully immersive, three-dimensional digital environment. Through the use of specialized VR headsets or goggles, individuals can explore and interact with this virtual world as if they were physically present within it. This simulated experience often incorporates various sensory inputs, such as visual, auditory, and sometimes tactile, to create a heightened sense of presence and immersion for the user.

One of the key aspects of VR is the creation of an immersive environment. VR headsets or goggles cover the user's entire field of view, effectively blocking out the physical world and fully immersing them in the simulated digital realm. This immersion is further enhanced by the ability to interact with and manipulate objects within the virtual environment using specialized input devices like motion controllers, gloves, or even hand gestures. This interactivity provides users with a sense of agency and presence, as if they are truly present within the simulated world. Advanced VR systems also incorporate sensory feedback mechanisms to further enhance the immersive experience. Spatial audio technologies create realistic soundscapes that envelop the user, while haptic feedback devices simulate tactile sensations such as vibrations or pressure, mimicking the sense of touch within the virtual environment. At the heart of VR technology lies powerful graphics processing units (GPUs) and real-time rendering techniques. These components work in tandem to generate high-quality, responsive visuals that maintain the user's immersion and minimize any potential discomfort or motion sickness. The ultimate goal of VR is to create a convincing sense of "presence," where users feel as though they are physically present within the virtual environment, even though they are aware of the simulated nature of the experience.

VR finds applications across a wide range of industries, from gaming and entertainment to education, healthcare, architecture, engineering, and training simulations. In the realm of communication technology, VR offers innovative solutions for facilitating interactions and connections in virtual environments. Platforms like Oculus Rooms, AltspaceVR, and VRChat allow users to meet, socialize, and collaborate in shared virtual spaces, hosting events, playing games, or simply engaging in casual conversation. Users can create custom avatars and explore unique virtual environments, fostering a sense of community and connection despite physical distances.

Companies like Spatial offer VR meeting and conferencing solutions that enable remote teams to collaborate seamlessly in virtual spaces. Participants can share presentations, whiteboards, and 3D models in real-time, enhancing communication and facilitating productive discussions. Social VR platforms like Facebook Horizon and Rec Room empower users to create their own virtual worlds, games, and experiences, fostering creativity and enabling immersive social interactions with friends and strangers alike. Educational and training platforms, such as Engage, leverage VR to provide immersive learning experiences, allowing users to attend virtual lectures, workshops, and conferences in simulated environments that mimic real-world settings.

While VR technology continues to evolve, challenges remain, including addressing potential motion sickness, ensuring hardware affordability and accessibility, creating compelling and engaging content, and optimizing performance to maintain seamless immersion without causing discomfort or fatigue for users.

Nonetheless, the potential of VR in communication technology is substantial, offering innovative ways for people to connect, collaborate, and socialize in virtual environments that transcend physical boundaries. As technology advances, VR is poised to reshape how we perceive and experience digital interactions, paving the way for new and immersive forms of communication and connection (Radianti, Majchrzak, Fromm, & Wohlgenannt, 2020).

3. HOW VIRTUAL REALITY DIFFERS FROM AUGMENTED REALITY IN COMMUNICATION TECHNOLOGY

Virtual reality (VR) and augmented reality (AR) are both immersive technologies employed in communication, yet they diverge in their approaches to presenting and interacting with digital content. Here's a more detailed explanation:

Immersive Experience: VR constructs a fully immersive digital realm that entirely supplants the user's physical surroundings. Donning VR headsets, users are transported into entirely virtual spaces where they can interact with digital objects and other users. In contrast, AR superimposes digital content onto the user's real-world environment, enabling them to perceive and interact with virtual elements alongside physical objects.

Isolation vs. Integration: VR typically isolates users from their physical surroundings, as they are fully immersed in the virtual environment. This isolation can be advantageous for certain applications, such as gaming or virtual meetings, where complete immersion is desirable. Conversely, AR integrates digital content into the user's real-world environment, allowing for seamless interaction with both virtual and physical elements. AR is often utilized in applications where users need to remain cognizant of their surroundings, such as navigation or remote assistance.

Table 1. Descriptive statistics of AR users

Years	AR users (in million)	VR users (in million) (non -headset)	VR users (in million) (headset)
2017	37.6	11.5	11
2018	59.5	19	15.6
2019	72.8	22.3	20.8
2020	83.7	28	22.2
2021	93.3	30.6	28.3
2022	101.6	32.7	31.3
2023	110.1	33.2	32.7

(Source: https://www.statista.com/statistics/1017008/united-states-vr-ar-users/)

Hardware Requirements: VR experiences typically necessitate specialized hardware, such as VR headsets or goggles, which completely obscure the user's view of the physical world. AR, on the other hand, can be experienced using devices like smartphones or AR glasses, which allow users to simultaneously perceive both the real world and virtual content.

Use Cases: VR is frequently employed for immersive gaming, simulations, virtual meetings, training, and entertainment experiences where users desire to be fully transported to a virtual environment. AR, conversely, finds applications in areas such as navigation, education, retail, remote assistance, and industrial training, where users need to interact with virtual content within the context of their real-world environment.

Furthermore, VR and AR differ in their level of immersion. VR offers a heightened sense of immersion by completely transporting users to a digital realm, while AR enhances the user's perception of the real world by overlaying virtual elements onto their physical surroundings (Aukstakalnis, 2016). The choice between VR and AR often depends on the specific requirements of the application and the desired level of immersion. VR is well-suited for scenarios where users need to be fully immersed in a virtual environment, while AR is more appropriate for situations where users need to maintain awareness of their physical surroundings while interacting with digital content. In the realm of communication technology, both VR and AR offer unique advantages and use cases (Olsson, Kärkkäinen, Lagerstam, & Ventä-Olkkonen, 2012). While VR provides a fully immersive experience for virtual meetings, training simulations, and entertainment, AR enables seamless integration of digital content into the user's real-world environment, enhancing communication, collaboration, and information exchange (Carmigniani & Furht, 2011).

4. ETHICAL ISSUES IN COMMUNICATION TECHNOLOGY

Ethics refers to the moral principles that guide an individual's behavior or the conduct of an activity. Ethics in technology, also known as tech ethics, focuses on the moral principles and values that steer the development, deployment, and utilization of technology. Ethical concerns in communication technology encompass a wide array of issues related to privacy, security, access, and societal impact. Some key ethical considerations include:

Privacy and Data Protection: With the increasing collection and use of personal data by communication technologies, concerns arise about safeguarding individuals' privacy rights and the potential for unauthorized access or misuse of sensitive information (Belkhir & Elmeligi, 2018). Ensuring proper data handling practices and protecting individual privacy is a critical ethical consideration.

Surveillance and Monitoring: The proliferation of communication technologies, such as social media platforms, raises concerns about surveillance and monitoring activities by governments, corporations, and other entities, potentially infringing on individuals' rights to privacy and freedom of expression. Striking a balance between security and individual liberties is an ethical challenge.

Cybersecurity: As communication technologies become more interconnected, the risk of cyberattacks, data breaches, and malicious activities increases, posing threats to individuals, organizations, and critical infrastructure. Maintaining robust cybersecurity measures and protecting against these threats is an ethical imperative (Lazer et al., 2018).

Digital Divide: Disparities in access to communication technologies, such as internet connectivity and digital devices, contribute to a digital divide that exacerbates inequalities in education, employment, healthcare, and other areas, creating social and economic barriers for marginalized communities. Addressing this divide and promoting equitable access is an ethical responsibility.

Misinformation and Disinformation: The rapid dissemination of misinformation and disinformation through communication technologies, particularly social media, undermines trust in institutions, distorts public discourse, and poses risks to democratic processes and social cohesion. Combating the spread of false and misleading information is an ethical obligation (Lazer et al., 2018).

Algorithmic Bias and Discrimination: Algorithms used in communication technologies can perpetuate biases and discrimination based on factors such as race, gender, and socioeconomic status, leading to unfair treatment and exclusion of certain groups. Developing and deploying ethical algorithms that promote fairness and inclusivity is a moral imperative.

Addiction and Mental Health: Excessive use of communication technologies, particularly social media and gaming platforms, can contribute to addiction, social isolation, and negative impacts on mental health, especially among vulnerable populations such as adolescents. Promoting responsible use and addressing potential harmful effects is an ethical concern.

Environmental Impact: The production, consumption, and disposal of communication technology devices and infrastructure contribute to environmental degradation, including carbon emissions, electronic waste, and resource depletion, raising concerns about sustainability and ecological impact. Adopting eco-friendly practices and minimizing the environmental footprint of communication technologies is an ethical responsibility.

Addressing these ethical issues requires concerted efforts from governments, technology companies, civil society organizations, and individuals to develop and implement policies, regulations, and practices that prioritize privacy, security, equity, and accountability in communication technology. Ethical considerations should be at the forefront of technological advancement to ensure that communication technologies are developed and utilized in a manner that upholds moral principles and promotes the well-being of individuals and society as a whole (Bélanger & Crossler, 2011).

5. MERITS AND DEMERITS OF INNOVATION IN COMMUNICATION TECHNOLOGY

It is imperative to recognize that communication technology, much like a coin, possesses two distinct facets. While its advancements have undoubtedly transformed the landscape of connectivity, collaboration, and access to information, they have also given rise to a myriad of challenges and potential drawbacks. The key lies in striking a judicious equilibrium, harnessing the transformative power of communication technology while simultaneously mitigating its adverse effects.

On the positive side, advancements in communication technology have facilitated unprecedented levels of connectivity, enabling individuals and communities to transcend geographical boundaries and forge connections with ease. From instant messaging and video conferencing to social media platforms, these innovations have empowered people to maintain relationships, collaborate on projects, and exchange ideas with unprecedented fluidity. Furthermore, these advancements have democratized access to information, providing individuals with a wealth of knowledge and resources at their fingertips, fostering informed decision-making, educational opportunities, and active participation in social and political discourse. Moreover, advancements in communication technology have catalyzed economic growth by facilitating trade, commerce, and entrepreneurship. The rise of e-commerce, online marketplaces, and digital platforms has created new avenues for businesses to reach global markets, spurring job creation and driving innovation in various sectors, including information technology, telecommunications, and digital media.

However, alongside these merits, communication technology advancements have also given rise to significant concerns and challenges. Privacy infringement is a paramount issue, as personal data may be collected, stored, and shared without consent, leading to surveillance, identity theft, and other privacy violations. The digital divide, too, persists, marginalizing individuals and communities with limited connectivity, skills, or resources, exacerbating inequalities in education, employment, and social participation. Additionally, the proliferation of communication technology advancements, such as social media and messaging platforms, has led to information overload, contributing to cognitive overload, distraction, and difficulty in discerning credible information from misinformation or disinformation. Cybersecurity threats, including cyberattacks, data breaches, and malware infections, pose risks to individuals, organizations, and critical infrastructure, jeopardizing privacy, security, and economic stability.

Furthermore, despite enhancing connectivity, communication technology advancements can paradoxically contribute to social isolation and loneliness, as people may prioritize virtual interactions over face-to-face communication, leading to decreased interpersonal relationships and potential mental health issues. Excessive reliance on these technologies, such as smartphones and social media, can also lead to dependency and addiction, resulting in negative impacts on productivity, relationships, and overall well-being. Lastly, the environmental impact of communication technology advancements cannot be overlooked. The production, use, and disposal of devices and infrastructure contribute to environmental degradation, including electronic waste, energy consumption, and pollution, raising concerns about sustainability and ecological impact.

Balancing the merits and demerits of advancements in communication technology necessitates a thoughtful consideration of ethical, social, and environmental implications. It requires collaborative efforts from stakeholders, including governments, technology companies, civil society organizations, and individuals, to develop and implement policies, regulations, and practices that prioritize privacy, security, equity, and accountability. In essence, while communication technology advancements have undoubtedly transformed our lives in remarkable ways, it is crucial to strike a balance between harnessing their potential and mitigating their adverse effects. By acknowledging and addressing the demerits, while simultaneously leveraging the merits, we can navigate the complexities of this ever-evolving technological landscape and shape a future where communication technology serves as a force for empowerment, progress, and the betterment of society.

6. CONCLUSION

Innovation serves as the driving force behind progress and lifestyle transformations in the realm of communication technology. Cutting-edge advancements, such as electric vehicles, smartphones, AI, and blockchain technology, continuously enhance existing concepts and introduce novel solutions, reshaping the way we interact and communicate. This rapid pace of innovation is propelled by advancements in semiconductor technology and the ever-increasing demand for faster and more reliable communication channels. Emerging technologies like the Internet of Things (IoT), AI, and 5G networking present fresh opportunities for innovation, influencing communication experiences and capabilities in unprecedented ways. AI, encompassing machine learning and natural language processing, replicates human intelligence in machines, augmenting communication technology through applications like chatbots, language translation, and speech recognition. Augmented reality (AR) further enriches communication by seamlessly overlaying digital content onto the real world, as exemplified by Snapchat lenses and Facebook Messenger AR effects. Virtual reality (VR) takes immersion a step further, enabling users to engage with digital content as if they were physically present, with applications ranging from social VR platforms to virtual meetings.

While these innovations in communication technology undoubtedly improve connectivity, collaboration, and access to information, they also raise ethical concerns surrounding privacy, security, access, and societal impact. Issues such as privacy infringement, cybersecurity threats, the digital divide, and the spread of misinformation necessitate collaborative efforts from various stakeholders to address these challenges effectively. Proactive measures must be taken to safeguard privacy, enhance cybersecurity, bridge the digital divide, and combat the dissemination of misinformation. Moreover, it is imperative to consider the ethical, social, and environmental implications of communication technology innovations, ensuring that progress is achieved in a responsible and sustainable manner. In essence, this section highlights the transformative power of innovation in communication technology, while acknowledging the potential pitfalls that must be addressed through collective efforts and a holistic approach to ethical, social, and environmental considerations.

Summary

- Communication technology has evolved from basic non-verbal forms to sophisticated systems, enabling efficient and effective communication globally. The history of communication technology spans millennia, from early writing systems to modern digital platforms, revolutionizing how people share information and connect. Ongoing innovations continue to transform the landscape of communication.

- Innovation in communication technology has revolutionized global connectivity and collaboration. Advancements in semiconductors, user-centric design, and emerging technologies like AI and 5G have enabled faster, more efficient, and reliable communication systems. This continuous innovation has transformed how people interact, fostering cultural exchange and economic growth worldwide.
- Artificial Intelligence (AI) enables machines to perform human-like tasks such as learning, reasoning, and problem-solving. AI's key subfields include machine learning, natural language processing, computer vision, and robotics. AI enhances communication technology through chatbots, translation tools, and personalized experiences, enabling more efficient, accessible, and seamless interactions across platforms.
- Augmented reality (AR) seamlessly blends digital content with the real world, creating immersive experiences. AR enables overlaying of digital information, real-time interaction with virtual objects, and contextual relevance. AR has applications in various industries and has transformed communication technology, adding interactive and engaging elements to video calls and messaging platforms.
- Virtual reality (VR) creates a fully immersive, three-dimensional digital environment that users can explore and interact with using specialized headsets. VR has applications across industries and enables innovative communication and collaboration solutions, allowing people to connect, socialize, and work together in virtual environments. However, challenges remain in ensuring accessibility and optimizing performance.
- Virtual reality (VR) and augmented reality (AR) differ in their approach to immersion. VR fully immerses users in a digital realm, while AR overlays virtual content onto the real world. The choice between VR and AR depends on the application's requirements, with VR suited for full immersion and AR for integrating digital content with the physical environment.
- This document discusses the ethical considerations in communication technology, including privacy, surveillance, cybersecurity, digital divide, misinformation, algorithmic bias, addiction, and environmental impact. It highlights the need for concerted efforts to develop policies, regulations, and practices that prioritize moral principles and promote the well-being of individuals and society.
- Communication technology has transformed connectivity and access to information, but also raised ethical concerns around privacy, digital divide, misinformation, cybersecurity, social isolation, addiction, and environmental

impact. Balancing the merits and demerits requires collaborative efforts to develop policies and practices that prioritize ethical principles and societal well-being.

REFERENCES

Aukstakalnis, S. (2016). *Practical augmented reality: A guide to the technologies, applications, and human factors for AR and VR*. New York Addison-Wesley Professional.

Bélanger, F., & Crossler, R. E. (2011). Privacy in the digital age: A review of information privacy research in information systems. *Management Information Systems Quarterly*, 35(4), 1017–1041. DOI: 10.2307/41409971

Belkhir, L., & Elmeligi, A. (2018). Assessing ICT global emissions footprint: Trends to 2040 & recommendations. *Journal of Cleaner Production*, 177, 448–463. DOI: 10.1016/j.jclepro.2017.12.239

Billinghurst, M., Clark, A., & Lee, G. (2015). others,". *A survey of augmented reality," Foundations and Trends R in Human–Computer Interaction, 8*(2-3), 73-272.

Binkley, M., Erstad, O., Herman, J., Raizen, S., Ripley, M., Miller-Ricci, M., & Rumble, M. (2012). Defining twenty-first century skills. *Assessment and teaching of 21st century skills*, 17-66.

Carmigniani, J., & Furht, B. (2011). Augmented reality: an overview. *Handbook of augmented reality, 10*(2), 3-46.

Grubert, J., Langlotz, T., Zollmann, S., & Regenbrecht, H. (2016). Towards pervasive augmented reality: Context-awareness in augmented reality. *IEEE Transactions on Visualization and Computer Graphics*, 23(6), 1706–1724. DOI: 10.1109/TVCG.2016.2543720 PMID: 27008668

Järvelä, S., & Renninger, K. (2014). *Designing for learning: Interest, motivation, and engagement.* London.

Lazer, D. M., Baum, M. A., Benkler, Y., Berinsky, A. J., Greenhill, K. M., Menczer, F., & Rothschild, D. (2018). The science of fake news. *Science*, 359(6380), 1094–1096. DOI: 10.1126/science.aao2998 PMID: 29590025

Olsson, T., Kärkkäinen, T., Lagerstam, E., & Ventä-Olkkonen, L. (2012). User evaluation of mobile augmented reality scenarios. *Journal of Ambient Intelligence and Smart Environments*, 4(1), 29–47. DOI: 10.3233/AIS-2011-0127

Radianti, J., Majchrzak, T. A., Fromm, J., & Wohlgenannt, I. (2020). A systematic review of immersive virtual reality applications for higher education: Design elements, lessons learned, and research agenda. *Computers & Education*, 147(7), 103–178. DOI: 10.1016/j.compedu.2019.103778

Chapter 4

Digital Engagement in Academic Libraries:
An Analysis of Short Video Platforms and Student Satisfaction

Adebowale Jeremy Adetayo
https://orcid.org/0000-0001-7869-5613
Adeleke University, Nigeria

Modupe Atinuke Otuyalo
https://orcid.org/0009-0007-4639-047X
Lagos State University of Science and Technology, Nigeria

Bosede Olufunmilayo Makinde
Lagos State University, Nigeria

ABSTRACT

The purpose of this chapter is to investigate students' satisfaction with the usage of short video platforms by libraries for marketing library services. A descriptive survey research design was employed, and the sample consisted of 245 {Anonymized} students. Descriptive statistics were used to analyze the data. Findings indicated that while students rarely create and upload videos on short video platforms, they are frequent watchers of short videos on platforms like TikTok, Instagram Reels, and YouTube Shorts. They have also seen various library services marketed on these short video platforms. The study concluded that patrons are satisfied with the usage of short video platforms to market library services and therefore recommended that libraries begin to employ short video platforms to market their services, with an emphasis on TikTok and Instagram reels.

DOI: 10.4018/979-8-3693-3579-6.ch004

INTRODUCTION

The academic library system in Nigeria is facing significant challenges in terms of patronage due to the increasing popularity of big tech companies and their ability to offer easy access to information from anywhere in the world (Adetayo, 2022). While this trend has the potential to negatively impact the viability of academic libraries, the concern is not necessarily about big tech replacing libraries, but rather a lack of awareness among patrons about the specialized services that libraries can offer (Okorafor et al., 2017). Many patrons believe that libraries are simply places to read books, but in reality, they offer a range of services that patrons may not be aware of. As a result, library administrators must work to promote and market these services effectively. One effective method of doing so is by leveraging big tech platforms and creating short videos to showcase the unique offerings of academic libraries.

Short video platforms, also known as short-video apps, refer to social media tools designed for the creation and sharing of short video content, typically less than 5 minutes in length. According to Smith (2022), these platforms are software or tools that aid in the creation and editing of short videos. The low cost of production, rapid spreadability, and blurring of lines between producers and consumers are all hallmark characteristics of short videos (Kong, 2018). Unlike video-sharing platforms like YouTube or social networking services like Facebook and Instagram, short-video apps are specifically focused on user-generated video content and engagement with this content rather than with each other (Chen et al., 2019)

Research has shown that short videos on mobile devices have a high level of effectiveness in capturing viewer attention and engagement (Omar & Dequan, 2020; Scherr & Wang, 2021). This can be attributed to the popularity of video sharing among adolescents and young people, which is one of the key features of short video platforms (McRoberts et al., 2017). Currently, TikTok is considered the dominant players in the short-form video market (Tolentino, 2019). However, in response to the popularity of TikTok, other tech companies have introduced similar capabilities in their own applications, such as Instagram's introduction of 'Instagram Reels.' While TikTok remains a popular platform, Instagram Reels has quickly emerged as a potential competitor, offering a similar user experience (Lerman, 2020).

Instagram Reels allows its users to become self-producing consumers by enabling them to create short videos using their smartphone camera, edit them with various inbuilt editing options, and share them with their friends and followers. The consumption of Reels videos, by fans or followers, can result in increased engagement through likes, comments, and shares. Facebook has also added the Reels feature to its platform.

Wright (2017) noted in previous research that user preferences for short video features on social media are driven by convenience and content curation, where an algorithm compiles a selection of posts for users to view. This content curation strategy has the potential to adapt social media material for consumers while also increasing traffic within a short time frame. However, there has been limited research to determine whether students would be satisfied with academic libraries using short video platforms to market their services. The aim of this current study is to investigate students' satisfaction with libraries utilizing short video platforms for marketing library services.

Objective of the Study

The objective of this chpater is to investigate students' satisfaction with academic libraries utilizing short video platforms for marketing their services. To achieve this broad objective, the chapter aims to achieve the following specific objectives:

- To examine the frequency of creating and uploading videos on short video platforms.
- To determine the frequency of watching videos on short video platforms.
- To analyze the library services viewed on popular short video platforms such as Reels, TikTok, and YouTube Shorts.
- To assess the level of satisfaction with short video platforms for marketing library services.

LITERATURE REVIEW

Short Video Platforms

The rise of short video platforms can be traced back to the creation of Vine in 2013 (Xiao et al., 2019). At the time, mobile internet technology was rapidly advancing, and Vine's concept of allowing users to make 6-second looping videos proved to be a hit. By 2015, Vine had amassed 200 million active members, making it one of the most popular social platforms at the time. However, its popularity waned, and it was eventually deactivated in 2017. Subsequently, Musical.ly rose to prominence as a platform that allowed users to produce up to 60-second lip-syncing videos. This platform, which was popularized by the use of hashtags to promote trends and global events, was one of the first instances where content creation was centered around a specific sound. By 2016, Musical.ly had become one of the fastest-growing social platforms, and in 2017, it was acquired by Bytedance Technology, Co. (Silvia, 2022).

The popular short video platform TikTok was created as a result of the merging of the applications Musical.ly and Douyin in 2018, by Bytedance. TikTok has become one of the most widely used apps globally, attracting hundreds of millions of users, including a significant number of children and teens, who engage in activities such as uploading, watching, and browsing videos and memes. The app allows users to create and publish up to 60-second lip-synched videos that incorporate a range of creative and interactive elements. TikTok videos can range from fun and entertaining dancing content (Vázquez-Herrero et al., 2020) to more serious topics such as social activism (Subramanian, 2020), public health information (Basch et al., 2020), and celebrity content (Abidin, 2020). To maximize the limited time frame, TikTok users can utilize a variety of tools such as music samples, filters, fast cuts, stickers, and other creative add-ons in their videos.

The widespread popularity of TikTok cannot be overstated, with the app becoming a global phenomenon in just a few short years. Initially launched in China as Douyin in 2016, it was later made available for iOS and Android in regions outside of China in 2017. The speed at which TikTok has grown is truly impressive, attracting over 1.5 billion active users in just three years and surpassing the likes of Netflix, YouTube, Snapchat, and Facebook (Chapple, 2020). In the first quarter of 2020 alone, TikTok recorded over 315 million installations on the App Store and Google Play, which is the most ever for any app in a single quarter. This led to a total of over 2 billion downloads globally since its inception (Chapple, 2020).

However, the emergence of Reels, a social networking platform within Instagram, has also captured the attention of users who are seeking to connect with others and expand their social capital (Scherr & Wang, 2021). With its roots in TikTok, Reels offers a similar experience with user-generated content and algorithmic curation (Meng & Leung, 2021). Unlike TikTok, however, Reels is not a standalone app, but rather a feature of the existing Instagram platform that users can access by scrolling to the bottom of the Instagram camera. Reels was introduced in 2020 and has quickly become a global phenomenon, with availability in 50 countries worldwide (Instagram, 2020). With its growing user base and the support of the larger Instagram community, Reels has the potential to become one of the most popular social networking platforms of our time especially as TikTok faces scrutiny from leaders across the world.

According to Nguyen and Mikolajczyk (2021), Instagram Reels provides users with a creative platform to create and share 15-60 second videos with others. The feature offers a range of creative options to help users produce high-quality content and engage with their audience in a quick and easy manner. Instagram Reels has the potential to serve as an effective instructional tool for fostering creativity and motivation among students.

Additionally, YouTube Shorts, another short video platform, has gained significant popularity. YouTube Shorts, introduced in India in 2020 and in the United States in 2021, has become a popular feature for short, vertical videos. As noted by Khan (2017), these videos are made on a smartphone and can be uploaded to YouTube through the YouTube app with ease. The videos are typically in 16:9 format and range from 15 to 60 seconds in duration. They allow for simple and immediate sharing, commenting, liking, disliking, and subscribing, which has led to its rapid growth, with daily views reaching over 6.5 billion globally (Beveridge, 2022).

Library Marketing With Short Video Platforms

Library marketing is crucial for ensuring that current and potential library users are well-informed about the resources and services available to them. To effectively promote the library's offerings and broaden its user base, effective marketing strategies must be employed. Martin (2012) highlights a successful marketing campaign directed towards the campus community, which serves as a blueprint for other institutions seeking to promote their library resources and services through the use of short video platforms such as Reels, TikTok, and YouTube Shorts (Adetayo et al., 2022).

The use of short video platforms as a marketing tool for libraries has been widely observed. According to Simon (2021), libraries and their staff have creatively utilized platforms to promote their services, share reader tips, and spread the joy of literacy to the public. Cascade Public Library, for instance, created a short video to advertise a program and included fun elements such as emojis and behind-the-scenes information in the commentary (Superlibrarymarketing, 2020). The Oceanside Public Library also created a lighthearted video to encourage patrons to adjust their clocks for Daylight Saving Time. Additionally, Glenthorne Library gained significant attention after reposting a video from an author, and Clifton Park-Halfmoon Public Library successfully leveraged a popular audio clip on Reels to reach their audience (Superlibrarymarketing, 2020). These examples demonstrate the potential of short video platforms as effective marketing tools for libraries.

Libraries can leverage short video platforms such as TikTok, Reels, and Youtube Shorts to promote their resources, services, and events. By creating short and engaging videos, libraries can attract and retain patrons, educate them about library resources and programs, and inspire them to take advantage of the library's offerings (King, 2022). As noted by Merga (2021), these platforms also offer an opportunity for libraries to enhance their reader advisory services. The creation of TikTok challenges and library-themed content can also be an effective way to attract users and introduce them to the library's spaces and services (Anderson, 2020). King (2018) provides seven easy steps for libraries to create visual marketing videos in American Libraries Magazine, emphasizing the importance of utilizing these platforms, where users are already watching videos, to effectively communicate what the library offers.

Student Satisfaction in Relation to Library Services

Student satisfaction with library services has been a critical area of focus in academic library research, particularly as libraries evolve to meet the changing needs of modern students. Recent studies have explored various aspects of library services and their impact on student satisfaction, highlighting the importance of both traditional and digital offerings. Misra et al. (2023) conducted a study on the continuous usage intention of digital library systems among postgraduate students in India. Their findings revealed that students' satisfaction and intention to continue using digital library systems were significantly influenced by the quality dimensions of these systems. This suggests that the overall quality of digital library services plays a crucial role in determining student satisfaction.

The importance of e-services in academic libraries was further emphasized by John (2023), who examined the influence of e-services on student satisfaction in higher learning institutions. The study found that students were satisfied with e-service provision when the systems were simple to use, reliable, efficient, secure, time-saving, and cost-effective. Specifically, students appreciated e-services that facilitated administrative tasks such as registration, tuition fee payment, and accessing online library resources. Udem and Ogungbeni (2023) focused on the use of mobile technology-based library services and its relationship to information needs satisfaction among undergraduates in South-West Nigeria. Their research revealed a highly positive correlation between the use of mobile technology-based library services and students' information needs satisfaction. This underscores the growing importance of mobile technologies in meeting students' library-related needs and enhancing their overall satisfaction.

However, not all studies have reported high levels of student satisfaction with library services. Agina-Obu and Evelyn (2023) examined postgraduate students' satisfaction with information resources at a university library in Nigeria. Their findings indicated that the majority of postgraduate students were not satisfied with the available library information resources. This highlights the need for libraries to continually assess and improve their resources to meet student expectations.

The quality of library services has been identified as a key determinant of student satisfaction. Butt et al. (2023) evaluated the service quality of Pakistani university libraries using a modified SERVQUAL model. While users were satisfied with the "Empathy" dimension, which related to staff behavior and the library environment, gaps were identified in other service quality dimensions such as "Tangible," "Reliability," "Responsiveness," and "Assurance." This suggests that libraries need to focus on multiple aspects of service quality to enhance overall student satisfaction. Sukarjono et al. (2023) investigated the relationship between the availability of sports literature, library service quality, student satisfaction, and learning motivation. Their study found that both the availability of relevant literature and the quality of library services had significant effects on student satisfaction, which in turn influenced learning motivation. This demonstrates the broader impact that library services and resources can have on students' academic experiences. In a comprehensive study of library services and facilities,

METHODS

Settings

The setting for this study is Adeleke University, a private institution located in Osun state, Nigeria. The University Library, situated on the campus, serves as a crucial educational resource for the students, with the goal of preparing them for success in life. The Library has a strong online presence, maintaining pages on several social media platforms including Facebook, LinkedIn, TikTok, YouTube and Instagram, where it promotes its services and interacts with users. The Library administration has taken proactive measures to increase students' engagement with its social media pages, including sharing posts on students' Whatsapp groups to increase visibility and accessibility.

Research Design

The study adopted a descriptive survey research design, with the aim of investigating students' satisfaction with the use of short video platforms by the library for marketing services. The descriptive survey approach was chosen because it provides a comprehensive understanding of the phenomenon being studied, reflecting the students' perceptions, attitudes and opinions about the subject of interest.

Population and Sampling Size

The target population for this study consisted of 245 undergraduate students who frequently visit Adeleke University Library. As the population size was relatively small, the entire population was included in the study.

Instrumentation

The study utilized a structured questionnaire as the primary data collection tool. The questionnaire was selected as the preferred method of data collection due to its ability to reach a large number of participants quickly and cost-effectively. The questionnaire consisted of both demographic and attitudinal questions and was designed using a Likert scale with 4 as the highest and 1 as the lowest. The questionnaire's face validity was assessed by the research supervisor to ensure that the components of the measure were relevant to what was being measured and that the measurement method was appropriate for capturing the variables under investigation. Based on the outcome of the face validity assessment, some questionnaire items were modified and clarified, while others with the same meaning were omitted.

Procedure for Data Collection

The approved questionnaire was sent to 245 students at Adeleke University and collected over a period of three weeks. The data collection process was conducted anonymously to ensure confidentiality. The students were assured that their responses would be kept private and used only for research purposes, and that such information would not be disclosed to a third party. Out of the 245 questionnaires distributed, 221 were collected, resulting in a success rate of 90.2%. The analysis was based on the data collected from the 221 respondents.

Data Analysis Technique

The collected data was analyzed using descriptive statistics. The study questions were answered using frequency counts, percentages, mean, and standard deviation scores. These descriptive statistics were utilized to provide a comprehensive understanding of the students' satisfaction with the use of short video platforms by the library to market its services.

RESULTS

This section uses descriptive statistics to present the study's findings. The specifics are shown in Tables 1-5.

Table 1. Demographics of respondents

Age Range	Frequency (221)	Percent
20 & Below	146	66.1
21-25	44	19.9
26-30	13	5.9
31 & above	18	8.1
Gender		
Female	120	54.3
Male	101	45.7

The table presents a summary of the demographic information of the study participants, including gender and age. The results reveal that the majority of the respondents were female (54.3%) with the majority being aged 20 and below (66.1%). This age group consists of young people, who have been demonstrated to use technology more than older generations (Adetayo & Williams-Ilemobola, 2021).

Table 2. Frequency of creating and Uploading Videos on Short Video Platforms

S/N	Shorts Video Platforms	Always	Often	Rarely	Never	Mean	Std Dev
i	Facebook reels	23(10.4%)	22(10.0%)	59(26.7%)	117(52.9%)	1.78	1.000
ii	TikTok	27(12.2%)	41(18.6%)	56(25.3%)	97(43.9%)	1.99	1.057
iii	Instagram reels	19(8.6%)	35(15.8%)	67(30.3%)	100(45.2%)	1.88	0.972
iv	YouTube Shorts	17(7.7%)	21(9.5%)	35(15.8%)	148(67.0%)	1.58	0.948

The table 2 provides an overview of the students' use of short video platforms for creating and posting videos. The results indicate that TikTok was the most popular platform with a mean score of 1.99, while YouTube Shorts was the least used platform with a mean score of 1.58. The findings suggest that, despite the popularity of short video platforms among young people, the students rarely use these platforms for creating and posting videos. These findings provide important insights into the students' preferences and habits when it comes to the use of short video platforms, which can inform the library's marketing strategies and efforts to engage with students more effectively.

Table 3. Frequency of watching videos on Short Video platforms

S/N	Shorts Video Platforms	Always	Often	Rarely	Never	Mean	Std Dev
i.	Facebook reels	44(19.9%)	47(21.3%)	69(31.2%)	61(27.6%)	2.33	1.085
ii.	TikTok	92(41.6%)	62(28.1%)	43(19.5%)	24(10.9%)	3.00	1.025
iii.	Instagram reels	73(33.0%)	76(34.4%)	48(21.7%)	24(10.9%)	2.90	0.988
iv.	YouTube Shorts	55(24.9%)	77(34.8%)	74(33.5%)	15(6.8%)	2.78	0.900

Table 3 presents the results of the students' usage of short video platforms for watching short videos. The findings indicate that TikTok emerged as the most frequently used platform for watching short videos.with a mean rating of 3.00, which represents a high level of utilization by students. This high level of utilization can be attributed to TikTok's unique features, such as its algorithmic feed that curates content based on users' interests and preferences, which make it highly engaging for users.

Furthermore, Instagram reels were found to be a close second, with a mean rating of 2.90, indicating that students also consider it to be a popular platform for watching short videos. On the other hand, Facebook reels received the lowest mean rating of 2.33, suggesting a lower level of utilization among students. This low level of utilization may be a reflection of the decline in popularity of Facebook among the younger demographic, as more engaging platforms, such as TikTok, emerge in the market.

These findings provide valuable insights into the usage patterns of short video platforms among undergraduate students, highlighting the popularity of TikTok and Instagram reels, as well as the lower utilization of Facebook reels. These insights can inform library marketing strategies by suggesting which short video platforms to utilize in order to reach and engage students effectively.

Table 4. Library Services viewed on reels, TikTok and YouTube shorts

S/N	Library Services	Always	Often	Rarely	Never	Mean	Std Dev
i	Library events	47(21.3%)	85(38.5%)	64(29.0%)	25(11.3%)	2.70	0.931
ii	New books arrivals	35(15.8%)	102(46.2%)	62(28.1%)	22(10.0%)	2.68	0.859
iii	Library news update	33(14.9%)	94(42.5%)	67(30.3%)	27(12.2%)	2.60	0.887
iv	Library users education	46(20.8%)	86(38.9%)	65(29.4%)	24(10.9%)	2.70	0.921
v	Electronic services available in the library	47(21.3%)	93(42.1%)	53(24.0%)	28(12.7%)	2.72	0.941

The findings in Table 4 provide insight into the impact that short video platforms have on students' perception of library services. By leveraging the power of social media, libraries have effectively been able to reach students and inform them about their offerings. This can be seen in the fact that most students have viewed library electronic services. Additionally, the promotion of library events and educational opportunities on these platforms has also resonated with students. This suggests that libraries are effectively utilizing the viral nature of short video platforms to increase students' engagement and exposure to their offerings. On the other hand, the lower number of students who have viewed information about new book arrivals on these platforms highlights a potential area for improvement for libraries in terms of their marketing efforts.

Table 5. Level of satisfaction with short video platforms for marketing library services

S/N	Shorts Video Platforms	Extremely Satisfied	Satisfied	Slightly Dissatisfied	Not Satisfied	Mean	Std Dev
i	Facebook reels	59(26.7%)	109(49.3%)	41(18.6%)	12(5.4%)	2.97	0.820
ii	TikTok	133(60.2%)	66(29.9%)	19(8.9%)	3(1.4%)	3.49	0.711
iii	Instagram reels	119(53.8%)	86(38.9%)	13(5.9%)	3(1.4%)	3.45	0.670
iv	YouTube Shorts	111(50.2%)	90(40.7%)	18(8.1%)	2(0.9%)	3.40	0.678

The findings in Table 5 reveal the students' level of satisfaction with the use of short video platforms by libraries for marketing library services. The results indicate that the majority of students view short video platforms as an effective method for promoting library services. TikTok emerged as the most favored platform, with a mean rating of 3.49, indicating high levels of satisfaction among students. This is followed by Instagram reels (mean = 3.45) and YouTube Shorts (mean = 3.40), with students expressing similar levels of satisfaction with these platforms. However, Facebook Reels received the lowest satisfaction rating (mean = 2.97), signifying that students are less impressed with its utilization by libraries for marketing purposes. This study highlights the importance of short video platforms as a viable medium for libraries to engage with and inform the younger demographic.

DISCUSSIONS OF FINDINGS

The popularity of short video platforms, particularly among younger generations, has been widely acknowledged. Ortega-Martínez et al. (2021) argue that libraries should adapt their policies to include social media platforms like Facebook, WhatsApp, Telegram, Twitter, Instagram, and TikTok for digital works. However, the current study's findings indicate that students rarely create and upload videos on short video platforms. This observation aligns with the findings of Preece et al. (2004), who noted that students tend to exhibit passive consumption behavior on these platforms, exemplified by Instagram Reels users who primarily share recycled content from other applications (Instagram, 2021). Consequently, it appears that students primarily utilize short video platforms for content consumption rather than active participation in its creation and dissemination.

Regarding video consumption, the findings of this study reveal that students frequently watch videos on short video platforms, with TikTok being the most commonly used platform for this purpose. This finding is supported by previous literature. Scherr and Wang (2021) note that TikTok gained instant global popularity, especially among millennials and Gen Xers, shortly after its launch in China in 2016. The app's unique user interaction features allowed it to surpass Facebook in terms of monthly usage minutes. Similar trends were observed among Generation Z in the United States, with TikTok being the most popular platform, followed by Instagram and Snapchat, as reported by Statista (2022). These findings emphasize the importance of incorporating short video platforms, such as TikTok, into libraries' marketing strategies to engage with students and promote library services effectively.

The appeal of TikTok among students in this study can be attributed to several factors identified by various research studies. Leander and Burriss (2020) highlight that the app's designers specifically targeted teenagers and employed a short video format to cater to their shorter attention spans. Additionally, the app's diverse set of effects and challenges fosters creativity, while its algorithm makes it easier for users to achieve viral status.

Scherr and Wang (2021) conducted research among Chinese TikTok users and identified four reasons for its popularity, including socially rewarding self-promotion, trendiness, escapist addiction, and novelty. Similarly, Meng and Leung (2021) conducted a study on TikTok engagement behaviors in China and found six traditional uses and gratifications (U&G) 1.0 motives, such as escape, entertainment, fashion, sociability seeking, money-earning, and information seeking, as well as three U&G 2.0 motives, namely navigability, interaction, and modality. The study concluded that platform-based affordances were the strongest predictors of TikTok involvement. Sundar and Limperos (2013) also employed a uses and gratification 2.0 framework based on the MAIN model in their research. All of these findings, as identified by Scherr and Wang (2021)Meng and Leung (2021), and Sundar and Limperos (2013), contribute to a deeper understanding of TikTok's appeal among students in the study area.

Overall, the findings suggest that while students may not actively create and upload videos on short video platforms, they heavily consume video content, with TikTok being the most popular platform. Libraries should recognize the significance of incorporating short video platforms, particularly TikTok, into their marketing strategies to effectively engage with students and promote library services. Understanding the factors that contribute to TikTok's appeal, such as its format, user interaction features, and the various motives underlying its usage, can aid libraries in tailoring their approach to reach and connect with their target audience more effectively.

The findings of this study indicate that a significant number of students have encountered library services being marketed on short video platforms, highlighting the growing use of these platforms by libraries to promote their offerings. This finding is supported by Anderson (2020), who provides evidence of early adoption of short video platforms, including TikTok, by libraries and librarians. The hashtag #librariansoftiktok, for example, leads to videos from various types of libraries, such as public, university, and school libraries, featuring short promotional videos and comprehensive tours of library spaces.

Moreover, the study reveals that students have also come across library user education services on short video platforms. While many libraries have yet to fully utilize short video platforms for user education, educators can draw inspiration from the implementation of TikTok in schools. Some teachers have already embraced the app and incorporated it into their coursework or established TikTok clubs (Lorenz, 2019). Basith and Syafi'i's research (2020) further supports this finding, demonstrating that Instagram can serve as an effective educational medium facilitating connections between students and teachers both inside and outside the classroom. Recent research on the use of TikTok among university students (Escamilla-Fajardo et al., 2021; Essel et al., 2021) also presents positive results, highlighting the potential for higher education institutions to adopt the platform for teaching and learning purposes.

The dissemination of information through short video platforms is a crucial factor contributing to their popularity, and this study underscores their potential for effectively marketing library services. The results indicate that students are satisfied with the use of short video platforms for marketing, with TikTok receiving the highest level of satisfaction. This can be attributed to its outreach potential, as libraries can create TikTok challenges to engage users and introduce them to their spaces and services (Anderson, 2020). The study by Merga (2021) on BookTok, a TikTok community focused on promoting reading activities, further supports the idea that libraries can leverage TikTok to enhance their reader advisory programs. Librarians who have adopted TikTok have reported that it offers a fresh medium for interacting with a new generation of users who may have become disillusioned with other social media platforms like Facebook, Instagram, and Twitter (Superlibrarymarketing, 2020).

It is important to note that this study's findings are limited to the perspectives of students and do not consider the viewpoints of library personnel. However, since students are a main target audience for academic library services, this study provides valuable insights that can assist library administrators in enhancing service recognition among users. Future research should explore the perspectives of library personnel to gain a comprehensive understanding of the challenges and opportunities associated with incorporating short video platforms into library marketing and user education strategies.

CONCLUSION

The results of this study provide a clear indication that students have a positive view of the use of short video platforms by libraries for marketing purposes. TikTok was found to be the most frequently used platform for creating and posting videos, while YouTube Shorts was the least. Despite this, regular usage of short video platforms for watching videos was found to be widespread among patrons, who frequently used TikTok, Instagram Reels, and YouTube Shorts. The majority of patrons indicated that they had seen various library services on these platforms, with electronic services being the most commonly viewed and news updates being the least. This highlights the importance of utilizing short video platforms to market library services effectively.

In light of these findings, the study recommends that libraries should consider incorporating short video platforms into their marketing strategies, with a focus on TikTok and Instagram Reels as the most popular platforms. To ensure success, library marketers should adhere to best practices for short videos, including keeping the videos short, using relevant background music, and incorporating a clear call to action. These strategies can help reach a wider audience and effectively communicate the library's services and offerings.

In conclusion, this study highlights the potential of short video platforms in marketing library services and the importance of incorporating them into library marketing strategies. It provides useful insights for libraries looking to enhance their outreach efforts and better engage with their patrons.

Implications of the study

The findings of this study carry significant implications for policy and practice in the library industry.

In terms of policy, this study highlights the impact of short video platforms on the younger generation, which is crucial for library policies. With the current library policy in Nigeria emphasizing the use of technology, but not specifying any specific technology, this study suggests that existing policies should be revised to prioritize short video platforms as an effective tool for libraries to reach their users and the world at large. This shift in focus towards short video platforms can be a major turning point for library services, which require a modern approach.

In terms of library practice, while social media has already been widely used by libraries to market their services, the use of short video platforms is still in its infancy. The lack of viral library videos on social media platforms could be due to the absence of a well-structured strategy for libraries to take advantage of the potential of short video platforms. Therefore, the findings of this study have significant implications for library service marketing practices. Libraries should make a concerted effort to incorporate more short videos into their marketing strategies, rather than relying solely on traditional social media features such as Facebook pages and Instagram posts. With TikTok being the most popular and effective short video channel for marketing library services, it should be given special attention by library management.

REFERENCES

Abidin, C. (2020). Mapping Internet Celebrity on TikTok: Exploring Attention Economies and Visibility Labours. *Cultural Science Journal*, 12(1), 77–103. DOI: 10.5334/csci.140

Adetayo, A. J. (2022). Harnessing Coopetition for the Survival of Nigerian Public Libraries: Role of Smart Technologies. *The International Information & Library Review*, •••, 1–15. DOI: 10.1080/10572317.2022.2070400

Adetayo, A. J., Komolafe, R. S., & Olalere, J. O. (2022). Virtual Learning and Library Usage during COVID-19 Pandemic: The Adeleke University Experience. *Journal of Library & Information Services in Distance Learning*, 16(1), 1–21. DOI: 10.1080/1533290X.2022.2066744

Adetayo, A. J., & Williams-Ilemobola, O. (2021). Librarians' generation and social media adoption in selected academic libraries in Southwestern, Nigeria. *Library Philosophy and Practice (e-Journal), 4984*. https://digitalcommons.unl.edu/libphilprac/4984

Agina-Obu, R., & Sawyerr-George, O. E. (2023). Evaluation of Users' Satisfaction of Information Resources in University Libraries in Nigeria: A Case Study. *Indian Journal of Information Sources and Services*, 13(1), 1–5. DOI: 10.51983/ijiss-2023.13.1.3378

Anderson, K. E. (2020). Getting acquainted with social networks and apps: It is time to talk about TikTok. *Library Hi Tech News*, 37(4), 7–12. DOI: 10.1108/LHTN-01-2020-0001

Basch, C. H., Hillyer, G. C., & Jaime, C. (2020). COVID-19 on TikTok: Harnessing an emerging social media platform to convey important public health messages. *International Journal of Adolescent Medicine and Health*. Advance online publication. DOI: 10.1515/IJAMH-2020-0111/MACHINEREADABLECITATION/RIS PMID: 32776899

Basith, A., & Syafi'i, A. (2020). Instagram as Media in Teaching Writing Recount Text for Senior High School Students. *SALEE: Study of Applied Linguistics and English Education*, 1(01), 11–20. DOI: 10.35961/salee.v1i01.66

Beveridge, C. (2022, January 12). *How to Make YouTube Shorts: Everything You Need to Know*. Hoostsuite. https://blog.hootsuite.com/youtube-shorts/

Butt, K., Arif, M., & Rafiq, M. (2023). Assessing University Students' Satisfaction with Web-Based Library Services Based on SERVQUAL Model: A Case of a Pakistani University. *Libri*, 73(2), 153–165. DOI: 10.1515/libri-2022-0050

Chapple, C. (2020, April). *TikTok Crosses 2 Billion Downloads After Best Quarter For Any App Ever*. Sensor Tower. https://sensortower.com/blog/tiktok-downloads-2-billion

Chen, Z., He, Q., Mao, Z., Chung, H. M., & Maharjan, S. (2019, May 17). A study on the characteristics of douyin short videos and implications for edge caching. *ACM International Conference Proceeding Series*. DOI: 10.1145/3321408.3323082

Escamilla-Fajardo, P., Alguacil, M., & López-Carril, S. (2021). Incorporating TikTok in higher education: Pedagogical perspectives from a corporal expression sport sciences course. *Journal of Hospitality, Leisure, Sport and Tourism Education*, 28, 100302. DOI: 10.1016/j.jhlste.2021.100302

Essel, H. B., Vlachopoulos, D., & Tachie-Menson, A. (2021). The relationship between the nomophobic levels of higher education students in Ghana and academic achievement. *PLoS One*, 16(6), e0252880. DOI: 10.1371/journal.pone.0252880 PMID: 34133434

Instagram. (2020, August 5). *Introducing Instagram Reels*. https://about.instagram.com/blog/announcements/introducing-instagram-reels-announcement

Instagram. (2021, January 5). *The Dos and Dont's for Reels*. https://www.instagram.com/p/CJrNqV6h3RK/

John, Y. (2023). Do Electronic Services Improve Students' Satisfaction among Public Sector Organizations? The Case of Tanzanian Higher Learning Institutions. *Perspectives on Global Development and Technology*, 22(1–2), 27–43. DOI: 10.1163/15691497-12341650

Khan, M. L. (2017). Social media engagement: What motivates user participation and consumption on YouTube? *Computers in Human Behavior*, 66, 236–247. DOI: 10.1016/j.chb.2016.09.024

King, D. L. (2018, November 1). *Making Library Videos*. American Libraries Magazine. https://americanlibrariesmagazine.org/2018/11/01/making-library-videos/

King, D. L. (2022). *TikTok And Your Library - It Might Be Time!* Princh Library Blog. https://princh.com/blog-tiktok-and-your-library/#.YoUCrcPMLIU

Kong, D. (2018, January 5). *2017 Short Video Industry Report*. http://www.199it.com/archives/672181.html

Leander, K. M., & Burriss, S. K. (2020). Critical literacy for a posthuman world: When people read, and become, with machines. *British Journal of Educational Technology*, 51(4), 1262–1276. DOI: 10.1111/bjet.12924

Lerman, R. (2020, August 5). *Instagram Reels, Facebook's new TikTok competitor, is released amid political scrutiny over the Chinese social media platform*. The Washington Post. https://www.washingtonpost.com/technology/2020/08/05/tiktok-instagram-reels-facebook/

Lorenz, T. (2019). *TikTok Clubs in High School? It Was a Thing*. The New York Times. https://www.nytimes.com/2019/10/19/style/high-school-tiktok-clubs.html

Martin, C. M. (2012). One-minute video: Marketing your library to faculty. *RSR. Reference Services Review*, 40(4), 589–600. DOI: 10.1108/00907321211277387

McRoberts, S., Ma, H., Hall, A., & Yarosh, S. (2017). Share first, save later: Performance of self through Snapchat stories. *Conference on Human Factors in Computing Systems - Proceedings, 2017-May*, 6902–6911. DOI: 10.1145/3025453.3025771

Meng, K. S., & Leung, L. (2021). Factors influencing TikTok engagement behaviors in China: An examination of gratifications sought, narcissism, and the Big Five personality traits. *Telecommunications Policy*, 45(7), 102172. DOI: 10.1016/j.telpol.2021.102172

Merga, M. K. (2021). How can Booktok on TikTok inform readers' advisory services for young people? *Library & Information Science Research*, 43(2), 101091. DOI: 10.1016/j.lisr.2021.101091

Misra, P., Chopra, G., & Bhaskar, P. (2023). Continuous usage intention for digital library systems among students at higher learning institutions: Moderating role of academic involvement. *Journal of Applied Research in Higher Education*, 15(5), 1752–1766. DOI: 10.1108/JARHE-06-2022-0185

Nguyen, L., & Mikolajczyk, K. (2021, July 21). *A Beginner-Friendly Guide to Using Instagram Reels for Business*. Hootsuite. https://blog.hootsuite.com/instagram-reels/

Okorafor, K., Arioie, I. A., & Anyalebechi, L. I. (2017). Awareness and Use of Reference and Information Services amongst Library Users in University Libraries in two States in Nigeria. *Journal of Applied Information Science and Technology*, 10(3), 109–116. https://www.jaistonline.org/10vol3/14.pdf

Omar, B., & Dequan, W. (2020). Watch, share or create: The influence of personality traits and user motivation on TikTok mobile video usage. *International Journal of Interactive Mobile Technologies*, 14(4), 121–137. DOI: 10.3991/ijim.v14i04.12429

Ortega-Martínez, E., Pacheco-Mendoza, J., García Meléndez, H. E., Ortiz-Díaz, E. M., & Saavedra-Alamillas, C. (2021). Digital services adapted by libraries in Mexico to COVID-19 pandemic: A critical review. *Digital Library Perspectives*, 37(1), 3–17. DOI: 10.1108/DLP-07-2020-0063

Preece, J., Nonnecke, B., & Andrews, D. (2004). The top five reasons for lurking: Improving community experiences for everyone. *Computers in Human Behavior*, 20(2), 201–223. DOI: 10.1016/j.chb.2003.10.015

Scherr, S., & Wang, K. (2021). Explaining the success of social media with gratification niches: Motivations behind daytime, nighttime, and active use of TikTok in China. *Computers in Human Behavior*, 124, 106893. DOI: 10.1016/j.chb.2021.106893

Silvia. (2022, January 14). *The Evolution & Future of Short Film Video*. Bbgg. https://bbggadv.com/insight-short-form-video/

Simon, L. (2021, March 18). *#LibraryTikTok Takes You Behind the Shelves*. I Love Libraries. https://ilovelibraries.org/article/librarytiktok-takes-you-behind-shelves/

Smith, M. (2022, February 12). *20 Best Short Video Apps*. Marketing Scoop. https://www.marketingscoop.com/short-video-app/?utm_source=rss&utm_medium=rss&utm_campaign=short-video-app

Statista. (2022, January). *Instagram: users by country*. https://www.statista.com/statistics/578364/countries-with-most-instagram-users/

Subramanian, S. (2020). Bahujan girls' anti-caste activism on TikTok. *Feminist Media Studies*, 21(1), 154–156. DOI: 10.1080/14680777.2021.1864875

Sukarjono, S., Nasrulloh, A., & Fakhrurozi, Z. A. (2023). Effect of Literature Availability in Sports and Quality of Service of Library towards the Satisfaction and Learning Motivation of the Students Major of Sports. *INTERNATIONAL JOURNAL OF MULTIDISCIPLINARY RESEARCH AND ANALYSIS*, 06(11). Advance online publication. DOI: 10.47191/ijmra/v6-i11-46

Sundar, S. S., & Limperos, A. M. (2013). Uses and Grats 2.0: New Gratifications for New Media. *Journal of Broadcasting & Electronic Media*, 57(4), 504–525. DOI: 10.1080/08838151.2013.845827

Superlibrarymarketing. (2020). *The 2022 Guide to Social Media for Libraries: Everything You Need to Know About Instagram – Super Library Marketing: Practical Tips and Ideas for Library Promotion*. https://superlibrarymarketing.com/2021/11/22/instagramforlibraries2022/

Tolentino, J. (2019, September 23). *How TikTok Holds Our Attention*. The New Yorker. https://www.newyorker.com/magazine/2019/09/30/how-tiktok-holds-our-attention

Udem, O. K., & Ogungbeni, J. I. (2023). Use of Mobile Technology-Based Library Services and Information Needs Satisfaction of Undergraduates in Universities in South-West Nigeria. *Folia Toruniensia*, 23, 15–33. DOI: 10.12775/FT.2023.001

Vázquez-Herrero, J., Negreira-Rey, M. C., & López-García, X. (2020). Let's dance the news! How the news media are adapting to the logic of TikTok. *Journalism*. Advance online publication. DOI: 10.1177/1464884920969092

Wright, C. (2017). Are beauty bloggers more influential than traditional industry experts? *Journal of Promotional Communications*, 5(3), 303–322. https://promotionalcommunications.org/index.php/pc/article/view/113

Xiao, Y., Wang, L., & Wang, P. (2019). Research on the Influence of Content Features of Short Video Marketing on Consumer purchase intentions. *Proceedings of the 2019 4th International Conference on Modern Management, Education Technology and Social Science (MMETSS 2019)*, 415–422. DOI: 10.2991/mmetss-19.2019.82

ADDITIONAL READING

Dey, S., & Satpathy, D. (2015). Marketing of Information Services in Libraries. *Social Science Research Network*, 295-307.

Liu, Y., Chiu, D., & Ho, K. (2023). Short-Form Videos for Public Library Marketing: Performance Analytics of Douyin in China. *Applied Sciences (Basel, Switzerland)*, 13(6), 3386. Advance online publication. DOI: 10.3390/app13063386

Minami, T. (2010). Towards Collaborative Library Marketing System for Improving Patron Satisfaction., 237-246. https://doi.org/.DOI: 10.1007/978-3-642-12101-2_25

Tyler, K., & Hastings, N. (2011). Factors Influencing Virtual Patron Satisfaction with Online Library Resources and Services. *The Journal of Educators Online*, 8(2), 1–210. DOI: 10.9743/JEO.2011.2.3

KEY TERMS AND DEFINITIONS

Facebook Reels: Facebook Reels is a feature within the Facebook app that enables users to create and share short-form videos. It is Facebook's answer to the growing popularity of short-form video content.

Instagram Reels: A feature within the Instagram app that allows users to create and share short-form videos, similar to TikTok. Reels can be up to 60 seconds long and can include audio, effects, and other creative tools.

Library Marketing: The process of promoting library resources, services, and events to current and potential users through various marketing channels, such as social media, email, print materials, and community outreach.

Library Services: The range of resources and assistance provided by a library to its users, including book lending, research support, reference services, computer and internet access, educational programs, and more.

Short Video Platforms: Social media platforms or features within existing platforms that are specifically designed for creating and sharing short-form videos, typically ranging from a few seconds to a couple of minutes in length. Examples include TikTok, Instagram Reels, Facebook Reels, and YouTube Shorts.

TikTok: A popular social media platform for creating and sharing short-form videos, typically up to 60 seconds long. Users can browse a personalized feed of videos, participate in viral trends and challenges, and engage with other users through likes, comments, and shares.

YouTube Shorts: A feature within the YouTube platform that allows users to create and share short-form videos, typically up to 60 seconds long. Shorts are designed to compete with platforms like TikTok and can be discovered through the YouTube app or website.

Chapter 5
Enhancing Communication Dynamics Through Interactive Activities in Distance Education Environments:
Instructional Design and Course Creation

Rachel Sale
South Texas College, USA

Sharon K. Andrews
University of Houston-Clear Lake, USA

Janice Newsum
https://orcid.org/0009-0001-4222-5043
University of Houston-Clear Lake, USA

Caroline M. Crawford
University of Houston-Clear Lake, USA

ABSTRACT

This multiple case study is a qualitative approach towards better understanding developing digital learning landscapes. Equally impactful are considerations to-

DOI: 10.4018/979-8-3693-3579-6.ch005

wards generative artificial intelligence, large language models, machine learning, and simulations integrated into course creation. The findings reflect twelve major themes that fall into categories of course design (course design with decisions based within environment, course design experience, course design), organizational philosophy of instructional supports (social learning, organization as collegiality and community supports, training and talent development, teaching complexity), student success and learner motivational supports (learner success as retention and competencies-capabilities, designing user experience and accessibility, learner motivation and ethical knowledge attainment), and innovation dynamics (flawed viability, viability of innovation).

INTRODUCTION

The continuously evolving digital learning landscape is ever-impacted by the latest technology trends that offer a multi-faceted understanding around the concept of designing instructional spaces within the digital environment. Clearly, the learning management system environment (LMS) is one consideration towards how learning is designed and delivered, while the more recent generative artificial intelligence (generative AI) and large language models (LLMs) as considerations towards machine learning (ML) and associated supports yield the efficient acceleration of the learning process. Within distance education learning environments, it is imperative to recognize and integrate the design of communication dynamics. This is recognized through Bandura's social learning theory (Bandura, 1969, 1971, 1972, 1986; Bandura & Walters, 1977; McLeod, 2011; Nabavi, 2012) wherein the sociocultural engagement of the learners is of paramount importance and impact, towards engaging in framing the knowledge for the learners within Vygotsky's understood Zone of Proximal Development as well as conceptual frameworks of understanding (Daniels, 1996; Vygotsky, 1962, 1978, 1987; Wertsch & Sohmer, 1995). The emphasis upon AI within the learning environment, supported through LLMs and ML as well as equally viable enhanced engagements with information learned within a digital environment of learning, supports the emphasis upon the digitalization of communication dynamics within learning environments.

Equally impactful, are communication dynamics within distance education environments that structurally support the inclination towards an evolved social learning embodiment of the learning environment, supporting Wenger-Trayner's community of practice (Wenger, 1998), as well as the evolution of instructionally-embedded learning in landscapes of practice (Wenger-Trayner, et al., 2014; Wenger-Trayner & Wenger-Trayner, 2015, Wenger-Trayner & Wenger-Trayner, 2020) that may support and enhance a reflection of:

...an understanding around value creation within the bounds of more socialized elearning spaces that are inherent within the hyperconnected world of the Internet, or the Web of Things (WoT) that has been embedded within the Digital Age's immediacy of information and the socialization that occurs as inherent within the learning process. (Crawford et al., 2022, p. 1)

The question posed, and desires serious ethical consideration is, towards what end? The ability to enhance communication dynamics through interactive activities is imperative, especially within a distance education environment. As such, the impact of interactive activities throughout the instructional design and course creation process is imperative to study and towards developing a stronger understanding of the associated interactive considerations that occur. The metaphoric shifting sands of teaching and learning within the higher education milieu emphasizes the desire for clarity as well as ethical guidance, associated with the concept of generative AI and LLMs. This is especially impactful within appropriate considerations towards ML and the impact upon the learning journey. To reflect upon deeply impactful recognitions within the real world of actual implementation versus theoretical ruminations, immersed higher education faculty coalesce, to consider these questions of significant impact.

To query the currently explosive topic within higher education environments is to take on a snapshot of understanding within the field. Yet to bring together thoughtfully impactful and dynamic faculty who are deeply immersed within the teaching and learning process that occurs within higher education courses is an effort and impact that heightens the discussion. Highlighting the strengths of experience, highlighting the strengths of extensive instructional knowledge and facilitative engagement, while equally emphasizing a recognition of subject matter expertise, evolves the discussion to one of impactful recognition. As such, to what extent does generative AI, LLMs, and ML, support the enhancement of communication dynamics through interactive activities within distance education environments?

A snapshot of current understanding associated with generative AI, LLMs, and ML, throughout the instructional design and course creation process is necessary and potentially of significant impact towards future considerations. To what extent might communication dynamics within distance education environments simulate and strengthen the potential for mirrored comfortably traditional learning experiences? Indeed, communication dynamics are continuously evolving and devolving through interrelationally designed interactive activities that include the following engagement considerations:

- Learner-content (Moore, 1989)
- Learner-interface (Hillman, Willis & Gunawardena, 1994)
- Learner-instructor (Moore, 1989)

- Learner-learner (Moore, 1989)
- Learner-self (Crawford, 2001)
- Learner-community (Burnham & Walden, 1997)
- Instructor-community (Crawford, 2001)
- Instructor-content (Crawford, 2001)
- Instructor-interface (Crawford, 2001)
- Instructor-self (Crawford, 2001, 2003)
- Learner-Web of Things (Crawford et al, 2021)
- Instructor-Web of Things (Crawford et al, 2021)
- Instructor-instructor colleagues (Crawford et al, 2021)

The interactive activities embedded two frames of communication dynamics within the digitalization of the learning environment's impact; specifically, through the learner's different forms of digitalization of communication dynamics and through the instructor's engaged and interactive digitalization of communication dynamics. The impact of the digitalization of communication dynamics through interactive activities as supporting and enhancing communication dynamics within the distance education environment, is the focus of this multiple participant qualitative approach towards more deeply understanding distance learning course instructional design and course creation. Through the lens of interactive activities, four higher education professionals will consider communication dynamics towards instructional design and course creation within digital spaces.

BACKGROUND

To embed more deeply within a discussion of the ability to enhance communication dynamics through interactive activities, especially within a distance education environment, a multiple case self-study approach was implemented. The focus upon the embedding reflective queries as critically reflective understandings, that may guide a grounded theory approach. This focus is towards understanding the current recognition of generative AI and LLMs, as well as potential associated with machine learning (ML), directly impact the feasibility of instructional design and course creation within LMSs and associated impacts upon the design and associated creation of digital spaces for learning engagement. The higher education professionals will respond to the following query prompts:

- How many years have you created content for implementation within learning management systems (LMSs)?

What has been your experience designing and creating digital spaces for learning engagement, within learning management system environments (WebCT, Moodle, BlackBoard, Canvas, etc.)?

- To what extent did you design and create the distance course for the learner experience:
- To what extent did you design and create the distance course for the instructional facilitator experience:
- To what extent have you integrated artificial intelligence (AI) resources, machine learning (ML), and/or simulations into your course design? In what way(s) have you integrated AI, ML, and simulations? Merely a few examples of AI learning support may be offered as adaptive learning, intelligent tutoring systems, virtual facilitators, chat bots, learning personalization, assessment and grading, anti-cheating detection (Online Education, n.d., LearnDash Collaborator, 2020)
- What is your current opinion on the viability of AI, ML, and simulations, towards communication dynamics and learning engagement within distance education environments?

Participants

This manuscript integrates three participant colleagues from various spaces and places within higher education, as well as within their academic journey. To more fully develop an understanding of the multiple case studies within this critically analytic self-study, it is appropriate to support the reader's understanding of each participant's descriptive professional journey in its entirety.

Participant One

Participant One is embedded within a college environment, actively engaged in an administrative leadership position that reflects her extensive knowledge and expertise within the teaching, learning, talent development, and instructional design realms. Equally impactful is the technical expertise that this collegial participant embeds within her emphasis upon service within her professional community. Participant One has almost a quarter century of experience, creating content for implementation within learning management systems (LMSs). This participant highlights five years developing K-12 teacher-focused professional development opportunities, that were presented via groups and shared file systems that were considered to be leading-edge environments at the time. Further, this participant highlights almost twenty years of course content design and development within higher education LMS environments.

This participant brings forward extensive experience within the realms of designing and creating digital spaces for learning engagement, specifically within LMSs. Participant One has led instructional design (ID) teams, with the focus upon the development of higher education courses that were implemented within multiple campus levels of engagement, including within LMSs that include ANGEL, early traditional Blackboard, Blackboard Ultra, Moodle, and Canvas. Further, as a consultant, this participant has developed courses for higher education as well as corporate client's learning environments, including Nursing RN-BSN, education master's degrees, teacher certification coursework, and technology-embedded professional development training and talent development events.

Participant Two

Participant Two has been engaged in the creation of content for implementation within the learning management systems (LMSs) for over fifteen years, yet this participant has been engaged in the professorial charge associated with high quality instructional experiences, for well over thirty years of a successful academic career. This participant is currently a professor, as well as taking on a leadership role within which the participant proactively develops and guides a recognized program of study. Throughout this time period, the participant brought forward extensive expertise and success, resulting in a nationally ranked graduate program of study.

An intriguing differentiation of course design and implementation within and between Blackboard and Canvas is articulated. The participant's experience with Canvas reflected a perception that its strengths lie in its simple and direct support for copying of course materials from other Canvas courses, in any level of granularity. Further, its ability to track each student's time in a course, pages viewed, and trends associated with course activity over time is useful. Canvas also has the advantage of allowing for modules to be made dependent on the successful completion of other modules. Canvas' weaknesses include an over simplification of its interface, with a loss of the ability to customize the look and feel for each page in the course. The lack of ability in Canvas to uniquely colorize course pages as was easily accomplished in BlackBoard has caused file upload mistakes, wherein updates to content were loaded into the wrong course due to each course having the same look and feel. Students have reported going into the wrong course and not immediately recognizing that they are trying to submit an assignment in the wrong course environment. This is a problem that never occurred in this participant's prior BlackBoard color-coded course designs.

Canvas does offer the opportunity to create rigidly structured courses that guide an undergraduate or K-12 student methodically through a linear training step-type course. If this linear course flow is needed, then Canvas matches this need well. For graduate level projects and courses, or seminars and research topics courses that rely on student-directed learning and requires high levels of innovation and creativity, BlackBoard's lack of imposed linear module structure is the better solution in my experience thus far.

Participant Three

Participant Three is a colleague deeply embedded within an understanding of information science, having developed a strong professional reputation within a large metropolitan regional area over decades of embedded elementary and secondary school leadership experience while equally engaged within higher education online course instruction, as well as course redesign and enhancements so as to engage the learners more fully. Participant Three reflects from a more personalized perception of the online course learner experience, emphasizing LMS as well as non-LMS course enhancements.

Creating content within a learning management system is a new endeavor for this participant. Before entering higher education, content creation was limited to such non-LMS platforms as Canva, NearPod, and animoto. These K-12 tech tools allowed for sharing and designing educational content for use by students. The difference between these systems and the LMS are report generation and data analysis.

After entering higher education in 2017, the content has primarily been pre-existing resources or archived content from prior courses. An understanding of online courses primarily focuses upon enhancing the pre-existing online course learning experience, while equally focusing upon imbuing the personality and philosophical beliefs associated with the learning and facilitation supports that are strengths of this participant. The value of the LMS is that it simplifies access to learning content and streamlining tasks. For the past seven years, the participant has modified and or adapted content within the LMS.

Concerning strengths of the online course digital space's LMS environment, this participant suggests that, while many of the design elements were already in place in the courses inherited in 2017, this participant's strength is in being able to focus on learner's needs and to create personalized learning experiences. As technology is so prevalent in school libraries, including technology in course assignments encourages students to effectively integrate technology into their daily practice. Participant Three also emphasized providing access to the most current accessible resources and researched, selected, and replaced print text with new e-textbooks for each course, thereby reducing costs for students and ensuring the most up-to-date course learning materials. Creating writable pdfs to replace paper-based forms facilitated digital signatures and made possible the organizing and uploading of the forms into the LMS. Additionally, the time that it takes Participant Three to get things accomplished is a definite weakness. Participant Three expressed uncertainty if this weakness is a form of perfectionism or procrastination. Either way it hinders forward movement and is quite concerning. Participant Three's=goal is to tackle this weakness during a development leave and return to campus with fully revamped courses.

The best opportunity associated with creating online course digital space within the LMS environment, is the time to work on course design with minimal interruptions of course assessment, standards alignment, data analysis, or other distractions to course enhancements. There is also a gap in the participant's knowledge of instructional design which can be addressed through professional development and additional training. Continued experience in higher education and guidance from other faculty members will assist in the development of the skills needed to improve in course design. Through such courses offered by the Association of College and University Educator (ACUE) Participant Three has been able to learn and implement practices in effective online instruction. However, the biggest threat at this time is an outside one. School libraries are undergoing extensive challenges. Some K-12 public school districts are eliminating school libraries. There are unprecedented challenges to library collections. Although libraries represent the most accessible sources of information, maintaining library preparation programs is becoming increasingly difficult. Course design for those who remain in programs must address these concerns with realistic options and forward-thinking solutions.

COURSE DESIGN FOR THE LEARNER EXPERIENCE

Each participant was prompted to respond to the question: To what extent did you design and create the distance course for the learner experience? Explicitly, the participants were offered a list of acknowledged interactive activities to consider, including:

- Learner-content (Moore, 1989)
- Learner-interface (Hillman, Willis & Gunawardena, 1994)
- Learner-instructor (Moore, 1989)
- Learner-learner (Moore, 1989)
- Learner-self (Crawford, 2001)
- Learner-community (Burnham & Walden, 1997)
- Learner-Web of Things (Crawford et al, 2021)

Each participant was offered the opportunity to respond as was considered to be most appropriate for their own professional experience and comfortability. The presentation of participant responses reflects the level of detailed narrative preferred.

Participant One

Participant One addressed four of the interactive activities, reflecting the extent to which this participant designed and created distance courses focused upon the learner experiences articulated as: learner-content; learner-interface; learner-learner; and, learner-Web of Things. The personalized narrative is reflective of the critically analytic engagement in the depth of the embedded instructional design efforts towards learner experience achievement.

Learner-Content

In my first higher ed position, I led a team focused only on moving traditional course content to the ANGEL LMS (2008-2009). The focus was on academic content with text-heavy readings rather than engagement. From an instructional design view, the team's challenge was creating clear instructions to help students identify and complete the reading content as preparation for a traditional exam or paper submission. Online courses were new to this institution, and the feeling of the departments was that students would automatically choose to read the content if the instructions were clear. After two semesters of launching this class type with minimal success, the faculty users became more open to considering the online class needs for engagement/interaction. With the data from our first year, change was welcome and enhanced by bringing a campus membership to the Quality Matters organization.

Years later, I now encourage my IDs to use the learning content as a springboard for investigating engagement, interactivity, and alternative formats. The content is a checklist that must be completed, but each item in that list is an opportunity for s different way of presenting the content.

Learner-Interface

The learner-interface engagement is developing into the biggest challenge for instructional designers over the next few years. The rush for online development during the Pandemic years has produced a plethora of third-party vendors and tools that can be embedded to connect directly with a course grade book with Learning Tools Interoperability (LTI), which is especially attractive to instructors. Extra professional development (PD) may be necessary to ensure proper settings are used by each instructor. If many such additions are made to the traditional interface, it may be impossible to adequately train a help desk system.

The student challenges are widespread. Even if the settings are correct, some tools work only on specific devices, omitting Chromebooks, certain tablets, or phone access. Device access is a standard part of choosing one's campus LMS, but a committee looking for LTIs may miss that step. In the active learning environment, students must be trained for these separate tools in addition to normal LMS use. Students encountering unresolved usage issues may drop a course, directly affecting retention.

Learner-Learner

Learner-learner engagement is another commonly accepted need for the online course to be successful. The past few years have produced societal and technology changes that enable the options while questioning the need for engagement from many students. Those who facilitate and/or develop Competency Based Education are quick to identify student resistance to waiting for others to engage in an activity. After all, these programs promise quick completion based on the experience and learner initiative. In my teaching experience with CBE, even when students do not score well on a competency pretest, they take sole ownership of the score and just want to study for the posttest, not engage with others.

Technology changes provide so many options to engage with other humans though cell phone or tablet options. Social media platforms thrive on this activity, but perhaps it all reduces benefit of scholarly activity. When even professional work activities are lessened by the "another meeting that could have been an email" mentality, are students willing to engage an learn via conversations?

Learner-Web of Things

Engagements between learners and the Web of Things will only increase as workforce programs impact the higher ed campus. In the health care field alone, think how many classes routinely expect the students to engage with interactive/online technology! It's almost impossible to imagine a job that does not use components of this area.

The design of the course and related training are impacted by the growing options and even upgrades to older options. Instructional designers who are skilled at chunking content into quick bits of timely training will indeed have value to any institution. In previous times, this group might have been well staffed by the ones on your team that enjoyed "tinkering" with new technology and seeing where it could be used. The overall growth of the Web of Things may require IDs to have a solid Information Technology or Ed Technologies training added to a program.

Participant Two

Participant Two addressed four of the interactive activities, reflecting the extent to which this participant designed and created distance courses focused upon the learner experiences articulated as: learner-content; learner-interface; learner-learner; and, learner-Web of Things. The personalized narrative is reflective of the critically analytic engagement in the depth of the embedded instructional design efforts towards learner experience achievement.

Learner-Content

My course design is centered around a set of topics. Each topic is supported by a lecture with PowerPoint slides, a set of reviews that are organized by a set of questions I have posted about the topics along with answers to that question that serve as additional viewpoints of the topic from that presented by the text book or other assigned readings. The reviews and PowerPoint reflect my own personally prepared content that extend external resources such as text book chapters, journals, and conference papers surrounding the topic, as well as a collection of videos that showcase conference presentations that address key present day topics relevant to the course topics.

Learner-Interface

The interface to the course I present to the learner is typically centered around phases in the software life cycle and the order of presentation is dependent upon whether or not a course is designed for a planned development process model (like Waterfall) or Incremental process model (Agile for example). For example, in my architecture course, the architectural concepts that are relevant to phases of the planned development process model, are presented followed by design phase concepts that require architectural considerations, and so one through testing and maintenance. Each of these concept topics will involve watching videos, reading assigned papers and PowerPoint slides, and attending the zoom session for live discussion. After such preparatory work, assignments are given to exercise concepts related to the topic. In BlackBoard these topics were presented as a set of hyperlinked html files which allowed students to move freely between the various elements of each topic and the different units as needed to prepare and formulate a solution to an assigned project, essay, test, quiz, or software specific knowledge application task. In Canvas these topics were turned into modules which have an imposed linear order. In Blackboard the order was not imposed and the html hyperlinked interface could be used to move through the course via a set of assigned topics depending on the phases of the lifecycle being emphasized.

Learner-Instructor

The course is designed so that students must access, read, and absorb the notes, reviews, lectures, and videos in order to understand the basic knowledge and then synthesize and apply this knowledge by performing the assignments. As the instructor, my role is to provide the information in several different communication styles and with multiple different viewpoints with different mediums of delivery. Examples of different viewpoints include my viewpoint as a subject expert, videos and papers that express other known expert's viewpoints of the same topic. I make it a point to provide different mediums of expression of these viewpoints such a textual as in narrative writings, graphical as in html pages with graphics that contain hyperlinked explanations as well as automation of animation that illustrates the process control and process steps including inputs and outputs that the learners need to understand.

Learner-Learner

Most of my courses are designed so that students can take the course and learn the material without a lot of interaction with other students. There is of course the usual exception of group assignments and group discussion and peer feedback, that I provide to support student's efforts at working with other students in the class. I vary the number and depth of group assignments from semester to semester, as well as the number and depth of required group discussion and peer feedback. In project classes, group projects are the main emphasis and students must primarily work together to specify, design, and implement a software project. In such classes group work can be both rewarding and frustrating depending on the group makeup. Students who are non-contributing to the group become a stress factor for the group and students typically will report such behavior. There is typically one student in each class that starts the class with this behavior. In most cases this student's behavior is remedied once they realize their grade is negatively affected by this.

Learner-Self

The course is designed so that a learner can hopefully find a mode of instruction most suited to the type of learning they find most beneficial. For example, if a student's learning styles responds well to multiple explanations of the same concept from different viewpoints and in different mediums then that is available. If the student finds attending a zoom session and bringing a set of questions to ask of the instruction and engage in a discussion with the instructor and or the class, that is an option as Question and Answer Zoom sessions are held for each online class. Additionally a set of self-check quizzes are provided for students to take up to three times each to test themselves on their understanding of basic fundamental concepts.

Learner-Community

My course design addresses all learners in my class as a community of learners most directly in student required topic presentations. In many of my classes the course design involves assigning each student a unique topic in the class. Each student must study that topic and understand the basics of the topic, its intersection with the class objectives, why the topic is important to the course, advantages, and disadvantages relative to the topic and other fundamental understandings important to the topic. The topic assigned will be very fine grained, to allow for all such ideas to be ascertained in a relatively short amount of time (4 weeks for example) and can be presented in about 20 to 25 minutes in an oral presentation. The goal is for the student to present the topic to the class and be able to answer questions I may pose to the student during the presentation. It is in effect a mini oral exam. This type of exercise has proven to hone the student's motivation for preparation, knowledge acquisition and study skills. Also during the presentation I will not only question the student but correct any misunderstanding or mis-statements and give advice on what the student needed to include or should have left out or should clarify and expand. This activity lets the other students in the class have a dynamic view of a detailed discussion of the topic. Other students learn from the interaction as much as the presenter. This is most effective but it is quite exhausting for both the student presenter and the instructor. It is requiring a lot of class time so is only feasible in classes of about 15 or less which fortunately is a typical graduate class size in my program.

Learner-Web of Things

In my course design I provide access to freely available tools the students may find helpful in both the course itself as well as outside of the course. Tools such as web site generators, word processors, spreadsheets, wikis, software measurement tools, software analysis tools. Also some course designs will require students to find available software tools and use these tools and compare and analyze such tools. Assignments to analyze AI tools is an example of some of the most current ones I have incorporated into the course design of some courses.

Participant Three

Participant Three addresses seven of the learner-focused interactive activities, reflecting participant designed and redesigned course experiences as well as course enhancements that develop a more personalized instructor approach to the learning and instructional facilitation experience. The personal narrative is well developed yet a succinct understanding associated with this type of online course experience that imbues a stronger sense of learning community and learner engagement within the online course environment.

Learner-Content

Some of these design changes seem minor, however, the students' responses to the changes were positive. Recommended instructional technologies were updated; redundant discussion forums were eliminated, multi-part assignments were consolidated, and time was extended for comprehensive assignments.

Learner-Interface

A very simple, but effective interface change was to update the images on the course landing page. The images did not reflect the diversity of the university and were dated. Opening the course and viewing a more modern version of a library or librarian made the interface aesthetically pleasing.

Learner-Instructor

If this relates to the learner and instructor engagement, then my experience is broad. So much is learned from my students and as a lifelong learner, I am always seeking opportunities to learn more. Students have been instrumental in increasing my engagement with social media, learning new instructional technologies they have discovered and are using, and in keeping abreast of trends in the various districts.

Learner-Learner

To update the standard "introduction" at the beginning of a course, students have another assignment in which they research, select, and use a comic creation software/application to introduce themselves to the class.

Learner-Self

While several learner-self assignments are presented in various courses, students indicate there is one that resonates with them more than the others. Students are asked to use presentation software to create infographics and video or audio only presentations of the people they would like to have with them if on a deserted island. The assignment does not limit real/unreal, living/deceased. But it does restrict the people to non-family members. Students frequently draw upon their knowledge of good literature or memorable movies to think of people to include.

Learner-Community

The learner community is the primary beneficiary of course design. The courses are designed with an eye towards efficacy and best practices. The learner community extends to the students served by our students. In order to serve the broader community, the learning experiences of our students must reflect the proven approaches to pedagogy and an emphasis on critical higher order thinking.

Learner-Web of Things

Design related to the Web of Things continues. As the ubiquitous nature of the web becomes more and more pervasive, the learner's experience should be effortless. The usability of the internet is a necessary literacy for users of multiple devices. Students are perhaps less aware of the representation of their devices on the internet, but a rudimentary understanding of information/data transfer, communication, and compatibility will facilitate ease of use. The WoT is not necessarily a design element in the course, but it is information that is shared with students.

COURSE DESIGN FOR THE INSTRUCTIONAL FACILITATOR EXPERIENCE

Each participant was prompted to respond to the question: To what extent did you design and create the distance course for the instructor facilitator experience? Explicitly, the participants were offered a list of acknowledged interactive activities to consider, including:

- Instructor-community (Crawford, 2001)
- Instructor-content (Crawford, 2001)
- Instructor-interface (Crawford, 2001)

- Instructor-self (Crawford, 2001, 2003)
- Instructor-Web of Things (Crawford et al, 2021)
- Instructor-instructor colleagues (Crawford et al, 2021)

Each participant was offered the opportunity to respond as was considered to be most appropriate for their own professional experience and comfortability. The presentation of participant responses reflects the level of detailed narrative preferred.

Participant One

Participant One addressed one of the interactive activities, reflecting the extent to which this participant designed and created distance courses focused upon the instructional facilitator experiences articulated as: instructor-content. The personalized narrative is reflective of the critically analytic engagement in the depth of the embedded instructional design efforts towards instructional facilitator experience achievement.

Instructor-Content

The premium design experience comes when the instructor as subject matter expert (SME) and I work together. In this process, they vet and choose the interactive/engagement options and tools. If they are unsure of how to use the tools as the facilitator we have time for the support team to provide quality instruction.

There are many reasons, institutions are turning to the use of master courses. They allow a significant reduction for the overall cost of development. Budgets are also reduced by using more adjuncts and fewer tenure track faculty. Occasionally, I hear the altruistic goal of offering the same quality instruction to all students.

Master courses can be created quickly with an instructional design team trained to work and support each other. My current team is composed of four IDs, 3 course developers, and a media expert. The team lead reviews a course request and the syllabus while one of the IDs creates a course map. We have a fairly standard *start here* module, and one of the course developers reaches out for any personalized content such as the instructor information and welcome statements. Prior to meeting with the instructor for any development time, the Blackboard AI tool uses the Syllabus to create a suggested course module pathway. If this outline matches the course map, the project is ready for team activity.

The team lead meets with the instructor and illustrates the process we have so far. They agree on the general look and feel of the class, and what elements of Bb and third -party tools will be used. If approved, each ID/course designer is assigned individual parts of each module to create. It is very common for this team approach to shorten a full course initial design to less than a week.

Revisions and changes are scheduled around the instructor's decisions. Each course must also be at least 95% accessible, and sometimes that takes extra work on the part of the design team. When the course is complete, an approval form showing the map and a link to the course for review is routed to the chair, the dean, and then finalized by my approval as the Dean of the Digital Learning Department. These approvals are valid for 5 years unless major changes are implemented.

Participant Two

Participant Two addressed instructor-focused interactive activities, including: instructor-community; instructor-content; instructor-interface; instructor-self; instructor-Web of Things; and instructor-instructor colleagues.

Instructor-Community

My online courses are designed for my use only and are not shared purposely by myself with the community of other instructors at large. Nor are my course materials placed in public sites for download. However, students are known to download course materials and distribute them. With the use of AI I believe students will be feeding all course materials into the AI tool sites so that most of the hard designed course materials will be used without attribution.

Instructor-Content

My online courses are all designed so that I can change the course easily from semester to semester in terms of the topics to be included, the types of assignments, the number and types of quizzes and the types of projects. For example, I have developed for each course an extensive database of quiz questions for a fine grained set of topics contained in each course. Each topic is contained in a quiz data base in BlackBoard, and in Question Banks for each topic in Canvas. Each topic contains between 30 – 60 questions with half essay and half of multiple choice. From this database a quiz in any course can be quickly created for a topic or set of topics. This is one of the more efficient aspects of my course design. The assignments are not as modular and cannot be generated from a set of topics but, instead, be hand designed or reused. I typically do not reuse assignments or quizzes unless several years have passed in between uses.

Instructor-Interface

Participant Two reflected upon: how do I design the interface in my course.

The interface in my course design is centered around a central content document which is an html document that links all topics to all content related to that topic. In addition to this, I provide a set of weekly assignments, quizzes, discussion boards, peer reviews and at times special presentation assignments. The navigation through the material from the content html allows the user to move freely from topic to topic as required in order to accomplish the assignments. In effect the content html removes the linear restriction of moving module to module that Canvas imposes. Canvas modules include assignment instructions, presentation instructions, and support files for assignments and course tasks.

Instructor-Self

Participant Two reflected upon how do I view myself in the course.

As instructor, my job is to guide the students to and through the set of topics and necessary topic details student must learn to meet the course objectives. The course materials I create deliver these topic details. It is my job to design a course that illustrates and explains the most important concepts needed to apply the skills learned. The selection of topics and the depth of them, what to include and what to leave out is most important. Too many details confuse where not enough detail is insufficient. Knowing what is most important about a topic is the hallmark of good instruction and of an effective course. All class assignments and activity must center and reflect this guiding principle.

Instructor-Web of Things

As instructor I do at times make use of freely available open source software to assign to students to study or modify or analyze.

Instructor-Instructor Colleagues

I have not participated in team teaching a lecture course with a colleague, but this is the most prominent example of direct colleague course interaction I can think of. I have given thought to participating in such an endeavor but it presents a reimbursement entanglement at my university such that I have not been willing to negotiate. I have had discussion regarding how such a course would be structured and I believe if not careful with topic distribution it could be more confusing to the students than helpful. Though I do think the instructors would enjoy the endeavor.

However, if the course were a seminar or project style course I think team teaching could be easily implemented and two or more instructors could be quite engaging and helpful to students.

Participant Three

Participant Three addressed the instructor-engaged interactive activities that directly addresses instructor-community, instructor-content, instructor-interface, instructor-self, instructor-Web of Things, and instructor-instructor colleagues.

Instructor-Community

We work in such an isolated environment that I give little consideration to instructor-community within the university. In the courses there is an emphasis upon the importance of stakeholders in the profession. Students are encouraged to consider the impact of all stakeholders on their professional practice and to be aware of the ways in which the community can support or hinder their work. Assignments are included that require students to engage in collaborative efforts with classroom teachers, parents, district librarians, and public libraries.

Instructor-Content

Content is constructed so that additional faculty can readily and easily utilize the course materials. This facilitates the use by adjunct faculty or other faculty who may teach the courses from time to time.

Instructor-Interface

The LMS interface is pre-set. There has been no motivation to alter or modify the interface for my benefit. The changes to the interface have been to make it more appealing to students. Other than ease of use, the design elements I focus on most are primarily for students.

Instructor-Self

Designing for the facilitator is less desirous to me because the course is meant to engage the students and promote their learning. Course design can go a long way towards helping students feel valued. According to an ACUE teaching module (2023) both employee retention and student success are bolstered by a sense of belonging.

Instructor-Web of Things

As an instructor I feel pretty much about my relationship to the Web of Things as I do about the students' WoT relationship. The Web of Things will continue to evolve, and use will continue. Thus, instructors and students must be adept at managing both the complexity and quantity of devices.

Instructor-Instructor Colleagues

Designing and creating distance learning experiences between colleagues is minimal. The primary consideration is whether or not the courses are easily and readily replicable for adjuncts or other faculty who may teach using the content.

COURSE DESIGN: INTEGRATION OF AI, ML, AND SIMULATIONS

The integration of more recent resources and associated tools within course design offers an intriguing snapshot of current understanding within the higher education milieu. The prompt to which the participants responded was a deep, rich approach to better understanding the more recent understandings associated with the implementation of artificial intelligence (AI), machine learning (ML), and the concept of differentiated simulation approaches into the learning environment. Comfortability and vulnerability associated with this developing understanding and engagement with development environments and associated resulting products from AI, ML, and simulations, offers a superior cross-sectional understanding of *where we are* in higher education during this moment in time. The prompt offered to the participants was:

> *To what extent have you integrated artificial intelligence (AI) resources, machine learning (ML), and/or simulations into your course design? In what way(s) have you integrated AI, ML, and simulations? Merely a few examples of AI learning support may be offered as adaptive learning, intelligent tutoring systems, virtual facilitators, chat bots, learning personalization, assessment and grading, anti-cheating detection (Online Education, n.d., LearnDash Collaborator, 2020).*

Participant responses viewed through this lens of understanding offers a query worthy of intrigue.

Participant One

Participant One responded to both prompts, offering well-rounded understandings of instructional design and course development efforts.

In the past year, my team has integrated AI into many courses. We began faculty PD by reviewing how much AI is already used. The Microsoft Editor tool can produce as many suggestions as any purchased tool. Instructors were excited to see how the Blackboard LMS AI feature will create a full course shell including learning content using your syllabus. This class provided the perfect opportunity to experience AI benefits and see how humans still need to fact-check and edit to correct errors. Collaborating with the LMS AI tool to create new quiz banks was not only exciting for most, but it also immediately supported our quest for additional Open Educational Resources (OER) across campus.

Merely a few examples of AI learning support may be offered as adaptive learning, intelligent tutoring systems, virtual facilitators, chat bots, learning personalization, assessment and grading, anti-cheating detection (Online Education, n.d., LearnDash Collaborator, 2020)

My design team has implemented several AI projects. We mastered the new LMS AI options and pass these on to faculty via online and face-to-face professional development. These options include Course module design, suggested images with alt txt, assessment question creation, and general page content creation.

Our faculty and student help desk services utilize a bot to encourage using the knowledge base and increase the value of service without extended personnel costs. We are implementing a bot in lesson modules that will allow students to ask up to 10 questions about each module and receive an answer. For example, what learning objectives are assessed by each graded item in this module? How many overall points or percentage of my grade is in this module? What chapters of the text are in this module? While this is information in the Overview, the bot may be of quicker service to the students.

Our learning engagement process is always important, but we just signed a contract to add H5P to our Blackboard Ultra courses. The H5P AI has research data showing that course creation time can be reduced by up to 80% by the use of the AI in their system.

Participant Two

AI has not been integrated into my course design except as a class project topic whereby I assign students to analyze, compare and asses current AI Tools and their implementation methods. I have yet to find a need or good reason to use current flawed AI tools to grade, detect cheating, or create my lectures. Software included with Blackboard (Safe Assign) and Canvas to detect cheating is minimally useful. If future plagiarism or other AI tools built into Canvas prove to be correct and reliable then I would test them and consider how I might use them to improve aspects of my course.

Participant Three

Participant Three responded to both prompts, articulating an intriguing understanding of AI, ML, and simulations within an articulated online course understanding as well as a *next step* consideration within a larger realm of understanding and consideration.

As concerns the integration of AI, ML, and simulations into this participant's course design, at this time, I have not integrated artificial intelligence, machine learning, or simulations into my course design. Learning the nuances of AI is just at the beginning stages for me and I have not located any simulations of the work that goes on in school libraries. The best we can do at this time is the use of video recordings. The LMS provided by the university has built-in anti-cheating detection. A recent conversation with one of my students revealed that Canvas has an AI detection feature.

COMMUNICATION DYNAMICS AND LEARNING ENGAGEMENT

A differentiation from the course design consideration, is the emphasis upon instructional engagement and learning community-building effort that is appropriately designed into a course experience. Specifically within distance education environment course experiences, the embedded emphasis upon interactive activities must be specifically designed into the course experience so as to support the social learning strengths of engagement. The participants responded to the following prompt: What is your current opinion on the viability of AI, ML, and simulations, towards communication dynamics and learning engagement within distance education environments? The differentiation in critical analysis from the course design and into the social learning aspects of learning communities, communities of practice, and perhaps even learning in landscapes of practice, more fully embed and embrace the participant perspectives and current philosophical understandings of the strengths associated with enhancing communication dynamics within distance ed

Participant One

Simulations are invaluable for many areas. Nothing prepares a nurse for the reality of daily life like working in a fully simulated hospital environment for their academic years. Workforce programs such as welding and auto repair or painting could not turn out the large number of workers needed without beginning with simulations before putting students in front of live equipment.

My feeling of AI is that it is going to allow teachers to focus on teaching, not requiring all of them to be technology users first. If we can help faculty to accept and grow with AI, they will bypass stressors that have grown with technology. I can remember having to master the mimeograph machine during teaching, and then years of learning the tips and tricks of a new departmental copier. Today's faculty have the ability to create and share content with nothing more than the ability to visualize and talk about the concepts. The key to this reality lies in our ability to lessen fear of change and help value the tools.

Participant Two

The use of AI as a vehicle for the student to study and compare software tools in terms of its limitations and strengths, what it can and cannot do well, is a worthy exercise of benefit to graduate software students. Students need to study such tools to be aware of the implementation methods employed and the extent to which each tool is effective. Learning to perform such an analysis and comparison assessment is a highly effective learning activity. The viability of each AI tool resides in the correctness of the large language model it is based upon as well as the correctness of the domain analysis and domain knowledge behind the model. The larger the domain the less likely the model will be correct. The direct use of current AI tools in a graduate software engineering class has little to no benefit as the class requires skills that are very hard to capture and automate in solving novel software problems. AI tools can be of benefit for the generation and reuse of well-known solution to common coding solution to known problems. Such can be used to free the software developer to concentrate on hard to solve novel aspects of system design problems.

Participant Three

This participant's current opinion of AI, ML, and simulations, as supports towards communication dynamics within learning engagement, distance education environments are already perceived to be embedded. Machine learning and simulations have been used for many years in medical, commercial, and private industries. There are potentially many learning applications for artificial intelligence. AI has the potential to reduce time on tasks for writers, developers, educators grading, and similar repetitive tasks. AI can also support the scaffolding a learning making the necessary adjustments based on students' success in learning new content. The danger is that students who have not yet developed the necessary skills in any content area will use AI to complete assignments without the depth of knowledge needed to grow as learners. The dilemma is the unethical use of this new tool in education. AI must be integrated into distance learning protocols because the benefits outweigh the risks.

FINDINGS: SOLUTIONS AND RECOMMENDATIONS

An analysis of the participant data was qualitatively analyzed through hand coding. The codes, categories, and themes were articulated by major topic area as designed within the chapter's flow of major headings and associated participant presentation of insight. Towards reflecting the major themes that arose from the analysis phase, Table 1 represents the thematic alignment, articulated within the four overarching thematic areas of Course Design, Organizational Philosophy of Instructional Supports, Student Success and Learner Motivational Supports, and Innovation Dynamics.

Table 1. Major thematic representation and alignment indication by manuscript section

	Course Design	**Organizational Philosophy of Instructional Supports**	**Student Success and Learner Motivational Supports**	**Innovation Dynamics**
Course Design for the Learner Experience	Course Design Decisions Based within Environment	Social Learning	Learner Success: Retention & Competencies-Capabilities	
Course Design for the Instructional Facilitator Experience	Course Design Experience	Organization: Collegiality & Community Supports	Designing User Experience & Accessibility	
Course Design: Integration of AI, ML, and Simulations	Course Design	Training and Talent Development		Flawed Viability
Communication Dynamics and Learning Engagement		Teaching Complexity	Learner Motivation & Ethical Knowledge Attainment	Viability of Innovations

The articulation of the four main areas of focus within this discussion, specifically:

- Course Design for the Learner Experience
- Course Design for the Instructional Facilitator Experience
- Course Design: Integration of AI, ML, and Simulations
- Communication Dynamics and Learning Engagement

Supported the presentation of the themes articulated within each data analysis undertaken. The articulation of the major manuscript sections with aligned themes are:

- Course Design for the Learner Experience
    - Course Design Decisions Based within Environment
    - Social Learning
    - Learner Success: Retention & Competencies-Capabilities
- Course Design for the Instructional Facilitator Experience
    - Course Design Experience
    - Designing User Experience & Accessibility
    - Organization: Collegiality & Community Supports
- Course Design: Integration of AI, ML, and Simulations
    - Course Design
    - Flawed Viability
    - Training and Talent Development
- Communication Dynamics and Learning Engagement
    - Learner Motivation & Ethical Knowledge Attainment
    - Teaching Complexity
    - Viability of Innovations

Table 2 represents the codes, categories, and themes that naturally developed through the data analysis of the participant data in the Course Design for the Learner Experience manuscript section.

Table 2. Course design for the learner experience codes, categories, and themes

Code	Category	Theme
Academic Content versus Engagement	Content	Course Design Decisions Based Within Environment
Content as a Springboard	Content	
Instructor-Developed and Instructor-Framed Content	Content	
Streamline Course Content and Engagement	Content	

continued on following page

Table 2. Continued

Code	Category	Theme
Progressive Phases of Course Content Presentation	Content	
Course Design Interface Decisions Based upon Capability of LMS	Design	
Initial Foray = Minimal Success	Design	
Present Information in Different Ways, thru Different Mediums	Design	
Co-Learning Opportunities Collegial Learning Environment	Design	
Course Updates Aesthetically Appeal	Design	
Learner Preferences	Design	
Question Learner Engagement Need	Course Community Experience	Social Learning
Over-Scheduled Overwhelmed	Course Community Experience	
Cognitive Modeling Cognitive Apprenticeship	Course Community Experience	
Real World Pedagogical Approaches	Course Community Experience	
School-to-Work Resource Implementation Learning in Landscapes of Practice	Course Community Experience	
Learner Initiative Interwoven Course Experience Mandates Quick Course Completion	Course Completion	Learner Success: Retention & Competencies- Capabilities
Focus on Course Completion, Not Community Engagement	Course Completion	
Third Party Solutions	Support	
Help Desk System Supports	Support	
Learner Training on Instructional Engagement Tools	Support	
Learner Personality Display	Learner Competency Attainment	
Learner-Reflective Practices Creativity	Learner Competency Attainment	
Skill and Knowledge Acquisition Competency	Learner Competency Attainment	

The initial theme brought forward is articulated as Course Design Decisions Based Within Environment. The emphasis upon content and design is clearly in the majority of the category designations. As regards the content category, one impactful quotation is pulled from Participant Three:

> *"Some of these design changes seem minor, however, the students' responses to the changes were positive. Recommended instructional technologies were updated; redundant discussion forums were eliminated, multi-part assignments were consolidated, and time was extended for comprehensive assignments." (Participant Three)*

The focus upon the learner-content emphasis upon streamlined course content and engagement revisits the course instructional design towards an acknowledged learner knowledge acquisition and associated flow of work. Yet an intriguing consideration, as offered by Participant One, does emphasize design considerations that support initial course design and the clearly articulated emphasis upon the need to offer a recursive reconsideration and potential redesign associated with the level of success associated with initial course *live* implementation with learners.

> *"After two semesters of launching this class type with minimal success, the faculty users became more open to considering the online class needs for engagement/ interaction. With the data from our first year, change was welcome and enhanced by bringing a campus membership to the Quality Matters organization." (Participant One)*

Equally associated with course design, is the recognition that course design interface decisions may be based upon the capability of the learning management system within which the course is designed and will be implemented:

> *"In BlackBoard these topics were presented as a set of hyperlinked html files which allowed students to move freely between the various elements of each topic and the different units as needed to prepare and formulate a solution to an assigned project, essay, test, quiz, or software specific knowledge application task. In Canvas these topics were turned into modules which have an imposed linear order. In Blackboard the order was not imposed and the html hyperlinked interface could be used to move through the course via a set of assigned topics depending on the phases of the lifecycle being emphasized." (Participant Two)*

The emphasis upon content and design is a natural occurrence throughout this discussion. However, the participants brought forward nuggets of impact and influence, that dove deeper into an understanding of the discussion.

The Social Learning theme brought forward the categories of course community experience, emphasizing intriguing aspects associated with a real world recognition that:

"When even professional work activities are lessened by the "another meeting that could have been an email" mentality, are students willing to engage an learn via conversations?" (Participant One)

The real world pedagogical approaches brought forward as a significant code reflects Participant Three's understanding of the Course Community Experience category:

"The learner community is the primary beneficiary of course design. The courses are designed with an eye towards efficacy and best practices. The learner community extends to the students served by our students. In order to serve the broader community, the learning experiences of our students must reflect the proven approaches to pedagogy and an emphasis on critical higher order thinking." (Participant Three)

Participant Two more fully develops this discussion through a *deep dive* into an engagement with bridging between school and workplace resource implementation, with a strong nod to learning in landscapes of practice (Wenger-Trayner, et al., 2014; Wenger-Trayner & Wenger-Trayner, 2015, Wenger-Trayner & Wenger-Trayner, 2020):

"In my course design I provide access to freely available tools the students may find helpful in both the course itself as well as outside of the course. Tools such as web site generators, word processors, spreadsheets, wikis, software measurement tools, software analysis tools. Also some course designs will require students to find available software tools and use these tools and compare and analyze such tools. Assignments to analyze AI tools is an example of some of the most current ones I have incorporated into the course design of some courses." (Participant Two)

The strength undergirding the learning theory umbrella that is Social Learning supports a larger understanding around the strength of community engagement within the learning environment, as well as the distance learning environment desire to create a depth of engagement and support that displays similar socio-cultural person-to-person relational engagement.

The third theme that naturally developed during the analysis process was Learner Success: Retention & Competencies-Capabilities, including the diverse categories of course completion, support, and learner competency attainment. Participant One initiated a course completion focus when articulated:

"Those who facilitate and/or develop Competency Based Education are quick to identify student resistance to waiting for others to engage in an activity. After all, these programs promise quick completion based on the experience and learner initiative." (Participant One)

Participant Two focused upon skill and knowledge acquisition of competencies, supporting the learner competency attainment category:

"My course design addresses all learners in my class as a community of learners most directly in student required topic presentations. In many of my classes the course design involves assigning each student a unique topic in the class. Each student must study that topic and understand the basics of the topic, its intersection with the class objectives, why the topic is important to the course, advantages, and disadvantages relative to the topic and other fundamental understandings important to the topic. ... This type of exercise has proven to hone the student's motivation for preparation, knowledge acquisition and study skills." (Participant Two)

More fully developing the concept of learner personality display and professional presentation within a course environment, also supporting the category of learner competency attainment:

"To update the standard 'introduction' at the beginning of a course, students have another assignment in which they research, select, and use a comic creation software/application to introduce themselves to the class." (Participant Three)

Table 3 represents the codes, categories, and themes that naturally developed through the data analysis of the participant data in the Course Design for the Instructional Facilitator Experience manuscript section.

Table 3. Course design for the instructional facilitator experience codes, categories, and themes

Code	Category	Theme
Instructor-Specific Course Design - Student Distribution of Course Materials without Instructor Approval (Intellectual Property Owner Approval)	Course Design	Course Design Experience
Real World - Learning in Landscapes of Practice	Course Design	
Ease of Course Revisions	Course Design	
LMS Design Enhancements, "Work Arounds"	Course Design	
Focus on Topic and Learning Objective Alignment	Course Design	
Instructor Interface is Not Important		
Open Source Instructional Resources	Course Complexity: Resources & Tools	

continued on following page

Table 3. Continued

Code	Category	Theme
Managing Complexity	Course Complexity: Resources & Tools	
Instructional Designer as Intermediary & Cognitive-Instructional Guide	Course Design Team	
Accessibility	Adaptive Accessibility	Designing User Experience & Accessibility
Master Courses	Course LMS Template	
Master Courses	Course LMS Template	
Learner Interface - User Design	Course LMS Template	
Instructor Isolation	Collegial Connections	Organization: Collegiality & Community Supports
Solo Effort - Replicable Courses	Collegial Supports	
Team Teaching	Organization Supports	

The initial theme brought forward is articulated as Course Design for the Instructional Facilitator Experience is the Course Design Experience. The categories articulated within this theme are course design, course complexity including resources and tools, as well as the course design team. Towards an acknowledgement of instructor philosophies of learning as well as concerns related to intellectual property:

> *"My online courses are designed for my use only and are not shared purposely by myself with the community of other instructors at large. Nor are my course materials placed in public sites for download. However, students are known to download course materials and distribute them. With the use of AI I believe students will be feeding all course materials into the AI tool sites to that most of the hard designed course materials will be used without attribution." (Participant Two)*

An outstanding insight was offered by Participant One, noting that:

> *"The premium design experience comes when the instructor (SME) and I work together. In this process, they vet and choose the interactive/engagement options and tools. If they are unsure of how to use the tools as the facilitator we have time for the support team to provide quality instruction." (Participant One)*

Participant Three equally emphasizes a course design that supports real world implementation and understanding of coursework, and equally includes a focus upon learning in landscapes of practice (Wenger-Trayner, et al., 2014; Wenger-Trayner & Wenger-Trayner, 2015, Wenger-Trayner & Wenger-Trayner, 2020):

> *"Assignments are included that require students to engage in collaborative efforts with classroom teachers, parents, district librarians, and public libraries." (Participant Three)*

The second theme that naturally arose from the data analysis was Designing User Experience and Accessibility. Through this lens, the categories articulated were adaptive accessibility, and course LMS template. Participant One noted the course LMS template as an altruistic goal that supports an underlay design of instructional quality and self-regulatory support for learners:

> *"There are many reasons, institutions are turning to the use of master courses. They allow a significant reduction for the overall cost of development. Budgets are also reduced by using more adjuncts and fewer tenure track faculty. Occasionally, I hear the altruistic goal of offering the same quality instruction to all students." (Participant One)*

Equally, Participant Three emphasizes her program's focus upon course design that easily flows between different course instructors:

> *"Content is constructed so that additional faculty can readily and easily utilize the course materials. This facilitates the use by adjunct faculty or other faculty who may teach the courses from time to time." (Participant Three)*

The third and final theme articulated is an Organization focus, looking towards Collegiality and Community Supports. Participant Three articulated that, "We work in such an isolated environment that I give little consideration to instructor-community within the university" (Participant Three). Equally intriguing is Participant Two's interest in organization supports:

> *"I have not participated in team teaching a lecture course with a colleague, but this is the most prominent example of direct colleague course interaction I can think of. I have given thought to participating in such an endeavor but it presents a reimbursement entanglement at my university such that I have not been willing to negotiate. I have had discussion regarding how such a course would be structured and I believe if not careful with topic distribution it could be more confusing to the students than helpful. Though I do think the instructors would enjoy the endeavor. However, if the course were a seminar or project style course I think team teaching could be easily implemented and two or more instructors could be quite engaging and helpful to students." (Participant Two)*

Table 4 represents the codes, categories, and themes that naturally developed through the data analysis of the participant data in the Course Design: Integration of AI, ML, and Simulations manuscript section.

Table 4. Course design integration of AI, ML, and simulation codes, categories, and themes

Code	Category	Theme
LMS course auto-creation	Course Design	Course Design
Course Creation Timesaver	Course Design	
AI Learning Supports	Self-Regulation Supports	
Human Fact-Check & Edit AI Errors Additional OER Resources	Errors	Flawed Viability
Lack of AI tool instructional viability (flawed AI tools)	Flawed AI Tools	
LMS AI professional development trainings	Professional Development	Training and Talent Development
Student Knowledge versus Instructor Knowledge	Novice versus Expert	

The initial theme brought forward is Course Design. Participant One noted that course design is an imperative consideration, including LMS focus, communicating that, "Instructors were excited to see how the Blackboard LMS AI feature will create a full course shell including learning content using your syllabus" (Participant One). A second theme that arose is the understanding associated with Flawed Viability. More specifically, Participant One reflected that:

> *"This class provided the perfect opportunity to experience AI benefits and see how humans still need to fact-check and edit to correct errors. Collaborating with the LMS AI tool to create new quiz banks was not only exciting for most, but it also immediately supported our quest for additional Open Educational Resources (OER) across campus." (Participant One)*

Alternatively, Participant Two noted that:

> *"AI has not been integrated into my course design except as a class project topic whereby I assign students to analyze, compare and asses current AI Tools and their implementation methods. I have yet to find a need or good reason to use current flawed AI tools to grade, detect cheating, or create my lectures. Software included with Blackboard (Safe Assign) and Canvas to detect cheating is minimally useful." (Participant Two)*

A final theme was articulated as Training and Talent Development. Participant One noted professional development as an imperative:

> *"My design team has implemented several AI projects. We mastered the new LMS AI options and pass these on to faculty via online and face-to-face professional development. These options include Course module design, suggested images*

with alt txt, assessment question creation, and general page content creation." (Participant One)

While Participant Three considered a novice versus expert category, coding the following quotation with an acknowledgement of student knowledge versus instructor knowledge:

"Learning the nuances of AI is just at the beginning stages for me and I have not located any simulations of the work that goes on in school libraries. The best we can do at this time is the use of video recordings. The LMS provided by the university has built-in anti-cheating detection. A recent conversation with one of my students revealed that Canvas has an AI detection feature." (Participant Three)

Table 5 represents the codes, categories, and themes that naturally developed through the data analysis of the participant data in the Communication Dynamics and Learning Engagement manuscript section.

Table 5. Communication dynamics and learning engagement codes, categories, and themes

Code	Category	Theme
Simulation Resource for Hands-on/ Technical Learning	Learner Engagement	Learner Motivation & Ethical Knowledge Attainment
Learner Analysis Assignment		
Knowledge Attainment in Learning Ethical Parameters of Usage	Knowledge Attainment	
Focus on Teaching versus Technology	Teaching Focus Ubiquitous Technology	Teaching Complexity
Freedom to Focus on Hard Problems	Complex Problems	
Change Management Diffusion of Innovations	Diffusion of Innovations Viability	Viability of Innovations
AI Viability	AI Viability	
Potential Strengths of AI	AI Viability	

Communication Dynamics and Learning Engagement brought forward three themes. The first theme is Learner Motivation and Ethical Knowledge Attainment. Participant One focused upon learner engagement, with a specific focus upon simulation resource for hands-on and technical learning:

"Simulations are invaluable for many areas. Nothing prepares a nurse for the reality of daily life like working in a fully simulated hospital environment for their academic years. Workforce programs such as welding and auto repair or painting could not turn out the large number of workers needed without beginning with simulations before putting students in front of live equipment." (Participant One)

Participant Two specifically focused upon learner analysis assignment focus:

"The use of AI as a vehicle for the student to study and compare software tools in terms of its limitations and strengths, what it can and cannot do well, is a worthy exercise of benefit to graduate software students. Students need to study such tools to be aware of the implementation methods employed and the extent to which each tool is effective. Learning to perform such an analysis and comparison assessment is a highly effective learning activity." (Participant Two)

While Participant Three was more fully embedded within knowledge attainment, associated with learning. Equally considered were ethical parameters of usage:

"The danger is that students who have not yet developed the necessary skills in any content area will use AI to complete assignments without the depth of knowledge needed to grow as learners. The dilemma is the unethical use of this new tool in education. AI must be integrated into distance learning protocols because the benefits outweigh the risks." (Participant Three)

The second theme brought forward was Teaching Complexity. Participant One highlighted not only a teaching focus but equally impactful was the concept of ubiquitous technology, noting that, "My feeling of AI is that it is going to allow teachers to focus on teaching, not requiring all of them to be technology users first" (Participant One). Equally impactful was Participant Two's emphasis upon complex problems; more specifically upon a freedom to focus on hard problems:

"AI tools can be of benefit for the generation and reuse of well-known solution to common coding solution to known problems. Such can be used to free the software developer to concentrate on hard to solve novel aspects of system design problems." (Participant Two)

The final theme was the Viability of Innovations. Participant One brought forward a category associated with diffusion of innovations viability or, more specifically, the dual understanding of change management and diffusion of innovations:

"Today's faculty have the ability to create and share content with nothing more than the ability to visualize and talk about the concepts. The key to this reality lies in our ability to lessen fear of change and help value the tools." (Participant One)

Participant Two focused upon AI viability, noting that:

"The viability of each AI tool resides in the correctness of the large language model it is based upon as well as the correctness of the domain analysis and domain knowledge behind the model. The larger the domain the less likely the model will be correct." (Participant Two)

Participant Three also focused upon AI viability, although more fully developed a consideration associated with the potential strengths of AI:

"There are potentially many learning applications for artificial intelligence. AI has the potential to reduce time on tasks for writers, developers, educators grading, and similar repetitive tasks. AI can also support the scaffolding a learning

making the necessary adjustments based on students' success in learning new content." (Participant Three)

The strength of the participant quotations merely undergirds and supports the depth as well as perceived breadth of discussion currently occurring around the concept of enhancing communication dynamics through interactive activities within distance education environments, with specific consideration towards instructional design and course creation.

FUTURE RESEARCH DIRECTIONS

The clear emphasis of this multiple case study qualitative approach, towards deepening the understanding of an enhanced communications dynamic that is implemented through an interactive activity lens, is vital to developing an understanding and recognition associated with the instructional design and course creation that supports a continuously burgeoning understanding of distance education environments. Yet the emphasis upon needs around more fully and more deeply understanding the nuances of successful distance education environments continues. The gap in the literature remains, with more fully recognized areas of embedded multiple case study approaches at the faculty level of consideration, the administrative depth of high-level understanding while equally approaching the reality of project-based flow of work towards attaining quality course products as well as quality instructional facilitation. The field of distance education environments is still in its relative youth, as the significant focus upon the viability and necessity of online learning efforts, virtual learning efforts, and the consideration towards the integration of generative AI, large learning models (LLMs), machine learning (ML), and simulations that embed the learner more deeply within the learning process, continue to develop at a lightning speed. An effort towards future research directions remains a focus upon the ethical implementation of resourceful tools that not only ease the flow of work within learning environments, but an equal nod towards the importance and potential impact upon the learning process and the learner.

CONCLUSION

The focus of this multiple case study approach was upon a critically analytic self-study that attempted to more deeply understand and potentially enhance the communication dynamics through interactive activities with a special focus upon distance education environments. The interrelationality of this multiple case study approach immersed the focus within participants who approached the support and emphasis upon the individualistic, highly personalized approaches. This understanding supports and is towards designing distance education environments, while focused upon the extent to which course design and associated learner experience had the potential consideration towards communication dynamics and learning engagement. Equally, while equally looking more deeply into the current integration of AI, LLMs, ML, and simulations into the learning environment experience.

The emphasis upon enhancing communication dynamics within distance education environments was framed through a focus upon interactive activities within learning environments, highlighting the importance and impact of the digitalization of communication dynamics within learning environments. More directly addressed is the emphasis upon the importance associated with designing communication dynamics into the distance education environment during the course design and course creation process. The shift from face to face learning experience into a distance learning experience through digitally supported environments has the opportunity to unequivocally enhance the impact of communication dynamics. Although the distance education environment is more carefully designed and created as a learning environment experience, the importance associated with the communication dynamics within learning environments is emphasized as social learning (Bandura, 1969, 1971, 1972, 1986; Bandura & Walters, 1977; McLeod, 2011; Nabavi, 2012) supports as well as engaging within sociocultural understandings associated with conceptual frameworks of understanding (Daniels, 1996; Vygotsky, 1962, 1978, 1987; Wertsch & Sohmer, 1995) enhancements.

REFERENCES

Association of College and University Educators (ACUE). (2023). Fostering a culture of belonging: Four insights you need to know. https://acue.org/blog/fostering-a-culture-of-belonging-four-insights-you-need-to-know/

Bandura, A. (1969). Social-learning theory of identificatory processes. In D. A. Goslin (Ed.), *Handbook of socialization theory and research* (pp. *213*- 262. Rand McNally & Company. Bandura, A. (1971). *Social learning theory*. General Learning Press.

Bandura, A. (1972). Modeling theory: Some traditions, trends, and disputes. In Parke, R. D. (Ed.), *Recent trends in social learning theory* (pp. 35–61). Academic Press. DOI: 10.1016/B978-0-12-545050-8.50008-0

Bandura, A. (1986). *Social foundations of thought and action: A social cognitive theory*. Prentice-Hall, Inc.

Bandura, A., & Walters, R. H. (1977). *Social learning theory* (Vol. 1). Prentice Hall.

Burnham, B., & Walden, B. (1997). Interactions in distance education: A report from the other side. In *Proceedings of the Adult Education Research Conference*. Oklahoma State University. http:// www.edst.educ.ubc.ca/aerc/1997/97burnham.html

Crawford, C. M. (2001). Developing webs of significance through communications: Appropriate interactive activities for distributed learning environments. *Campus-Wide Information Systems*, 18(2), 68–72. DOI: 10.1108/10650740110386675

Crawford, C. M. (2003). Emerging learning environments: Enhancing the online community. *Academic Exchange Quarterly*, 7(4), 131–135.

Crawford, C. M., Andrews, S., & Wallace, J. K. Y. (2021). Co-creative collegial communities of instructional engagement. [IJHIoT]. *International Journal of Hyperconnectivity and the Internet of Things*, 5(2), 38–56. DOI: 10.4018/IJHIoT.2021070103

Crawford, C. M., Andrews, S., & Young Wallace, J. K. (2022). Implicit cognitive vulnerability through nudges, boosts, and bounces. *International Journal of Hyperconnectivity and the Internet of Things*, 6(1), 1–14. DOI: 10.4018/IJHIoT.285588

Dai, C., & Ke, F. (2022). Educational applications of artificial intelligence in simulation-based learning: A systematic mapping review. *Computers and Education: Artificial Intelligence*, 3, 1–17. DOI: 10.1016/j.caeai.2022.100087

Daniels, H. (Ed.). (1996). *An introduction to Vygotsky*. Routledge.

Hillman, D., Willis, D. J., & Gunawardena, C. (1994). Learner-interface interaction in distance education: An extension of contemporary models and strategies for practitioners. *American Journal of Distance Education*, 8(2), 30–42. DOI: 10.1080/08923649409526853

LearnDash Collaborator. (2020). *How is AI being used in e-learning in 2020?*https://www.learndash.com/how-is-ai-being-used-in-e-learning-in-2020/

McLeod, S. (2011). *Albert Bandura's social learning theory*. Simply Psychology.

Moore, M. (1989). Three types of interaction. *American Journal of Distance Education*, 3(2), 1–7. DOI: 10.1080/08923648909526659

Nabavi, R. T. (2012). Bandura's social learning theory & social cognitive learning theory. *Theory of Developmental Psychology*, 1(1), 1–24.

Online Education. (n.d.). *Artificial intelligence innovations in online learning*.https://www.onlineeducation.com/features/ai-in-distance-learning

Vygotsky, L. S. (1962). *Thought and language*. MIT Press. DOI: 10.1037/11193-000

Vygotsky, L. S. (1978). *Mind in society: The development of higher psychological processes*. Harvard University Press.

Vygotsky, L. S. (1987). Thinking and speech. In R.W. Rieber & A.S. Carton (Eds.), *The collected works of L.S. Vygotsky, Volume 1: Problems of general psychology (pp. 39–285)*. Plenum Press. (Original work published 1934.)

Wenger, E. (1998). *Communities of practice: Learning, meaning, and identity*. Cambridge University Press., DOI: 10.1017/CBO9780511803932

Wenger-Trayner, E., Fenton-O'Creevy, M., Hutchison, S., Kubiak, C., & Wenger-Trayner, B. (2014). *Learning in Landscapes of Practice: Boundaries, Identity, and Knowledgeability in Practice-Based Learning*. Routledge., DOI: 10.4324/9781315777122

Wenger-Trayner, E., & Wenger-Trayner, B. (2015). Communities of practice: A brief introduction. Wenger-Taylor. https://wenger-trayner.com/introduction-to-communities-of-practice/

Wenger-Trayner, E., & Wenger-Trayner, B. (2020). *Learning to Make a Difference: Value Creation in Social Learning Spaces*. Cambridge University Press., DOI: 10.1017/9781108677431

Wertsch, J. V., & Sohmer, R. (1995). Vygotsky on learning and development. *Human Development*, 38(6), 332–337. DOI: 10.1159/000278339

KEY TERMS AND DEFINITIONS

Communication Dynamics: This is a term that describes the ways through which persons within an environment interact, with the specific focus herein around the distance education environment topic.

Course Design: This is the process through which a learning environment is designed and developed and may takes the form of distance education environments within this focused discussion, although face to face, blended, flipped, and other types of learning environments may also be included in this learning environment design and development process.

Distance Education Environments: A space within which learning is meant to successfully occur, at a distance from the instructor and most likely also at a distance from the other student learners associated within the experience.

Higher Education: A post K-12 educational opportunity, that includes community colleges, colleges, and universities.

Instructional Facilitation: The role of the course teacher who is assigned to support the student learner's progression through a learning experience and within which specific learning objectives are meant to be successfully achieved, with an eye towards attained competencies as well as forward-leaning capability achievements.

Interactive Activities: Different forms of engagement that occur within different community experiences. Within a distance education environment, the focus is upon learner motivation, self-regulation, as well as knowledge acquisition understanding with an equal emphasis upon developing appropriate knowledge contexts and conceptual frameworks of understanding.

Online Learning: This is another term for distance education environments, with a specific focus upon Internet-supported instructional learning opportunities.

Chapter 6
Exploiting the Opportunities of 5G for Streamlined Virtual Event Experiences From Connection to Engagement

Sonali Vijay Shinkar
https://orcid.org/0000-0002-8873-4708
SCTR's Pune Institute of Computer Technology, India

Kiran Patil
G.H. Raisoni College of Engineering and Management, India

Archana Dongardive
G.H. Raisoni College of Engineering and Management, India

Sabyasachi Pramanik
https://orcid.org/0000-0002-9431-8751
Haldia Institute of Technology, India

ABSTRACT

This abstract looks at how enhanced connection and immersive interactivity may be used by 5G technology to change virtual event experiences. The essay examines how 5G networks may provide more network capacity, very low latency, and quicker data rates. These developments make it possible to utilize virtual reality apps, communicate

DOI: 10.4018/979-8-3693-3579-6.ch006

in real time, and transmit content smoothly. Moreover, it underscores the impact of 5G on several sectors such as business, education, and entertainment, showcasing its ability to create virtual worlds that faithfully replicate the live event experience.

INTRODUCTION

With its profound digital revolutions across industries, 5G is set to unleash immense social and economic value. Its arrival marks a historic milestone in the development of connection. 5G, the fifth generation of cellular network technology, outperforms 4G in several measures, including dependability, network efficiency, peak data speeds (20Gbps vs. 1Gbps for 4G), latency (1 ms vs. 60 ms), and capacity for enormous device connection. These advantages highlight its potential as the ultimate connection fabric that opens the door to breakthroughs ranging from smart cities and intelligent automation across manufacturing networks to precision medicine and self-driving cars. By the end of 2022, there were over 780 million 5G subscribers worldwide, indicating a growth rate that exceeded that of preceding cellular generations. By 2027, over 4.4 billion 5G connections are expected, which will account for more than 50% of all mobile subscriptions globally. One of the main drivers is the widespread availability of 5G-compatible devices, such as industrial sensors and smartphones. Currently, there are 5G networks operational in more than 70 countries throughout the globe, mostly in the Americas, Asia Pacific, and Western Europe. Nonetheless, by 2025, 5G spectrum allocations are expected for more than 100 more nations, and commercial infrastructure installations suited for 5G are already under way. The majority of subscriber growth is anticipated to come from developing nations with fast growing populations of youthful, tech-savvy urban customers in the next years. Because of their lower rates of fixed broadband penetration, 5G's mobile internet accessibility is essential for promoting digital inclusion. Deploying 5G is facilitating corporate and consumption-driven digital change, which is significant. According to economic impact analyses, revolutionary 5G use cases might generate up to $13.2 trillion in value for the world economy by 2035, having a cross-cutting effect on many important industries. For example, it is anticipated that 5G would inspire smart city solutions that enhance sustainability, safety, and citizen experiences, generating over $2 trillion in value. The next stage of Industry 4.0 will be driven by real-time supply chain coordination and manufacturing automation, which are expected to generate over $700 billion in value. The high-bandwidth, low-latency connection of 5G has the potential to create a $4.5 trillion market for immersive video and virtual reality. Other trillion-dollar industries include autonomous driving technology ($1.2 trillion) and linked healthcare ($2.5 trillion).

Research Objective

This study's main goal is to multidimensionally map the environment of corporate change driven by 5G in order to expedite the production of fair value across industries while controlling risks. In particular, we aimed to: 1. To classify the application areas with the greatest potential effect, identify and size important technical use cases across sectors where 5G connection offers distinct benefits over previous networks, satellite, or fixed broadband solutions. 2. Using combined data-driven forecasts based on these use cases' whole value chains, estimate the size of the total addressable market and the possible income from them through 2025. 3. While weighing options, estimate broader economic benefits and employment multipliers from scaling 5G-led transformation beyond direct company effect. 4. Describe the issues with technology, regulations, skills availability, societal acceptability, and policy that are preventing policy interventions from being implemented. 5. Examine early deployment data to simulate differences in 5G access and adoption trends across urban and rural populations in order to evaluate the hazards associated with the digital divide. 6. Combine viewpoints from a variety of stakeholder groups, such as members of the community, civil society organizations, industry leaders, and regulators, to discuss how to balance inclusive innovation. 7. Suggest methods for recognized stakeholder groups to work together to expedite 5G-led corporate transformation in a fair and responsible manner.

OVERVIEW OF VIRTUAL EVENTS WITH 5G TECHNOLOGY

Even when the pandemic limitations are gone, COVID-19 has caused a change in innovation events, having a persistent focus on digital platforms (Jauhiainen, 2021). Because they provide a seamless and interesting experience for attendees, virtual event platforms are quickly replacing traditional in-person events. These technologies make use of 5G technology's ability to increase connection, enabling real-time communication and immersive experiences. By using the potential of 5G, virtual event experiences have advanced from basic connection to full immersion, offering a multitude of opportunities to planners and attendees alike. One should not undervalue the importance of 5G technology in enhancing virtual event experiences. Leading the way in these technological advances is augmented and virtual reality. Elbamby et al. predict that virtual reality will play a significant role in 5G networks and that it will revolutionize the way that people enjoy virtual events. According to Park and Bennis (2018), it provides users with immersive and realistic surroundings in which they may interact with virtual items, hence activating many senses. Furthermore, emerging technologies like augmented reality and virtual reality would have much more potential because to the advantages of 5G networks (Onwuegbuzie, 2022). Technologies like as augmented reality and virtual reality rely on the 5G network's quick transmission rates, large signal transmission capacity, and minimal network latency (Liang et al., 2023).

An Overview of Trends in Connectivity: An Elucidation of 5G

The upcoming release of 5G, the most recent version of wireless communication technology, is expected to significantly alter connection and enable a wide range of innovative applications and services. 5G provides faster transmission, more bandwidth, and higher throughput, which lays the groundwork for seamless virtual event experiences. 5G technologies deliver faster connectivity, more capacity, and higher throughput, according to Ahmad et al. (2021).

5G's Importance for Virtual Events

5G is very important for virtual events since it makes it easier for participants to get immersive experiences. 5G facilitates efficient data transmission and processing by using the capacity to virtualize and distributes network operations, ensuring low latency and consistent connectivity (Tang, 2022). Virtual events rely heavily on robust and continuous connection. "5G technology facilitates improved mobile broadband connectivity, minimal delay in data transmission, and dependable communication. Additionally, it supports extensive communication services for various applications." In order to ensure that attendees may actively engage and interact in the virtual event environment, 5G technology provides the necessary foundation.

The Progression of Evolution from Connectivity to Immersion

Merely being connected is not enough when it comes to virtual events; immersion is essential. Participants may fully immerse themselves in the event experience thanks to virtual reality and augmented reality (VR/AR) technologies. These technologies immerse users in amazingly realistic virtual environments and give them a sense of being present. With the arrival of 5G technology, the possibilities for perfect virtual event experiences are fully realised. 5G deployment will be very beneficial for emerging technologies like augmented and virtual reality. Users may explore and engage with previously inconceivable virtual worlds thanks to the immersive and amazing experiences that virtual reality and augmented reality technology generate.

These experiences may include things like taking a virtual tour of historical sites, going to the solar system, or seeing a live concert from the greatest seat in the house.

However, the transmission of large amounts of real-time data required for VR/AR technologies may exceed 5G networks' capacity. To tackle this issue, efforts are being made to improve the network design and ensure data delivery without interruption. The virtualization of network services, which allows for more efficient resource allocation and real-time capacity adjustments to meet the demands of virtual events, is an example of advancements in this subject. Through the use of virtualization, the network may more efficiently distribute resources and improve data transfer to provide engaging user experiences. This guarantees dependable, quick connections with minimal latency and sufficient capacity for virtual events.

Using 5G to Improve Immersion Experiences

Virtual event planners may take use of 5G's cutting-edge capabilities to provide attendees more immersive experiences.

Large-scale machine-type communications, trustworthy communication, lower latency, and quicker communication are all made possible by 5G's increased bandwidth, throughput, and speed of communication.

Along with real-time interaction and high-quality video streaming, participants may also enjoy other immersive elements like accurate spatial audio and haptic feedback. Two popular services that are used in 5G networks are virtual and augmented reality (Kitanov et al., 2021). However, given the potential to exceed capacity and latency restrictions, current 5G network limitations could not be sufficient for certain applications, such haptic technologies, virtual conference rooms, and remote surgery. However, ongoing efforts in the field of 5G research and development aim to surmount these limitations and fully realize the potential of virtual event interactions. Virtual event experiences may transcend traditional constraints and create an environment that is really immersive by using 5G's capabilities. Virtual event planners have the opportunity to drastically change how guests connect and interact with their surroundings by using 5G's capabilities. Participants may experience the virtual world via tactile sensations in addition to their visual and aural senses by using haptic technology. A high degree of immersion like this may greatly improve the whole experience and make virtual events more fascinating and lifelike. Furthermore, 5G's low latency and fast data transfer capabilities allow attendees to engage in virtual events without interruption or delay. Furthermore, the integration of augmented reality into virtual event experiences is made possible by the use of 5G technologies (Chamola et al., 2020). Through the integration of 5G and augmented reality, users will be able to overlay virtual things on top of the real world, creating a mixed reality experience that blurs the lines between the real and virtual worlds.

This can present virtual event planners with countless chances to create unique and enthralling experiences, like virtual classrooms where attendees actively engage in interactive learning activities or virtual exhibition halls where visitors can explore and interact with virtual booths and products. Additionally, immediate collaboration and communication in virtual events is made possible by 5G's large bandwidth and quick data transmission rate (Liang et al., 2023).

ADVANTAGES

The provision of high-speed connection that is reasonably priced, dependable, and really omnipresent is the most basic potential that 5G networks enable. Baseline peak download rates of 20 Gbps and upload speeds of 10 Gbps are required by 5G specification requirements, which is more than 20 times higher than current average 4G speeds. Actual carrier installations show that these rates are often exceeded. Crucially, the increased capacity allows a large number of users and devices inside congestion-free coverage zones to consume high bandwidth concurrently. Beyond raw bandwidth power, reliability in the promised experience is just as important. Latency, or the interval of time between transmitting and receiving data, is less than 5–10 ms for 5G compared to 30–60 ms for 4G. Additionally, jitter, or variations in delay, decreases tenfold to 1-2 ms. This significant improvement comes from 5G's new core architecture, which lowers signaling and backhaul traffic. Consistent connectivity is further aided by support for up to 1 million device connections per square kilometer and effective antenna beam forming that maintains links even at cell borders or inside. In some networks, 99.999% reliability corresponds to fewer than 6 minutes of yearly downtime. This potent capacity coupled with steadfast, consistent dependability opens up a world of high-bandwidth applications that were previously unreliable to execute. The viability of any solution that depends on moving or processing massive amounts of mission-critical data in real-time is greatly increased. With 5G providing ubiquitous high-speed backbone connection, application cases such as fleet coordination of autonomous vehicles, industrial automation with various robotic devices, and pervasive VR collaboration platforms or metaverse settings become more viable. By doing this, connection is successfully elevated from a barrier impeding innovation to a facilitator opening up new value chains. Importantly, because of the improved infrastructure, the ubiquity equally encompasses conventional rural and urban geographies. Mid-band spectrum and small cell boosters are ideal for cost-effectively propagating across large regions. Developing nations using 5G technology stand to benefit much from practically jumping ahead of transitional 2G–4G investments to revolutionary mobile broadband. More than 70% of these increases are expected to come from Latin America, Africa, and

Asia by 2025. The widespread availability of inexpensive gadgets, cellphones, and Internet of Things sensors increases the options available to those who have not yet been granted digital access. Personalized experiences still need, however, to balance enough capacity by adding more macrocells in congested areas. Large metro areas already have carriers enhancing mmWave small cells operating in high-frequency bands like as 26/28Ghz. Despite having a much smaller range of propagation, they are well suited for packed areas. Delivering ubiquitous connection is optimized when infrastructure layers are synchronized with activity density. But for there to be any success, cooperative policy approaches between telecoms and municipalities are needed to get beyond deployment obstacles related to site clearances and fiber backhaul constraints. It also still depends on providing equitable access to resources, expertise, and operator trust for underprivileged populations. Thus, a well-defined plan that takes use of 5G's much increased capacity and reliability in offering consistently low-cost high-speed connectivity across the board includes: • Optimal placement of tiered infrastructure in accordance with site density • Multi-domain policy coordination that makes approvals easier; • Localized capacity boosting using tiny cells and Wi-Fi offloading • Cost-effective, inclusive device and data plans • Promoting digital literacy to generate informed demand

Real-time, low latency analytics Beyond its massive bandwidth capacity, 5G's ultra-reliable low latency is undoubtedly its most transformational quality. An completely new class of use cases that depend on real-time analytics and quick data-driven decision automation are made possible by guaranteed latency of less than 10 milliseconds. A 5G researcher sums it by saying, "Timeliness matters more than speed." First, reduced latency opens up new possibilities for remote control over physical processes that were previously unautomatable because of response delays. Surgeons may now perform tele-operated robotic surgery without being constrained by fractional delays, moving the technology from science fiction to practical reality. The immediate tactile input required for precise tool manipulation is not available over 4G connections. Sub-10 ms 5G latency, however, is now proving viable in lag-sensitive operations such as stent placement for the trachea. The device mitigates gaps in physical closeness by augmenting immersive video feeds with sensor data to improve situational awareness. Patients in need of specialised treatment may have easier access if such tele-surgical programs are expanded. Parallel uses in dangerous settings also appear, such as autonomous drones for disaster response that evaluate damage in locations too dangerous for people to enter. Over large distances, precise navigation and coordination of activities become possible. Similar to this, achieving the productivity increases that industrial automation promised has long been hampered by time lag margins. Thanks to quicker reaction times, automated guided vehicles (AGVs) working in warehouses find obstacles over 80% more often when utilizing 5G connection than when using WiFi or 4G. Reliability in quickly communicating sensor inputs and coordinating reactions becomes critical as manufacturing and transportation facilities make the shift to fleets of collaborating robots. It allows for the choreography of device networks at a higher size and complexity. Smooth Cross-Device Interactions Experience consistency across contexts and devices is a key expectation that consumers and workers have of contemporary digital technologies. Coherence is required whether it be in the form of multi-screen unified visual interfaces or the smooth transfer of work started on a phone to a desktop. This is significantly aided by 5G's increased dependability, which allows most devices to have consistent high bandwidth access across locations. Stable connections are specifically made possible by the ability to slice virtual channels free from broader network congestion, sustain connection densities of up to a million devices per square kilometer, and stably maintain high speeds across larger regions. This eliminates delays and dropouts as users move processes across displays and endpoints. Superior engagement over 5G is anticipated from immersive systems like as AR/VR, which combine multi-modal inputs from sensors, wearable, and touch screens in a seamless integration. Practically speaking, unlike inconsistent Wi-Fi, video conversations remain crisp and responsive while transferring between cellphones, computers, and conference room projection systems. This kind of adaptable mobility also helps when implementing technological interfaces in situations where permanent terminal access points are inappropriate. Cost-effective aerial mesh networks with low latency may be used to relay data and extend 5G coverage to rural areas. Using spectrum pooling to provide fast failover between terrestrial and satellite communications guarantees dependability and permits governor access from moving vehicles. Wearable technology that doesn't need hands-on manipulation makes life easier for frontline workers who need to be connected across numerous places. Seamless handovers expand use case options, from continuous phone conversations in corporate campuses to multi-player immersive gaming. Crucially, however, it is the responsibility of hardware and service providers to guarantee that device ecosystems seamlessly interact via standardized interfaces. Presently, proprietary authentication systems and fragmented platform signatures pose a danger to smooth roaming. Handoffs may be slowed down by an inadequate signaling architecture, which can potentially overburden network resources. While pursuing commercial interests alone risks suboptimal fragmentation, these observations highlight the importance of industry cooperation for sharing radio stack implementations, open API access facilitating application portability, and mutual roaming/spectrum sharing arrangements between telcos to aid seamless connectivity. Furthermore, businesses need to include user experience continuity into their workforce transformation plan from the very beginning, when they rethink their digital processes. Workflows may avoid persistent legacy rigidity by being structured on user journeys rather than devices or connections. Device-neutral experiences are aided by the use of cloud orchestration, process automation, and omnichannel content delivery. Acceptance is aided by staff acclimatization via digital upskilling.

Revolution of the Internet of Things

Vast Increase in the Number of Networked Sensors and Devices Enabling huge machine-type communications to rapidly grow the Internet of Things (IoT) ecosystem is one of 5G's most revolutionary features. 5G provides the potential for high densification in sensors and endpoints, with over 1 million device connections per square kilometre as opposed to only 2,000 on 4G. Manufacturing and healthcare industries are to undergo a transformation thanks to the enormous size of hyper-connectivity and its unlocking of collective data flows and automation possibilities. According to market projections, over 65 billion IoT devices worldwide—a 3X increase from 2018—will be linked by 2025 in both consumer and industrial sectors. This growth is accelerated by 5G's ability to simultaneously serve dense device clusters at high speeds. Small, inexpensive 5G chipsets make possible Internet of Things applications that depend on the widespread availability of low-powered sensors for more accurate real-time monitoring, such as smart power metering, infrastructure tracking, or environmental mapping. Their quantity may increase without endangering the user's experience. The portfolio's coverage of both indoor and outdoor endpoints is significant. At a very little additional expense, small cells in warehouses transmit signals to allow businesses to monitor inventory using RFID tags attached to each pallet or box. In logistics, tracking the origin of individual shipments across transit hubs facilitates last-mile delivery by allowing traceability across whole supply chains. Advanced proximity applications such as context-aware alerts, safety warnings, and precise asset monitoring in public areas including parking lots, malls, and public transportation are also made possible by dense sensor arrays. Wireless sensors will replace outdated instrumentation in industrial and utilities, enabling higher-fidelity, downtime-free telemetry capture from aged equipment. Without the limitations of old device retrofitting or cabling, their numbers can expand linearly. Instead of relying just on individual sampling, proactive maintenance prognostics and diagnostics are enhanced by larger historical datasets spanning several aligned assets. Increases in operational resilience and safety for large machinery, dams, power plants, etc., provide controlled productivity improvements long until the end of equipment life. Additionally, sophisticated real-time automation flow coordination emerges. Fleets of autonomous vehicles depend on real-time optical, positional, and lidar data to safely communicate with one another. Similarly, for navigation, analytics, and control, swarms of drones inspecting catastrophe areas or collaboratively building platforms in midair need trustworthy, low latency device interconnection. The crucial dependability of sensors, controllers, and links 5G promises to open these kinds of applications that profit from coordinating large numbers of robots. However, in addition to facilitating the number of devices, connection quality is crucial for maximising the benefits of IoT installations. For time-sensitive automation use cases spanning smart factories, linked healthcare clinics, or intelligent transportation systems, telemetry that is sluggish or inconsistent is insufficient. Here, the application scope is significantly increased by 5G's speed, low latency, and consistency. Longer battery lifespan also equate to more time between terminal recharges, which makes maintenance easier. These affordability and dependability considerations, together with growing IoT chipset ecosystems, help drive down sensor prices, enabling additional scaling. Realizing the anticipated surge in connected devices will, however, also need carefully weighing the possible trade-offs in relation to e-waste, data privacy, and security threats from exponential attack surfaces. However, over the next years in the 5G era, society can balance these intricately connected IoT potential and problems for shared prosperity with the help of commercial innovation addressing ethical issues and socially responsible policies. Improved Supply Chain Data Flows and Intelligent Infrastructure Identifying sensors and endpoints across infrastructure opens up a wealth of opportunities, one of which is the significantly increased insight it provides into operational performance, utilization patterns, and external risk exposures. The analytics and information derived on asset health, project lifecycles, and cascading interdependencies become exponentially more complex with deeper telemetry and richer embedded data flows. In order to successfully manage the surge in data quantities transferred from multiplying monitoring nodes, 5G is essential. Urban planners may dynamically simulate use patterns to optimize routes, schedules, and first- and last-mile connection by monitoring citizen mobility via linked automobiles and public transportation access. The reduction of congestion and emissions occurs when travels are more intelligently coordinated with demand. In order to make land use choices that combine sustainability and convenience, city managers may also employ sensor arrays that track traffic patterns, parking space occupancy, air and noise pollution levels, and other relevant data. These decisions are made based on factual data rather than conjecture on customer behavior. Similarly, equipment telemetry allows for fine-grained monitoring, which helps infrastructure operators—private utilities and grid corporations in particular—improve their predictive maintenance skills. When early symptoms may be identified from performance changes, warning indications surrounding ageing assets like as transformers, switchgear, underground cables, or turbine engines appear long before catastrophic breakdowns. However, thorough equipment coverage is crucial for the dependability of diagnostics, which presents a costly barrier when using wired instrumentation. 5G mitigates this by allowing businesses to skip the expensive cable retrofitting and go straight to widespread wireless coverage. Digital twin models with previously unheard-of fidelity are produced when scaling equipment sensors does not negatively impact the availability or performance of vital infrastructure. These models visualize asset conditions. Instead of using clumsy failover processes that affect whole neighborhoods, grid operators may do asset health assessments, run simulations to evaluate vulnerabilities, and respond precisely when problems arise. As a result, even during disturbances, outage durations decrease as reactions become more precise. As a result, reliability indicators surpass restrictions, drawing more funding for improvements and expenditures. Real-time monitoring of material flows is facilitated by ubiquitous, low-cost sensors enabled by 5G, particularly in complicated production settings and warehouse inventory management. Supply chain analysts can better estimate supplier concentration-related risks with the use of automated risk signals. Demand forecasting algorithms are balanced with depletion notifications to optimize inventory levels. In order to reduce under- or overstocking losses, shipping warnings automatically initiate just-in-time replenishments depending on lead times. According to McKinsey, when combined, these data-driven improvements may save global supply chains over $500 billion in costs and restore capacity via transparent, integrated processes. However, the widespread connection that the 5G deployment has made inexpensive at scale is necessary for their sustainability. But moving from fragmented information flows, data interoperability standards, and holes in legacy analytics capabilities is necessary to move towards intelligent, data-enhanced decision structures. Even after adding sensors and 5G networks, change management that focuses on human obstacles surrounding trust, in-house tool literacy, and cross-functional collaboration is essential. Furthermore, maintaining sustainable value requires conscientious governance via the ethical management of privacy sensitivities and cyber dangers that may arise from expanding data collection footprints if neglected. Overall however, supply chains and heavily instrumented smart infrastructure clearly increase resilience, visibility, and responsiveness, yielding significant societal and economic benefits from the scaling of 5G connection.

Capabilities for Intelligent Automation

Although an increasing number of hyperconnected sensors and multiplying endpoints provide enormous amounts of very fine-grained data on system performances, analyzing and reacting in real-time to new insights rather than just compiling historical data yields the most advantage. This is where 5G's dependable intelligent automation features come into play. Thanks to much decreased latency, processing data instantly enables algorithms to sense surroundings, perform contextual actions, and continually change behaviors without waiting for human interactions. To complete the automation loop, feedback must be sent smoothly so that controllers can effectuate changes. Again, 5G's better dependability helps here. In utilities, transportation networks, and distributed energy resources, this combination opens up advanced predictive diagnostics and self-healing systems for essential infrastructure. For example, an intelligent water distribution system may detect unexpected flow deviations and pressure losses and autonomously isolate distant leaks. Then, by adjusting pressure control valves downstream to lower outflow limits before circulating gels to seal ruptures, it stops pipe damage from spreading. Instead of manually locating faults across km of subterranean mains, crews get accurate spatial coordinates for targeted repairs. Despite interruptions, service continuity continues. Similarly, wind turbines improve asset utilization by more than 60% when they use area sensors to detect approaching lightning activity and begin automatically modifying blade angles to lower hazards seconds before actual hits. Unlike sluggish WiFi, the gesture recognition algorithms that drive automatic visitor check-ins at workplaces instantly gather data from a plethora of motion and biometric sensors via 5G. Instead of wasting time seeking for open spaces, smart parking systems use data about available bay occupancy from linked cameras to route cars to available places. Delivery drones avoid wasting fuel by loitering unnecessary long and use the locations from the swarm's sensors for navigation, obstacle avoidance, and coordinated package dropping. The opportunities for machine cooperation increase in industrial settings. By precisely mapping the locations of surrounding robots, inventory, and human staff members using positional data transmitted over 5G's dependable low latency networks, connected AGVs prevent collisions and arrange optimized paths. When environmental sensors signal that slower human operations are preventing customers from receiving their orders on time, automated guided carts take over the duty with ease. The quick correction of 3D print nozzle alignments during test runs by quality assurance sensors that identify minute errors in outputs stops whole production batches from failing requirements. Tailored adjustments maximize uptime rather than halting machines with blanket overrides. Networked sensors even help processes that are presently purely reliant on human expertise and experience by managing repeated operations and offering helpful insights from gathered data that reduce mistakes. Expanded instrumentation streams processed intelligently over 5G enhance human capability rather than threaten replacement. Examples include augmenting sous chefs with nutrition recommendations, warning technicians to best practices from maintenance histories, or automating crop disease diagnoses for targeted early interventions boosting farm productivity. The convergence of growing numbers of 5G-capable endpoints with AI has the ability to unleash enormous potential for social good via responsible, ethical automation that balances worker welfare with economic efficiency and safety.

5G's Revolutionary Potential to Revolutionize Virtual Events

Furthermore, 5G may make it easier to integrate augmented reality into virtual event experiences. This connection makes it possible to superimpose virtual objects on the actual world, creating a mixed reality experience that makes it difficult to distinguish between the real and virtual worlds. The capacity for participants to interact in real time with virtual environments and objects raises the overall degree of immersion and engagement throughout the event. Furthermore, 5G makes it easier to transmit haptic feedback, which enhances the immersive experience. Haptic technology provides tactile input for attendees at virtual conferences or exhibits, enabling them to feel sensations that closely mimic those in the real world in addition to seeing and hearing virtual content. By using 5G technology, virtual event planners can create a very seamless and engaging experience for participants.

Unlocking the Potential with Seamless Virtual Events and 5G Technology

To ensure seamless virtual event experiences, it is essential to thoroughly evaluate several critical elements in order to optimize the potential of 5G. First and foremost, the infrastructure required to support 5G connection must be provided. This includes putting in place the necessary network infrastructure and 5G base station installations to provide wide coverage and consistent connection.

Moreover, in order for virtual event platforms and apps to be compatible with 5G networks, they must perform better.

In order to take use of 5G's low latency and fast data transfer capabilities, the code and content must be optimized. It also involves making use of technologies like content delivery networks and edge computing to reduce latency and ensure smooth streaming of virtual event material. In addition, collaboration between content creators, virtual event organizers, and network providers is essential to ensuring that virtual events are built with 5G capabilities in mind. This will make it possible to fully use 5G's promise and integrate immersive technologies with ease. Moreover, user devices play a crucial role in guaranteeing a seamless virtual event encounter over 5G networks. For immersive technologies like virtual reality and augmented reality to function properly, these devices need to have the necessary hardware and software capabilities. They should also be able to support 5G networks' high data transfer speeds and low latency requirements. In order to optimize the potential of 5G for seamless virtual event experiences, event planners need to consider these aspects. According to Chamola et al. (2020), 5G has the potential to revolutionize virtual event experiences by enabling dependable connection and immersive technologies like as augmented reality and virtual reality.

Case Studies: Effective 5G Implementations in Virtual Events

Let's now look at a few case studies that show how 5G technology has been successfully incorporated into virtual events. 1. A unique opportunity to showcase the possibilities of 5G technology in virtual event interactions was presented by the Tokyo Olympics 2021.

The organizers made use of 5G's capability to provide immersive watching experiences to spectators all around the globe. By giving viewers a feeling of total immersion in the action, virtual reality video streaming significantly improved the virtual event experience. By means of 5G wireless connections, viewers were provided with a network connection that was both fast and large enough to allow for continuous streaming of virtual reality video content.

1. An important annual event in the mobile industry, the Mobile World Congress (MWC) attracts a large number of attendees from all over the world. The promoters of the 2022 edition worked with a well-known telecommunications company to integrate 5G technology across the whole event. The MWC effectively improved attendees' virtual event experiences by deploying a specialized 5G network. This included the real-time transmission of virtual reality content, interactive augmented reality experiences, and the usage of teamwork tools for business networking and meetings. These case studies demonstrate how 5G technology may be used to simulate events in an efficient manner.

They demonstrate how 5G may transform virtual event experiences by providing attendees with more engagement, immersive technology, and constant connection. Because virtual reality video can provide an immersive watching experience while meeting low-latency, high-bandwidth network needs, it has emerged as the perfect use case for 5G wireless connection in virtual events.

Challenges and Benefits of Implementing 5G for Virtual Events:

While 5G technology has great potential to provide perfect virtual event experiences, there are still challenges that need to be solved in order to put it into practice. One significant challenge is 5G networks' capacity to transport large amounts of real-time data, which is required for immersive technologies like augmented and virtual reality. This might be a barrier to fully using 5G's benefits for virtual events that are immersive, since these apps could be able to use more bandwidth than 5G networks can now support. Moreover, employing 5G for virtual events may provide challenges due to the need for very low latency in applications like haptic technology and remote surgery. However, these challenges also provide opportunities for creativity and advancement in the field of 5G. To fully use the potential of 5G for virtual events, technologies and solutions that increase network capacity and reduce latency are essential. In addition, there are a lot of advantages to using 5G technology for virtual events. The advantages include: - Better connectivity: 5G networks provide faster data transfer speeds and lower latency, which means better connectivity and an overall better virtual event experience. - Uninterrupted streaming: 5G's fast data speeds allow virtual events to stream in high definition without any hiccups, guaranteeing that viewers can fully immerse themselves in the immersive viewing experience. - Immersive and captivating experiences: 5G's low latency and small latency allow for real-time interaction and cooperation, which allows users to actively participate in virtual events and increases the overall immersive experience. - Improved accessibility: 5G networks are widely available, so virtual events can reach a wider audience.

CONCLUSION

Virtual event experiences with 5G have a lot of interesting potential. 5G has the potential to revolutionize virtual events by providing smooth connection, improved interaction, and immersive watching experiences with technologies including immersive technology, high-speed data transfer, and low latency. Virtual event planners may provide guests with engaging and uninterrupted experiences by using 5G networks. Significant real-time data transfer is made easier by 5G technology, which also satisfies stringent network standards for high bandwidth and low latency. As such, it facilitates the emergence of virtual reality video as the ultimate application in virtual events. Furthermore, the development of 5G technology might encourage the invention of original use cases and applications for virtual events. Potential applications include virtual conference rooms, haptic devices, and remote surgery. With its enhanced connection, lower latency, and immersive technology, 5G promises to completely transform virtual event experiences. In order to accomplish a smooth integration of 5G technology into virtual event experiences, issues related to capacity, latency, and real-time data transmission must be resolved. In order to go above these barriers and take use of 5G, virtual event planners might provide participants with really seamless and immersive experiences. In conclusion, 5G has enormous promise for delivering seamless virtual event experiences. 5G has features like low latency and fast data transmission that may support applications for virtual and augmented reality that are very immersive. This has the potential to drastically alter how virtual events are conducted.

REFERENCES

Ahmad, H., Islam, M. Z., Ali, R., Haider, A., & Kim, H. S. (2021). Intelligent stretch optimization in information centric Networking-Based tactile internet applications. *Applied Sciences (Basel, Switzerland)*, 11(16), 7351. DOI: 10.3390/app11167351

Chamola, V., Hassija, V., Gupta, V., & Guizani, M. (2020). A Comprehensive Review of the COVID-19 Pandemic and the Role of IoT, Drones, AI, Blockchain, and 5G in Managing its Impact. *IEEE Access : Practical Innovations, Open Solutions*, 8, 90225–90265. DOI: 10.1109/ACCESS.2020.2992341

Jauhiainen, J. S. (2021). Entrepreneurship and Innovation Events during the COVID-19 Pandemic: The User Preferences of VirBELA Virtual 3D Platform at the SHIFT Event Organized in Finland. *Sustainability (Basel)*, 13(7), 3802. DOI: 10.3390/su13073802

Kitanov, S., Petrov, I., & Janevski, T. (2021). 6G MOBILE NETWORKS: RESEARCH TRENDS, CHALLENGES AND POTENTIAL SOLUTIONS. *Journal of Electrical Engineering and Information Technologies*, 6(2), 67–77. DOI: 10.51466/JEEIT2162186067k

Liang, X., Liu, F., Wang, L., Zheng, B., & Sun, Y. (2023). Internet of Cultural Things: Current research, challenges and opportunities. *Computers, Materials & Continua*, 74(1), 469–488. DOI: 10.32604/cmc.2023.029641

Liu, Z., Li, Q., Chen, X., Wu, C., Ishihara, S., Li, J., & Ji, Y. (2020). Point Cloud Video Streaming in 5G Systems and Beyond: Challenges and Solutions. *IEEE Network*. https://doi.org/DOI: 10.36227/techrxiv.13138940.v1

Onwuegbuzie, I. U. (2022). 5G: next generation mobile wireless technology for a fast pacing world. [JPAS]. *Journal for Pure and Applied Sciences*, 1(1), 1–9. DOI: 10.56180/jpas.vol1.iss1.57

Park, J., & Bennis, M. (2018). URLLC-eMBB Slicing to Support VR Multimodal Perceptions over Wireless Cellular Systems. *IEEE Conference on Global Communications (GLOBECOM)*. https://doi.org/DOI: 10.1109/GLOCOM.2018.8647208

Tang, Q., Yu, F. R., Xie, R., Boukerche, A., Huang, T., & Liu, Y. (2022). Internet of Intelligence: A survey on the enabling technologies, applications, and challenges. *IEEE Communications Surveys and Tutorials*, 24(3), 1394–1434. DOI: 10.1109/COMST.2022.3175453

Chapter 7
COVID-19 Communication Messaging and Culture:
A Survey of Latino and Non-Latino Americans

Anne Pesavento
Touro University Worldwide, USA

Caroline Howard
Touro University Worldwide, USA

Kathleen Hargiss
Touro University Worldwide, USA

ABSTRACT

Major disasters and public health crises require effective communication strategies to disseminate vital information and save lives. During the recent COVID-19 pandemic, use of digital communication to communicate information became critical as governments enacted measures to protect public health, including COVID-19 curtailment strategies intended to save lives. To increase response effectiveness to critically important messaging, it is imperative to understand how to communicate with different communities to best achieve cooperation and compliance. Studies have identified culture as being a strong determinant of how people reacted to COVID-19 messaging. The following chapter describes a survey of 184 Hispanic and Latino Americans and non-Hispanic and Latino Americans to determine their attitudes towards COVID-19 mitigation protocol compliance and effective messaging strategies. The study found that significant differences between Hispanic and Latino and

DOI: 10.4018/979-8-3693-3579-6.ch007

non- Hispanic and Latino culture and preferences for specific messaging strategies during the pandemic.

INTRODUCTION

During major disasters and crises, governments and public health institutions need effective communication strategies to disseminate life-saving information and minimize potential consequences. To reduce the impacts of COVID-19 throughout the pandemic, governments of countries throughout the world and major public health organizations developed digital communications and social media strategies designed to educate and gain cooperation with measures aimed at lessening the spread of coronavirus infections (Brammer, Branicki, & Linnenluecke, 2020; Plohl & Musil, 2021).

The primary measures most countries put into place were lockdowns and enforcement of quarantine requirements (Bennett, Freckelton, & Wolf, 2023). Mandatory closings of educational institutions from pre-school classes to graduate programs resulted in significant learning gaps for affected students of all ages (Meinck, Fraillon, & Strietholt, 2022). Businesses were required to enforce COVID-19 mitigation protocols including social distancing, masking, and mandating customers' provide proof of vaccination to enter the businesses (Stephenson, 2021). COVID-19 mitigation protocols and government mandates were shown to reduce the spread of coronavirus (Walker, Whittaker, Watson, Baguelin, Ainslie, Bhatia, Dom & Ghani, 2020). Simultaneously, the severity of the mitigation efforts and mandates severely affected businesses, causing closures and bankruptcies (Purwaningsih & Pernamasari, 2024).

Organizations, such as the World Health Organization (WHO), have laid out guidelines throughout the pandemic that are aimed at slowing the virus's spread, but compliance with the guidelines has not been universal (Plohl & Musil, 2021). Due to the consequences of government mitigation measures, many people reacted negatively to COVID-19 protocols (Dillard, Tian, Cruz, Smith, & Shen, 2023; He, Cai, Geng, & Klug, 2023; Morales & Paat, 2024), and businesses faced resistance and sometimes defiance from customers when attempting to implement COVID-19 mandates (Wood, & Schulman 2021).

In response to the government prevention measures, that many believed were excessive, resistance was common, and there were protests in reaction to government mandates throughout the United States (Burruss, Jaynes, Moule Jr., & Fairchild, 2021). The public reaction reduced the ability of governments to curtail the spread of the virus. According to the research done by Plohl and Musil (2021), slowing the growth of COVID-19 cases required compliance and identifying the determinants of cooperation. Plohl and Musil (2021) distributed their survey through social media sites to 525 participants to measure their level of compliance and related factors, including their perception of COVID-19, political ideals, religious background, conspiracy ideation, and education level.

One source of both individual and community resistance to compliance with mandated health safety requirements during the pandemic is cultural differences. (Demirtas-Madran, 2021; Morales & Paat, 2024). With the large Hispanic and Latino American community of almost 60 million within the United States during the pandemic (US Census, 2021), it was important to understand the community culture and identify which digital messaging strategies resulted compliance with COVID-19 mandates. Based on the public health data from New York City and 16 states included in their meta-analysis, Tirupathi, Muradova, Shekhar, Salim, Al-Tawfiq, & Palabindala (2020) reported that Hispanic and Latino persons were at a substantially higher risk of serious illness and death from COVID-19 when compared to Non-Hispanic Whites. With the pandemic's severe impacts in the Hispanic and Latino American community (Baxter et al., 2023), it is essential to better understand both individual and community attitudes and preferences towards digital strategies for communication with the Latino and Hispanic population. This increased understanding will enable best practices for communicating crisis information and requests for compliance with government health mandates. While studies have examined the impacts of culture on resistance to COVID-19 mitigation protocols (Bazzi, Fiszbein, & Gebresilasse, 2020; Biddlestone, Green, & Douglas, 2020: Frey, Chen, & Presidente. 2020; Germani, Buratta, Delvecchio., & Mazzeschi, 2020; Maaravi, Levy, Gur, Confino, & Segal, 2021), research has been needed to understand the differences in attitudes and preferences between Hispanic and Latino Americans and non- Hispanic and Latino Americans towards COVID-19 and the most effective strategies for increasing their compliance with government mandates.

The primary objective of the chapter and research reported was to address the lack of understanding of the cultural differences between Hispanic and Latino Americans and non-Hispanic and Latino Americans in attitudes and preferences towards COVID-19 mitigation protocol compliance and how to maximize the effectiveness of digital messaging strategies. Plohl and Musils's validated survey, *Compliance with COVID-19 Prevention Guidelines Scale,* was completed by a sample of 92 Hispanic and Latino Americans and 92 non- Hispanic and Latino Americans. The chapter reports the results of the study along with descriptions of the prior research and underlying reasons for the study along with the need for research to better facilitate communications that result in greater compliance with COVID-19 and other health related protocols from different cultural groups. With the review of publications describing research and theory on COVID-19 and culture along with reporting the authors' research, the chapter aims to enable better understanding Hispanic and

Latino attitudes and preferences for digital messaging strategies that align with best practices for communicating crisis information and requests for compliance with government health mandates.

BACKGROUND

In December 2019, the first case of COVID-19 was reported in Wuhan, China. Since then excess mortality statistics have led to estimates as high as 4.8 million deaths worldwide (Temsah, Jamal, & Al-Tawfiq, 2023). As COVID-19 became a global pandemic, government mandates forced people into lockdown, businesses were closed, and many firms struggled to survive. A 2020 survey by Goldman Sachs found that 96% of business owners already felt the effects of the pandemic, and many business leaders predicted that they would face further financial consequences (Liguori & Pittz, 2020).

In 2022, COVID-19 was still affecting businesses, causing closures, especially for small businesses (Statista, 2022). In February 2022, McKinsey reported that responding to COVID-19 requires that businesses have innovative strategies and increased expenditures on technology and infrastructure to survive the challenges. Mishrif, & Khan (2023) reported that technology adoption was a key survival strategy for small and medium enterprises during COVID-19.

In August 2020, Main Street American surveyed 1,414 small business owners about how the pandemic affected them. The survey reported that 30% had not had to close their doors during the pandemic, but 53% had to close only temporarily at the beginning of the pandemic and have since re-opened (Powe, 2020). Participants described how they added a variety of measures to ensure the safety of their employees and customers, including adding delivery services, having employees work from home, and providing hand sanitizer stations. Many described how they had implemented additional protocols, including requiring staff to be masked and limiting the number of guests allowed in their business (Powe, 2020).

COVID-19 and the Latino/Hispanic Population in the United States

While only 18% of the United States population identifies as Hispanic or Hispanic and Latino (US Census Bureau, 2020), they comprise almost 33.8% of the cases of COVID-19 (Moore, 2021). Within the community, individuals are more likely than Non-Hispanic Whites to contract COVID-19, require hospitalization, be admitted to the intensive care unit, and die from COVID-19 than non-Hispanic-Whites (Baxter, Shenoy, Lee, Griffith, Rivas-Baxter, & Park, 2023; Moore, 2021; Tirupathi et al., 2020; Figueroa et al., 2020; Holtgrave et al., 2020; Izzy et al., 2020; Tirupathi et al., 2020). According to the Centers for Disease Control (CDC), the disparity in health outcomes was caused by lack of access to proper healthcare, as well as increased exposure due to jobs that require employees to work closely together (Noe-Bustamante, Krogstad, Lopez, 2021). Hispanics and Latinos had more front-line jobs, that did not allow for the spacing between people recommended to reduce infections, resulting in greater exposure to the virus than other populations. This meant more infections within the community (Calo, Murray, Francis, Bermudez, & Kraschnewski, 2020). 45% of jobs held by Hispanics require them to work outside the home (Noe-Bustamante, Krogstad, Lopez, 2021), and the Hispanic community is more likely to live in multigenerational households, raising the risk of exposure further (Calo, et al. 2020).

LITERATURE REVIEW

The literature focusing on COVID-19 prevention highlights the importance of understanding the relationship between culture and COVID-19 prevention measures and appropriate communication strategies for different audiences. The proposed research addresses the gap in the literature by focusing on adding to the knowledge about the cultural difference between Hispanic, Latina, and Latino and non- Hispanic, Latina, and Latino Americans in attitudes towards compliance with COVID-19 protocols and what communication strategies can be taken to obtain greater compliance.

Curtailing COVID-19 required maximally effective and impactful digital communications strategies. Kalam, Hossain, Jahan, and Kim (2021) surveyed 264 participants to understand what steps could help keep businesses open and the public safe in the vulnerable economy of heavily populated Bangladesh. Kalam, et al. (2021) describe how protective materials and social distancing had a positive effect on stopping the viruses from spreading, allowing customers to reduce risks while shopping. They found that the awareness of symptoms, understanding social distancing, and learning how to use protective equipment was positively related to business survival during the pandemic. They report that if an individual was able to understand the actual symptoms of the disease and how to use protective equipment, they were more likely to take protective measures designed to prevent the spread of the virus. Specifically, Kalam, et al. (2021) found that the more the population is educated in the symptoms and preventative measures, the more they are likely to self-isolate. Effective media messaging included using a variety of platforms to raise awareness of the symptoms, the effectiveness of social distancing, and knowledge on disease reduction resulting from the use of protective equipment was key to customer retention and resilience for businesses (Kalam, et al., 2021).

Culture, Communication, and Compliance

A number of studies have found that peoples' culture is a determinant in how they respond to the COVID-19 pandemic and how likely they are to comply with COVID-19 prevention protocols. Research indicates that cultural factors are related to a community's receptiveness to messaging, and their findings suggest that differences in country culture require that different approaches be taken to implementing protocols to reduce the spread of COVID-19. (Maaravi, et al, 2021). For example, a study of 111 countries by Chen et al. (2021) found that culture is important in determining a country's success in implementing COVID-19 policies. To better understand the link between culture and compliance with COVID-19 mitigation protocols, Maaravi et al. (2021) analyzed data from 69 countries, comprising 75% of the world's population and found evidence that the spread of COVID-19 can be linked to "societies' individualistic or. collectivistic orientation (Maaravi, et al, 2021). They based their research on Hofstede's cultural dimensions which was used to define country culture, including determining whether a company is high in individualism or collectivism (Hofstede, 2021). Like Chen et al. (2020), Maaravi et al. (2021) found evidence that the spread of COVID-19 can be linked to "societies' with individualistic or. collectivistic orientations (Maaravi, et al, 2021). The authors found a significant correlation between populations that favor collectivism and adherence to government COVID-19 mitigation protocols. Populations with more individualistic orientations were found to be less likely to comply with the prevention measures. The findings indicate that government policy implementation is more difficult in countries ranking higher in individualism, with its emphasis on personal freedom, than those ranking high in collectivism, emphasizing the well-being of the group. The results reported by Chen et al. (2020) support the importance of culture in determining how populations respond to COVID-19 and prevention protocols.

Similarly, studies have found that there is a relationship between attitudes towards COVID-19 vaccinations and culture. Leonhardt, & Pezzuti (2021) conducted a largescale multilevel analysis of responses from a survey with over 400,000 participants. They found higher COVID-19 vaccination intentions in countries that are higher in collectivism. Mo, Yu, Luo, Wang, Zhao, & Lau. (2021) surveyed university students in 165 classes in five provinces of China. They found that collectivistic culture was a significant determinant of COVID-19 vaccination.

Luo, Chen, Cui, & Liao (2021) performed semantic network analysis on two social media platforms in China and the United States, two countries chosen for the study due to substantial differences in the cultural characteristics defined by Hofstede (2011). They found that Chinese users of social media expressed more positive feelings about the COVID-19 vaccine than did American users. Specifically, Luo et al. (2021) found that compared to the Americans the Chinese were less likely to express negative sentiments towards taking the COVID-19 vaccine.

Culturally appropriate COVID-19 information and messaging platforms are required to best communicate messages about COVID-19 prevention (Ala, Edge, Zumla, & Shafi, 2020). Effective media messaging through a variety of platforms can raise awareness of the symptoms spread, as the effectiveness of social distancing and disease reduction resulting from the use of protective equipment is key for businesses retaining customers and staying resilient (Kalam, et al., 2021). Lagman, Vergara, & Sarmiento (2021) stress the critical importance of culture in determining how to communicate about and respond to COVID-19. For a community to be receptive, messaging must also be clear, consistent, easy to understand, and culturally appropriate (Ala, Edge, Zumla, & Shafi,, 2020). This is especially true in minority populations, like the Hispanic, Latina, and Latino community.

Studies have found that communication strategies for mitigating COVID-19 are best received when guided to suit the minority audience (Ala, et al., 2021; Shelus, Frank, Lazard, Higgins, Pulido, Richter, & Hall, 2020). Messaging to minority populations has less effective due to complacency, skepticism, and mistrust of politicians and politicians when compared to community leaders for whom there is greater trust within communities of color (Ala et al., 2021). Ala et al. (2020), also, recommend involving in COVID-19 messaging those, who have recovered from COVID-19, along with religious leaders, and members of faith-based organizations.

The selection of the venue used to communicate messages is, also, important (Ala et al., 2020). A variety of venues and messengers may be useful for communicating messages through video and documents including community and social organizations, social media, community champions, local pharmacies, and primary care facilities (Ala et al., 2020). The social media site, TikTok, is especially useful for communicating to adolescents and young adults (Basch, Mohlman, Fera, Tang, Pellicane, & Basch, 2021). Based on the relationship between the Uncertainty Avoidance cultural dimension developed by Hofstede (n.d.) and social distancing behavior, Huynh (2020) recommends messages that emphasize uncertainty when trying to persuade individuals to social distance in cultures high on the Uncertainty dimension.

One of the main issues that may cause confusion in compliance and messaging is the difference in subjective cultures, or the beliefs or norms that are associated with a certain culture. This idea means that the two different cultures examined, Hispanic and Latino and non-Hispanic and Latino, have different ideals that relate to them. Three subjective cultural ideals need to be examined in learning how to discuss healthcare, including self-construal, ethnic identity, and cultural health attributions (Oetzel, Vargas, Ginossar, & Sanchez, 2007).

Self-construal is one self's image, feelings, and values regarding relationships in their life. This involves face-to-face interactions, family relationships, and media. Unlike the typical American relationship, Hispanic culture typically relies on interdependent relationships (Oetzel, et al., 2007). Ethnic identity may also be considered when connecting with patients of different cultures. People may relate more to their culture of origin, they may lean more towards the society they are currently living in, or they may have mixed orientations. Differences in ethnic identity may influence communication preferences. If one consumes a lot of mainstream media, they may have increased knowledge of the country they live in. If they spend time consuming media from the culture of their origin, they may have more of a connection to their original country and ethnic group. The last subjective cultural idea is cultural health attribution. The Hispanic culture does not always lean toward traditional medicine, sometimes they will lean on alternative options of healthcare. This would include santeros or religious leaders, herbal remedies, and folk remedies (Oetzel, et al., 2007).

Studies have found that communication strategies for mitigating COVID-19 were best received when guided to suit the minority audience (Ala, et al. 2021; Shelus, Frank, Lazard, Higgins, Pulido, Richter, & Hall, 2020). Within minority communities, messaging was less effective when received from politicians due to skepticism, and mistrust in comparison to messages from community leaders for whom there is greater trust within communities of color (Ala et al., 2021). Ala et al. (2020) recommended communications from people who recovered from COVID-19, along with religious leaders, and faith-based communities. The selection of the venue used to communicate messages was, also, found to be important (Ala et al., 2020). A variety of venues and messengers appeared useful for communicating messages through video and documents including community and social organizations, social media, community champions local pharmacies, primary care facilities (Ala et al., 2020). The social media site, TikTok, was especially useful for communicating to adolescents and young adults (Basch, Mohlman, Fera, Tang, Pellicane, & Basch, 2021). Based on the relationship between the Uncertainty Avoidance cultural dimension developed by Hofstede (n.d.) and social distancing behavior, Huynh (2020) recommends messages that emphasize uncertainty when trying to persuade individuals to social distance in cultures high on the Uncertainty dimension.

To determine how the Hispanic community viewed the COVID-19 in the early days of the pandemic, the Pew Research Center conducted a survey (Calo et al., 2020) that showed that two-thirds of Hispanic adults saw that the COVID-19 was a threat to the whole population. The same survey showed that only half of the respondents in the United States similarly viewed COVID-19 as being a threat to the population (Calo, et al., 2020).

In focus group studies, Shelus et al. (2020) found that Hispanic and Latino participants did not feel that the current COVID-19 messaging resonated with them and favored messages that included the word, "together", and those, which emphasized that "we're all in this together" were well received. Their findings reflect the culture of Latin American counties, which score high on In-Group Collectivism (Dorfman, Javidan, Hanges, Dastmalchian, & House, 2012), especially when compared to the United States, which is highly individualist (Wu, 2006).

Similarly, Airhihenbuwa, Iwelunmor, Munodawafa, Ford, Oni, Agyemang, & Okosun (2020), Munodawafa Ford, Oni, & Agyemang (2020) acknowledge the importance of communicating individual risk but describe that for communities that were hard hit by COVID-19, the engagement of people living within the community is more effective, particularly in communities, like the Hispanic and Latino community, that was most heavily hit by the pandemic. Similarly, based on their analysis of lessons learned during the pandemic, Alberti, Castaneda, Castrucci, & Harrison (2020) recommended formulating messages that emphasize community problems rather than a public health problem. In addition, messages focusing on doing it for others, community, faith, and family appealed to the Hispanic and Latino participants (Shelus et al., 2020). After interviewing Indiana doctors, Guerra Rodriguez & Hudson (2020), determined that trust needs to be built to communicate with their Hispanic patients. The message needs to not just be translated, but also needs to be tailored to the community when coming from a healthcare professional (2020).

In small Hispanic communities, research showed that a personal touch would be recommended when communicating information on COVID-19, like text messages from the universities and their healthcare professionals. Also because of the language barrier, some participants had to get their news from Spanish-languages sources like Univision instead of getting it from local sources, so having local news sources translated would help communities like these understand local issues regarding the virus (Moyce, et al., 2021). Based on their findings from an online experiment based on the cultural theory of risk, Yuan, & Chu (2021) emphasize the importance of understanding the worldviews of the audience.

Hispanic Community, Historical Medical Abuse, and Attitudes of Skepticism

Since the winter of 2020, there have been three vaccines that have been approved by the FDA to help prevent COVID-19, but many of the citizens of the United States are not planning on getting vaccinated. There are several reasons that some of the population has had anti-vaccination attitudes. This includes skepticism against the medical community, fear of long-term side effects, distrust of political powers, and issues based on historical injustices that happened to minority communities (Peteet, Belliard, Abdul-Mutakabbir, Casey, & Simmons, 2021).

Fears of the medical community are prominent when speaking about minority populations. Experiments on the community have created distrust in medicine today. In the 1920s, the Supreme Court legalized unethical eugenic sterilization (Olagunju, 2021). Compulsory sterilization was included in H.H. Laughlin's publication of Eugenical Sterilization in 1922, which introduced the idea of forced sterilization throughout the country. It was thought that the establishment of the law would prevent hereditary degeneration (Farber, 2008). Women who were Native Americans, African Americans, and Puerto Ricans were given hysterectomies against their will. Over 60,000 people of color, poor, disabled, and institutionalized were forced to be sterilized (Olagunju, 2021). The process was reviewed by the Carnegie Institute in 1935, showing no medical merit, and the experiment lost its funding in 1939 (Farber, 2008).

One of the most well-known experiments that lost the trust of minority communities would be the Tuskegee Syphilis Study. It is reported that researchers wanted to study the course of the disease (Alsan & Wanamaker, 2018). The study included 600 poor black men from Alabama, 399 of who tested positive for syphilis (Olagunju, 2021). The participants were told they had "bad blood" and were given blood draws and painful spinal taps to help "treat" them. Researchers and physicians were told to treat patients for all ailments except for syphilis as the study was to study the victim's bodies after they died. (Alsan & Wanamaker, 2018). Even though it was examined by multiple medical communities, the study did not end until the 1970s, when many of the men involved were already deceased (Olagunju, 2021).

The unethical experiments and procedures did not end with the Tuskegee study. In the 1990s, an unethical experiment was done on 6 to 11-year-old boys who had Hispanic or Black backgrounds. The boys had to have an older sibling who were labeled delinquents or show Attention-Deficit/ Hyperactivity disorder traits (Olagunju, 2021). 34 young men in New York City were given the drug Fenfluramine to test criminal or violent behavior when levels of brain chemicals are messed with (Hilts, 1998). The drug was later linked with valvular heart disease and pulmonary arterial hypertension (Olagunju, 2021). Many in the scientific community worry that the study exposed children to substantial risk, with no medical benefits to the participants (Hilts, 1998).

Views on Vaccinations

With the introduction of the COVID-19 vaccine, it is even more important to know why this fear exists and how to improve compliance in the protocols involving the COVID-19 virus. Some employers have required the COVID-19 vaccination for their workers, but have been met with rejection and refusal (Smith, & Nagele-Piazza, 2022). Rejection of vaccinations is nothing new. In 2009, a study conducted by Ulrich, Thompson, Livaudais, Espinoza, Cordova, and Coronado (2013) was conducted in the Lower Yakima Valley of Washington State to gain more of an understanding about the willingness to take part in biomedical research, attitudes towards the HPV vaccine, and characteristics of the community. 117 Hispanic households qualified for the study and were interviewed and surveyed about their trust of doctors, the medical field, and other barriers that they might have against biomedical research (Ulrich, et al., 2013). Several ideas were reported as barriers, including cultural belief (65%), family care (82%), having family with the disease (80%), and lack of trust in the medical and research field (71%) (Ulrich, et al., 2013).

In 2020 another study was performed by Padilla, Frietze, Shenberger-Trujilo, Carrillo, & Loya (2020) to attain information on the views of vaccines, mainly influenza. A survey was distributed, adapted from the National Flu Survey, and it was concluded that health-care providers need to focus more on cultural sensitivity and targeted education (Padilla, et al., 2020). This information foreshadowed the issue with compliance in the CDC's recommendation to get the COVID-19 vaccine.

The CDC reports that only 38% of Hispanic individuals received the influenza vaccine in 2020. This is compared to 41% of the Black population and over half of the non-Hispanic whites. Many minorities fear that vaccines are unsafe. This fear is because of medical abuse that has been conducted on these groups. Kricorian and Turner (2021) conducted a study in January 2021 to gain more understanding of the COVID-19 and beliefs about the COVID-19 vaccine. By surveying 1950 adults in America, the researchers were able to gain more of an understanding of who is willing to get the COVID-19 vaccination and who has fear regarding it. 28.2% of Hispanics reported wanting to get the shot immediately, whereas 38.5% of non-Hispanic whites shared the same views. 27% of Hispanics did share that they did not trust the vaccination though. Having trust in the vaccine is very important as many workplaces and events are wanting those to comply with their vaccination requirements (Kricorian & Turner, 2021).

Those who are pro-vaccine need to be empathic towards the fears of others, make personal connections, and deliver clear and facts. It is also important to avoid ignoring historical wrongdoings. One way to reach this part of the population and hope for compliance in vaccinations is by delivering free virtual webinars to address fears and allow facts on the vaccine to be shared (Peteet, et al., 2021). A seminar was conducted in December 2020. The event hosted more than 500 Black and Latina/o community members in Southern California, and later gained more than 7000 views on Facebook live. The event had COVID-19 experts on the panel to help address fears, describe the health disparities, and educate people on the vaccine. Although the results of the event are still forthcoming, it did allow those in the medical field to address concerns and hopefully mend some relationships with those who have fears about the vaccine.

Barriers to Quality Medical Treatment

Many people complain about not knowing what the truth is when it comes to COVID-19. This can be an issue when asking people to comply with the protective protocols. Part of the hesitancy in the Hispanic community seems to be the limited language-appropriate services, including a shortage of medical professionals who speak Spanish and the lack of medical interpreters available at facilities. Between 2010 and 2016, North Dakota, West Virginia, and South Dakota have seen massive growth in their immigrant population, the majority being Hispanic or Latina/o. Those areas may have more of a demand for Spanish-speaking doctors and they may not be prepared for it, especially because of the pandemic (Ortega, Shin & Perez-Corodon, 2020). Since over 40 million American citizens speak Spanish in the home, it is important that the medical community has exposure and first-hand experiences with the culture to understand their needs and how to get them to understand why compliance is so important when dealing with COVID-19 (Ortega, et al., 2020).

The medical field has also shown to be affected because of the pandemic, not allowing medical students to have exposure to different cultures. This means that they have less opportunity to gain Spanish skills. With the pandemic, students have not been able to have exposure to minority communities, go study abroad, or do case studies which allow them to examine other cultures.

Since the pandemic started, colleges and universities have included more virtual options to their students, allowing them to gain experience in dealing with different minority populations. Virtual access can help because it allows students all over the nation to have access to education in Spanish classes without the geographic barriers previously put up. With online courses, rural areas can be reached to educate those who are not able to communicate with their local healthcare professionals (Ortega, et al., 2020).

Medical school needs to take advantage of the new normal to give the opportunity of linguistic and cultural courses to educate the medical students on different parts of the US patient population. This method will help those students learn more about the cultural ideals, as well as the specific language aspects that are needed to communicate medical needs, not just learning vocabulary and grammar, but instead patient-centered communication. These ideas are not taught in a classroom but need interactions and real-life experience to be taught. Classes can include virtual role-playing, real-time feedback, and peer to peer review. They also will allow medical students to use non-verbal cues, instead of just relying on language to express ideas and concerns. By doing this, communication with patients will become easier and allow for the medical community to express why compliance is so important (Ortega, et al., 2020).

When dealing with compliance to COVID-19 protocols, it is important for the patients to feel satisfaction in their medical care. Hispanic and Latino/a patients have shown low satisfaction in their healthcare compared to White and English-speaking individuals (O'Brien & Shea, 2011). To gain more understanding on the views of the Hispanic community on the compliance of these protocols, it is important not only to look at the historical impact of abuse, but also on how this part of the population deals with other aspects of healthcare. One of the most prominent causes of death among Hispanic women in the United States is cancer. Oetzel, et al. (2007) researched the subject to gain more understanding of the preferences that Hispanic women have when dealing with healthcare and how doctors can communicate the importance of complying with the COVID-19 protocols. While white women are being educated on breast health, including self-screening. Hispanic women feel ignored because there is no familiarity with their culture and the differences that need to be addressed when dealing with healthcare. Women with a Hispanic background, have less breast health messaging or informative sources than other races (Oetzel, et. al, 2007).

Diabetes has also hit the Hispanic population much harder than the non-Hispanic white population, 12% compared to 7%. Moreira, Hernandez, Scott, Murillo, Vaughan, & Johnston (2018) researched the cultural beliefs of medicine and disease to understand the impact it has on the treatment of diabetes in the Hispanic population. The main three ideas that were examined are *susto, coraje*, and *fatalismo*.

Susto, or fright, is the idea that after a traumatic event, one that "makes the soul leave the body (Moreira, et al., 2018)," can cause physical ailments. It has been reported that people have experienced gastrointestinal issues, as well as cold sweats. Those who believe in *susto* use prayer, herbs, and healers to combat the symptoms instead of going to a traditional doctor. This means that someone in the community could be ill with a disease like diabetes but use non-traditional methods to try to heal it instead of reaching out. *Coraje* is what some think may cause diabetes in an individual. A long-term struggle, anger, or family violence can be part of *coraje.* The thought is that the physical, verbal, or emotional abuse that comes from those events are a causal factor in one developing diabetes (Moreira, et al., 2018). There is also a belief that things happen because of fate, also known as fatalismo. Fatalismo causes the patient to believe there is nothing that can be done to change the outcome. The belief that they are going to be sick no matter what will shut down any medical advice given (Moreira, et al., 2018). The literature focusing on COVID-19 prevention highlights the importance of understanding the relationship between culture and COVID-19 prevention measures and appropriate communication strategies for different audiences. The research addressed the gap in the literature by focusing on adding to the knowledge about the cultural difference between Hispanic and Latino and non- Hispanic and Latino Americans in attitudes towards compliance with COVID-19 protocols and what communication strategies can be taken to obtain greater compliance.

Healthcare Outreach

When examining the Hispanic communities' ideas on healthcare, especially when dealing with cancer and diabetes, there were clear ideas that were brought to light. To work on communicating with Hispanic women, using peer counseling to deliver the message will help spread the message as there is more of a strong sense of community in the Hispanic culture. There is already trust built, so the peer counseling will be more believed and hopefully have more women learn about breast health screening (Oetzel, et al., 2007). The same idea can be used regarding COVID-19. By using peer counseling, the trust does not need to be fully built, so the message may be received more easily. To help combat the non-traditional medical ideas, healthcare providers need to gain the trust with patients. One way to do this would be to engage family members in the process. The family of the patient may shed some light on the background and beliefs of the patient. Asking questions to the patient and the family about the beliefs and any barriers or distrust they have will help to find a way forward with treatment. This may also allow a patient to have more confidence in their relationship with their doctor or medical professional (Moreira, et al., 2018).

The National Association of Hispanic Nurses, founded in 1975, is the leading organization that wants to help the Hispanic, Latino/a community. The Hispanic,Latino/a population represented 10.2% of almost 4 million licensed RNs in 2020. They have been pressing the COVID-19 protocols recommended by the CDC to the vulnerable community in the Hispanic population (Cuellar, Aquino, Dawson, Gracia-Dia, Im, Jurado, & Toney, 2020). Having these nurses spread the message is important, but there would also be benefits with compliance in COVID-19 protocols from an increase investment in healthcare centers and hospitals in low-income areas to help serve those communities and undocumented immigrants. These centers would need to have multilingual services, and free encrypted platforms to help healthcare professionals to communicate with their patients about COVID-19-related issues. Also allowing Medicaid for those who have lost their insurance because of loss of a job would help, as well as not pursuing debt collection for COVID-19 related health services (Lopez, Hart, & Katz, 2021).

Besides medical professionals understanding how to speak to their patients, there have been programs set up to help the community outreach and push compliance of the COVID-19 protocols. Programs like REACH, founded in 1999, have been set up to understand the Hispanic community. They started by addressing health concerns, like promoting walking daily to help improve overall wellbeing. By researching communities in parts of Pennsylvania, which had an estimated 970,000 Hispanic communities in 2020, researchers were able to understand the fear and anxiety that came from the language barrier and the limited health care information that has been spread around the community. The REACH program has tried to reach the community by offering programs to support COVID-19 pandemic efforts (Calo, et al, 2020). The organization wanted to figure out how to address Hispanic families need to stay informed, as well as how to test and prevent the spread of COVID-19 (Calo, et al, 2020). One way REACH was able to achieve communication and education with rural communities was by offering Zoom classes. Since many have internet and smartphones, this was an ideal way to promote healthy lifestyle changes in this community. The first class was called "Preparing your household for COVID-19 (Calo, et al, 2020)." These lessons were later published on Youtube so the public could use them as well.

Rural Community of Hispanic and Latino/a

The rural community should not be overlooked in examining how the Hispanic/ Latino community has dealt with COVID-19. One main reason is that there is not a lot of access to health care due to communities being physically isolated (Moyce, Velazquez, Claudio, Thompson, Metcalf, Aghbashian, & Sisson, 2021). In rural communities, spreading information throughout the community can be difficult. One reason is the language barrier. Some patients may not be able to understand the messaging provided or be able to communicate with their healthcare providers that they need testing or more information (2021).

To collect data, Moyce, et al.,(2021), used snowball sampling to gain more understanding of the area. The researchers conducted 14 semi-structured interviews in April 2020. recorded and transcribed in Spanish, translated to English. Five themes resulted from the data analysis of the interviews including news sources, changing behavior, emotional responses, coping mechanisms, and the use of natural remedies. Most respondents in the area stated that they received their news about COVID-19 from local, national, and international sources. To be widely understood, messages had to be broadcasted in Spanish due to a large number of people having difficulty understanding the English news networks. Many of those interviewed discussed their wariness of trusting the news posted on social media. Traditional English speaking American news broadcasts proved hard for many in the community to understand due to the language barrier (Moyce, et al., 2021).

When asked what would help to spread the correct information in the community, the recommendation was made to have local information about the pandemic translated into their native language. Because of the language barrier, some participants had to get their news from Spanish-language sources like Univision instead of local sources. The participants also stated that they would prefer more personal touches when getting information, like text messages from the universities and their healthcare professionals (Moyce, et al., 2021).

RESEARCH QUESTIONS

The research questions addressed by the study were:

1. What is the relationship between Hispanic, Latina, and Latino and non- Hispanic, Latina, and Latino culture and attitudes toward compliance with COVID-19 mitigation protocols?
2. What is the relationship between Hispanic, Latina, and Latino and non- Hispanic, Latina, and Latino culture and attitudes towards messaging strategies about COVID-19 mitigation protocols?

METHODOLOGY

The validated survey on Compliance with COVID-19 prevention guidelines by Plohl & Musil (2020) was used to measure the differences in attitudes towards compliance resistance and strategies for overcoming compliance. Variables incorporated into the study included perception of COVID-19, political ideals, religious background, ideation, and education level.

Data Collection

The population for the research study was adults over the age of 18, who live in the United States of America. Since it is not feasible to survey the entire population, a sample of 184

Hispanic and Latino and non- Hispanic and Latino Americans were selected to be representatives of the population. A large enough number of people need to be surveyed though to ensure a significant and accurate conclusion, and reduce the likelihood of error (Kaliyadan, & Kulkarni, 2019). According to the United States Census Bureau, the population of individuals over the age of 18 is 258.3 million (Ogunwole, Rabe, Roberts, & Caplan, 2021), meaning a sample size of 184 Hispanic and Latino Americans and non- Hispanic and Latino Americans was necessary for the study. The sample size calculator on SurveyMonkey.com also identified 184 participants as appropriate for a population that size (Survey Monkey, 2022). The survey was conducted using Survey Monkey's survey tool, surveying 92 Hispanic and Latino Americans and 92 non- Hispanic and Latino Americans from the Survey Monkey Audience. Survey Monkey allows a researcher to design their survey for a fee and sends it out to the target population.

Because Survey Monkey is online, the survey was completed quickly, completely anonymous, and helped to eliminate bias (Abd Halim, Foozy, Rahmi, & Mustapha, 2018). No personal information was collected within the survey. Each participant was be provided a letter of consent before they answered the survey. The researcher obtained the Touro University Worldwide IRB approval before the survey was distributed.

Data Analysis

The study used the *Compliance with COVID-19 prevention guidelines scale* developed and validated by Plohl and Musil (2021) to measure the relationship between individual characteristics and compliance with COVID-19 prevention guidelines. authors reported that the scale exhibited acceptable internal consistency and reliability (Plohl & Musil, 2021).

Limitations

The study had several limitations. First, the results are dependent on the respondent's understanding, responding to the questions honestly, and providing complete answers. While the questions appear to be clear and straightforward, the method of distribution through Survey Monkey and closed questions will not allow participants to ask for clarification or other information to better understand the questions. Murdoch, Simon, Polusny, Bangerter, Grill, Noorbaloochi, and Partin (2014) describe anonymous respondents to surveys as more likely to honestly disclose sensitive information. Therefore, the use of Survey Monkey to distribute the surveys to participants, who will remain anonymous, is more increases the likelihood of receiving honest answers

SURVEY RESULTS

Confidentiality

The online survey platform, Survey Monkey, allows anonymous data to be collected. Researchers can choose to ensure all that collected are anonymous responses. Once the data is collected, the information will be password protected. The information on Survey Monkey will also be deleted after three years (Making Responses Anonymous, 2021).

RESULTS

The survey was sent out to 471 people through the Survey Monkey database. Only 218 finished the survey. Out of those responses, 34 were disqualified. The resulting sample population ended up being 92 Hispanic, Latino/a, and 92 non-Hispanic Latino/a responses that qualified, collected, and used for this research.

Demographical Data

Table 1 displays the percent of Hispanic/Latino along with Non-Hispanic/Latino participants, who were male and female. Surprisingly the gender ratio for both groups was the same, with 43% of those surveyed being male, 55% being female, and 1% either not wanting to answer or being intersex.

Table 1. Gender

	Total	Hispanic and Latino	Non-Hispanic and Latino
Male	43%	43%	43%
Female	55%	55%	55%
Intersex	1%	1%	1%

As shown in the Table 2 below, the majority of those who participated were between the ages of 18 to 44, showing most being 25 to 34 of age. There were less than 10% over the age of 65 that answered the survey's data.

Table 2. Age

Age	Total	Hispanic and Latino	Non-Hispanic and Latino
18-24	16%	20%	12%
25-34	41%	40%	40%
35-44	22%	27%	17%
45-54	9%	7%	12%
55-64	6%	3%	9%
65+	5%	2%	9%

The results from the income levels of those surveyed varied much more than age or gender, showing most made between $50,001 to $100,000, but there were more than 10% in each income level, other than those making more than $150,000.

Table 3. Income

Income	Total	Hispanic and Latino	Non-Hispanic and Latino
Less than $25,000	12%	10%	14%
$25,000 to $50,000	26%	27%	25%
$50,001 to $100,000	39%	43%	34%
$100,001 to $150,000	15%	15%	15%
More than $150,000	7%	4%	9%
Prefer not to answer	2%	0%	3%

In both Hispanic and Latino, and Non-Hispanic and Latino, the age and income did not vary between the two groups, but politically, there was a more of a drastic difference in their answers. Politically the views leaned more towards liberal and neutral, but there was 10% that were either unsure or preferred not to answer. Hispanic and Latino wanted to state their political views, having the non-Hispanic and Latino either state they were neutral, unsure, or preferring not to answer.

Table 4. Political views

Political Views	Total	Hispanic and Latino	Non-Hispanic and Latino
Very Conservative	7%	9%	4%
Conservative	13%	15%	11%
Neutral	33%	28%	37%
Liberal	28%	29%	27%

continued on following page

Table 4. Continued

Political Views	Total	Hispanic and Latino	Non-Hispanic and Latino
Very Liberal	9%	11%	8%
Unsure	6%	8%	5%
Prefer not to answer	4%	0%	8%

Hypothesis Testing

The following hypotheses were tested to examine if there is a relationship between Hispanic and Latino and non- Hispanic and Latino culture and attitudes toward compliance with COVID-19 mitigation protocols and with attitudes towards messaging.

Hypothesis One: There is a significant relationship between Hispanic and Latino and non- Hispanic and Latino culture and attitudes toward compliance with COVID-19 mitigation protocols

Table 6. Hypothesis 1 regression

Regression Statistics								
Multiple R	**0.001917**							
R Square	3.68E-06							
Adjusted R Square	-0.00549							
Standard Error	0.502739							
Observations	184							
ANOVA								
	Df	*SS*	*MS*	*F*	*Significance F*			
Regression	1	0.000169	0.000169	0.000669	0.979393			
Residual	182	45.99983	0.252746					
Total	183	46						
	Coefficients	*Standard Error*	*t Stat*	*P-value*	*Lower 95%*	*Upper 95%*	*Lower 95.0%*	*Upper 95.0%*
Intercept	1.49505	0.194917	7.670187	1E-12	1.110463	1.879638	1.110463	1.879638
X Variable 1	8.45E-05	0.003269	0.025865	0.979393	-0.00636	0.006534	-0.00636	0.006534

The null hypothesis was not rejected at 0.05 level of significance (β=8.45E-05, p = 0.979393). The hypothesis testing results indicated that *Hispanic and Latino and non- Hispanic and Latino culture* did not significantly predict *attitudes toward compliance with COVID-19 mitigation protocols.* (See tables X, y z11, 12, 13) The relationship between *Hispanic and Latino and non- Hispanic and Latino culture*, the independent variable, and *attitudes toward compliance with COVID-19 mitigation protocols*, the dependent variable, was not significant at an alpha level of 0.05. The regression indicates only a very small percent (R Square < .001) of the variation of the variation in *attitudes towards compliance with COVID-19 mitigation protocols* can be explained by *Hispanic and Latino and non- Hispanic and Latino culture* (R Square= 3.68E-06, Adjusted R Square= -0.00549), and that the model was not significant.

Hypothesis Two: There is a significant relationship between Hispanic and Latino and non- Hispanic and Latino culture and preferences for specific messaging strategies about COVID-19 mitigation protocols.

Table 7. Hypothesis 2 regression

Regression Statistics								
Multiple R	**0.301398**							
R Square	0.090841							
Adjusted R Square	0.085845							
Standard Error	0.479362							
Observations	184							
ANOVA								
	Df	*SS*	*MS*	*F*	*Significance F*			
Regression	1	4.178679	4.178679	18.18497	p <.001			
Residual	182	41.82132	0.229787					
Total	183							
	Coefficients	*Standard Error*	*t Stat*	*P-value*	*Lower 95%*	*Upper 95%*	*Lower 95.0%*	*Upper 95.0%*
Intercept	1.425676	0.039403	36.18165	p <.001	1.34793	1.503422	1.34793	1.503422
Preference For Specific Communication Method	0.37988	0.089082	4.264384	p <.001	0.204114	0.555646	0.204114	0.555646

The null hypothesis was rejected at 0.05 level of significance (B = 0.37988, p <.001). The hypothesis testing results indicated that *Hispanic and Latino and non- Hispanic and Latino culture* did not significantly predict *attitudes toward messaging strategies about COVID-19 mitigation protocols*. The relationship between *Hispanic and Latino and non- Hispanic and Latino culture*, the independent variable, *preferences for specific messaging strategies about COVID-19 mitigation protocols*, the dependent variable, was significant at an alpha level of 0.05. The regression indicates that only approximately nine percent of the variation of *attitudes towards messaging strategies about COVID-19 mitigation protocols* the variation in can be explained by *Hispanic and Latino and non- Hispanic and Latino culture* (R Square = 0.090841, Adjusted R Square = 0.085845), but the model was significant ($p < .001$).

DISCUSSION

This study was designed to increase the understanding of cultural differences between relationship between Hispanic and Latino and non-Hispanic and Latino. The first hypothesis examined the attitudes towards compliance with COVID-19 mitigation protocols. The regression indicated a very small percent of variation, showing the model was not significant. This finding was different than discussed in the literature review. A possible reason is that most of the articles discussed earlier were from 2020 and 2021. At that time people were still confused about COVID-19 and were receiving conflicting information about COVID-19. These articles different views than were found in this study and found with the sample population in this study, conducted in May of 2022.

Because the study took place later in the pandemic, people have been living with the virus for over two years. This means that there have been new less harmful variants, a vaccine, more experience living with a pandemic, and COVID- 19 fatigue that has settled in. Omicron, or SARS- CoV-2, was first discovered in November 2021. Even though it is more transmissible, the variant has shown to be less harmful, having people show milder symptoms. The variant is less likely to cause lung infection because it infects the upper respiratory tract (Fan, Li, Zang, Zhou, 2022). Due to the milder infection, many did not take the surge seriously in December 2021, causing the nation to hit a new high of daily cases, 300,886. Even during this surge of cases, the CDC changed their recommendations to shorter time in quarantine, saying people who were asymptomatic should only isolate for five days and those who are sick should quarantine for ten days (Elamroussi, 2021). Another change that could have caused the results to not be significant is the introduction of the vaccine. Once the vaccine was widely available, many stopped seeing COVID-19 as an issue once vaccinated. There were many who celebrated getting their vaccination, thinking that the vaccine meant they could go back to living a "normal" life (Jones, Barajas, 2021). These changes to life with the virus could have caused some of the answers to the survey to change from the beginning of the pandemic to now.

One of the largest variables that could have affected the answers received in this survey would have been 'COVID-19 Fatigue.' People are over the social distancing, the missing of events, and the wearing masks. During the summer of 2022, there has been another surge, but many have been ignoring it. Gathering indoors without masks, people seem to be living carefree. They weary of wearing masks and exhausted by the number of cases going up and then down and up again. Many are not even reporting their exposures or infection because of home testing being so easy and symptoms being so mild. At the beginning of the pandemic, people acted cautiously out of fear, but now the feeling has changed drastically, having the thought of going back to 'normal life being the driving force (Schneid, Perez-Moreno, & Branson-Potts, 2022).

The second hypothesis examined the attitudes towards messaging strategies about COVID-19 mitigation protocols with Hispanic and Latino, as well as non-Hispanic and Latino people in the United States. The survey did show that the model was significant, meaning the null hypothesis was rejected. One of the questions examined the type of communication does the participant use to receive their news on COVID-19. When running the dummy regression on this question, there was a significant difference between the two groups.

Approximately half of each group received news from American news sources but only about a third of each group would choose to receive news from those sources if given a choice. Similarly, the portions of each group choosing messaging from doctors or medical professionals and word of mouth decreased when given a choice of communication messaging. While preference for messaging from American news sources, the CDC website and doctors or medical professionals decreased more for non-Hispanics, preference for word of mouth had the largest decrease for both groups. Again, the confusion and conflicting information about COVID-19 and drop in trust of certain sources for accurate information could explain these differences. Since COVID-19 was new in 2020, much of the initial information was later revised. That could have created the impression that information provided by American news sources, CDC and doctors or medical sources was not reliable. A drop in trust for word-of-mouth communications could be explained by the variety of individual opinions and proliferation of misinformation potentially spread by word of mouth.

SOLUTIONS AND RECOMMENDATIONS

Health and other public crises necessitate understanding of how to effectively disseminate emergency information and communicate mandates. Research has shown that culture is an important factor in determining the receptiveness a community will have to receiving digital messages. In this study the most effective messaging strategies was significantly different for the Latino and Hispanic versus Non-Latino and Hispanic Americans. Prior research has shown similar findings from other cultures and communities regarding the attitudes and preferences towards digital communications. To facilitate maximally effective emergency health and crisis messaging, the people communicating vital information need to be aware of cultural and community attitudes and preferences in order to develop the best methods and messages.

FUTURE RESEARCH DIRECTIONS

When determining messaging strategies, greater understanding of the attitudes and preferences of community receiving communications is needed. Prior research along with the study reported here, demonstrated significant differences in the types of message sources preferred by the Latino and Hispanic versus Non-Latino and Hispanic Americans and with the choices each group would make in the future about messaging. Future studies and data analysis is needed to examine the relationships between Hispanics, Latina, and Latinos and Non-Hispanics, Latinas, and Latinos on the specific messaging strategies and changes in messaging preferences if given a choice. In addition, studies on other cultures and communities needs to be completed to develop better messaging strategies during emergencies and crises. Further research is needed to determine the reasons for the preferences and changes needed in digital communication strategies.

CONCLUSION

In conclusion, based on this study, there was no significant difference between the relationship the different cultures, Hispanics, Latina, and Latinos and Non-Hispanics, Latinas, and Latinos, and the way compliance of COVID-19 protocols are viewed. As mentioned previously, these results may be because of the length the pandemic has gone on, as well as the exhaustion people have from the inconsistent messaging by the medical community. The study did show that there is more of a difference in terms of specific communication preferences. Receiving their news from American news sources seemed to be popular with about half of each group, but at least a third of those stated they would choose to change where they receive their news from. Since cultural differences have been shown to be significant indicators of message preferences and attitudes, further research is needed to better understand which messages are most effective for the communities studied here along with other communicates.

REFERENCES

Abd Halim, M., Foozy, C. F. M., Rahmi, I., & Mustapha, A. (2018). A review of live survey application: SurveyMonkey and SurveyGizmo. *JOIV: International Journal on Informatics Visualization, 2*(4-2), 309-312.

Airhihenbuwa, C. O. (1989). Perspectives on AIDS in Africa: strategies for prevention and control. *AIDS Education and Prevention.* Airhihenbuwa, C. O. (1989). Perspectives on AIDS in Africa: strategies for prevention and control. *AIDS Education and Prevention.*

Airhihenbuwa, C. O., Iwelunmor, J., Munodawafa, D., Ford, C. L., Oni, T., Agyemang, C., Mota, C., Ikuomola, O. B., Simbayi, L., Fallah, M. P., Qian, Z., Makinwa, B., Niang, C., & Okosun, I. (2020). Peer reviewed: Culture matters in communicating the global response to COVID-19. *Preventing Chronic Disease*, 17, 200245. https://www.ncbi.nlm.nih.gov/pmc/articles/PMC7367065/. DOI: 10.5888/pcd17.200245

Akbari, J., Salamzadeh, Y., & Farzad, F. S. (2021). Collaborative Governance, a Tool for Enhancing Compliance with COVID-19 Prevention Guidelines. A Study in Iranian Context.

Ala, A., Edge, C., Zumla, A., & Shafi, S. (2021). Specific COVID-19 messaging targeting ethnic minority communities. *EClinicalMedicine*, 35, 100862. https://www.thelancet.com/journals/eclinm/article/PIIS2589-5370(21)00142-5/fulltext. DOI: 10.1016/j.eclinm.2021.100862 PMID: 33997747

Alberti, P. M., Castaneda, M. J., Castrucci, B. C., & Harrison, L. M. (2020). Engaging With Communities—Lessons (Re) Learned From COVID-19. *PCD Staff, 17*, E65. https://www.cdc.gov/pcd/collections/pdf/US-Public-Health-Response-to-COVID-19.pdf#page=7

Alsan, M., & Wanamaker, M. (2018). Tuskegee and the health of black men. *The Quarterly Journal of Economics*, 133(1), 407–455. DOI: 10.1093/qje/qjx029 PMID: 30505005

ANNALS. (2021, July).. . *AAPSS*, 696, 223. DOI: 10.1177/00027162211050007

Apuke, O. D. (2017). Quantitative research methods: A synopsis approach. *Kuwait Chapter of Arabian Journal of Business and Management Review*, 33(5471), 1–8.

Basch, C. H., Mohlman, J., Fera, J., Tang, H., Pellicane, A., & Basch, C. E. (2021). Community mitigation of COVID-19 and portrayal of testing on tiktok: Descriptive study. *JMIR Public Health and Surveillance*, 7(6), e29528. https://publichealth.jmir.org/2021/6/e29528. DOI: 10.2196/29528 PMID: 34081591

Baxter, T., Shenoy, S., Lee, H. S., Griffith, T., Rivas-Baxter, A., & Park, S. (2023). Unequal outcomes: The effects of the COVID-19 pandemic on mental health and well-being among Hispanic/Latinos with varying degrees of 'Belonging'. *The International Journal of Social Psychiatry*, 69(4), 853–864. DOI: 10.1177/00207640221140285 PMID: 36573293

Bazzi, S., Fiszbein, M., & Gebresilasse, M. (2020). Rugged Individualism and Collective (In)action During the COVID-19 Pandemic (No. w27776; p. w27776). National Bureau of Economic Research. https://doi.org/DOI: 10.3386/w27776

Bekalu, M. A., Dhawan, D., McCloud, R., Pinnamaneni, R., & Viswanath, K. (2021). Adherence to COVID-19 mitigation measures among American adults: The need for consistent and unified messaging. *Health Education Research*, 36(2), 178–191. DOI: 10.1093/her/cyab002 PMID: 33496331

Bennett, B., Freckelton, I. R., & Wolf, G. (2023). *COVID-19, Law and Regulation: Rights, Freedoms, and Obligations in a Pandemic*. Oxford University Press. DOI: 10.1093/oso/9780192896742.001.0001

Biddlestone, M., Green, R., & Douglas, K. M. (2020). Cultural orientation, power, belief in conspiracy theories, and intentions to reduce the spread of COVID-19. *British Journal of Social Psychology*, 59(3), 663–673. DOI: 10.1111/bjso.12397 PMID: 32592420

Brammer, S., Branicki, L., & Linnenluecke, M. K. (2020). COVID-19, societalization, and the future of business in society. *The Academy of Management Perspectives*, 34(4), 493–507. DOI: 10.5465/amp.2019.0053

Briefing, E. February 9, 2022. https://www.mckinsey.com/business-functions/risk-and-resilience/our-insights/COVID-19-implications-for-business

Burruss, G. W., Jaynes, C. M., Moule, R. K.Jr, & Fairchild, R. E. (2021). Modeling Individual Defiance of COVID-19 Pandemic Mitigation Strategies: Insights From the Expanded Model of Deterrence and Protection Motivation Theory. *Criminal Justice and Behavior*, 48(9), 00938548211010315. DOI: 10.1177/00938548211010315

Chen, C., Frey, C. B., & Presidente, G. (2021). Culture and contagion: Individualism and compliance with COVID-19 policy. *Journal of Economic Behavior & Organization*, 190, 191–200. DOI: 10.1016/j.jebo.2021.07.026 PMID: 34566218

Cuellar, N. G., Aquino, E., Dawson, M. A., Garcia-Dia, M. J., Im, E. O., Jurado, L. F. M., Lee, Y. S., Littlejohn, S., Tom-Orme, L., & Toney, D. A. (2020). Culturally congruent health care of COVID-19 in minorities in the United States: A clinical practice paper from the National Coalition of Ethnic Minority Nurse Associations. *Journal of Transcultural Nursing*, 31(5), 434–443. DOI: 10.1177/1043659620941578 PMID: 32794439

Demirtas-Madran, H. A. (2021). Individual differences in accepting restrictions and compliance with recommended preventive behaviors for COVID-19: A discussion based on the key approaches and current research on fear appeals. *Frontiers in Psychology*, 12, 2084. https://www.frontiersin.org/articles/10.3389/fpsyg.2021.558437/pdf. DOI: 10.3389/fpsyg.2021.558437

Dillard, J. P., Tian, X., Cruz, S. M., Smith, R. A., & Shen, L. (2023). Persuasive messages, social norms, and reactance: A study of masking behavior during a COVID-19 campus health campaign. *Health Communication*, 38(7), 1338–1348. DOI: 10.1080/10410236.2021.2007579 PMID: 34879752

Donthu, N., & Gustafsson, A. (2020). Effects of COVID-19 on business and research. *Journal of Business Research*, 117, 284–289. DOI: 10.1016/j.jbusres.2020.06.008 PMID: 32536736

Dorfman, P., Javidan, M., Hanges, P., Dastmalchian, A., & House, R. (2012). GLOBE: A twenty year journey into the intriguing world of culture and leadership. *Journal of World Business*, 47(4), 504–518. DOI: 10.1016/j.jwb.2012.01.004

Elamroussi, A. (2021) Omicron surge is 'unlike anything we've ever seen,' expert says. *CNN*https://www.cnn.com/2021/12/30/health/us-coronavirus-thursday/index.html

Fan, Y., Li, X., Zhang, L., Wan, S., Zhang, L., & Zhou, F. (2022). SARS-CoV-2 Omicron variant: Recent progress and future perspectives. *Signal Transduction and Targeted Therapy*, 7(1), 1–11. DOI: 10.1038/s41392-022-00997-x PMID: 35484110

Farber, S. A. (2008). US scientists' role in the eugenics movement (1907–1939): A contemporary biologist's perspective. *Zebrafish*, 5(4), 243–245. DOI: 10.1089/zeb.2008.0576 PMID: 19133822

Figueroa, J. F., Wadhera, R. K., Lee, D., Yeh, R. W., & Sommers, B. D. (2020). Community-level factors associated with racial and ethnic disparities in COVID-19 rates in Massachusetts. *Health Affairs*, 39(11), 1984–1992. DOI: 10.1377/hlthaff.2020.01040 PMID: 32853056

Frey, C. B., Chen, C., & Presidente, G. (2020). Democracy, Culture, and Contagion. *Political Regimes and Countries Responsiveness to COVID*, 19, 20.

Galdas, P. (2017). Revisiting bias in qualitative research: Reflections on its relationship with funding and impact. *International Journal of Qualitative Methods*, 16(1), 1609406917748992. DOI: 10.1177/1609406917748992

Germani, A., Buratta, L., Delvecchio, E., & Mazzeschi, C. (2020). Emerging Adults and COVID-19: The Role of Individualism-Collectivism on Perceived Risks and Psychological Maladjustment. *International Journal of Environmental Research and Public Health*, 17(10), 3497. DOI: 10.3390/ijerph17103497 PMID: 32429536

Goldstein, I., Koijen, R. S., & Mueller, H. M. (2021). COVID-19 and its impact on financial markets and the real economy. *Review of Financial Studies*, 34(11), 5135–5148. DOI: 10.1093/rfs/hhab085

Guerra Rodriguez, Y., & Hudson, B. L. (2020). [Impact on the Hispanic Community: How Understanding Culture Can Improve Outcomes.]. *COVID*, •••, 19s.

He, W., Cai, D., Geng, G., & Klug, D. (2023). Factors influencing wearing face mask in public during COVID-19 outbreak: A qualitative study. *Disaster Medicine and Public Health Preparedness*, 17, e141. DOI: 10.1017/dmp.2022.52 PMID: 35241205

Hills, S., & Eraso, Y. (2021). Factors associated with non-adherence to social distancing rules during the COVID-19 pandemic: A logistic regression analysis. *BMC Public Health*, 21(1), 352. DOI: 10.1186/s12889-021-10379-7 PMID: 33581734

Hilts, P. J. "Experiments On Children Are Reviewed." *The New York Times*, The New York Times, (15 Apr. 1998,) https://www.nytimes.com/1998/04/15/nyregion/experiments-on-children-are-reviewed.html

Holtgrave, D. R., Barranco, M. A., Tesoriero, J. M., Blog, D. S., & Rosenberg, E. S. (2020). Assessing racial and ethnic disparities using a COVID-19 outcomes continuum for New York State. *Annals of Epidemiology*, 48, 9–14. DOI: 10.1016/j.annepidem.2020.06.010 PMID: 32723697

Huynh, T. L. D. (2020). Does culture matter social distancing under the COVID-19 pandemic? *Safety Science*, 130, 104872. https://www.ncbi.nlm.nih.gov/pmc/articles/PMC7284251/. DOI: 10.1016/j.ssci.2020.104872 PMID: 32550745

Ikuomola, O. B., & Airhihenbuwa, C. O. An Application of the PEN-3 Model. https://www.globaldiasporanews.com/an-application-of-the-pen-3-model/

Insights, H. 2021. Compare countries. Retrieved from https://www.hofstede-insights.com/product/compare-countries/

Insights, H. (n.d.). Country comparison between China and the United States. Retrieved from https://www.hofstede-insights.com/country-comparison/china,the-usa/

Izzy, S., Tahir, Z., Cote, D. J., Al Jarrah, A., Roberts, M. B., Turbett, S., Kadar, A., & Smirnakis, S.

Jiang, S., Wei, Q., & Zhang, L. (2020). Impacts of Cultural Difference on the Transmission of COVID-19Individualism vs. Collectivism (SSRN Scholarly Paper ID 3646229). Social Science Research Network. https://doi.org/DOI: 10.2139/ssrn.3646229

Jones, C., & Barajas, J. (2021). 'It feels like freedom,' 8 people describe getting their COVID-19 vaccine. *PBS*. https://www.pbs.org/newshour/health/it-feels-like-freedom-8-people-describe-getting-their-covid-19-vaccine

Jovancic, N. (2021) 5 Types of Bias in Research and How to Make Your Surveys Bias-Free.

Kalam, A., Hossain, M. A., Jahan, N., & Kim, M. (2021). New Normal and Business Sustainability in the Age of Global pandemic. *Asia-Pacific Journal of Business, 12*(1), 71-86. https://www.koreascience.or.kr/article/JAKO202113150643397.pdf

Kaliyadan, F., & Kulkarni, V. (2019). Types of variables, descriptive statistics, and sample size. *Indian Dermatology Online Journal*, 10(1), 82. DOI: 10.4103/idoj.IDOJ_468_18 PMID: 30775310

Kricorian, K., & Turner, K. (2021). COVID-19 Vaccine Acceptance and Beliefs among Black and Hispanic Americans. *PLoS One*, 16(8), e0256122. DOI: 10.1371/journal.pone.0256122 PMID: 34428216

Lagman, J. D. N., Vergara, R. J. D., & Sarmiento, P. J. D. (2021). Culture as a contributory factor in combatting the COVID-19 pandemic. *Journal of Public Health (Oxford, England)*, 43(2), e289–e290. https://academic.oup.com/jpubhealth/article/43/2/e289/6102664?login=true. DOI: 10.1093/pubmed/fdaa280 PMID: 33454771

Lead Quizzes. (July 4, 2021). https://www.leadquizzes.com/blog/types-of-bias-in-research/

Leonhardt, J. M., & Pezzuti, T. (2021). EXPRESS: Vaccination Acceptance across Cultures: The Roles of Collectivism, Empathy, and Homophily. Journal of International Marketing, 1069031X211073179.

Liguori, E. W., & Pittz, T. G. (2020). Strategies for small business: Surviving and thriving in the era of COVID-19. *Journal of the International Council for Small Business*, 1(2), 106–110. DOI: 10.1080/26437015.2020.1779538

Liu, N., Zhang, F., Wei, C., Jia, Y., Shang, Z., Sun, L., Wu, L., Sun, Z., Zhou, Y., Wang, Y., & Liu, W. (2020). Prevalence and predictors of PTSS during COVID-19 outbreak in China hardest-hit areas: Gender differences matter. *Psychiatry Research*, 112921, 112921. https://www.ncbi.nlm.nih.gov/pmc/articles/PMC7102622/. DOI: 10.1016/j.psychres.2020.112921 PMID: 32240896

Lopez, L.III, Hart, L. H.III, & Katz, M. H. (2021). Racial and Ethnic Health Disparities Related to COVID-19. *Journal of the American Medical Association*, 325(8), 719–720. DOI: 10.1001/jama.2020.26443 PMID: 33480972

Lotfollahzadeh, A., Rastgoo, L., Shirinzadeh, I., Kharghani Moghadam, S. M., & Ebrahimi, H. (2021). Investigating the compliance of COVID-19 protocols in the workplaces of Ardabil, Iran. *Work*, (Preprint), 1. 1 Jan. 2021: 1031 – 1037.7.

Luo, C., Chen, A., Cui, B., & Liao, W. (2021). Exploring public perceptions of the COVID-19 vaccine online from a cultural perspective: Semantic network analysis of two social media platforms in the United States and China. *Telematics and Informatics*, 65, 101712. https://www.ncbi.nlm.nih.gov/pmc/articles/pmc8429027/. DOI: 10.1016/j.tele.2021.101712 PMID: 34887618

M., Feske, S. K., Zafonte, R., Fishman, J. A., & El Khoury, J. (2020). Characteristics and outcomes of Hispanic and Latino patients with COVID-19 in comparison with other ethnic and racial groups. *Open Forum Infectious Diseases*, 7(10), ofaa401. DOI: 10.1093/ofid/ofaa401

Maaravi, Y., Levy, A., Gur, T., Confino, D., & Segal, S. (2021). "The Tragedy of the Commons": How Individualism and Collectivism Affected the Spread of the COVID-19 Pandemic. *Frontiers in Public Health*, 9, 627559. Advance online publication. DOI: 10.3389/fpubh.2021.627559 PMID: 33643992

Malecki, K. M., Keating, J. A., & Safdar, N. (2021). Crisis communication and public perception of COVID-19 risk in the era of social media. *Clinical Infectious Diseases*, 72(4), 697–702. https://www.ncbi.nlm.nih.gov/pmc/articles/PMC7337650/ . DOI: 10.1093/cid/ciaa758 PMID: 32544242

McClaughlin, E., Vilar-Lluch, S., Parnell, T., Knight, D., Nichele, E., Adolphs, S., Clos, J., & Schiazza, G. (2023). The reception of public health messages during the COVID-19 pandemic. *Applied Corpus Linguistics*, 3(1), 100037. DOI: 10.1016/j.acorp.2022.100037 PMID: 37521321

McLaughlin, K. (2022). *COVID-19: Implications for business February 9, 2022*. McKinsey.

Mishrif, A., & Khan, A. (2023). Technology adoption as survival strategy for small and medium enterprises during COVID-19. *Journal of Innovation and Entrepreneurship*, 12(1), 53. DOI: 10.1186/s13731-023-00317-9

Moore, K. S. (2021). The Impact of COVID-19 on the Hispanic and Latino Population: A Scoping Literature Review. *Public Health Nursing*. https://www.ncbi.nlm.nih.gov/pmc/articles/PMC8251024/

Morales, D. X., & Paat, Y. F. (2024). Hesitancy or resistance? Differential changes in COVID-19 vaccination intention between black and white Americans. *Journal of Racial and Ethnic Health Disparities*, 11(1), 23–35. DOI: 10.1007/s40615-022-01494-1 PMID: 36547772

Morales-Vives, F., Dueñas, J. M., Ferrando, P. J., Vigil-Colet, A., & Varea, M. D. (2022). Compliance with pandemic Commands Scale (COCOS): The relationship between compliance with COVID-19 measures and sociodemographic and attitudinal variables. *PLoS One*, 17(1), e0262698. https://journals.plos.org/plosone/article?id=10.1371/journal.pone.0262698. DOI: 10.1371/journal.pone.0262698 PMID: 35045104

Moreira, T., Hernandez, D. C., Scott, C. W., Murillo, R., Vaughan, E. M., & Johnston, C. A. (2018). *Susto, Coraje, y Fatalismo*: Cultural-Bound Beliefs and the Treatment of Diabetes Among Socioeconomically Disadvantaged Hispanics. *American Journal of Lifestyle Medicine*, 12(1), 30–33. DOI: 10.1177/1559827617736506 PMID: 29541008

Moyce, S., Velazquez, M., Claudio, D., Thompson, S., Metcalf, M., Aghbashian, E., Vanderwood, K., & Sisson, N. (2021). Exploring a rural Latino community's perception of the COVID-19 pandemic. *Ethnicity & Health*, 26(1), 126–138. DOI: 10.1080/13557858.2020.1838456 PMID: 33126820

Munodawafa, D., Ford, C., Oni, T., & Agyemang, C. (2020). Culture Matters in Communicating the Global Response to COVID-19. *PCD Staff, 17*, E60. https://www.cdc.gov/pcd/collections/pdf/US-Public-Health-Response-to-COVID-19.pdf#page=30

Murdoch, M., Simon, A. B., Polusny, M. A., Bangerter, A. K., Grill, J. P., Noorbaloochi, S., & Partin, M. R. (2014). Impact of different privacy conditions and incentives on survey response rate, participant representativeness, and disclosure of sensitive information: A randomized controlled trial. *BMC Medical Research Methodology*, 14(1), 1–11. https://bmcmedresmethodol.biomedcentral.com/articles/10.1186/1471-2288-14-90he. DOI: 10.1186/1471-2288-14-90 PMID: 25027174

Noe-Bustamante, L., Krogstad, J. M., & Lopez, M. H. (2021, July 16). *For U.S. Latinos, COVID-19 Has Taken a Personal and Financial Toll.* Race & Ethnicity. https://www.pewresearch.org/race-ethnicity/2021/07/15/for-u-s-latinos-COVID-19-has-taken-a-personal-and-financial-toll/

O'Brien, M., & Shea, J. (2011). Disparities in patient satisfaction among Hispanics: The role of language preference. *Journal of Immigrant and Minority Health*, 13(2), 408–412. DOI: 10.1007/s10903-009-9275-2 PMID: 19626440

O'Neil, A. (2021, December 10). Latin America- Statistics & Facts. Statista. Retrieved from https://www.statista.com/topics/3287/latin-america/

Oetzel, J., De Vargas, F., Ginossar, T., & Sanchez, C. (2007). Hispanic women's preferences for breast health information: Subjective cultural influences on source, message, and channel. *Health Communication*, 21(3), 223–233. DOI: 10.1080/10410230701307550 PMID: 17567254

Ogunwole, S., Rabe, M., Roberts, A., & Caplan, Z. (2021). Population Under Age 18 Declined

Olagunju, A. (2021). The Impact of Race and History on COVID Vaccination Acceptance.

Ortega, P., Shin, T. M., Pérez-Cordón, C., & Martínez, G. A. (2020). Virtual Medical Spanish Education at the *Corazón* of Hispanic/Latinx Health During COVID-19. *Medical Science Educator*, 30(4), 1661–1666. DOI: 10.1007/s40670-020-01058-0 PMID: 32839674

Padilla, M. E., Frietze, G., Shenberger-Trujillo, J. M., Carrillo, M., & Loya, A. M. (2020). Influenza and intentions to vaccinate in an underserved Hispanic population: The role of theoretically derived constructs. *Journal of Pharmacy Practice*, 33(3), 326–333. DOI: 10.1177/0897190018810595 PMID: 30453817

Pérez, E., HyunJeong Lee, J., Oaxaca Carrasco, A. L., Matthews, C., & Ritsema, M. (2023). Unexpected, but consistent and pre-registered: Experimental evidence on interview language and Latino views of COVID-19. *Research & Politics*, 10(2), 20531680231168736. DOI: 10.1177/20531680231168736

Perreira, K. M., Allen, C. D., & Oberlander, J. (2021). Access to health insurance and health care for Hispanic children in the United States. *The Annals of the American Academy of Political and Social Science*, 696(1), 223–244. DOI: 10.1177/00027162211050007 PMID: 37123537

Peteet, B., Belliard, J. C., Abdul-Mutakabbir, J., Casey, S., & Simmons, K. (2021). Community-academic partnerships to reduce COVID-19 vaccine hesitancy in minoritized communities. *EClinicalMedicine*, 34, 34. DOI: 10.1016/j.eclinm.2021.100834 PMID: 33880441

Plohl, N., & Musil, B. (2021). Modeling compliance with COVID-19 prevention guidelines: The critical role of trust in science. *Psychology Health and Medicine*, 26(1), 1–12. DOI: 10.1080/13548506.2020.1772988 PMID: 32479113

Powe, M. (2020) New Survey Results: How Small Businesses are Managing the Pandemic and Social Unrest. *Main Street America*. Retrieved from https://www.mainstreet.org/blogs/national-main-street-center/2020/09/28/new-survey-results-how-small-businesses-are-managi

Purwaningsih, S., & Pernamasari, R. (2024). Financial Performance Analysis in Predicting Corporate Bankruptcy: Before and During the COVID-19 Pandemic. *Journal of Business and Management*, 2(2), 36–48.

Rajkumar, R. P. (2021). The relationship between measures of individualism and collectivism and the impact of COVID-19 across nations. *Public Health in Practice (Oxford, England)*, 2, 100143. https://www.sciencedirect.com/science/article/pii/S2666535221000689. DOI: 10.1016/j.puhip.2021.100143 PMID: 34494009

Schneid, R., Perez-Moreno, H., & Branson-Potts, H. (2022). 'I'm over it.' Many in L.A. shrug off COVID-19 wave despite super-infectious subvariants. *L.A. Times*. https://www.latimes.com/california/story/2022-07-21/la-me-covid-mood-mask-mandate

Shelus, V. S., Frank, S. C., Lazard, A. J., Higgins, I. C., Pulido, M., Richter, A. P. C., Vandegrift, S. M., Vereen, R. N., Ribisl, K. M., & Hall, M. G. (2020). Motivations and barriers for the use of face coverings during the COVID-19 pandemic: Messaging insights from focus groups. *International Journal of Environmental Research and Public Health*, 17(24), 9298. https://www.mdpi.com/1660-4601/17/24/9298/pdf. DOI: 10.3390/ijerph17249298 PMID: 33322672

Sitto, K., & Lubinga, E. (2020). A Disease of Privilege? Social Representations in Online Media about COVID-19 among South Africans during Lockdown. *Papers on Social Representations*, 29(2), 6–1.

Smith, A., & Nagele-Piazza, L. "Employers React to Workers Who Refuse a COVID-19 Vaccination." *SHRM*, SHRM, (3 Feb. 2022), https://www.shrm.org/resourcesandtools/legal-and-compliance/employment-law/pages/if-workers-refuse-a-covid-19-vaccination.aspx

Statista (2022) Share of small businesses that have closed due to the COVID-19 pandemic in the United States from October 2020 to January 2022. Statista.com. https://www.statista.com/statistics/1222202/us-COVID-19-closings-small-businesses/

Stephenson, J. (2021, January). While Waiting for Vaccination, Most US Adults Will Wear Masks, Follow Social Distancing Guidelines, Poll Finds. In *JAMA Health Forum* (Vol. 2, No. 1, pp. e210024-e210024). American Medical Association.https://jamanetwork.com/journals/jama-health-forum/fullarticle/2775257

Survey Monkey. (2021). Making Responses Anonymous. (2021). *Survey Monkey.* Retrieved from https://help.surveymonkey.com/articles/en_US/kb/How-do-I-make-surveys-anonymous

Survey Monkey. (2022). Sample Size Calculator. *Survey Monkey*. Retrieved from https://www.surveymonkey.com/mp/sample-size-calculator/

Survey Monkey. (n.d.) Non-Probability Sampling. *Survey Monkey.* https://www.surveymonkey.com/mp/non-probability-sampling

Tai, D. B. G., Shah, A., Doubeni, C. A., Sia, I. G., & Wieland, M. L. (2020). The disproportionate impact of COVID-19 on racial and ethnic minorities in the United States. *Clinical Infectious Diseases*, 72(4), 703–706. DOI: 10.1093/cid/ciaa815 PMID: 32562416

Theofanidis, D., & Fountouki, A. (2018). Limitations and delimitations in the research process. Perioperative nursing, 7(3), 155-163. https://www.spnj.gr/articlefiles/volume7_issue3/pn_sep_73_155_162b.pdf

Tirupathi, R., Muradova, V., Shekhar, R., Salim, S. A., Al-Tawfiq, J. A., & Palabindala, V. (2020). COVID-19 disparity among racial and ethnic minorities in the US: A cross sectional analysis. *Travel Medicine and Infectious Disease*, 38, 101904. https://www.ncbi.nlm.nih.gov/pmc/articles/PMC7603979/. DOI: 10.1016/j.tmaid.2020.101904 PMID: 33137491

Ulrich, A., Thompson, B., Livaudais, J. C., Espinoza, N., Cordova, A., & Coronado, G. D. (2013). Issues in biomedical research: What do Hispanics think? *American Journal of Health Behavior*, 37(1), 80–85. DOI: 10.5993/AJHB.37.1.9 PMID: 22943104

Uriel, E. (2013). Hypothesis testing in the multiple regression model. Introduction to Econometrics, 1-49. https://d1wqtxts1xzle7.cloudfront.net/34695537/4_Hypothesis_testing_in_the_multiple_regression_model-with-cover-page-v2.pdf?Expires=1643629129&Signature=RExpdIJ0dE2MAqguFrUwoxWAw9WXvfG-3~PGfl-ddw~91sbjF5tprG6STN0pNCNTIpD7uO5qvHR184rs6LTkp~xB5886sMY98GsielVy7L88cA4iiDfj94aPfpX9Gil3Q5Mp9xqiMcOHdPZOYguWZI3CYt0FjQ8otQRlsShII4t8VNlWZnaAAweuKlwnTlRhg3bYiZ1TIaI9Y3Kg9eh4jJLyAmmZMy1393B1kbKTeSJD56yiUI1LEHxV9iYIomA1Q2G85YTFbmGo3bUtj2iiEXfimT1IJRc8V0qsWnVvwgcGfA5aebHjrpQIzqNHIlw66q7qCwZyFUQB0aHZ7Ao1vA__&Key-Pair-Id=APKAJLOHF5GGSLRBV4ZA

USCensusBureau. (2020). Quick Facts United States._https://www.census.gov/quickfacts/fact/table/US/RHI725219

Van Wees, J. D., van der Kuip, M., Osinga, S., Keijser, B., van Westerloo, D., Hanegraaf, M., & van Furth, M. T. (2021). SIR model for assessing the impact of the advent of Omicron and mitigating measures on infection pressure and hospitalization needs. *MedRxIv*. DOI: 10.1101/2021.12.25.21268394

Webster, G. D., Howell, J. L., Losee, J. E., Mahar, E. A., & Wongsomboon, V. (2021). Culture, COVID-19, and collectivism: A paradox of American exceptionalism? *Personality and Individual Differences*, 178, 110853. DOI: 10.1016/j.paid.2021.110853 PMID: 36540785

Whembolua, G. L., & Tshiswaka, D. I. (2020). Public trust in the time of the Coronavirus Disease 2019 (COVID-19): the case of the DR Congo. *The Pan African Medical Journal, 35*(Suppl 2).

Wood, S., & Schulman, K. (2021). Beyond politics—Promoting COVID-19 vaccination in the United States. *The New England Journal of Medicine*, 384(7), e23. https://www.nejm.org/doi/full/10.1056/nejmms2033790. DOI: 10.1056/NEJMms2033790 PMID: 33406324

Wu, M. (2006). Hofstede's cultural dimensions 30 years later: A study of Taiwan and the United States. *Intercultural Communication Studies*, 15(1), 33. http://www.homecont.ro/pitagora/Hofstede-04-Ming-Yi-Wu.pdf

Yuan, S., & Chu, H. (2021). Vaccine for yourself, your community, or your country? Examining audiences' response to distance framing of COVID-19 vaccine messages. *Patient Education and Counseling*. https://www.ncbi.nlm.nih.gov/pmc/articles/PMC8384529/

KEY TERMS AND DEFINITIONS

COVID-19 Compliance (or Compliance): Cooperation and enforcement of the COVID-19 mitigation protocols required by governments and organizations throughout the United States.

COVID-19 Mitigation Protocols: Governmental and organizational requirements that people adhere to COVID-19 prevention mandates such as including social distancing, masking, and mandating customers' proof of vaccination to enter the businesses.

Herd Immunity: The idea that once a high enough portion of the population had a disease, immunity would result for that population.

Hispanic and Latino Participants: For the study and chapter, Hispanic and Latino participants were residents of the United States.

Individualistic or Collectivistic Orientation: Refers to Hofstede's cultural dimension describing whether the culture tends to more highly value individual or communal rights.

Messaging Strategies: Refers to the different types of digital communications methods used to communicate about COVID-19.

Protective Protocols: Synonymous with COVID-19 mitigation protocols.

Chapter 8
Crisis Communication Strategies in the Age of Social Media:
The Case of Volunteered Geographic Information

Munir Ahmad
https://orcid.org/0000-0003-4836-6151
Survey of Pakistan, Pakistan

Muhammad Farrukh
https://orcid.org/0000-0002-0741-5394
The University of Lahore, Pakistan

H. M. Atif Wafik
https://orcid.org/0000-0002-6032-3800
University of Scholars, Bangladesh

Popoola Adeolu Gabriel
Federal College of Education (Special), Nigeria

ABSTRACT

Crisis communication in today's digital age, shaped by social media and Volunteered Geographic Information (VGI), has revolutionized response strategies. Social media-based VGI can enable rapid dissemination of information and foster two-way communication, empowering communities to report incidents, seek help, and share experiences in real-time. This transparency builds trust between authorities and the public, dispels rumors, and ensures adherence to safety measures. VGI can

DOI: 10.4018/979-8-3693-3579-6.ch008

amplify the voices of those affected, humanizing crises and mobilizing support for relief efforts. However, using VGI requires careful verification and ethical considerations like privacy protection and combating misinformation. Techniques such as cross-referencing with credible sources and utilizing geolocation data enhance reliability. Collaborative platforms can streamline information exchange, facilitating efficient crisis response and volunteer coordination. Actionable updates with visual aids can improve public engagement.

INTRODUCTION

Social media presents a new landscape that has revolutionized how information is shared during crises (Bukar et al., 2020). The earlier form of media that was widely used during crises were television, radio, and newspapers. However, social media has made this process open to anyone who owns a smartphone and has access to the internet to share information with the public in real-time. Some of the tools that seem to be in operational use include micro-blogging sites such as X (formally Twitter), and Facebook, and map-based crisis applications such as OpenStreetMap among others. This shift has led to what has been termed as Volunteered Geographic Information (VGI) which is information on geographical location by common citizens (Goodchild, 2007). Experience has shown that VGI is an effective technology that can be applied for crisis response and its management (Ahmad, 2023, 2024; Anderson-Tarver, 2015; Haworth et al., 2012; Minghini et al., 2017).

VGI through social media platforms offers the possibility of frequent information dissemination and frequency of updates is particularly important in the course of a crisis as shown in Figure 1. Through social media channels such as Twitter and Facebook, and specialized crisis mapping applications it is possible to get frequent updates and may even surpass traditional media in speed. Figure 2 shows key social media platforms that can be used for crisis communication. Citizens on the scene may share what they witness, post photos and videos, and give an on-the-spot description of the situation in progress which is broader and more real-time. Secondly, the approach of VGI is decentralized, which means that more people can share their opinions and ideas about the crisis. This inclusivity also ensures that the information received is not only from the usual channels but also from people who are directly experiencing the crisis. Personal testimonies can reveal facets of the story that might not be covered by television or newspapers, which creates a more holistic picture. Furthermore, stand-alone crisis mapping applications such as OpenStreetMap or Ushahidi can help to develop dynamic maps that provide further insight into the development of the event (Jessica Pechmann & Claudio de los Reyes, 2023; Pánek et al., 2017). These maps can also display the positions of significant occurrences, resources, and secure zones which all the responding personnel and citizens would benefit from. The incorporation of social media data into these maps further improves the efficiency of the maps.

Figure 1. Crisis communication process using social media

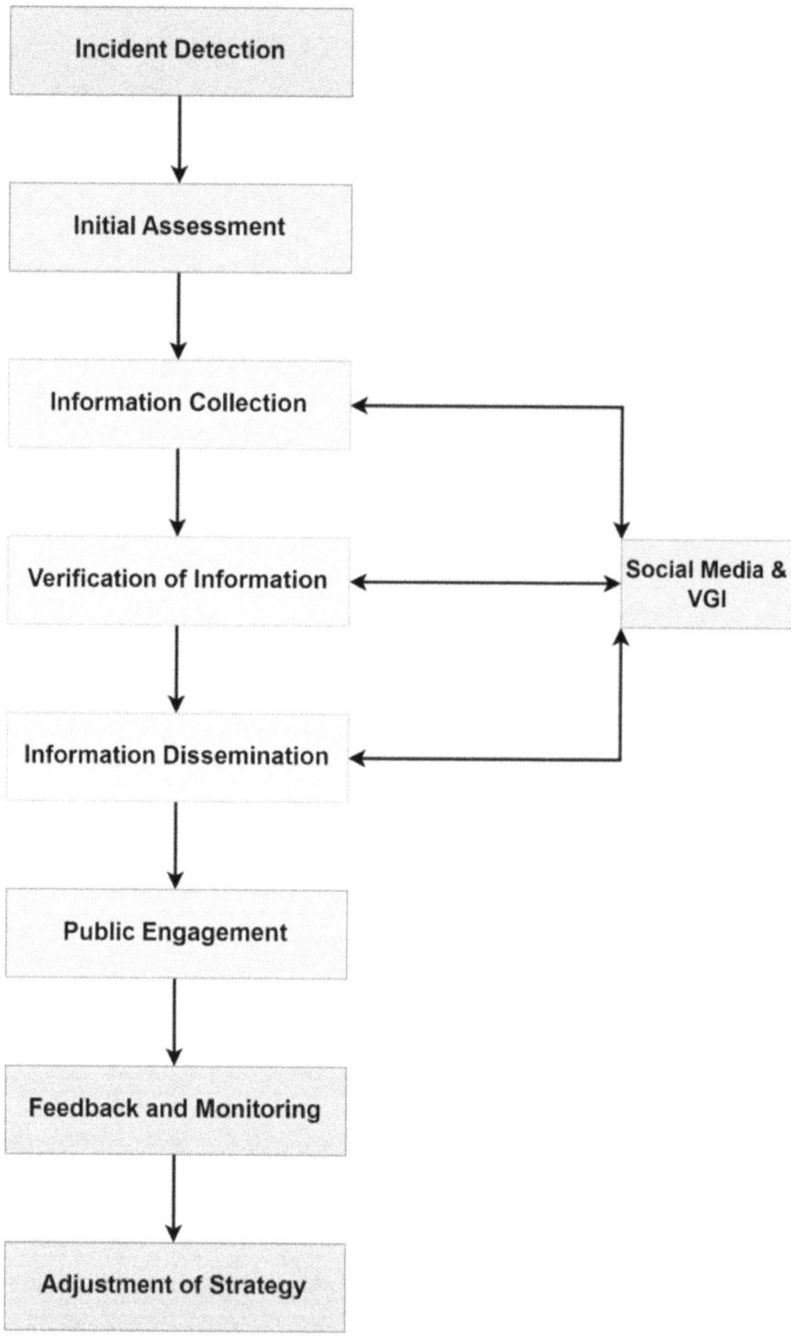

However, the issues that accompany the increase of VGI have also emerged. Extraordinary amounts of information can be created during a crisis situation, which makes it challenging to distinguish relevant and reliable information for response. Based on the fact that VGI is inherently decentralized, information is personalized and comes from multiple sources most of which can be considered unreliable or uncreditable. Under these circumstances, the content of such information is not always accurate and may contain distortions, which can result in the sharing of erroneous data.

Figure 2. Key social media platforms for crisis communication

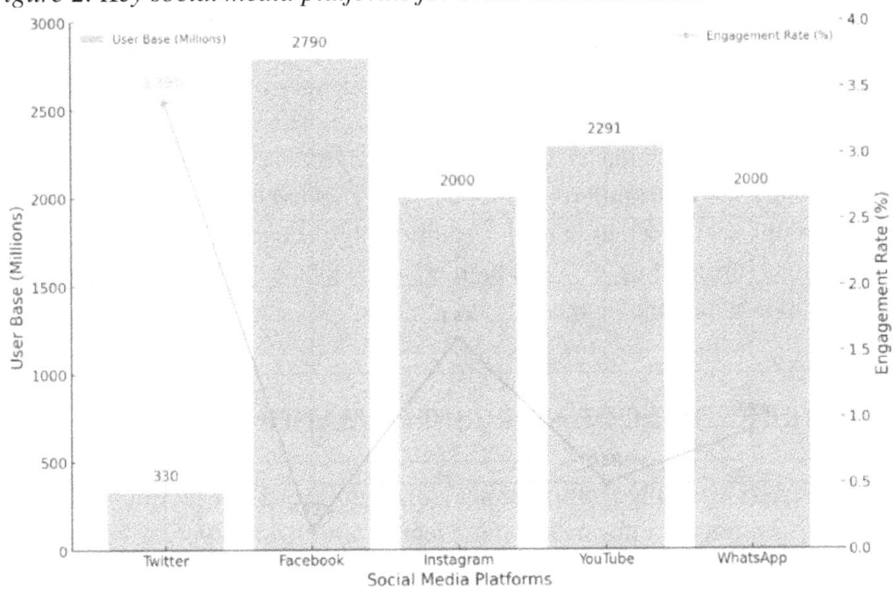

Some of the measures that have been implemented to tackle the above challenges are strict verification procedures that can be used to determine the validity of the information passing through the system. Many techniques and procedures have been designed to reduce noise and work out the sources that can be trusted to cope with such a huge amount of information. Nevertheless, human intervention is still crucial here to ensure nuanced judgment and understanding within the particular context. Crisis response teams utilize real-time monitoring and analyses in their operations as part of their daily routine practice. These teams have what it takes to monitor and assess constantly information flow in VGI so they can be able to counter negative information and spread relevant and accurate information in the shortest time.

There is also a very strong focus on the interaction between official agencies and organizations, non-governmental agencies, and the public (Haupt, 2021). Well-developed communication lines and reliable networks make the information flowing between nodes more reliable (Firestone, 2020). This problem-solving perspective allows multiple stakeholders, who can provide and use the VGI, to contribute and improve the outcomes of the crisis responses. Another type of existing strategy is public education campaigns. These campaigns are geared towards informing the public on the need to practice healthy skepticism before reposting or passing on any information, especially through social media platforms. This serves to reduce the prevalence of fake news and encourages enhanced awareness in the population.

Understanding how to leverage VGI while mitigating its potential pitfalls is crucial for successful crisis communication. This chapter aim to explore this critical intersection, examining the opportunities and challenges presented by VGI in the age of social media. The next section explores the existing strategies for utilizing VGI in crisis communication by describing methods for verifying information, fostering collaboration with online communities, and effectively disseminating official updates through social media channels. Additionally, the chapter address the ethical considerations surrounding VGI and highlight best practices for responsible crisis communication in the digital age.

VOLUNTEERED GEOGRAPHIC INFORMATION

Volunteered Geographic Information (VGI) is the broad category of voluntarily contributed geographic information (Goodchild, 2007). This type of information emerges from people who consciously contribute their insights and even real-time logs of what they see around them. VGI could be made of various types and formats and all of them prove useful in the time of crises and other momentous occurrences. There are various benefits of VGI with text description being the most common form of volunteered geographic information. This can comprise tweets, status updates, blog posts, and comments that people post on social media sites (Adelfio et al., 2020; Amirkhanyan & Meinel, 2016; Issa et al., 2017; Nguyen et al., 2016; Wu et al., 2018). While text-based VGI can be considered incomplete when compared with other forms of media like images and videos, it may contain firsthand reports of an event, descriptions of a venue, or situational reports that may be invaluable in constructing an event timeline.

Photos make up another very important element of VGI (Antoniou et al., 2010; Gunter & Önder, 2021; Yuan & Medel, 2016). Some of what is happening in different parts of the world can be seen through photos that common people post on social media. They can give information on the state of the infrastructure, the magnitude of disasters in the region, or the position of rescuers or threats. Photos can be provided with the geographical reference of the location in which the photos were captured, which makes it possible to incorporate this data into any emergency response measures. Video is an especially powerful type of additional information because it gives more dynamic and detailed coverage than words and pictures can fail to deliver. They can illustrate how one thing leads to another, record people's immediate response and depict and setting or mood. Live videos posted on sites like YouTube, Facebook, and X (formally Twitter) can sometimes be of much help in analyzing complicated situations, and give an approximate measure of the requirements of the area that has been affected.

Information about the location linked to shared social materials is the other significant component of VGI. It is also known that many social networks let users put their posts with geographic parameters, which can then be used for mapping and tracking of events. Such spatial information enables the responders to then locate and differentiate between the hot areas, observe the trends and rates of crises, and even distribute resources. Besides these formats, VGI may encompass data drawn from niche apps and services primarily geared towards supporting crisis mapping and operations. Figure 3 depicts information flow in a VGI system.

Figure 3. Information flow in a VGI system

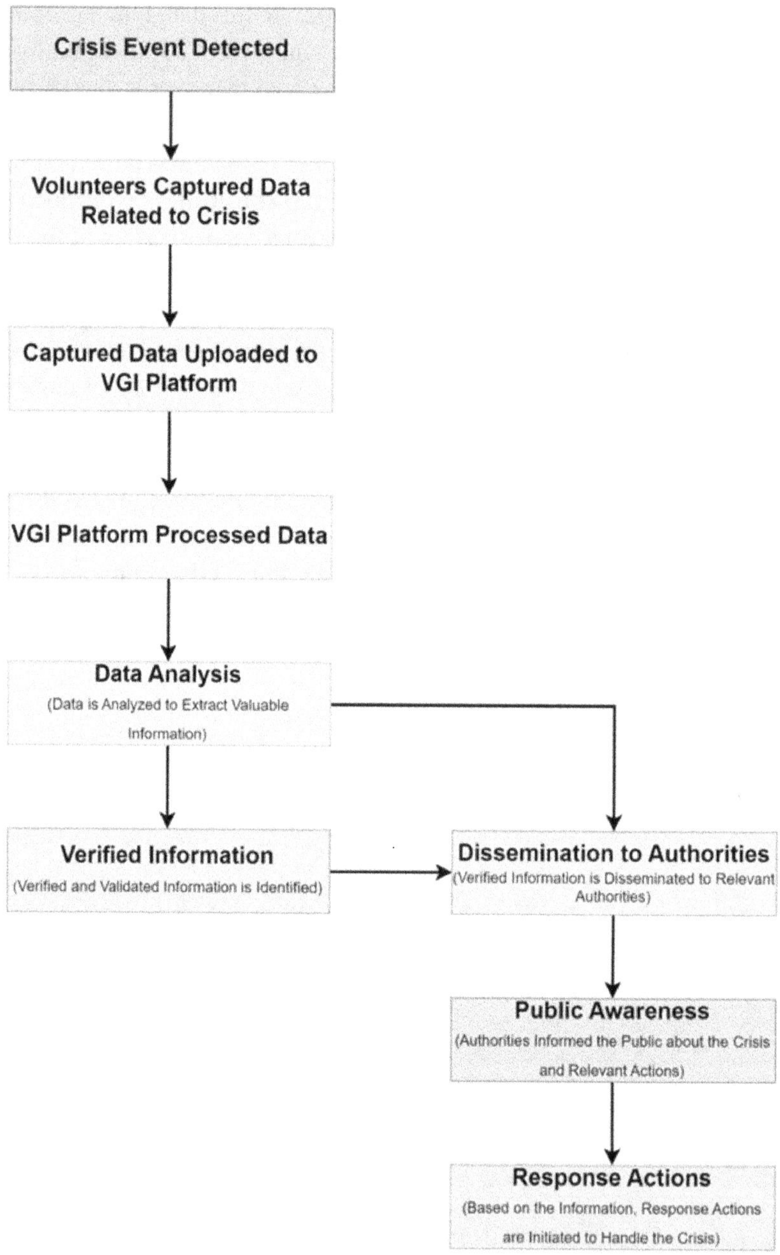

VGI has emerged as a powerful tool in crisis response, transforming the way information is gathered and disseminated during critical events. The concept of VGI can be traced back to the early days of citizen science, where individuals contributed observations to scientific studies. However, the rise of social media and smartphones has significantly accelerated the growth and impact of VGI. Today, with internet connectivity readily available and location-based services integrated into most mobile devices, anyone can become a VGI contributor. This real-time flow of information has revolutionized crisis communication and disaster response efforts.

VGI comes from a variety of sources, including social media platforms Twitter, Facebook, and other social media platforms are major sources of VGI. Users can share eyewitness accounts, photos, and videos tagged with their location, providing valuable insights into unfolding events. Moreover, several mobile applications are specifically designed for VGI collection during crises. These apps allow users to report damage, request assistance, and share critical information with response teams. Websites like Ushahidi and CrisisMappers allow users to submit VGI reports via web forms or SMS messages. These platforms can then aggregate and visualize the data, providing a comprehensive overview of the situation.

VGI offers a multitude of benefits for crisis response and management. VGI has the ability to provide real-time updates on the situation on the ground, allowing response teams to make informed decisions quickly. The vast amount of data from VGI helps create a more comprehensive picture of the crisis, encompassing affected areas, resource needs, and emerging issues. Moreover, VGI facilitates communication and coordination between different response agencies, volunteers, and affected communities. VGI can also empower the public to become active participants in crisis response, fostering a sense of community and collective action (Ahmad, 2023, 2024).

Despite its advantages, VGI also presents some challenges. VGI reports can be inaccurate or misleading. Careful verification procedures are necessary to ensure the credibility of the information. Moreover, as already described the sheer volume of VGI data can be overwhelming, making it difficult to identify reliable information and prioritize response efforts. Further, sharing VGI can raise privacy concerns. Strategies are needed to anonymize data and protect the privacy of individuals affected by the crisis. Not everyone has access to the internet or smartphones, potentially limiting the reach of VGI and exacerbating existing inequalities during a crisis.

SIGNIFICANCE OF CRISIS COMMUNICATION

The significance of crisis communication in contemporary society can be understood through several key points.

Speed and Reach

The speed and reach of social media play a crucial role in crisis communication. Platforms like Twitter and Facebook enable the rapid dissemination of information to a vast audience, allowing authorities to swiftly inform and instruct affected populations. For instance, during natural disasters like hurricanes, authorities can use social media to provide real-time updates on storm paths, evacuation orders, and shelter locations. The immediacy of this communication can be lifesaving, as it ensures that people receive critical information quickly enough to take necessary precautions.

A poignant example of this was seen during Hurricane Harvey in 2017. The Houston Police Department and other local agencies used Twitter to provide constant updates on flooding, rescue operations, and safety advisories. This real-time communication helped coordinate relief efforts and keep residents informed, significantly contributing to the overall response strategy (Zhou et al., 2022; Zou et al., 2023). Similarly, during the COVID-19 pandemic, health organizations like the World Health Organization (WHO) and the Centers for Disease Control and Prevention (CDC) used social media to share timely updates about virus transmission, safety guidelines, and vaccine availability, reaching millions of people globally and influencing public behavior in real-time (Ma et al., 2023; Yin et al., 2024).

Engagement and Empowerment

Another critical aspect of crisis communication is engagement and empowerment. Social media platforms facilitate two-way communication, allowing individuals not only to receive information but also to share their experiences, report incidents, and seek assistance in real-time. This level of engagement empowers communities to play an active role in the response effort, turning passive recipients of information into active participants. For example, during the 2010 Haiti earthquake, social media and SMS services enabled people affected by the disaster to report their locations and conditions directly to relief organizations (Fauzi, 2023; Li et al., 2023). Platforms like Twitter and Facebook were used to crowdsource real-time data on the most urgent needs and areas requiring immediate assistance. This grassroots-level input significantly enhanced the situational awareness of decision-makers and emergency responders, enabling them to allocate resources more effectively and prioritize interventions based on actual conditions reported by those on the ground.

Similarly, during the Australian bushfires in 2019-2020, social media played a crucial role in crisis communication. Residents used platforms like Facebook and Twitter to share updates about fire locations, road closures, and safe evacuation routes (Gardiner et al., 2023; Zander et al., 2023). This real-time information sharing helped authorities and emergency services to stay informed about rapidly changing conditions and to communicate directly with the public, offering timely advice and instructions. The ability of individuals to share their firsthand experiences and observations enriched the overall understanding of the situation, allowing for a more responsive and adaptive crisis management approach. This interactive dynamic of social media helps bridge the gap between authorities and the public, fostering a collaborative environment where everyone contributes to crisis management. By actively engaging with the public, authorities can build stronger relationships and trust within the community. For instance, police departments and emergency management agencies often use social media to answer questions, dispel rumors, and provide personalized responses to individuals seeking help. This direct interaction not only helps in disseminating accurate information but also reassures the public that their concerns are being heard and addressed.

Transparency and Trust

Transparency and trust are fundamental components of effective crisis communication. Transparent communication practices build and maintain trust between authorities and the public, especially during crises when accurate information is paramount. Social media plays a crucial role in facilitating this transparency by enabling real-time updates, directly addressing public concerns, and clarifying any misconceptions. When authorities communicate openly and honestly, it reassures the public and enhances the credibility of the information being shared. This transparency helps to reduce uncertainty and fear, which are often prevalent during emergencies. For example, during the early stages of the COVID-19 pandemic, transparent communication from public health officials about the nature of the virus, safety protocols, and the development of vaccines was critical. Regular updates through social media channels from organizations like the World Health Organization (WHO) and the Centers for Disease Control and Prevention (CDC) helped to inform the public and build trust in the recommended guidelines and precautions.

Social media platforms allow for immediate and direct interaction with the public, making it easier for authorities to address concerns and answer questions in real-time. For instance, during natural disasters like hurricanes or wildfires, emergency management agencies can use social media to provide up-to-the-minute information on evacuation orders, shelter availability, and safety measures. By engaging with the public directly, authorities can dispel rumors and misinformation that may arise during the chaos of a crisis. This direct line of communication helps to establish a reliable source of truth, which is essential for maintaining public trust. A good example of this practice was seen during the 2011 earthquake and tsunami in Japan (Siawsh et al., 2023). The Japanese government and various international organizations used social media to provide continuous updates on the disaster response and radiation levels from the Fukushima nuclear plant. This transparency helped to alleviate public fears and provided a clear, consistent source of information amidst widespread confusion and concern.

In addition, transparent communication can help to humanize authorities and make them more relatable to the public. By showing empathy, understanding, and a willingness to engage openly, officials can foster a stronger connection with the community. This human element was evident during the Australian bushfires, where local leaders and firefighters used social media to share personal stories, express solidarity with affected communities, and provide heartfelt reassurance (Gardiner et al., 2023; Zander et al., 2023). This approach not only conveyed important information but also built a sense of unity and shared purpose. Trust is essential for ensuring that people follow guidelines and take necessary precautions during emergencies. When the public trusts the information provided by authorities, they are more likely to adhere to safety recommendations and cooperate with emergency measures. For example, trust in public health advisories was crucial during the Ebola outbreak in West Africa, where misinformation and mistrust initially hampered response efforts (Cénat et al., 2023). Transparent, consistent communication from health officials eventually helped to build trust and improve compliance with health directives, contributing to the containment of the virus.

Amplification of Voices

Furthermore, social media serves as a powerful amplifier of voices, providing a platform for affected individuals to share their stories and needs with a broader audience. During crises, whether natural disasters, pandemics, or social upheavals, social media enables those directly impacted to convey their experiences in real time. For example, in the aftermath of Hurricane Katrina in 2005, survivors used platforms like Twitter and Facebook to share harrowing tales of evacuation, loss, and resilience. These personal narratives not only raised awareness but also humanized the crisis, prompting widespread empathy and support from around the world. The amplification of voices on social media goes beyond personal stories; it also facilitates the mobilization of communities and organizations for collective action. During the Syrian refugee crisis, for instance, social media played a pivotal role in galvanizing global humanitarian efforts (Crisp, 2024). Images, videos, and firsthand accounts shared on platforms like Instagram and YouTube highlighted the plight of refugees, sparking public outrage and prompting governments and NGOs to increase aid and support initiatives.

Moreover, social media amplifies advocacy efforts by allowing affected individuals and advocacy groups to reach a wider audience with their messages. Activists fighting for social justice, environmental protection, or human rights utilize platforms such as Twitter campaigns and Facebook groups to raise awareness, organize protests, and advocate for policy change. This amplification of voices has been instrumental in driving social movements, where social media catalyzed widespread awareness, solidarity, and action against systemic injustices. In addition to amplifying individual voices and advocacy efforts, social media enhances community resilience by fostering solidarity among those affected by crises. Online platforms enable people to connect, share resources, and offer support in times of need. For example, during the COVID-19 pandemic, community groups on Facebook emerged to coordinate mask-making efforts, deliver groceries to vulnerable populations, and provide emotional support to isolated individuals. This grassroots mobilization demonstrated the power of social media in uniting communities and leveraging collective strength to address challenges collaboratively.

STRATEGIES FOR UTILIZING VGI IN CRISIS COMMUNICATION

This section provides details on specific tactics for harnessing the power of VGI during a crisis.

Verification and Validation

The deluge of information during a crisis can be overwhelming, and VGI is no exception. Given the potential for misinformation, robust methods for verifying the accuracy and credibility of VGI reports are essential. This section provides a brief on key techniques crisis communication teams can employ to ensure they are basing their actions on reliable information.

Cross-Referencing Information from Multiple Sources

Effective verification involves triangulating information from various sources. Emergency response agencies, government bodies, and established humanitarian organizations are credible sources of information during a crisis. It is required to regularly check their websites and social media channels for official updates and compare them with VGI reports. Moreover, reputable news organizations with a track record of fact-checking and on-the-ground reporting can provide valuable insights. Looking for news articles that corroborate VGI reports or offer additional details about the unfolding situation can be beneficial in this context. Further, many journalists and verified accounts from official organizations utilize social media platforms to share updates during crises. It is important to cross-reference VGI reports with information shared by these verified accounts to assess credibility. If a particular VGI report seems isolated and contradicts the overall narrative emerging on social media, it might require further investigation.

Utilizing Geolocation Data

Many social media platforms and VGI applications embed geolocation data with user-generated content. This data can be a powerful tool for verification to support effective crisis communication. In this context, analyze the geotagged location of a VGI report in relation to the reported event. For instance, during a flood event, reports with photos geotagged near overflowing rivers hold more weight than those originating from faraway locations (Antoniou et al., 2010; Fan et al., 2020; Sathianarayanan et al., 2024). Consider the timing of the geolocation data. Does the timestamp on the VGI report align with the reported event? Inconsistencies in timing can raise red flags, especially for reports claiming to depict events that are still unfolding. Moreover, utilize publicly available satellite imagery or real-time mapping tools to verify the visual content of VGI reports, particularly those containing photographs or videos. This can help identify inconsistencies or potential manipulations.

Collaborating with Trusted Online Communities

Establishing online communities with a history of responsible information sharing during crises can be invaluable partners in the verification process. Look for online communities with a strong track record of fact-checking information and fostering constructive dialogue during previous crisis events (HOTOSM, 2023a, 2023b). Often, these communities will have established protocols for verifying user-generated content. Moreover, collaborating with these communities to develop a system for flagging potentially inaccurate VGI reports and working together to verify the legitimacy of information. This can involve cross-referencing reports with local knowledge and expertise within the community. The collective knowledge and experience of online communities can be a powerful asset. By harnessing the collective intelligence of these groups, crisis communication teams can gain a more nuanced understanding of the situation on the ground and identify potential blind spots in VGI data.

Fostering Collaboration

Social media offers a powerful platform for forging connections and empowering the public during a crisis. This section explores strategies for fostering collaboration with online communities, transforming them from passive observers to active participants in crisis response.

Establishing Dedicated Hashtags

The hashtag facilitates the easy tracking of relevant information. By searching the hashtag, crisis response teams can quickly identify updates, requests for assistance, and emerging issues. Further, the hashtag acts as a communication hub for affected individuals, volunteers, and official channels. Individuals can share updates, volunteers can offer assistance, and official channels can disseminate critical information in a centralized location. In this context, a dedicated hashtag fosters a sense of community and shared purpose among those involved. It allows individuals to connect with others facing similar challenges and creates a platform for mutual support and collaboration.

Creating Online Forums for Information Exchange

Utilizing existing social media platforms or establishing dedicated online forums to create a central hub for information exchange can boost crisis communication. The forum allows the public to share critical updates, including eyewitness accounts, photos, and videos from the affected area. This provides valuable real-time information for crisis response teams. Individuals in need can utilize the forum to request specific types of assistance, such as medical aid, search and rescue, or temporary shelter. This allows for targeted resource allocation and ensures crucial support reaches those who need it most. Crisis communication teams can leverage the forum to disseminate official updates, safety instructions, and resource information. This ensures the public has access to accurate and timely information directly from official sources.

Leveraging the Expertise of Volunteers

Online communities often anchorage individuals with a wealth of valuable skills and knowledge. Crisis communication teams can tap into this expertise through various means to strengthen crisis communication. They can utilize social media platforms to recruit volunteers with specific skill sets relevant to the crisis. This could include doctors, engineers, translators, or individuals with experience in crisis response. Match volunteers with tasks that align with their expertise. For instance, data analysts can assist with processing VGI reports, translators can bridge communication gaps in multilingual communities, and individuals with experience in crisis response can help coordinate virtual volunteer networks. They can empower volunteers by providing clear instructions, training materials, and ongoing support. Publicly acknowledge the contributions of volunteers to foster a sense of ownership and encourage continued participation.

Effective Dissemination of Official Updates

In the midst of a crisis, timely and accurate communication from official sources is critical for managing public anxieties, ensuring safety, and facilitating a coordinated response. This section explores practices for disseminating official updates through social media channels.

Clear and Concise Messaging

Crafting messages that are easy to understand, even for audiences under duress can help in effective crisis communication. Avoid technical jargon and acronyms. Use simple, direct language that everyone can understand. Focus on providing clear instructions and actionable information. Tell people what they need to do to stay safe, seek assistance, or access resources. Utilize bullet points and numbered lists to break down complex information into easily digestible chunks.

Utilizing Visuals and Data Effectively

People process visuals more efficiently than text. Therefore, utilize clear and concise maps to illustrate the affected areas, evacuation routes, or shelter locations. Infographics can effectively communicate complex data points related to the crisis, such as the number of casualties or the spread of a disease. When possible, share high-quality photos and videos that accurately depict the situation on the ground. This can help people understand the severity of the crisis and the response efforts underway. Also, ensure that all visuals include alt-text descriptions for users with visual impairments.

Engaging in Two-Way Communication

Social media is a two-way street. Therefore, actively monitor social media platforms and respond promptly to comments and questions from the public. Address concerns directly and provide accurate information. Be transparent about the situation, even when there are unknowns. Acknowledge challenges and provide updates on ongoing response efforts. Honesty fosters trust and public confidence in official information. Showcase the human face of the response effort. Share stories of first responders, volunteers, and individuals coming together to help each other.

ETHICAL CONSIDERATIONS

The influx of VGI raises important ethical considerations and this section explores these considerations.

Privacy Concerns

Protecting the privacy of individuals affected by the crisis is crucial. Strategies for anonymizing data and respecting user privacy may be as follows: -

Anonymizing Data

When sharing VGI reports that include personal details like usernames or locations, consider removing or obfuscating this information to protect user privacy. This can be done in several ways. Blurring faces in photographs is a common practice for protecting the identities of individuals in crises. Techniques like pixelation or masking faces can be employed. Instead of sharing specific street addresses, consider using broader location descriptions like city districts or neighborhoods. This provides valuable information for crisis response while maintaining individual privacy. If VGI reports contain details like phone numbers or email addresses, completely remove this information before sharing the report publicly.

Obtaining Consent

If possible, seek consent from individuals before sharing their VGI reports, especially if the information is particularly sensitive. This demonstrates respect for user privacy and builds trust during a crisis. For situations where volunteers are actively collecting VGI data (e.g., damage assessments), develop clear and concise consent forms that explain how the information will be used and anonymized. When utilizing online platforms for VGI collection, implement opt-in mechanisms where users can explicitly choose whether to share their information publicly. Even when explicit consent is not feasible, ensure clear communication about how VGI data will be handled. Inform users about anonymization practices and their right to request the removal of their information.

Misinformation and Disinformation

The potential for false information to spread rapidly presents a significant challenge in the age of VGI. This section explores methods for identifying and debunking misinformation, as well as promoting media literacy among the public.

Fact-Checking and Verification

Utilize credible fact-checking websites and collaborate with journalists to verify the accuracy of VGI reports. Look for inconsistencies, suspicious language, and a lack of verifiable sources. If a VGI report includes a particularly alarming image, conduct a reverse image search using online tools. This can help determine the origin of the image and expose potential manipulations. Consider the timing of a VGI report. Does it align with other reports from the crisis zone? Does the content seem plausible given the unfolding events? Inconsistencies in timing can raise red flags. Always compare VGI reports with official statements from emergency response agencies, government bodies, and reputable news outlets.

Flagging Misinformation

Many social media platforms allow users to flag posts containing misinformation. Encouraging the public to utilize these tools and report any suspicious content can be helpful in this context. Collaborate with credible social media personalities and community leaders to debunk misinformation and promote reliable sources of information. When addressing misinformation, focus on providing accurate information in a clear and easy-to-understand manner. Avoid inflammatory language and focus on correcting factual inaccuracies.

Promoting Media Literacy

Develop educational campaigns that teach the public how to critically evaluate information online. This can include workshops on identifying bias, recognizing manipulative tactics, and verifying sources. Educate the public on how to assess the credibility of online sources. This includes examining website domains, author credentials, and potential biases of the information provider. Misinformation often appeals to emotions like fear or outrage. Train the public to recognize emotionally charged language and manipulative tactics used to spread false information. Equip the public with basic verification skills, such as reverse image search and fact-checking websites. This empowers individuals to become discerning consumers of information online.

CONCLUSION

The evolving landscape of crisis communication in the digital age, shaped significantly by social media and VGI, underscores several critical facets essential for the effective response and management of crises. Beyond speed and reach, social media-led VGI can foster two-way communication, empowering communities to actively participate in crisis response. People can share experiences, report incidents, and seek help in real time. This transparency builds trust between authorities and the public, who can address concerns, dispel rumors, and provide updates to reassure people and encourage adherence to safety measures. Social media-based VGI can also amplify the voices of those affected by crises, raising awareness, humanizing situations, and galvanizing support for relief efforts.

However, utilizing Volunteered Geographic Information during crises requires careful consideration. While VGI offers valuable real-time insights, verification of information and addressing ethical concerns like privacy and misinformation are crucial. Techniques include cross-referencing information with credible sources, utilizing geolocation data, and collaborating with trusted online communities. VGI can further enhance crisis communication by fostering collaboration. Dedicated hashtags create communication hubs, while online forums centralize information exchange. People can share updates, request and receive help, and access official information. Additionally, social media allows for recruiting volunteers with specific skills and matching them with relevant tasks.

Effective dissemination of official updates is also crucial. Clear and concise messaging with actionable information, along with visuals like maps, infographics, and photos (with descriptions for visually impaired users), helps people understand the situation and response efforts. Two-way communication, where authorities actively monitor social media, respond to inquiries, and showcase the human side of the response effort, is essential for building trust. Ethical considerations regarding VGI include anonymizing data, obtaining consent before sharing reports, and utilizing opt-in mechanisms on online platforms. Combating misinformation involves fact-checking, collaborating with journalists, flagging misinformation, and promoting media literacy campaigns.

Looking ahead, future efforts in crisis communication should focus on harnessing the full potential of VGI while addressing its inherent challenges. Enhanced strategies for verifying the accuracy of VGI reports are essential to ensure reliable information amidst the deluge of data. Collaborating with trusted online communities, refining ethical guidelines for data privacy, and advancing media literacy initiatives will be critical. Moreover, integrating emerging technologies like artificial intelligence and machine learning can further enhance the capabilities of crisis communication teams in processing and utilizing VGI effectively.

REFERENCES

Adelfio, M., Serrano-Estrada, L., Martí-Ciriquián, P., Kain, J. H., & Stenberg, J. (2020). Social Activity in Gothenburg's Intermediate City: Mapping Third Places through Social Media Data. *Applied Spatial Analysis and Policy*, 13(4), 985–1017. Advance online publication. DOI: 10.1007/s12061-020-09338-3

Ahmad, M. (2023). Connecting People and Places: How Citizen Diplomacy and VGI Are Strengthening Disaster Response and Community Development. In Kavoğlu, S., & Köksoy, E. (Eds.), *Global Perspectives on the Emerging Trends in Public Diplomacy* (pp. 195–226). IGI Global., DOI: 10.4018/978-1-6684-9161-4.ch009

Ahmad, M. (2024). From Crisis to Opportunity: How VGI Is Transforming Disaster Response and Recovery Efforts. In Siyal, S. (Ed.), *Using Crises and Disasters as Opportunities for Innovation and Improvement* (pp. 215–232). IGI Global., DOI: 10.4018/978-1-6684-9522-3.ch013

Amirkhanyan, A., & Meinel, C. (2016). Analysis of the value of public geotagged data from twitter from the perspective of providing situational awareness. *Lecture Notes in Computer Science (Including Subseries Lecture Notes in Artificial Intelligence and Lecture Notes in Bioinformatics), 9844 LNCS*. DOI: 10.1007/978-3-319-45234-0_48

Anderson-Tarver, C. (2015). Crisis Mapping the 2010 Earthquake in OpenStreetMap Haiti. In *ProQuest Dissertations and Theses*. http://search.proquest.com/docview/1690497435?accountid=8330\nhttp://jn8sf5hk5v.search.serialssolutions.com/?ctx_ver=Z39.88-2004&ctx_enc=info:ofi/enc:UTF-8&rfr_id=info:sid/ProQuest+Dissertations+%26+Theses+A%26I&rft_val_fmt=info:ofi/fmt:kev:mtx:dissertati

Antoniou, V., Morley, J., & Haklay, M. (2010). *Web 2.0 geotagged photos: Assessing the spatial dimension of the phenomenon*. Geomatica.

Bukar, U. A., Jabar, M. A., Sidi, F., Nor, R. N. H. B., Abdullah, S., & Othman, M. (2020). Crisis informatics in the context of social media crisis communication: Theoretical models, taxonomy, and open issues. *IEEE Access : Practical Innovations, Open Solutions*, 8, 185842–185869. Advance online publication. DOI: 10.1109/ACCESS.2020.3030184

Cénat, J. M., Broussard, C., Darius, W. P., Onesi, O., Auguste, E., El Aouame, A. M., Ukwu, G., Khodabocus, S. N., Labelle, P. R., & Dalexis, R. D. (2023). Social mobilization, education, and prevention of the Ebola virus disease: A scoping review. In *Preventive Medicine* (Vol. 166). DOI: 10.1016/j.ypmed.2022.107328

Crisp, J. (2024). The Syrian Emergency and Its Impact on the Evolution of Global Refugee Policy. In *Urban Displacement*. DOI: 10.2307/jj.9891587.15

Fan, C., Wu, F., & Mostafavi, A. (2020). A Hybrid Machine Learning Pipeline for Automated Mapping of Events and Locations from Social Media in Disasters. *IEEE Access : Practical Innovations, Open Solutions*, 8, 10478–10490. Advance online publication. DOI: 10.1109/ACCESS.2020.2965550

Fauzi, M. A. (2023). Social media in disaster management: review of the literature and future trends through bibliometric analysis. In *Natural Hazards* (Vol. 118, Issue 2). DOI: 10.1007/s11069-023-06079-7

Firestone, S. (2020). Importance of Communication During a Crisis. In *Biblical Principles of Crisis Leadership*. DOI: 10.1007/978-3-030-44955-1_7

Gardiner, S., Chen, J., Abreu Novais, M., Dupré, K., & Castley, J. G. (2023). Analyzing and Leveraging Social Media Disaster Communication of Natural Hazards: Community Sentiment and Messaging Regarding the Australian 2019/20 Bushfires. *Societies (Basel, Switzerland)*, 13(6), 138. Advance online publication. DOI: 10.3390/soc13060138

Goodchild, M. F. (2007). Citizens as sensors: The world of volunteered geography. In *GeoJournal*. DOI: 10.1007/s10708-007-9111-y

Gunter, U., & Önder, I. (2021). An Exploratory Analysis of Geotagged Photos From Instagram for Residents of and Visitors to Vienna. *Journal of Hospitality & Tourism Research (Washington, D.C.)*, 45(2), 373–398. Advance online publication. DOI: 10.1177/1096348020963689

Haupt, B. (2021). The Use of Crisis Communication Strategies in Emergency Management. *Journal of Homeland Security and Emergency Management*, 18(2), 125–150. Advance online publication. DOI: 10.1515/jhsem-2020-0039

Haworth, B., Bruce, E., & Middleton, P. (2012). Emerging Technologies for Risk Reduction : Assessing The Potential Use of Social Media and VGI for Increasing Community Engagement. *Australian Journal of Emergency Management*, 30(3), 36–41.

HOTOSM. (2023a). *HOT'S RESPONSE TO THE TURKEY & SYRIA EARTHQUAKE*. https://www.hotosm.org/projects/join-the-turkey-and-syria-earthquake-response/

HOTOSM. (2023b). *What We Do | Humanitarian OpenStreetMap Team*. https://www.hotosm.org/what-we-do

Issa, E., Tsou, M. H., Nara, A., & Spitzberg, B. (2017). Understanding the spatio-temporal characteristics of Twitter data with geotagged and non-geotagged content: Two case studies with the topic of flu and Ted (movie). *Annals of GIS*, 23(3), 219–235. Advance online publication. DOI: 10.1080/19475683.2017.1343257

Li, L., Bensi, M., & Baecher, G. (2023). Exploring the potential of social media crowdsourcing for post-earthquake damage assessment. *International Journal of Disaster Risk Reduction*, 98, 104062. Advance online publication. DOI: 10.1016/j.ijdrr.2023.104062

Ma, L., Wang, Y., & Kim, J. (2023). How health organizations communicate about COVID-19 on social media: A comparative content analysis. *Journal of Communication in Healthcare*, 16(1), 83–92. Advance online publication. DOI: 10.1080/17538068.2022.2103334 PMID: 36919810

Minghini, M., Delucchi, L., Sarretta, A., Lupia, F., Napolitano, M., & Palmas, A. (2017). Collaborative mapping response to disasters through openstreetmap: The case of the 2016 Italian earthquake. In *Geoingegneria Ambientale e Mineraria* (Vol. 151, Issue 2).

Nguyen, Q. C., Kath, S., Meng, H. W., Li, D., Smith, K. R., VanDerslice, J. A., Wen, M., & Li, F. (2016). Leveraging geotagged Twitter data to examine neighborhood happiness, diet, and physical activity. *Applied Geography (Sevenoaks, England)*, 73, 77–88. Advance online publication. DOI: 10.1016/j.apgeog.2016.06.003 PMID: 28533568

Pánek, J., Marek, L., Pászto, V., & Valůch, J. (2017). The Crisis Map of the Czech Republic: The nationwide deployment of an Ushahidi application for disasters. *Disasters*, 41(4), 649–671. Advance online publication. DOI: 10.1111/disa.12221 PMID: 28059447

Pechmann, J., & de los Reyes, C. (2023, March 17). *Using OSM Data in the Turkey and Syria Earthquake Response*. https://www.hotosm.org/updates/using-osm-data-for-the-turkey-and-syria-earthquake-response/

Sathianarayanan, M., Hsu, P. H., & Chang, C. C. (2024). Extracting disaster location identification from social media images using deep learning. *International Journal of Disaster Risk Reduction*, 104, 104352. Advance online publication. DOI: 10.1016/j.ijdrr.2024.104352

Siawsh, N., Peszynski, K., Vo-Tran, H., & Young, L. (2023). Toward the creation of disaster-resilient communities: The Machizukuri initiative – The 2011 Tōhoku Great East Japan Earthquake and Tsunami. *International Journal of Disaster Risk Reduction*, 96, 103961. Advance online publication. DOI: 10.1016/j.ijdrr.2023.103961

Wu, X., Huang, Z., Peng, X., Chen, Y., & Liu, Y. (2018). Building a spatially-embedded network of tourism hotspots from geotagged social media data. *IEEE Access : Practical Innovations, Open Solutions*, 6, 21945–21955. Advance online publication. DOI: 10.1109/ACCESS.2018.2828032

Yin, S., Chen, S., & Ge, Y. (2024). Dynamic Associations Between Centers for Disease Control and Prevention Social Media Contents and Epidemic Measures During COVID-19: Infoveillance Study. *JMIR Infodemiology*, 4(1), e49756. Advance online publication. DOI: 10.2196/49756 PMID: 38261367

Yuan, Y., & Medel, M. (2016). Characterizing international travel behavior from geotagged photos: A case study of Flickr. *PLoS One*, 11(5), e0154885. Advance online publication. DOI: 10.1371/journal.pone.0154885 PMID: 27159195

Zander, K. K., Garnett, S. T., Ogie, R., Alazab, M., & Nguyen, D. (2023). Trends in bushfire related tweets during the Australian 'Black Summer' of 2019/20. *Forest Ecology and Management*, 545, 121274. Advance online publication. DOI: 10.1016/j.foreco.2023.121274

Zhou, B., Zou, L., Mostafavi, A., Lin, B., Yang, M., Gharaibeh, N., Cai, H., Abedin, J., & Mandal, D. (2022). VictimFinder: Harvesting rescue requests in disaster response from social media with BERT. *Computers, Environment and Urban Systems*, 95, 101824. Advance online publication. DOI: 10.1016/j.compenvurbsys.2022.101824

Zou, L., Liao, D., Lam, N. S. N., Meyer, M. A., Gharaibeh, N. G., Cai, H., Zhou, B., & Li, D. (2023). Social media for emergency rescue: An analysis of rescue requests on Twitter during Hurricane Harvey. *International Journal of Disaster Risk Reduction*, 85, 103513. Advance online publication. DOI: 10.1016/j.ijdrr.2022.103513

KEY TERMS AND DEFINITIONS

Crisis Communication: Involves the strategies and methods used to convey information and manage public perception during a crisis or emergency situation. Effective crisis communication aims to provide accurate, timely, and clear information to stakeholders, including the public, media, and authorities, to mitigate harm, manage rumors, and facilitate coordinated response efforts.

Online Communities: Groups of individuals who interact and communicate with each other through digital platforms and internet-based technologies. These communities are typically formed around shared interests, goals, or activities and can be found on social media platforms, forums, discussion boards, and other online spaces.

Spatial Data: Also known as geospatial data, refers to information that describes the location, shape, and relationships of geographic features and phenomena on the Earth's surface. This data can include coordinates, maps, and attributes that define the properties of physical objects, such as buildings, roads, rivers, and land use.

Social Media: Refers to digital platforms and applications that enable users to create, share, and interact with content online. Examples include Facebook, Twitter, Instagram, LinkedIn, and TikTok. Social media platforms facilitate communication, networking, and information dissemination, allowing users to connect with others, share experiences, and stay informed about current events.

Volunteered Geographic Information (VGI): A type of geospatial data that is collected, contributed, and shared voluntarily by individuals, often through the use of web-based platforms and mobile applications. VGI encompasses a wide range of data types, including location coordinates, maps, and descriptive information about geographic features, events, or phenomena.

Chapter 9
Digital Activism in Portugal: Strategies for Communication and Social Mobilization

Joana Ramalho
UNIDCOM, IADE, Universidade Europeia, Portugal

Maria João Silva
IADE, Universidade Europeia, Portugal

ABSTRACT

In recent years, several social movements have emerged on the international scene, some of which have a global reach, in part due to the digital environment, and in particular to social media. Digital activism describes different forms of activism that use digital technology and it has gained importance, as it enables users, activists and institutions to share content that has the potential to reach a very significant number of people. It is against this backdrop that this research was developed. Our main objective is to contribute to the understanding of how digital activism works in Portugal, in particular how activist organizations use digital media to communicate with their audiences. We try to understand whether these communication channels really represent an added value to publicize the mission, causes, campaigns and work carried out by the institutions, and if they can effectively mobilize the members of the public, so that they participate in activist actions proposed and organized by the institutions.

DOI: 10.4018/979-8-3693-3579-6.ch009

INTRODUCTION: WHY TRY TO GRASP DIGITAL ACTIVISM?

In recent years, several social movements have emerged on the international scene, some of which have a global reach, partly due to the digital environment, particularly social media. Digital activism – a term that describes different forms of activism that use digital technology (Gerbaudo, 2017) – has gained importance, as it enables users, activists, and institutions to share content that has the potential to reach a very significant number of people. That is because the digital medium allows users to share different content formats on various digital platforms, in a quick, easy, and effective way.

In the Digital Era, social networks play a fundamental role in organizing, publicizing, and mobilizing events, actions, and activist organizations. Thanks to the functionalities and characteristics of the digital medium, the role of user-producer of content has become very present in the media, since any user can create their content and publish it online, making it available for several people to see. This phenomenon makes it easier, and faster, for people to know about an important and substantial event, such as a social movement, that happened in a particular country and that movement can easily spread to other regions of the globe (Castells & Cardoso, 2005; Castells, 2010; Van Dijck, 2013; Tilly, Castañeda & Wood, 2020). In this case, social media play a huge role in developing and disclosing some social movements, being that users participate in awareness not only online, but have also been mobilized for offline protests (Tilly, Castañeda & Wood, 2020).

Nevertheless, it is important to know the differences between social movement and collective action. First, social movements are individuals' collective actions focused on issues such as human rights, better living conditions, or the emancipation of social groups, and it is not possible to achieve them only through an isolated person (Castañeda, 2020). Second, collective action is defined by the behaviors of a group of individuals who share the same principles and values, although they may have individual intentions during the practice of this action (Silva, 2018). Paulo Silva argues that collective actions happen because the individual action does not meet the objectives and the intended results. Collective action is the result of the discontent of a group of individuals with a given experienced situation, and it is thought to be the best way to act so that individuals can obtain their goals and publicize their opinions against the systems of power (Silva, 2018).

It is against this backdrop that the research here presented aims to contribute to the understanding of how digital activism works in Portugal, in particular, how activist organizations use digital media to communicate with their audiences. It aims to understand whether these communication channels represent an added value to publicize the mission, causes, campaigns, and work carried out by the institutions, and if they can effectively mobilize the members of the public so that they participate in activist actions proposed and organized by the institutions.

The relevance of the present work arises from its exploration of the different communication approaches of different organizations, which also have varied missions, proposals, projects, and campaigns. Moreover, it remains important as it contributes to demonstrating the differences, in terms of human and financial resources, between large and better-known organizations and small and more recent ones. This chapter also intends to contribute as it helps make known the realities and experiences lived by minorities – which are often not described or represented in most academic literature – and how they use the digital medium to carry out their activism, as well as the associations that represent them.

NOTES ON OUR PATH: QUESTIONS AND METHODOLOGY

Against this context, the present chapter seeks to answer the following research questions: (1) How do activist organizations use digital media, specifically their websites and social media, to reach the public?; (2) How does the digital medium contribute to the dissemination of activist ideals?; (3) What practices are carried out to ensure that digital activism contributes to offline activism?; (4) Is the number of online followers reflected in the number of participants in offline activist activities?

The organizations chosen have different missions and purposes, and they all operate in Portugal: UNICEF Portugal, Amnesty International Portugal, Associação ILGA Portugal – Intervenção Lésbica, Gay, Bisexual, Trans e Intersex [Lesbian, Gay, Bisexual, Trans and Intersex Intervention], EX Aequo [Support network for young lesbian, gay, bisexual, trans, intersex and supporters], APAV – Associação Portuguesa de Apoio à Vítima [Portuguese Victim Support Association], Quercus [Association for the conservation of natural resources and the environment], and IRA – Intervenção e Resgate Animal [Animal Intervention and Rescue].

For each organization, when exploring the results of the analysis carried out below, its objectives will be briefly presented. All of them are non-profit organizations, with no association with political parties. Their presence in Portuguese territory is more or less extensive, depending on the case. But for example, IRA – the most recent among them – is only present in the metropolitan areas of Lisbon and Porto and São Miguel (Azores). Two of them also operate worldwide, namely UNICEF – a United Nations organization founded after World War II – and Amnesty International – founded in London in the 60s. The rest were founded mainly in the 80s and 90s: Quercus in 1985, APAV in 1990, ILGA in 1995, Rede Ex Aequo (from ILGA's Decentralize project) in 2003, and IRA in 2016.

All the chosen organizations have developed work and action in the physical environment over the years. However, they were also chosen because they all have a presence in the digital environment: They have social media accounts, where they share various publications with users, and an official website, where information is made available in more depth.

The methodology of the research carried out was bipartite and sought to obtain an external point of view of communication in organizations, based on online mapping – content analysis of social media profiles – and an internal point of view, through semi-structured interviews with members of activist organizations.

Posts from three social media platforms – Instagram, Facebook, and Twitter –were analyzed for February, March, April, and May 2023. The online mapping of social media has as its reference date the 3rd of June 2023, so that changes that occurred in the organizations' profiles after that date, such as in the number of followers, likes, comments, shares, and posts, had no impact on the content analysis.

During the process of mapping the social media of these associations, not only the images of the posts were analyzed, but also their descriptions. After an initial analysis of the organizations' accounts, different categories were created to define each post. In addition to categorizing social media posts, the organizations' profiles on Facebook, Instagram and Twitter were also analyzed in terms of (1) the number of followers, (2) the average number of likes, (3) comments, (4) shares and retweets (5) the frequency of publication.

To complement the online mapping of organizations and associations, we carried out interviews, during June and July 2023. With this tool, we learned more about the organizations themselves, by hearing from people who are directly involved in the projects and, in particular, gained better knowledge of how the communication area of each organization works. Out of the seven organizations contacted, only five agreed to do the interview – APAV, ILGA, Ex Aequo, Amnesty International, and Quercus. All semi-structured interviews followed the same script, to obtain information about the organizations on the same topics, and then, based on the same analysis grid, observe the points in which they are similar and different.

It is important to take into account that the interviewees are not representative of the organization as a whole and the results obtained through the interviews are the view that the person interviewed has of the organization, through the involvement they have in the projects and the work they carry out within it.

THEORETICAL BACKGROUND: WHAT IS DIGITAL ACTIVISM AND HOW DO ORGANIZATIONS USE IT?

According to some authors, the term "activism" can involve various meanings. Chon and Park (2020) define activism as collective action by like-minded people aiming to change society, politics, or an organization on controversial issues. Jordan (2002) believes that activism is made and organized by a group of people. The only difference between "normal" people and activists lies in activists pursuing and advocating for a change in society, recognizing the necessity to change lifestyles.

However, activism is not a recent behavior that has spread and escalated with the usage of the Internet. Throughout the years, people have been active in expressing discontent about the conditions of the society they live in and acting to make a change. There have been numerous occasions where individuals joined forces to advocate for their rights and beliefs. Examples include the suffragette movement, the Storming of the Bastille, and various war protests (Hayward, 2018). These events have significantly impacted history and laid the foundations for modern activism.

Nowadays, with the Internet, a significant part of activism occurs online, resulting in digital activism. Also, the Internet and information technology have not only facilitated traditional activist practices but also changed them, making digital activism a constant aspect of online users' lives, due to the widespread connectivity it offers across the globe.

For Sandor Vegh (2003), online activism is a movement, which is dependent on the Internet and contains political motives. Another point of view about the term is from Paolo Gerbaudo (2017), who argues that digital activism is used to describe different ways of activism that use digital technology and since its appearance, with the beginning of the World Wide Web, has suffered a lot of changes.

As mentioned before, for many years, several activism practices and social movements did not involve the use of the Internet as people relied on traditional activist practices to contact and connect with others. However, as technology evolved and became widespread, the organizing process of these movements changed and developed. Therefore, many activists began to take advantage of the features and capabilities of technology and the Internet for activist purposes.

Thus, before the advent of Web 2.0, organizing activist movements faced limitations, but with the evolution of technology, the processes and practices of activism and the activists themselves have changed (Sandoval-Almazan & Gil-Garcia, 2014). Announcing a demonstration and reaching a greater number of people through social media, getting in touch more easily with the entities and/or organizations necessary to expose a certain social issue, and sharing information directly with the public without having any intermediary, have become easier processes to carry out (Viana & Sarmento, 2021).

With the emergence and the rise of this new form of activism – digital activism – there have been two distinct phases, according to Gerbaudo (2017). The first one began when users' development and mass use of the Internet also began. Observing this phenomenon, activists thought it would be the right time to create different projects and initiatives through technology. And digital activism accompanied the growth of the Internet, adapting to its functionalities. The activists soon saw the opportunities and benefits of the Internet and technology to reach a broader audience, achieve their goals, and defend the causes that they believed in (Vegh, 2003).

The second phase coincided with the rise of social media and world-famous hacker collectives. In this phase, activists used social media as a means of mass mobilization, when protests and social movements were organized and aimed to fill the streets with a lot of people (Gerbaudo, 2017).

Activists, to achieve their objectives, use, according to Vegh (2003), two types of strategies: one is to increase traditional techniques of activism through the Internet, and the other is to use the functionalities and capabilities of the Internet itself to do so.

For users to be able to do digital activism, digital technological devices and digital platforms are needed and the activists have to explore its features and goals. Through digital technologies, people can use them to better engage in digital activism. Joyce (2011) presents seven activist functions that users attribute to digital technology when these are used for the practice of activism. The functions are Document, Broadcast, Mobilize, Co-Create, Protect, Synthesize and Transfer Resources. These seven functions demonstrate the multiple ways that technology can be a benefit and a need for the good practice of digital activism.

Activists play a crucial role in the practice of activism because without people it would not be possible to execute activist practices and actions. Many activists affiliate with organizations to join other like-minded individuals who share common goals and interests, aiming to influence others to improve the world for everyone who lives in it. Using the Internet's features, activist groups, and their organizations can reach a wider audience and establish a closer relationship with them, to create connections and get support for the causes they defend.

In today's global society, there are several activist organizations dedicated to changing society's mentality: some are larger and globally recognized, while others are smaller and less renowned. The oldest, most renowned, and most resourceful organizations were founded when digital opportunities were less advanced. Nonetheless, thanks to technological advancements, the rise of the Internet, and digital possibilities, these organizations, sometimes operating across multiple countries simultaneously, were able to become closer to the public.

Moreover, over the years, new organizations have emerged and joined existing ones. While several organizations share the same purpose, each must have a different strategy to engage the public, and many have found the appropriate tools on the Internet. Many of the new organizations lack the funds and resources that more established organizations have. Therefore, the Internet serves as a means to increase and improve resources available to activists. Furthermore, independence from traditional news media is advantageous because it allows activists to connect directly with the public and have better control over the treatment and dissemination of their messages.

Simultaneously, the digital field allows audiences from different backgrounds and life experiences to come together and discuss a particular topic, as it helps create a dialogue between various individuals. It may be easier for users to talk about certain topics in the digital world than it would be in the real world. Social media allow any user to express their opinion on a variety of topics. When a certain subject begins to gain special attention from a certain number of people, a group is formed in which individuals have common interests. Thus, on digital platforms, active communities are created to express their opinions and share content produced by users, available in different formats.

Another Side of Digital Activism: Is Fake Activism Real?

As mentioned before, the Internet and digital media offer great opportunities for organizations and activists to engage with the public and raise awareness for their projects and causes. With this in mind, Web 2.0 and information technologies helped the improvement of activist practices. However, some activist practices spark debate about their effectiveness. Thus, it is also not certain whether they represent a negative side to activism, as they also have their advantages for some activists and organizations.

There are several practices deemed as fake activism, such as hacktivism, slacktivism, and performative activism. According to some authors, these practices are seen as not-so-correct practices of digital activism.

Therefore, hacktivism is a practice made possible by characteristics and features allowed in the virtual medium and utilized by its users. For many, this practice is an alternative to traditional activism practices, as it also causes effects and disruptions.

Hacktivism, according to Jordan (2002), occurs due to political and social reasons. With this practice of digital activism, hacktivists – the people who practice hacktivism – can do several actions, but the most known is the attack on websites to shut them down (Campos, Pereira & Simões, 2016). Hence, hacktivists found on the Internet two kinds of uses of hacktivism. Jordan (2001, 2002) identifies these uses as mass virtual direct action and individual virtual direct action.

For mass virtual direct action to occur, it requires a large number of people to simultaneously utilize the Internet to create civil disobedience digitally and monitor peaceful protests on the streets. While these protests occur offline, and online, hacktivists are pursuing their objectives (Jordan, 2001, 2002). One of the most common practices of mass virtual direct action hacktivism involves attacks aimed to shut down slower websites, mostly known as denial-of-service attacks, which often require multiple users to produce a significant impact, although these attacks can also be made by a single user, utilizing a specific program (Jordan, 2002).

Regarding the second use of the Internet by hacktivists – individual virtual direct action, which Jordan defines as "using technology to make a political point" (Jordan, 2011, p.11) – they use hacking techniques, primarily attacking websites for political proposes. Contrary to what its name may imply, these actions are not carried out individually by hacktivists but can be made individually – due to the nature of the actions – which do not rely on mass participation.

Another example of fake activism is slacktivism. As the name implies, this form of activism is done in a lazy, careless, and slacked way (Glenn, 2015). This form of activism, normally practiced in the digital realm, is characterized as an action with symbolic value in support of social, political, and feminist causes, among others. However, the user lacks the "intention or willingness to put forth significant effort in pursuit of social change" (Noland, 2019, p.2). The big difference between slacktivism and traditional activism is in the effort and determination that people invest in supporting a social cause (Noland, 2019) whereas, in the first case, this effort is debatable, because it occurs online and often does not transpose to the offline.

Slacktivism often carries a negative connotation, viewed as an activist practice in which users only practice it just to gratify themselves and to show to their peers that they are involved and preoccupied with some social cause, without investing time and actively supporting the cause (Glenn, 2015; Noland, 2019), such as through participation in movements or volunteering. According to Noland (2019), these slacktivism actions are motivated by laziness, a desire to show others a socially engaged image online, and impression management - the process individuals and organizations use to guide others' perceptions of them (Cuofano, 2023). However, these activist actions, for many organizations and activists, help increase interactions with other users, with minimal financial costs (Noland, 2019; Glenn, 2015).

At the same time, the main objective of activists is to change the mentality of society in several aspects and make it more involved and participatory in social causes, aiming for a world where all people can live in harmony. Thus, when users rely on slacktivism as their only activist action and the only way to participate in movements and political-social issues, they are not making an effort to achieve the change they want. In slacktivism, users express their support for causes simply by "liking" or sharing posts on social media, yet they do not show the feeling and emotion that occurs when trying to do something to cause a change (McCafferty, 2011) in the offline realm. While those engaging in slacktivism are more likely to support causes online, they are less inclined to take action offline (Noland, 2019).

Finally, the last practice to be explored is performative activism. As mentioned above, slacktivism is a practice more focused on the online image that is transmitted to others, without giving too much thought about participating and doing activist actions in the offline realm, performative activism reinforces this idea. Additionally, when performative activism is practiced, performative solidarity also takes place, concerning the supported cause, since there is a false sense of support from users. Performative activism and performative solidarity take place mainly in social media and can be done by individual users or by organizations and companies so that they maintain a good image in the digital medium.

As seen on many occasions, a society has several power dynamics. There are the majorities and the minorities. The first ones are privileged and have power whilst the second ones do not have that many privileges and search and fight for a place in society (Kutlaca & Radke, 2022). Furthermore, when there is the intention to draw society's attention to a particular subject or to an event that directly affected minorities and generated controversy and concern for global society, social movements are the events that move more people and organizations, both online and offline. Both these parties – people and organizations – try to find a way to participate in the cause, raise awareness of other individuals, and ensure that the subject does not fall into oblivion. Due to the features of the Digital Age, the digital medium, particularly social media, is the place where people express their opinions (Kutlaca & Radke, 2022).

Most of the time, users express their solidarity and support for a cause and show their position concerning a movement through social media. However, many users simply publish, share, and engage with posts about the situation, without giving much thought to the cause or participating in offline activist actions. When situations like this arise, it is referred to as performative activism.

Performative activism occurs when users from privileged groups express solidarity and support for minority groups, without effectively addressing their difficulties and injustices (Wellman, 2022). While interaction and activist participation in the digital medium is necessary, this practice does not help the cause since it floods platforms with superficial support that fails to inform or benefit minorities. By practicing performative activism, privileged users redirect the focus away from the pressing social issues at hand.

Digital activism is crucial for activism's growth and for the social mobilization of individuals. However, according to Abdalla et al. (2022), the majority of the time, the users who participate in activism in the digital medium see this action as participation in a trend and do not take the necessary measures to lead the change of mentalities in society. Hence, performative activism is perceived as superficial and ineffective, and those who practice it don't demonstrate the desire to promote change (Wellman, 2022).

Although performative activism actions and performative solidarity may attract attention and support, they still are practices considered selfish and insincere. Instead of genuinely helping those in need, they're more focused on improving the image that others have of the person who performs these actions (Wellman, 2022; Abdalla et al., 2022; Kutlaca & Radke, 2022).

Social media and the Internet have contributed to the development of these and other practices perceived as fake activism, while also facilitating the publication of one's opinion and support towards a cause or movement. Even though these practices are often viewed as examples of fake activism by many scholars, in some cases, they help raise awareness and catch the public's attention. Therefore, it is challenging to determine what is fake activism and whether it is detrimental to digital activism.

Although many scholars deem these practices as fake activism and a threat to digital activism, our research exposes another side of these practices. Through the interviews, a different perspective emerged, revealing how some actions seen as fake activism play an important role. Interviewees expressed that these actions effectively spread information on crucial topics related to their organizations, while also serving as a means to engage diverse audiences and promote meaningful dialogues.

Another important aspect brought by a few interviewees is that fake activism practices might be the only way for some people to engage in activism. Limited by factors such as time, energy, or privilege many find traditional forms of activism inaccessible. Therefore, they turn to digital platforms as their only way to practice advocacy, moved by genuine intentions to make a change in society. This is valuable for those who feel marginalized or unsafe in offline spaces, where expressing different opinions can cause rejection. Consequently, the digital realm becomes their safe space to express themselves, enabling them to voice opinions without fear.

Online Mapping: Main Empirical Notes

For all organizations, posts on Instagram and Facebook tend to be roughly the same on both social media platforms. In almost all organizations, it is Facebook that has the most followers (except for ILGA, where it is Instagram) and Twitter tends to be little used – often more for retweets or comments. However, the platform that systematically receives the most interactions from followers tends to be Instagram.

UNICEF's mission is to defend and ensure children's rights around the world (UNICEF, s.d.). In this sense, posts on social media seek to give visibility to their involvement in solving the problems that children face, especially in places affected by wars, natural disasters, and climate change, among others. In the three platforms analyzed, informative posts are in greater number, followed by fundraising posts. The first are posts that transmit information regarding a specific topic aligned with the organization's purpose; or that inform followers of the reality of individuals who seek help from the organization; or about an event and the organization's involvement. The latter specifically targets users with requests for financial help. In this case, it is due to the events that marked the analyzed months (e.g. the earthquake that affected Syria and Turkey, and the first anniversary of the invasion of Ukraine by Russia) and the posts served the purpose of informing users about the work that the organization was carrying out, as well as to ask for financial assistance to help the affected children.

Amnesty International's main objective is to defend and fight for human rights, doing everything possible to denounce and combat injustices and make more people aware of them to end them (Amnesty International, 2023; Aministia Internacional, s.d). The category that has the most publications on the organization's social networks is informative posts, followed by commemorative posts, which are those that commemorate periodic or specific events, such as a holiday, or a specific day that is important for the organization, or an event achieved due to the effort of the organization. In several publications, Amnesty International Portugal calls for the signing of petitions in the digital environment, so that certain freedoms and rights are guaranteed, both nationally and internationally.

These two organizations – UNICEF Portugal and Amnesty International – present a strong and constant graphic identity, which probably makes the users quickly recognize posts, mainly because they always have the logo and the same color palette. When looking at their social media, both seem to have more economic resources for communication, when compared to other organizations, and appear to have a consistent publication plan.

Concerning publication frequency, IRA is the organization that shares the most, publishing daily, and doing it up to six or more posts per day. IRA's work is limited to rescuing animals that are neglected or mistreated, based on complaints received from users or specific cases in which organization members testify (Ribeiro & Rodrigues, 2018). Through their social media platforms, the organization shows some rescue processes. They share updates on rescued animals, to show followers their work, raise awareness about animal abuse, and raise monetary funds to help with the treatment and feeding of rescued animals. Informative posts are the majority, followed by posts with requests for help and fundraising.

Anyone who follows IRA's social media has the feeling of following the work and daily life of someone they know since the images used are quite spontaneous and, above all, the language used in the descriptions is not formal or very descriptive. The way the descriptions are constructed involves the user in the event itself, telling the stories of the animals and how the organization intervened. In this way, users' attention is captured by the situations that occur. In contrast to the two previous organizations, IRA does not have a communications team or specific financial resources for this area. The organization does not have a well-defined graphic identity, however, it demonstrates consistency in the type of narrative and publication.

In the case of ILGA, the association publishes frequently on the three social media platforms, although the consistency of the publication varies depending on the month. ILGA wants LGBTI+ members and their families not to be targets of discrimination and promotes their integration into society without any type of prejudice (ILGA, s.d.). To be able to promote this type of mentality in society, the association spreads information and creates gatherings at the Lisbon LGBTI+ Center, where members of the LGBTI+ community and supporters can coexist in a safe space and, thus, meet more people with the same values. Informative posts and posts promoting offline actions are more present.

Through social media, the association transmits information about current issues that directly affect the Portuguese LGBTI+ community and uses these means of communication mainly to promote the various events it organizes throughout the year, such as the presentation of the Arco-Íris [Rainbow] Awards, which celebrate people and entities that stood out in the fight against discrimination. During the period under analysis were also important, for example, the Lesbian Visibility Week, the National and International Day for the Fight Against Homophobia, Transphobia, Biphobia, and Interphobia, and the festival Arraial Lisboa Pride, which always attracts many people. ILGA's online presence makes it noticeable that it has someone in charge of communication and social media management since posts have a graphic identity and follow a line of communication that is very identifiable with the association, with the logo present in most posts.

Regarding Ex Aequo's publication frequency, the association does not publish frequently and does so in an inconsistent manner. Facebook is used more to publicize online and offline actions, as they use its features to create "events" and promote the gatherings and conversations they organize. Twitter is also used to announce various events they organize. Ex Aequo, like ILGA, is an association that seeks to fight against discrimination and integrate LGBTI+ young people into society so that they do not feel alone in the process of personal discovery (Rede Ex Aequo, s.d.). To this end, the association organizes activities, gatherings, and conversations on topics related to the LGBTI+ community, both digitally and physically. Therefore, the categories more present are posts promoting offline actions, posts promoting online actions, and Informative publications.

Although there were not a large number of posts in the four months of the social media mapping, there seems to be attention to having a cohesive graphic image, which reflects the identity of the association. It seems clear that there is a communications team, which may not be so focused on social media, but rather on other Ex Aequo projects. It is also worth mentioning the fact that this is a nationwide association with several centers across the country, and it is necessary to have a team that organizes and publishes the various initiatives belonging to the different centers.

APAV publishes frequently and consistently on Facebook, Instagram, and Twitter, and all of these social media platforms have a large number of publications. The purpose of APAV is to inform, protect, and monitor individuals who are or who may become victims of any type of crime (APAV, s.d.). In this sense, in the three social media platforms, informative posts are the category with the most posts, followed by the fundraising posts category. In the months under review, campaigns against violence in dating, on the occasion of Saint Valentine's Day, and against crimes on the Internet, such as online romantic scams, stood out. The IV Braga Conference against Violence was also highly promoted, which is a very important event for the association.

On social media, the association presents and publicizes the campaigns it promotes throughout the year, developed through patronage by advertising agencies. These campaigns help reach a greater number of users and inform them that there is an entity that is available to support victims of different types of crime. They also help raise awareness of the different types of crimes.

APAV publications present a consistent and identifying graphic image, with the logo present. Their social media follow a publication plan and there seems to be an investment in these communication channels by the responsible team. The association may not have as many economic resources to carry out advertising campaigns and publicize them in traditional media, but it works with advertising agencies and magazines through patronage, managing to have these campaigns in the physical environment and on websites.

Quercus' mission is to fight for the defense and conservation of the environment, animals, and nature, and its work is based on volunteering and disseminating information about situations with a negative impact on the environment (Quercus, s.d.). In addition, they have wildlife protection centers, where injured animals recover and are then returned to their natural habitat. The categories that have a higher number of publications are informative posts and posts promoting offline actions. It is through these publications that they can reach more people and raise the human resources and financial help they need. Some of the publications presented over the four months promote activities to return recovered animals to nature, requests for animal sponsorship, publicity about volunteering actions, and the section they have in partnership with the national public broadcasting network RTP ("Minuto Verde").

In some aspects, Quercus's posts are similar to IRA's, since many of them are spontaneous photographs, namely of the actions they carry out in the physical environment, of the animals that are recovering and their return to nature, of the activities they carry out with schools and companies, among others. However, it has other publications of a more organizational nature, such as commemorative posts or calls for the General Assembly. As with Ex Aequo, Quercus' action extends across the country and the work of the various centers is publicized on social media, as well as environmental situations and occurrences throughout Portugal.

The Point of View of Some of Those in the Trenches

In the institutions analyzed, there are two different types of communication teams: professional and volunteer teams. As the name suggests, the latter are formed by individuals who are doing voluntary work and do not necessarily have training in the area before joining the team. On the contrary, in professional teams, members have specific training in the area in which they work. Some organizations depend on the work of volunteers to operate and carry out their functions, while in others, in addition to having voluntary work, certain positions and functions are carried out by professionals. On the other hand, some organizations mix the two modalities.

A common point among all interviewees is the purpose of the institution to be present in the digital medium. They all mentioned that it is extremely important because that's where people are and that way they can reach as many people as possible. Through the digital, they intend to publicize their work, as well as the activities they organize, raise awareness among users about the causes they defend, invite more people to collaborate with the institution, and make themselves known to users so that they know who they are, what they do, where you can find and contact them. They also seek to inform users of current situations, related to the causes they defend.

Another factor that favors activist organizations being present in the digital environment is being able to talk about any type of subject without having intermediaries approve the published content, which happens in traditional media. On the Internet, they can talk about topics that do not receive as much attention and demand on television, radio, and newspapers. One interviewee mentioned that, when it comes to the media, there is no positive response when asked to expose problems and issues related to the cause of their association. When promoting themselves online, there is no obstruction, and their content is not denied sharing. The digital environment is also much more financially accessible to all organizations, especially for those who do not have the financial resources to promote themselves in the physical environment. This line of thought was reinforced by the testimonies of the interviewees.

Another advantage of organizations being closer to the public online is the ability to create a closer relationship with them. This proximity and involvement can translate into awareness and mobilization for causes and make them grow, both digitally and physically. In the digital environment, it is possible to create communities of individuals who share common experiences and interests, creating safe spaces where they can talk openly.

These online communities can also meet offline. Regarding the participation of individuals in online and offline actions, the interviewees' general opinion is that there are not as many people in the activist actions in the physical environment as in the ones in the digital environment. For example, when an association organizes an event and asks for registration, it observes that there are always more people registered than those who eventually attend. However, the opposite happens with some events, namely large-scale events expected by many individuals, such as the LGBTI March or the festival Arraial Lisboa Pride. On the other hand, there are many online awareness campaigns, which are not easy to convert from digital to physical, in the sense that it is difficult to measure the success that the campaign had with the public. Overall, it appears that word of mouth, because of the trust factor, is still the best way to convince someone to attend activities.

Considering the extent of individuals' participation in online and offline actions, online allows people from different parts of the country to come together more easily than trying to bring them together physically, especially when in organizations spread across the country. It seems clear that the participation, both in the physical and digital environments, of individuals depends on the subject of the events and actions, as well as the time of year itself. Knowing that social networks have facilitated the sharing of content with activist content, organizations are also aware that there are people who practice false activism, but they can do little to protect themselves or prevent these practices, and it is not even clear whether this may not come with advantages.

Seven Organizations: Some Similarities and Differences

Based on the results obtained in online mapping and interviews, it is possible to verify that the organizations and associations analyzed are, in fact, distinct from each other, whether in their purpose and objective, in their internal organization, or the structure of each communication team, among other aspects. Each of these differences represents the diversity that exists among all organizations in the world, each of which has a specific mission, different methods of operation, and a different target audience.

However, despite the differences, it is important to note that there are similar points between the institutions covered, bringing them closer to one another. Firstly, and very importantly, there is a similarity in the level of dedication to the mission, that is, the commitment that organizations have concerning the causes they defend. Although the purposes and objectives may be different, the organizations and associations analyzed all have in common great dedication to their causes. In our empirical research, it was obvious that the organizations and people involved believe in what they defend, doing their best to reach more people so that they can act and promote change.

Secondly, despite communication teams and planning strategies not being equal, all institutions recognize how important communication is to achieve their specific objectives and transmit their work to the public. This way, it is possible to keep the public up to date with what happens in the organization, the topics it promotes, and the events that are directly linked to its mission.

Thirdly, following the previous point, when analyzing the posts on social media and websites, it was found that organizations actively use these platforms for a variety of purposes, including raising public awareness, disseminating information, promoting activist actions, fundraising, and facilitating contact with the public, among others. Taking into account that by effectively using these digital platforms, the reach can increase considerably, they are a fundamental means of promoting organizations and involving users, both online and offline.

Fourthly, organizations need to use digital tools to mobilize the public. This stands out as an objective, as without people supporting the work of institutions, especially those that have a voluntary work base, it would be difficult to fulfill their purpose. By involving the public in causes, interacting with them, promoting their participation, and encouraging them to do part of the organization, whether through volunteering or by becoming members, aims to participate and continuously support causes, activist actions, campaigns, support groups, gatherings, and activities developed.

Fifthly, and more specifically, the social media platform chosen by the interviewees (with a single exception) as the best for interacting with the public was Instagram, due to its functionalities and characteristics. From the perceptions of the interviewees and the online mapping, it was verified that this network is also the one that users prefer to interact with institutions and stay up to date with what they are doing.

And finally, sixthly, the online world has proven to be a place where you can promote various campaigns and publicize institutions without spending a lot of money, since it is cheaper to have ads in this medium than in the physical environment, and reach effectively a global audience. On the other hand, by having their social media and website, organizations can promote themselves for free.

Although the organizations mentioned have characteristics in common, they also have other aspects in which they differ. In addition to the aspect that primarily and most distinguishes them from one another, which is the mission and purpose they defend, it is necessary to consider the fact that some organizations operate at an international level while others are exclusively national. Amnesty International Portugal and UNICEF Portugal, compared to other institutions, have a long history of working in the field, a great impact, and visibility in society, and were also founded when the Internet was far from being a reality. It was observed that these organizations have a global reach that allows them to get to a large number of people and show that they have more financial resources to invest in communication online and offline, than other institutions operating at the national level. Unlike these two organizations, the remaining institutions analyzed - APAV, ILGA, IRA, Rede Ex Aequo, and Quercus - are national, more recent, and founded either when the Internet was beginning to take its first steps or when it was already well-established. However, despite being quite new and not having as many financial resources, compared to previous organizations, these are institutions that are quite nationally recognized and have a very broad field of action, seeking to enter the front line in promoting change in issues and mentalities in Portuguese society.

Another aspect focuses on the different approaches that organizations choose to display on social media, in terms of how they address and communicate with users. In the cases of UNICEF Portugal and Amnesty International Portugal, when one looks at their social media, one notices that the photographs in the publications are professional and carefully chosen. UNICEF Portugal uses images also present on the organization's international profile. Amnesty International, in addition to photographs, also has posts with other more creative elements. They both present a strong graphic image, making it possible to recognize publications and identify that they belong to the respective organizations. The language used is direct and institutional.

Institutions such as IRA and Quercus also make posts with photographs, although these do not appear to receive the same treatment and editing as the photographs used by the organizations mentioned above. They seem to be more spontaneous and informal, sometimes used to publicize an event. In the case of Quercus, in addition to photographs, there are also posts with more creative elements. Like UNICEF Portugal and Amnesty International Portugal, the language used is also formal and institutional. In this aspect, the IRA is the organization that most diverges from all the others, presenting an informal and, at times, colloquial language, addressing users directly. There are times when it appears to be telling a story about the events and rescues that happen in the organization's daily life.

Unlike the previous organizations, ILGA, APAV, and Rede Ex Aequo do not have many photographs on social networks. Instead, they have slightly more creative publications, with elements other than photographs. Some of these publications are shared on ILGA and Rede Ex Aequo profiles. Their language tends to be a little formal, but accessible to everyone, trying to bridge the distance to the public. Regardless of the more informal or informal nature of the language used, all seven organizations sometimes use hashtags.

FINAL REMARKS: POSSIBLE ANSWERS TO OUR QUESTIONS

The main objective of this study was to contribute to understanding how activist institutions reach the public through the digital medium and how they use it to disseminate, communicate, and interact with the public in Portugal. After collecting, analyzing, and reflecting on the data obtained through theoretical and empirical research, it was possible to try and answer the posed research questions.

How do activist organizations use the digital medium, specifically their websites and social networks, to reach the public? Based on the literature, social media platforms are means through which users can freely express themselves, exchange opinions and information, interact, and connect. Upon realizing that the public is increasingly present on these digital platforms, organizations have also begun to use their services to effectively reach and mobilize a broader audience. Thus, from the analysis of the different organization's social media accounts, it was observed that posts can serve various purposes, but the posts that were more consistently present were informative ones.

On social media, differences in the objectives of each organization and different forms of practicing activism are evident, with adaptations according to the mission and the causes advocated. The website serves as a comprehensive source of information about the institutions, offering reliable and updated information. It also makes it easier to contact the institutions. Essentially, it serves as the primary source of information. Interviews also highlight the importance of the digital realm for organizations in reaching more individuals and mobilizing them for action.

How does the digital medium contribute to the dissemination of activist ideals? It does so in diverse ways. This research has emphasized that the global reach enabled by the Internet makes it possible for advocated causes and launched campaigns to reach various countries and, consequently, more people. Also, through social media and websites, it is easier to share information, posts, images, and videos of a specific event related to activist causes, with the possibility of gaining attention from others who were not present or who were unaware of what was happening. These platforms allow users to interact directly with organizations and seek clarification and information.

As everything happens swiftly in the digital realm, mobilization for activist actions can be publicized on websites and social media with full information, and users quickly become aware of what will happen on a particular day and time, just needing to show up. This aspect was evident in the analysis of social media, where the dissemination of various online and offline actions was carried out. It also allows users easy access to information about the organization, especially on the website, so they can understand what it stands for.

Furthermore, it also gives voice to minority groups, who otherwise may not have the opportunity to share their experiences and perspectives with a broader audience, since in the digital realm, anyone can create their content and make it available to others. Thus, public awareness is essential for activist organizations, and through projects and activities presented online, it is possible to spread their ideals.

What practices are carried out to ensure that digital activism contributes to offline activism? According to the interviews, there isn't a fixed and correct practice developed by organizations that leads the public to view activist content in the digital realm and then certainly engage in offline activist actions. Some organizations offer offline actions, such as signing petitions, and even in those cases, there is no certainty that all users who came across the petition's promotion would take the time to sign it. Similarly, the interviewees are aware of the conveniences that social media has provided for activism, but also of the disadvantages they have brought, in the sense that if a user shares, comments, or likes a post, they feel like they've done their part and don't seek to take action offline. Thus, practices of fake activism are present in the Portuguese activist landscape – probably on a lower scale than in other countries –, as organizations are aware of these practices.

Is the number of online followers reflected in the number of participants in offline activist activities? The number of followers an institution has on social media may not have a significant correlation with the interactions it receives on those platforms. In the social media mapping and analysis, it was observed that although organizations have many followers, the interactions they receive are sometimes not of great volume. The interactions may also be inconsistent, with certain posts garnering many interactions unlike others, thus depending on the post's content. Only one organization – IRA – showed significant interactions on Facebook and Instagram.

As mentioned earlier, organizations are not certain that those who interact with the post are mobilized and participate in activist actions. Several factors may lead a person to not participate in offline actions but engage online. Occasionally people may simply forget that an action was going to occur or they didn't feel like going at the time; they may also not have noticed the content of the post and only interacted to support the organization's social media. However, there are other more serious reasons regarding this aspect. According to one interviewee, certain individuals do not feel comfortable exposing themselves and being publicly associated with a cause, either because they belong to minorities or live in precarious situations, preferring to do their activism part in the digital realm.

In conclusion, thanks to this study, it was possible to confirm that the digital medium is indeed an advantage for activist practices and that it proves to be an asset for organizations, in the sense that the typical processes of activism – communicating, interacting, mobilizing, and raising awareness – have become facilitated. One of the main objectives of organizations is to reach a large number of people, so that they become aware of the cause, sympathize with it, and support it, and with the digital medium it has become possible to reach citizens worldwide.

After analyzing the online mapping and conducting the interviews, it became clear that organizations being present in the digital medium and its assistance in their everyday activist practices is crucial. Both social media accounts and the website aim to reach the public, whether through offline and online actions they promote or through the vast information and campaigns they make available on these platforms. The best thing organizations can do is to continue using social media to promote what they do, developing quality content so that if a post is shared, it captures someone's interest and possibly motivates them to be interested in the issue and be mobilized for activist action, both online and mainly offline. Social media and websites are shown to be communication channels that allow users to obtain information about organizations and to contact them from anywhere in the world. Overall, the digital and its possibilities are a relevant tool for the spread of activist ideals and actions.

REFERENCES

Abdalla, A., & et al. (2022). "Social Media as a Stage: A Behind the Scenes Analysis of Performative Activism, "Cancel Culture," and Effective Allyship". McMaster Undergraduate Journal of Social Psychology, Vol. 1, n.º1, pp. 83-122.

Amnesty International. (29 de maio de 2023). "Who We Are". Available: https://www.amnesty.org/en/who-we-are/

Amnistia Internacional (s.d.). "Como fazemos a mudança acontecer?" Available: https://www.amnistia.pt/como-atuamos/ (last access 30/05/2023).

APAV. (s.d.). "Quem Somos". Available: https://apav.pt/apav_v3/index.php/pt/aapav/quem-somos (last access 11/06/2023).

Campos, R., Pereira, I., & Simões, J. A. (2016). "Ativismo Digital em Portugal: Um Estudo Exploratório". Sociologia, Problemas e Práticas, n.º82, pp. 27 - 47. DOI:. DOI: 10.7458/SPP2016826977

Castañeda, E. (2020). Analysing Contemporary Social Movements. In Tilly, C., Castañeda, E., & Wood, L. (Eds.), *Social Movements* (Vol. 1768-2018, pp. 167–176). Routledge.

Castells, M. (2010). *The Rise of the Network* (2. ª ed., Vol. 1). Wiley-Blackwell.

Castells, M., & Cardoso, G. (2005). *A Sociedade em Rede: Do Conhecimento à Acção Política*. Imprensa Nacional - Casa da Moeda.

Chon, M., & Park, H. (2020). Social Media Activism in the Digital Age: Testing an Integrative Model of Activism on Contentious Issues. *Journalism & Mass Communication Quarterly*, 97(1), 72–97. DOI: 10.1177/1077699019835896

Cuofano, G. (2023). "What Is Relationship Marketing and Why It Matters in Business". https://fourweekmba.com/what-is-relationship-marketing/

Gerbaudo, P. (2017). From Cyber-Autonomism to Cyber-Populism: An Ideological History of Digital Activism. *TripleC*, 15(2), 477–489. DOI: 10.31269/triplec.v15i2.773

Glenn, C. L. (2015). Activism or "Slacktivism?": Digital Media and Organizing for Social Change. *Communication Teacher*, 29(2), 81–85. DOI: 10.1080/17404622.2014.1003310

Hayward, N. (2018). "A Short History of Activism". Available: https://www.womenshistory.org/resources/general/short-history-activism

ILGA. (s.d.). “Porque existimos”. Available: https://ilga-portugal.pt/associacao/porqueexistimos/ (last access 14/06/2023).

Jordan, T. (2001). Mapping Hacktivism: Mass Virtual Direct Action (MVDA), Individual Virtual Direct Action (IVDA) And Cyber-wars. *Computer Fraud & Security*, 2001(4), 8–11. DOI: 10.1016/S1361-3723(01)00416-X

Jordan, T. (2002). *Activism! Direct Action, Hacktivism and the Future of Society*. Reaktion Books.

Joyce, M. (12 de abril de 2011). “The 7 activist uses of digital tech: the case of popular resistance in Egypt”. Available: https://www.nonviolent-conflict.org/the-7-activist-uses-of-digital-tech-the-case-ofpopular-resistance-in-egypt/ (last access: 17/02/2023).

Kutlaca, M., & Radke, H. R. M. (2022). Towards an understanding of performative allyship: Definition, antecedents and consequences. *Social and Personality Psychology Compass*, •••, 1–12. DOI: 10.1111/spc3.12724

McCafferty, D. (2011). Activism Vs. Slacktivism. *Communications of the ACM*, 54(12), 17–19. DOI: 10.1145/2043174.2043182

Noland, A. (2019). “Like, Share, Retweet: Testing Competing Models of the Theory of Planned Behavior to Predict Slacktivism Engagement”. Journal of Nonprofit & Public Sector Marketing, 1 - 22, DOI: .DOI: 10.1080/10495142.2019.1589626

Quercus (s.d.). “Quem Somos – Apresentação”. Available: https://quercus.pt/apresentacao/ (last access 14/06/2023).

Rede Ex Aequo (s.d.). “A Rede Ex Aequo”. Available: https://www.rea.pt/quemsomos/ (last access 11/06/2023).

Ribeiro, L., & Rodrigues, S. (16 de novembro de 2018). “Animais maltratados: Conheça o IRA, o grupo que faz justiça pelas próprias mãos”. Available: https://visao.sapo.pt/atualidade/sociedade/2018-11-16-animais-maltratados-conheca-oira-o-grupo-que-faz-justica-pelas-proprias-maos/ (last access 30/05/2023).

Sandoval-Almazan, R., & Gil-Garcia, J. (2014). “Towards cyberactivism 2.0? Understanding the use of social media and other information technologies for political activism and social movements”. Government Information Quarterly, Vol. 31, n.º3, pp. 365-378. DOI:DOI: 10.1016/j.giq.2013.10.016

Silva, P. (2018). A Ação Coletiva: O Desafio da Mobilização. *Revista Movimentos Sociais e Dinâmicas Espaciais*, 7(2), 62–87. DOI: 10.51359/2238-8052.2018.238618

Tilly, C., Castañeda, E., & Wood, L. J. (2020). *Social Movements, 1768 - 2018* (4.ª ed.). Routledge. UNICEF (s.d.). "A UNICEF". Available: https://www.unicef.pt/unicef/a-unicef/ (last access 12/12/2022).

Van Dijck, J. (2013). Engineering Sociality in a Culture of Connectivity. In *J. Van Dijck, The culture of connectivity: a critical history of social media* (pp. 3–23). Oxford University Press. DOI: 10.1093/acprof:oso/9780199970773.003.0001

Vegh, S. (2003). Classifying Forms of Inline Activism: The Case of Cyberprotests against the World Bank. In McCaughey, M., & Ayers, M. D. (Eds.), *Cyberactivism - online activism in theory and practice* (pp. 71–95). Routledge.

Viana, L., & Sarmento, R. (2021). "Ativismo Digital em Âmbito Regional: Um Estudo Sobre a Presença Online de Movimentos Sociais no Interior de Minas Gerais#. Revista Teoria & Pesquisa, Vol.30, n.º1, 2-26. DOI: DOI: 10.31068/tp.30101

Wellman, M. L. (2022). #Black Squares for Black Lives? Performative Allyship as Credibility Maintenance for Social Media Influencers on Instagram#. *Social Media + Society*, 8(1), 1–10. DOI: 10.1177/20563051221080473

Chapter 10
Future Cyber Physical System Security

Naren Kathirvel
Anand Institute of Higher Technology, India

Kathirvel Ayyaswamy
https://orcid.org/0000-0002-5347-9110
Department of Computer Science and Engineering, Tagore Engineering College, Chennai, India

K. Gunasekaran
Department of Computer Science and Engineering, Panimalar Engineering College, Chennai, India

ABSTRACT

Over the past few decades, the manufacturing industry has become a hotbed of research on the rapidly developing trend of smart technology. Numerous cutting-edge technologies, including the internet of things, industrial internet of things (IOT), cyberphysical production systems, artificial intelligence, block chains, and systematic data analysis, were included in smart technology (ST). In order to close the gap between humans and machines, we give an overview of various smart technologies that can be used to regulate the sophisticated characteristics of computer and sensor technologies. The entire literature review on ST to improve machinery equipment production and maintenance, as well as prospective threats to ST, is presented in this research article. Additionally, the survey report comparing the adoption rate currently to that of the previous five years is included in this article in order to reduce the amount of human labor required in the industrial industry through the use of various sensors, cost-effective methodologies, and real-time outcomes.

DOI: 10.4018/979-8-3693-3579-6.ch010

1. INTRODUCTION TO CPS

Cyber-physical systems are undergoing significant change due to several trends: A greater degree of Internet of Things (IoT) integration. It is becoming more and more integrated with the Internet of Things (IoT), which is the process of connecting physical items to the internet through the use of sensors and other devices (Khan, Rehman, Zangoti, Afzal, Armi, & Salah, 2020)

(Mittal, Khan, Purohit, Menon, Romero, & Wuest, 2020). The use of artificial intelligence (AI) and machine learning (ML) technologies is one of the major developments in cybersecurity going forward. These intelligent algorithms are able to analyze large amounts of data and identify trends or anomalies that could indicate a potential cyberthreat. Cyber-physical systems connect physical infrastructure and things to the Internet and one another by integrating sensing, computation, control, and networking. Leading the way in promoting developments in the essential skills and knowledge needed to realize cyber-physical systems is NSF (Lu, Liu, Wang, Huang, & Xu, 2020)(Kusiak, 2019)(Hedberg, Bajaj, & Camelio, 2019).

Blockchain is a tamper-proof distributed digital ledger. This digital ledger is safe, secure, transparent, and decentralized, which simply means that it is not controlled by a single authority. In Blockchain, there is no central authority, and the ledger runs on multiple computers and doesn't require any single person to authenticate or settle transactions.

Absolutely! Blockchain technology indeed offers a distributed ledger system that allows multiple participants, or peers, to collaborate in creating a decentralized network. By leveraging a consensus algorithm, these peers can communicate and share information securely and transparently.

Instead, it relies on cryptographic techniques and distributed consensus mechanisms to ensure the integrity and trustworthiness of transactions. By removing the need for intermediaries or central control, Blockchain brings about increased security, immutability, and transparency. This makes it an ideal solution for various industries such as finance, supply chain management, healthcare, and more.

Cyber-Physical Systems (CPS) refer to engineered systems that integrate computation, networking, and physical processes. They are characterized by a tight integration of cyber (software, computation, data) and physical (mechanical, electrical, environmental) components, often interacting in real-time. These systems are foundational in many modern technologies and applications.

Key Features of CPS:

1. Integration of Physical and Cyber Elements:
    - Sensors and actuators gather and influence real-world data.
    - Embedded systems process and analyze data for decision-making.

2. Real-Time Operation:
   - Systems operate under strict time constraints to ensure synchronized physical responses.
3. Connectivity:
   - Often integrated into networks like the Internet of Things (IoT), allowing distributed systems to work collaboratively.
4. Adaptability and Autonomy:
   - Many CPS can self-adjust and optimize based on environmental conditions or operational data.

Applications of CPS:

1. Healthcare:
   - Smart medical devices (e.g., pacemakers, insulin pumps).
   - Telemedicine systems for real-time patient monitoring.

2.Transportation:

- Autonomous vehicles and intelligent traffic management systems.
- Rail system automation and smart aviation networks.

3. Manufacturing:

- Industry 4.0 technologies like smart factories.
- Robots and automated systems for production efficiency.

4. Energy Systems:

- Smart grids that monitor and optimize energy distribution.
- Renewable energy integration systems.

5. Infrastructure:

- Smart cities with adaptive lighting, water systems, and waste management.
- Building automation systems.

Challenges in CPS Development:

1. Security and Privacy:
   - Ensuring data integrity and protecting against cyberattacks.
2. Scalability:
   - Designing systems that can handle large-scale deployment.
3. Interoperability:
   - Seamless integration of diverse components and standards.
4. Real-Time Constraints:
   - Guaranteeing performance under strict time and resource limitations.

5. Safety and Reliability:
    - Minimizing risks in critical applications like healthcare or transportation.

CPS forms the backbone of many emerging technologies and is essential for future advancements in automation, artificial intelligence, and sustainable systems. Would you like to delve deeper into any specific application or aspect of CPS?

1.1 Cyber System Components

A distributed cyber system, a networking and communication component, and a physical system make up a CPS's three primary parts as shown in the Figure 1. Distributed hardware, software, and network components integrated in physical systems and environments make up the design of CPSs. They blend physical (sensors and actuators) and cyber (computation and communication) capabilities. CPS is used in almost every industry and setting, including manufacturing, aerospace, building systems, electric power grids, medical, and autos. The three Cs of optimal security are collaborative, consolidated, and comprehensive. Cybercriminals are always coming up with new strategies to take advantage of small and medium-sized enterprises, governments, and large organizations. Cybersecurity is the defense against cyberthreats for internet-connected devices, including data, software, and hardware (Yao, Zhou, Lin, et al., 2019)(Kusiak, 2018)(Kang, Lee, Choi, et al., 2016) (Qi & Tao, 2019)(Tao, Qi, Liu, & Kusiak, 2018).

Figure 1. Distributed cyber system components

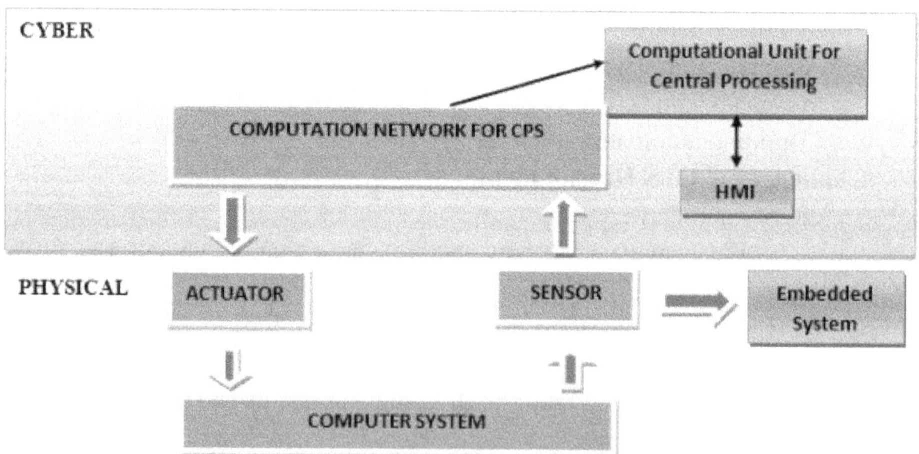

Both individuals and businesses employ this technique to guard against illegal access to data centers and other digital systems. In order to create a culture that is cognizant of cyber risk, the event is bringing together the four Ps of cyber security: people, process, policy, and partners. With CPS, reaction times will be faster, early failure detection will be possible, and bandwidth will be used properly. As a part of cloud computing, CPS can also give users resources according to their needs. Traditionally, cybersecurity and physical security measures have been seen as distinct endeavors (Wang, Ma, Zhang, Gao, & Wu, 2018)(Brougham & Haar, 2018)(Zhang, Xu, Liu, Lu, Xu, & Tao, 2019)(Uhlemann, Lehmann, & Steinhilper, 2017)(Yao, Zhou, Lin, et al., 2019). On the other hand, combining cybersecurity with physical security can lower costs, strengthen overall security posture, and enhance threat detection and response capabilities. Information systems can be infiltrated or destroyed by physical attacks; however cyberattacks carry serious dangers and may lead to malfunctioning physical components.

1.2 CPS Security

As a result, CPSs are required to follow and put into action policies that safeguard the physical and digital environments of their foundations for vital infrastructure. Cloud computing and cyber security grew more integrated during 2023, indicating a developing dependency. The increased use of cloud services has increased the potential and threats associated with digital security.

Millions of people's personal information could be compromised by a single security breech. Customers' trust is lost, and the organizations suffer severe financial consequences as a result of these breaches. Therefore, cyber security is crucial for shielding people and companies from spammers and other online criminals. Cyber-Physical Systems (CPS) are groups of computer and physical parts that work together to securely and effectively run an operation. Smart grids, robotics systems, water systems, and industrial control systems are a few examples of CPS.

Cyber Physical Systems Security (CPSs) It could be highly harmful for the CPS to not provide appropriate protection, depending on the circumstances (Rojas & Rauch, 2019)(Khan, Khan, & Haleem, 2020)(Tao, Qi, Liu, & Kusiak, 2018)(Mittal, Khan, Romero, & Wuest, 2018)(Wan et al., 2016). For instance, if a nuclear plant's usage of CPS has compromised its safety, the result could be a threat to the whole world.

1.3 Categories of Cybercrime

Application security guards against illegal usage and unauthorized access of linked data & apps.

Data security; cloud security; endpoint security; Internet of Things (IoT) security; mobile security; network security; critical infrastructure security; and so on.

1.4 Cyber Security Achieved in Ten Steps

• A system of dealing with risks. Evaluate the threats to the data and systems of your company by implementing a suitable risk management plan.

• Configured securely.

Monitoring, malware prevention, incident management, user education and awareness, network security, managing user privileges, and so on.

2. CYBER PHYSICAL PRODUCTION SYSTEMS

Cyber Physical Production System (CPPS) is a Smart Technology that is proposed for the artificial intelligence, this system is collection of collaborative technologies with this global world and its on-going process related to the further development of Computer and Communication technologies (CCT), this become the cause of fourth generation industry named as 4.0 (Pérez, Carou, Rubio, & Teti, 2020)(Zhu, Rodrigues, Leung, Shu, & Yang, 2018)(Xu, Yu, Griffith, & Golmie, 2018)(Liao, Rocha Loures, & Deschamps, 2018). The heap of raw data that is collected for the industrial manufacturing from different sensors is very helpful for detection of fault or to prognosticate equipment wear. There are different five level architectures of CPPS that produce an efficient productive system. The five C levels of architecture is described in the following diagram given in the Figure 2.

Figure 2. 5C architecture of cyber physical production system

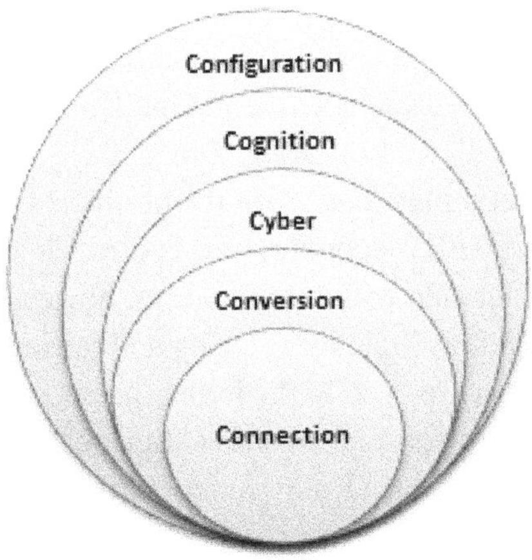

This five 5C surface level of CPS shows the close bonding from 1st level to 5TH level. All levels start from C, CPPS starts from the connection of self-senses and collection of data that has to be obtained from this system to network implementation to produce cost reducing products. The main architecture of Cyber physical production system is given in Figure 3.

Figure 3. How CPPS system works

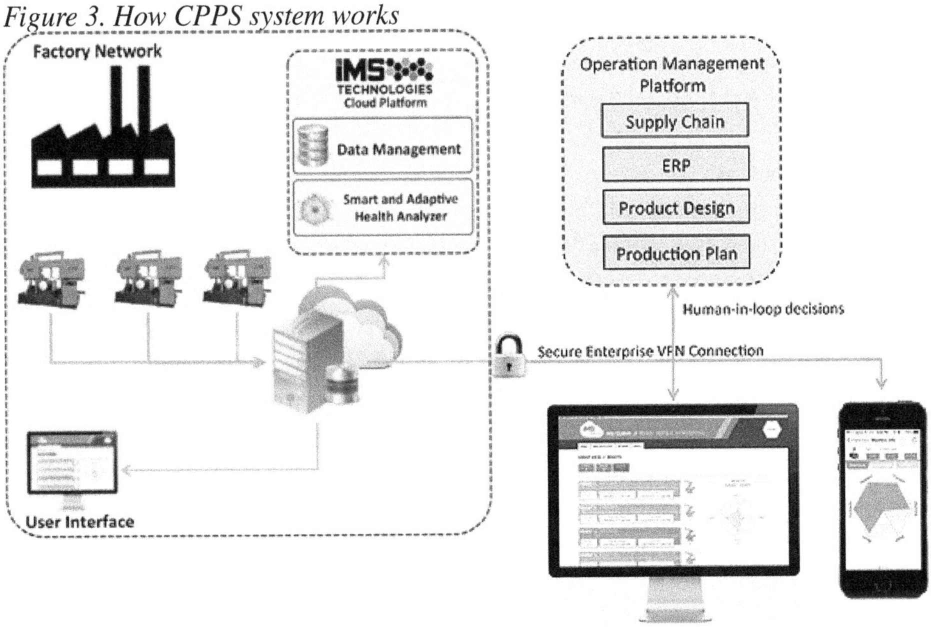

3. SENSORS AND AUTOMATIC IDENTIFICATION

Different type of sensors that are used in automatic identification of different factors like cost, temperature, motion, environmental, however most popular sensor that is used in automation called RFID (Radio Frequency Identification), theses sensors are used to meet the complete quality of the products in manufacturing sectors .The tags are used in the RFID technology to control the wireless technological devices, every product is labeled with different tags, however these tags are used to perform all particular activities including storing and retrieving of data in Figure 4. There are many sensors that are used in the industrial and automation industry like temperature sensor, pressure sensors, MEMs (Micro-electro-mechanical Sensors, motion sensors and torque sensors (Kathirvel & Naren, 2024).

Figure 4. Type of sensors

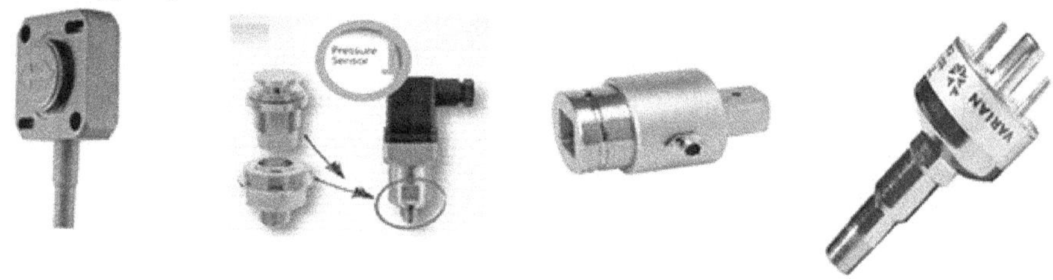

3.1 Sensors and Actuators

IoT and robotics rely on two fundamental technologies: sensors and actuators. These components are crucial for IoRT systems to function properly, as they provide essential functionalities and well-defined interfaces for identification and reaction. In contrast to IoT, the practical functionality of IoRT building blocks is based on sensors and actuators. Robotic Interaction Services (RoIS) define HRI functions and the usage of external building blocks, as well as abstracting the hardware in service robots. This standard allows for the development of HRI apps that can be used on both devices and gateways (Kathirvel, 2024a). Actuation is a crucial aspect of robotics that enables the manipulation of items, movement of people and products, and even the functioning of automated doors. Numerous actuation approaches, including automated strategy and execution by multiple robots, have been developed in the field of robotics.

3.2 Emerging IORT Technologies

Cost effective a solid-state semiconductors (CMOS) photographic sensors with proactive illumination are needed for IoRT applications to operate correctly in a variety of weather situations, including sunlight, darkness, rain, fog, and dust (Ayyaswamy, 2024). These sensors must have high accuracy and resolution for both vertical object identification and horizontal road surface scanning. Most sensors now only provide 2D sensing data, therefore sensor fusion is primarily concerned with 2D representation.

However, more height data, 3D visualization, and the combining of actuators and sensors are needed to better future IoRT activities. Robotic objects require a customized environment model that utilizes existing and new sensor technology in order to generate an ideal 3D environmental model that balances resource consumption and maximized performance. LIDAR systems, which use a revolving, scanning mirror to provide a complete picture, give autonomous robotic objects and cars 360-degree vision.

Figure 5. Yearly shift from non IoT to IoT devices

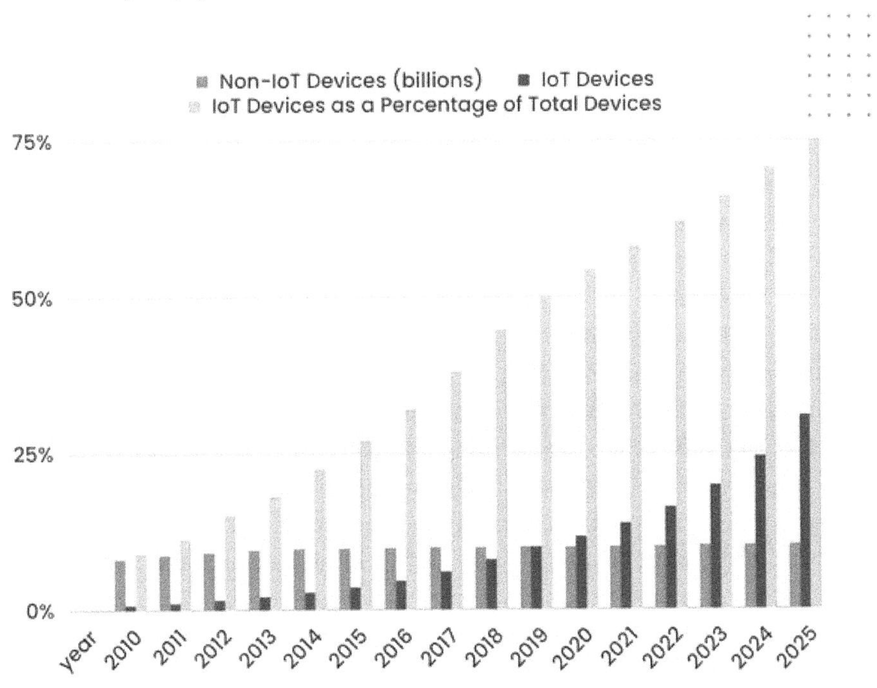

For self-driving, autonomous robotic things, LIDAR systems provide precise 3D data on the immediate surroundings that can be utilized for obstacle avoidance, motion vector analysis, object recognition, and collision prediction. However, LIDAR systems are ineffective for close-range control, so radars must be added to autonomous robotic objects and vehicles. The 76-81 GHz frequency range is typically used for radar because of its RF propagation characteristics and necessary resolution (Kathirvel, 2024c).

IoT devices around the world are taking over non IoT devices. The US, China, as well as the developed countries in the European Union and the Asia Pacific region, are leading IoT markets. The region is predicted to lead in expenditure and uptake in the worldwide market, in line with other APAC-related trends (Kathirvel & Naren, 2024d)(Kathirvel, Rithik, & Naren, 2024e). According to market predictions, China will account for two-thirds of the industrial IoT market in the APAC region by 2025 as shown in the Figure 5.

3.3 Communication Technologies

The communication infrastructure of IoRT necessitates novel methods in order to allow virtualization of operations on current computing engines and facilitate the usage of such infrastructures in diverse fields. These strategies should make it possible to exchange data streams (important for 3D-awareness and imaging systems), communicate internally, and use edge computing. Time-critical communication is essential for collision avoidance, which significantly lowers accidents and collisions.

IoRT commonly operates local robots using networking technology and remote robots using designated white spectrum frequencies (Kathirvel & Gobinath, 2024f). IoT employs machine-to-machine communication and implements emerging protocols like LoRa and SIGFOX, as well as 4G, WiFi, and Bluetooth.However, this level of interoperability and service development is more challenging and requires semantic data from other disciplines.

Due in part to the dynamic nature of IoRT conditions and their dependence on applications, contexts, and use cases, the ability to discover and classify services of objects often presents significant challenges. IoRT systems are built on communication protocols, which enable connectivity to networks and application interaction.

In order to exchange data over the network, edge devices and autonomous devices use a variety of communication protocols that define exchange of data formats, data encryption, device addressing schemes, and packet transportation from origin to destination. Various Wireless Local Area Network (WLAN) communication standards, such as 802.11a, 802.11b and 802.11g, 802.11n, 802.11ac, and 802.11ad, are included in the 802.11 - Wi-Fi protocols that are used.

These standards offer communication ranges of between 20 meters (indoors) and 100 meters (outdoors) with data speeds ranging from 1 Mb/s to 6.75 Gb/s. Various IoRT communication techniques can be used to achieve shared real-time computation, data exchange, and internal communication (Gobinath, Kathirvel, Kanna, & Annamalai, 2024g).

3.4 Voice Recognition, Voice Control

Nowadays, chatbots and devices with microphones are commonly used for communication. As the IoRT advances, more endpoints are created for people and robots to interact (Kathirvel & Maheswaran, 2023a). This leads to the emergence of a new digital experience where robotic objects and humans collaborate, resulting in IoRT mesh development through mutual contact between robotic entities (Pérez, Carou, Rubio, & Teti, 2020).

People should be able to communicate with robot fleets in more natural ways for IoRT applications that include tour-guiding, elder care, recovery, search and rescue, monitoring, education, general support in everyday settings, and in workplaces, factories, and homes. Using knowledge of the robot's movements and postures, powerful noise-reduction techniques can be used for speech recognition and voice control (Zhu, Rodrigues, Leung, Shu, & Yang, 2018).

To reduce background noise, the microphone's quality is crucial for automatic speech recognition (Kathirvel, Maheswaran, Subramaniam, & Naren, 2023b) (Kathirvel, Naren, Nithyanand, & Santhoshi, 2024h). A speech recognition automation system for robots can accurately identify voices from adults as well as kids in crowded spaces thanks to the reduction of reverberation, disruption, and noise in a portable multiple channels of communication technique with an outlier-robust broader side-lobe canceller method and a feature-space noise reduction criterion (Xu, Yu, Griffith, & Golmie, 2018).

3.5 Machine Learning as Enabler for Adaptive

The Internet of Things (IoT) community recognizes the significance of incorporating Machine Learning (ML) techniques into IoT devices for effective management of the vast and diverse sensory data generated by network nodes (Kathirvel, Gopinath, Naren, Nithyanand, & Nirmaladevi, 2024i). This integration enables IoRT solutions to adjust to different environments, while IoRT apps can learn from their surroundings and experiences. The ML service needs to be intelligent and distributed enough to encompass every IoRT node, including those at the network's edge. These forecasts might be used as inputs for an innovative method of learning on an alternative network node, which would then carry out additional forecasts and data fusion operations (Gobinath, Kathirvel, & Kathirvel, 2024). This would result in the development of a smart network of machine learning algorithms that would carry out progressive compilations of the sensed data.

3.6 Marketplace for an IoRT Ecosystem

Mechanisms for the monetization of service components and data are required to encourage participation and the development of an IoRT ecosystem.A market place must be established as the center of an IoRT ecosystem. A marketplace facilitates registration and the discovery of offerings, as well as the data or attributes that services provide. These services may act on robots or objects directly or as separate components accessible through IoRT platforms. Selling and buying of products and services take place in the marketplace, which acts as a central location for transactions.

- Offering authentication, where a supplier of an offering can provide a metadata description for the marketplace to ingest and index to help with discovery.
- Offering discovery, where a customer searches for offerings using a marketplace search interface. To facilitate data registration and discovery, common ontologies for the semantic description of data must be established.
- Role and privileges management could be used to secure access to the market. Authentication and authorization can also be used to secure access to the marketplace.
- Reputation management, which involves consumer evaluations of suppliers and their goods; these evaluations may be taken into consideration when selecting search outcomes and during investigation.

4. BIG DATA ANALYTICS

There is colossal amount of data also require a highly productive computing system, the demand of high amount of data that is collected than analyzed how to handle, how to work with this data, some specific data collection techniques are used in the smart manufacturing technique to reduce the cost, time constraints and budget issue, there is also need of collection of correct raw data, Big Data analytic system play as a key role in the manufacturing sector. Made research on the big data analytic industries and their challenges in his research he also defined the term Manufacturing Internet of Things (MIOT). Perform the big analytic data technologies and give comprehensive reviews of different big analytic data in smart manufacturing and give enhanced briefing on how to get productive results.

5. BLOCKCHAIN TECHNOLOGY

In the latest smart technologies, Blockchain has a key role in IIOT leading towards the next generation in the manufacturing industrial sector. Block chain Technology does not involve not only in manufacturing sector but also in healthcare, finance supply chain and car insurance. Ethereum virtual Machine is also established, mainly the nodes that are present in ork.Latest research shows how Blockchain can be developed for DT(Digital Twins) to produce the authentic, efficient and securing manufacturing.

5.1 Artificial Intelligence

Artificial Intelligence is a key element of ST, due to AI the Industry 4.0 come into exist, without artificial intelligence there may be no concept of ST, the success of any smart technology depends upon the rate at which level of AI is involved on that technique, the term AI and ML(machine learning) are related to each other. The motivation behind the AI and ML to create ST as to increase the fast process of analysis and decision making, SNA (Social Network Analysis) derived from the Social Network theories. In 2018, MHI and his team take a survey of different ST that are used in the whole world, this survey take long time of five years given in Table 1.

Table 1. Survey report of MHI in 2018

Name of ST	Rate of Adoption in Manufacturing sector
Cloud Computing	57%
Inventory and Network Optimization	44%
Sensors and Automatic Detection	45%
Predictive Analysis	20%
Internet of things	22%
Robotics and Automation	34%
Block chain	6%
Driverless Vehicle and Drones	11%
3D Printing	16%
Artificial Intelligence	6%

5.2 Serverless Computing

As serverless providers find ways to get around some of the problems with serverless computing, the technology continues to advance. Cold beginnings are one of these disadvantages.

The provider typically deactivates a serverless service that hasn't been used for a while in order to save energy and avoid over-provisioning. When a user launches an application that calls it later, the serverless provider will have to start it up again and host that function. Term "cold start" refers to the substantial lag that this initial period contributes.

Once the function is operational, following requests will be delivered significantly more quickly (warm begins); but, if the function is not used again for some time, it will eventually go for dormant.

In order to solve this issue, Cloudflare Workers has spun up serverless tasks ahead of time, during the TLS handshake. The outcome is a FaaS platform with no cold starts since Workers functions spin up at the edge in much less time—even less time than it takes to finish the handshake. Check out our Developer guide to learn how to use Cloudflare Workers.

We should anticipate a greater use of serverless architecture when the disadvantages of doing so are resolved and edge computing gains traction.

Simplified Backend Code: With FaaS, programmers can create basic applications that each perform a specific function, such initiating a request to an API, independently.

Faster Turnaround: Serverless architecture has an opportunity to significantly shorten the time to market. to release corrections for bugs and fresh functionality, developers don't need to go through a laborious deploy procedure; instead, they may add and edit code as needed.

5.2.1 Example of Serverless Computing

Global leaders in business have been using serverless computing to provide their clients with high-availability, high-performance online services. Among the noteworthy instances are:

5.2.1.1 Major League Baseball Advanced Media (MLBAM)

The Major League Baseball Association is one of the larger and most established athletic associations in the USA. The company runs a program called Statcast that gives customers access to precise and up-to-date sports analytics. You may perform intricate searches on the Statcast website using information such as pitch velocity, pitch type, season type, and individual player names. It can provide accurate data and help users make decisions about baseball games by utilizing serverless computing.

5.2.1.2 Autodesk

For the bandwidth-intensive and mission-critical construction, architecture, and engineering sectors, Autodesk provides robust software. It released a new tool named Tailor that could swiftly generate unique Autodesk accounts for businesses with all the required setups. With just two FTES handling the solution, Autodesk was able to launch Tailor in just two weeks by utilizing a serverless architecture.

5.2.1.3 Netflix

Netflix, one of the biggest over-the-top (OTT) video companies in the world, has long supported serverless computing. Since 2017 and before, it has been utilizing serverless to create a platform that can manage thousands of modifications every day. The only thing that Netflix developers have to do is define the adapter code, which controls how the platform reacts to user requests and system requirements. At the core of Netflix's exclusive Dynamic Scripting Platform is serverless architecture, which manages platform modifications, provisioning, and end-user delivery.

6. BLOCKCHAIN NETWORKS

A Blockchain network is a technical infrastructure that allows applications to access ledger and smart contract services. Smart contracts are primarily used to originate transactions, which are then transmitted to each peer node in the network and recorded immutably on their copy of the ledger. End-users using client applications or Blockchain (Iansiti & Lakhani, 2017) network administrators are examples of app users (Sudha & Kathirvel, 2023a)(Sudha & Kathirvel, 2023b) (Sudha & Kathirvel, 2022a)(Naveneethan, Madhan, & Kathirvel, 2022)(Kathirvel, Subramaniam, Naveneethan, Das, & Kirubakaran, 2022a).

Orders, accounts, payments, production and much more may be tracked using a Blockchain network. You can see all facts of a transaction end to end since members share a single view of the truth, providing you with greater confidence and additional efficiencies and opportunities.

6.1 Types of Blockchain

There are different types of Blockchain that are being used currently as shown in the Fig 6.

Figure 6. Types of block chain

	Public	Private	Hybrid	Consortium
Permissioned/ Permissionless	Permissionless	Permissioned	Permissioned & Permissionless	Permissioned
Control	No control by a central authority	Control by a central authority	Control by a central authority	Control by multiple central authorities
Main Advantages	✓ Independence ✓ Transparency	✓ Performance ✓ Scalability	✓ Performance ✓ Low Cost	✓ Performance ✓ Security
Main Disadvantages	✕ Performance ✕ Scalability Issues	✕ Security ✕ Trust	✕ Transparency ✕ Upgrading	✕ Transparency
Examples	Bitcoin Litecoin	Hyperledger Fabric	XRP token	Corda Quorum

A. Public Blockchain

A public Blockchain is completely decentralized and does not have a single entity that controls the network. In this type of Blockchain, anyone with an internet connection can join the network and participate in reading, writing, or auditing within the Blockchain. Hence anyone can review anything at a given point of time on a public Blockchain.

Simply put, anyone can use public Blockchain (Kathirvel, Subramaniam, Naveneethan, & Sabarinath, 2021a) from anywhere to input data and transactions and review them as long as they are connected to the network.

Public Blockchain can also be called Permissionless Blockchain because it allows any user to create a personal address and join the Blockchain network, i.e., become a 'node' of the network. Also, the public Blockchains do not restrict the rights of the nodes on the Blockchain network.

These types of Blockchain are more transparent and secure than permissioned Blockchains because there are many nodes to validate transactions. Thus, it would be difficult for bad actors to collude on the network. Also, it is not possible to modify or alter the data once it has been validated on the Blockchain.

(i) To be part of a public Blockchain, one needs to download the code without needing any permission from anyone and start running a public node on their local device. Once this is done, he can then validate transactions in the network, thus participating in the consensus process — the process for determining what blocks can be added to the Blockchain.

(ii) Anyone in the world can send transactions through the network and can expect them to be included in the Blockchain if they are valid.

(iii) In a public Blockchain, anyone in the world can access and read Blockchain transactions using a block explorer, which is a program or a website that allows a user to search and navigate the blocks of a Blockchain, their contents, and relevant details.

(iv) Speed of transactions in the public Blockchain is slower than in the case of private Blockchain.

Examples: To date, public Blockchains are primarily used for exchanging and mining cryptocurrency. A few examples of cryptocurrency public Blockchain platforms include Bitcoin, Ethereum, Litecoin, etc. Ethereum is actually the most popular public Blockchain at present.

B. Private Blockchain

The second type of Blockchain is Private Blockchain (Subramaniam, Kathirvel, Sabitha, & Basha, 2021b)., where the network participants have control over who can join the network and who can participate in the consensus process.

Access controls in the network will vary for every participant, and the Regulatory Authority or central organization would decide on various activities inside the network starting from joining the network until executing any of the functionalities in this closed network.

In this platform, the digital identities of the participants need to be managed and monitored by the Regulatory Authority.

Therefore, it is not necessarily decentralized even among its members but is a centralized system that uses distributed ledger technology.

Private Blockchain offers several advantages over public Blockchain:

(i) In the case of the public Blockchain, the credentials of nodes are not known; therefore, nobody can decipher who these validators are, which increases the risk of malicious activities. But in the case of a private Blockchain (Xu, Yu, Griffith, & Golmie, 2018), the credentials of participating nodes are present on the Blockchain; therefore, it is much easier to track where the fraud happened.

(ii) Additionally, since private Blockchains are generally smaller than public Blockchains and have far fewer participants, therefore it takes less time for the network to reach a consensus. As a result, more transactions can take place. Private Blockchains can process thousands of transactions per second. Also, they take up a lot less energy and power to validate transactions. On the contrary public Blockchain networks, which usually have thousands of computers to verify transactions, consume tons of energy.

(iii) Private Blockchain provides exciting opportunities for businesses to leverage trustless and transparent activities for internal and business-to-business use cases. For instance, banks and financial institutions can use private Blockchain. With the help of smart contracts, this technology could eventually replace many centralized businesses.

C. Consortium or Federated Blockchain

The third type of Blockchain is consortium Blockchain that tries to remove the sole autonomy which gets vested in just one entity as in private Blockchain.

There is a group of companies or representative individuals making decisions in the best interest of the whole network. Such groups are called consortiums or federations (Kathirvel & Naren, 2024).

These nodes have the authority to read or write transactions, and they can also allow or restrict participants on the network. But none of the nodes alone can add a block to the Blockchain.

To add a block, every single node has to approve the block. The mechanism of reaching a decision to add a block is based on the voting system, also referred to as the Proof of Vote mechanism. The purpose of this mechanism is to follow up on the selected nodes.

Here every node will need to vote in order to validate a block. The number of votes required will be pre-determined in order to reach a decision of adding the block to the Blockchain. This means that it has been decided in the code that for fifteen selected nodes, it might take ten nodes' votes or even fifteen nodes' votes to validate a block. By using the Proof of Vote mechanism, the network helps to ensure that no one is misusing their power (Kathirvel, Subramaniam, Naveneethan, & Sabarinath, 2021a).

D. Permissioned Blockchain

Both Private and Federal Blockchains are types of Permissioned Blockchain. Permissioned Blockchain maintains an access control layer to allow certain actions to be performed only by certain identifiable participants.

On such Blockchain platforms, there is a need of special permissions to read, access, and write information on them. The platform will allow anyone to join the Blockchain network after suitable verification of their identity, and then permissions are allocated and designated to perform only certain activities on the network.

The roles of each participant who can access and contribute to the Blockchain are defined.

Permissioned Blockchain has many other applications like payments, KYC validation, escrow, insurance claims, charity, music publishing, and much more. In a nutshell, the permissioned Blockchain offers a secure business environment with customization that allows wider industry adoption across multiple enterprises.

E. Hybrid Blockchain

The Blockchain which combines the best of both private and public Blockchains. Simply put, a hybrid Blockchain provides controlled access and freedom simultaneously. In other words, the hybrid Blockchain lets organizations set up a private, permission-based system alongside a public permissionless system.

With such a system, organizations can control -

- who can access what data on the Blockchain,
- what data can be opened up publicly,
- what data needs to be kept confidential on the private network?

Fusing different features from private and public Blockchains ensures that an organization can work with its stakeholders in the best possible way.

Like a consortium Blockchain, a hybrid Blockchain has the privacy benefits of a permissioned Blockchain.

But, unlike a consortium Blockchain with multiple participants collectively helping to maintain the network, a hybrid Blockchain can have a single entity network administrator.

Figure 7. Hybrid blockchain has features of both public and private blockchains

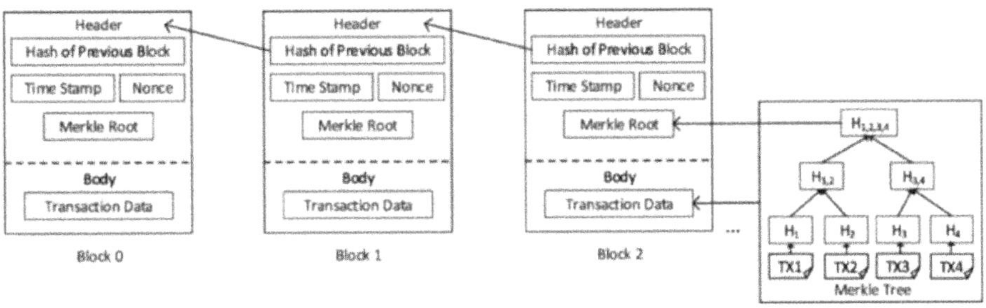

A transaction in a private network of a hybrid Blockchain is usually verified within that network. But users can also release it in the public Blockchain to get it verified.

In the hybrid Blockchain, there are two different types of users/nodes based on the level of information that they access:

(i) The first type of users/nodes are those who are part of the private Blockchain and have all the control over the Blockchain and can decide the level of security permissions for a particular user.
(ii) And the other type of users/nodes are those who are part of the public Blockchain and can just access the data released on the Blockchain.

Even though a set of individuals control the hybrid Blockchain network, they cannot change the transactions' immutability and security. They can only control which transactions are made public and which are not.

6.2 Block in the Blockchain

Blockchain Structure: Indeed, a Blockchain is a sequential chain of blocks, where each block contains a list of transactions and other pertinent information.
Content of Blocks: Blocks include transaction data, timestamps, and other metadata crucial for network operation and consensus.
Linear Connectivity and Security: Blocks are linked together in a linear fashion, with each block containing a cryptographic hash of the previous block. This linking ensures the integrity and immutability of the Blockchain.
Block Header: The header of each block contains vital information such as the hash of the previous block, the current block's hash, a nonce (a number used in mining), the Merkle root (a hash of all transactions in the block), and other relevant details.
Linear Connection via Previous Block Hashes: As you correctly noted, the hash of the previous block is used to compute the hash of the current block, creating a chain-like structure.
Genesis Block: The first block in a Blockchain is referred to as the Genesis Block. It serves as the foundation of the Blockchain and does not reference any previous block as shown in the Figure 7.
Consensus Mechanisms for Adding Blocks: Consensus mechanisms such as Proof of Work (PoW), Proof of Stake (PoS), and others are employed by Blockchain networks to facilitate agreement on the addition of new blocks. These mechanisms ensure that the network reaches consensus on the validity of transactions and the order of blocks in the Blockchain.

conceptual image of a block with reference to a ledger of transactions

Figure 8. Visualization of simpler conceptual block

The above image will create simpler conceptual block visualization in your head. However, the actual block contains a lot more information than the ledger image above.

6.2.1 Elements of a Block

- **Block Height:** The sequence number of the block in the Blockchain. The genesis block has a block height of 1, and subsequent blocks increment this value.
- **Block Size:** A 4-byte field indicating the size of the block in bytes. It helps determine the storage and transmission requirements of the block.
- **Block Reward:** The amount rewarded to the miner for successfully adding the block to the Blockchain. This incentive encourages miners to participate in the network and validate transactions.
- **Tx Count:** The number of transactions contained within the block. It provides information about the volume of transactions processed in the block.
- **Block Header:** An 80-byte field containing metadata about the block. It includes various essential details, such as:
- **Time:** The timestamp indicating when the block was mined, used for validating transactions and maintaining the Blockchain's chronological order.
- **Version:** A 4-byte field representing the protocol version used in the block. It helps ensure compatibility and upgradeability of the Blockchain protocol.

- **Previous Block Hash:** A 32-byte field containing the cryptographic hash of the previous block. This establishes the sequential linking of blocks in the Blockchain, ensuring its immutability.
- **Bits:** A 4-byte field indicating the difficulty level for adding the block. Miners adjust their computational effort (hash rate) based on this difficulty to maintain the Blockchain's security.
- **Nonce:** A 4-byte field containing a variable number used in the Proof of Work (PoW) consensus mechanism. Miners modify the nonce to find a valid block hash that satisfies the difficulty requirement.
- **Merkle Root:** A 32-byte field representing the root hash of the Merkle tree constructed from individual transaction hashes within the block. It efficiently summarizes all transactions in the block, facilitating quick verification.

6.2.2 Transactions

It's a variable-size field that includes the list of all transactions contained in the block. Each **bitcoin block** contains about 2000 transactions. The size of each block is approx 1MB. The size and number of transactions in a block vary in Blockchains. It's decided based on network congestion and communication overhead.

6.2.2.1 Hash

1. **Unique Identifier**: Just like a fingerprint uniquely identifies an individual, a hash uniquely identifies a block and all its contents. No two blocks can have the same hash, making it an excellent tool for distinguishing between different blocks in the Blockchain.
2. **Tamper Detection**: Any modification to the contents of a block, no matter how small, will result in a completely different hash. This property makes hashes invaluable for detecting tampering or unauthorized changes to block data. If the hash of a block changes, it indicates that the block has been altered, and its integrity has been compromised.
3. **Simultaneous Generation**: When a block is created, its hash is generated simultaneously. This ensures that the hash accurately represents the current state of the block's data. Any subsequent changes to the block will require regenerating the hash to reflect those alterations.
4. **Security Tool**: Hashes serve as a powerful security tool in Blockchain networks. They provide a tamper-evident seal for each block, making it computationally infeasible for malicious actors to alter block data without being detected.

In summary, hashes are essential components of Blockchain technology, providing a secure and immutable record of transactions. They enable tamper detection, ensure data integrity, and contribute to the overall security of the Blockchain network.

6.2.2.2 Timestamp

The fourth essential element that every block contains is the timestamp. Timestamp simply means "a proof that some data existed at a particular date and time." In other words, the timestamp can be referred to as "Proof of existence." Any digital data can be timestamped. The hash of the block containing data and transactions is timestamped and is then published on the network. By doing so, it is ensured that the transactions have existed at this point in time. Implementing a timestamp on the block also makes the block impossible to be repeated in the future since, in addition to the time, the date of creation of the block is also stored. Therefore, there is no possibility that any block in the future can be assigned a repeated hash that was given to any of the previous blocks a week, two months, or a year ago (Zhu, Rodrigues, Leung, Shu, & Yang, 2018).

6.2.2.3 Nonce

The fifth element of a block is nonce. The nonce is an abbreviation for "number only used once." The Nonce is an integer number that, along with the block number, data, and previous hash, serves as an input for the hashing algorithm to calculate the valid hash for the block. A valid hash for the block is a hash that meets a certain difficulty i.e. contains a number of predefined zeros at the beginning of the hash.

A **nonce** is basically a random number; it is important to find a nonce value as it helps to generate a valid hash containing 4 leading O's to make the specific block valid. One way to obtain a nonce value is by changing the nonce manually. Thus, it may take several iterations until the desired hash with four leading O's is generated.

Once the block becomes valid, it becomes a part of the Blockchain, and new blocks are added to this block.

Since the nonce value can only be used once, it plays an important role to keep the Blockchain immutable. Suppose the data is changed on Block 2, because of which the hash changes as data is used to calculate the hash. Also, Block 2 becomes invalid because its hash no longer has four leading 0's. Block 3's hash changes because Block 2's hash was used to calculate Block 3's hash. Also, Block 3 becomes invalid because its hash no longer has four leading 0's. The same is the case with other blocks in the Blockchain.

The only way to mutate Block 2 is to mine the block again to find a valid hash with four leading 0's and then mine all the blocks after to find their valid hash because all the blocks are linked together with the hash of the previous block. Since new blocks are always being added, it's nearly impossible to mutate the Blockchain.

6.3 Blockchain Structure

The Blockchain is a proficient combination of two hash-based data structures

6.3.1 Linked List

This is the structure of the Blockchain itself, which is a linked list of hash pointers. A regular linked list consists of nodes. Each node has 2 parts- data and pointer. The pointer points to the next node. In the Blockchain, simply replace the regular pointer with a hash pointer.

6.3.2 Merkle Tree

A Merkle tree is a binary tree formed by hash pointers, and named after its creator, Ralph Merkle as shown in the Figure 8.

Figure 9. Merkle tree

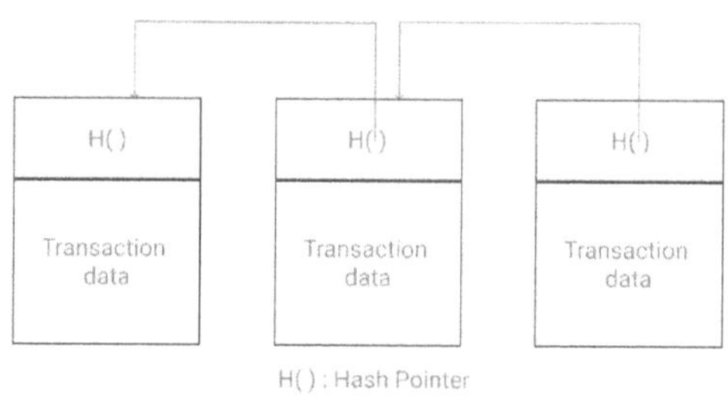

6.3.3. Block Structure

Block header: The header data contains metadata of the block, i.e information about the block itself see Figure 9. The contents of the block header include:

- Hash of the previous block header
- Hash of the current block

- Timestamp
- Cryptographic nonce
- Merkle root

6.3.4 Merkle Tree

1. **Hash Pointer-based Linked List**: One approach to storing transactions within a block is to create a hash pointer-based linked list of transactions and store this linked list in the block. However, this approach becomes impractical for large numbers of transactions, as it can lead to inefficiencies when searching for a specific transaction.
2. **Merkle Tree**: To address the inefficiencies of the hash pointer-based linked list approach, Merkle trees are used. A Merkle tree is a binary tree formed by hash pointers, where each leaf node represents an individual transaction. The interior nodes of the tree are computed by hashing the concatenation of the hashes of their children. This process continues until a single root hash is computed, which represents all transactions in the block.
3. **Efficient Verification**: The Merkle tree allows for efficient verification of transactions within a block. Instead of traversing the entire block and its linked list of transactions, a proof of membership can be generated using a logarithmic number of hash operations. This proof consists of a subset of hashes from the Merkle tree, enabling nodes to efficiently verify whether a particular transaction is included in the block without needing to access all transactions.
4. **Blockchain Structure**: In the context of the Blockchain, each block consists of a header and transactions. The transactions are organized in a tree-like structure, known as the Merkle tree, within the block. The block header contains essential metadata, including the Merkle root hash, which serves as a summary of all transactions in the block.

The Merkle tree plays a crucial role in optimizing the storage and verification of transactions within a block, enhancing the efficiency and scalability of the Blockchain network.

Figure 10. Blockchain structure

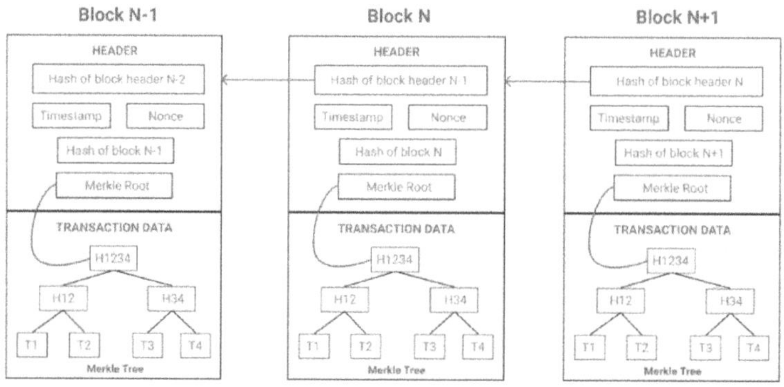

6.3.5 Merkle Tree Structure

Figure 11. Merkle tree structure

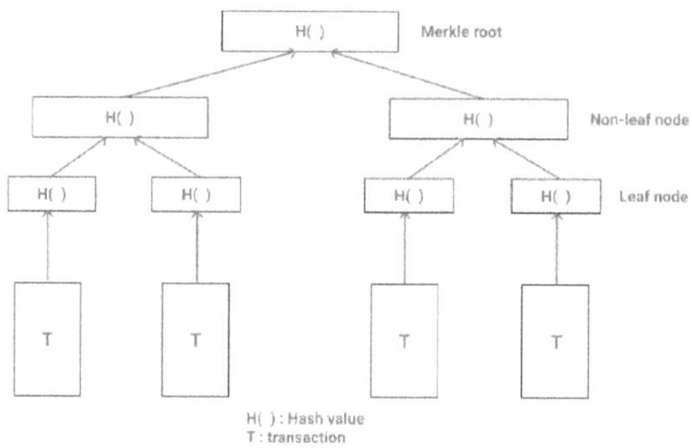

1. A Blockchain can potentially have thousands of blocks with thousands of transactions in each block. Therefore, memory space and computing power are two main challenges.
2. It would be optimal to use as little data as possible for verifying transactions, which can reduce CPU processing and provide better security, and this is exactly what Merkle trees offer.
3. In a Merkle tree, transactions are grouped into pairs. The hash is computed for each pair and this is stored in the parent node. Now the parent nodes are grouped into pairs and their hash is stored one level up in the tree shown in the Figure 10. This continues till the root of the tree.

The different types of nodes in a Merkle tree are:

Root Node: The root of the Merkle tree is known as the Merkle root and this Merkle root is stored in the header of the block.
Leaf Node: The leaf nodes contain the hash values of transaction data. Each transaction in the block has its data hashed and then this hash value (also known as transaction ID) is stored in leaf nodes.
Non-Leaf Node: The non-leaf nodes contain the hash value of their respective children. These are also called intermediate nodes because they contain the intermediate hash values and the hash process continues till the root of the tree.

4. Bitcoin uses the SHA-256 hash function to hash transaction data continuously till the Merkle root is obtained.
5. Further, a Merkle tree is binary in nature. This means that the number of leaf nodes needs to be even for the Merkle tree to be constructed properly. In case there is an odd number of leaf nodes, the tree duplicates the last hash and makes the number of leaf nodes even.

6.3.6 How Do Merkle Trees Work?

A Merkle tree is constructed from the leaf nodes level all the way up to the Merkle root level by grouping nodes in pairs and calculating the hash of each pair of nodes in that particular level. This hash value is propagated to the next level. This is a bottom-to-up type of construction where the hash values are flowing from down to up direction.

Hence, by comparing the Merkle tree structure to a regular binary tree data structure, one can observe that Merkle trees are actually inverted down as shown in the Figure 11.

Figure 12. Work flow of Merkle trees

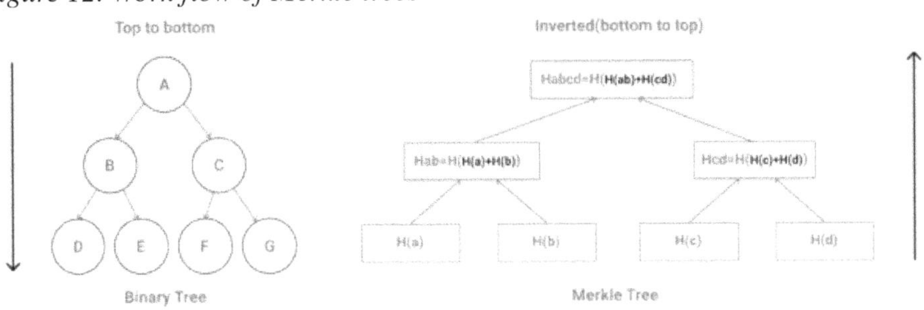

Example: Consider a block having 4 transactions- T1, T2, T3, T4. These four transactions have to be stored in the Merkle tree and this is done by the following steps-

Step 1: Compute the hash of each transaction. For example, H1 = Hash(T1).
Step 2: Store the hashes of individual transactions in the leaf nodes of the Merkle tree.
Step 3: Form non-leaf nodes by pairing leaf nodes from left to right and computing the hash of each pair. For example:

- Compute H12 = Hash(H1 + H2).
- Compute H34 = Hash(H3 + H4).
- These computed hashes, H12 and H34, become parent nodes.

Step 4: Continue pairing and computing hashes until a single hash, the Merkle root, is obtained. For example:

- Compute H1234 = Hash(H12 + H34).
- H1234 becomes the Merkle root, representing all transactions in the block.

By following this process, a Merkle tree efficiently summarizes all transactions in a block into a single root hash. This root hash can then be used to verify the integrity of transactions in the block without needing to access and hash each individual transaction separately.

Figure 13. Work flow of Merkle roots

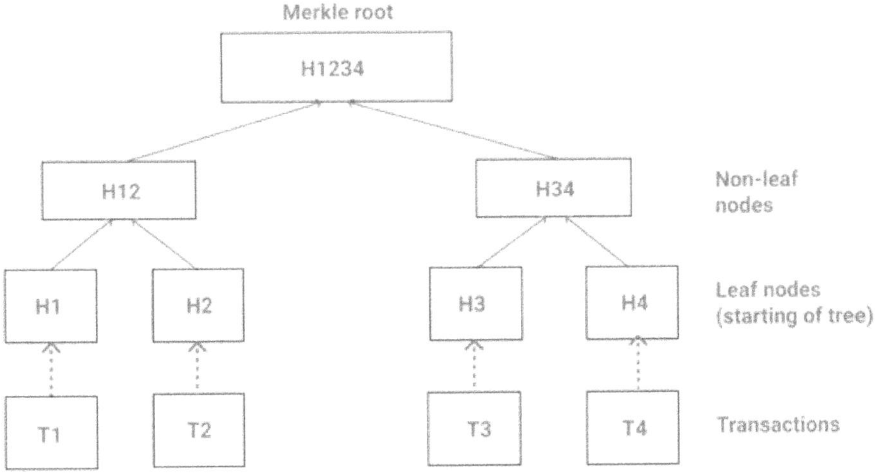

1. In order to check whether the transaction has tampered with the tree, there is only a need to remember the root of the tree (refer Figure 12).
2. One can access the transactions by traversing through the hash pointers and if any content has been changed in the transaction, this will reflect on the hash stored in the parent node, which in turn would affect the hash in the upper-level node and so on until the root is reached.
3. Hence the root of the Merkle tree has also changed. So Merkle root which is stored in the block header makes transactions tamper-proof and validates the integrity of data.
4. With the help of the Merkle root, the Merkle tree helps in eliminating duplicate or false transactions in a block.
5. It generates a digital fingerprint of all transactions in a block and the Merkle root in the header is further protected by the hash of the block header stored in the next block.

7. DISCUSSION

7.1 Present and Future Challenges

Due to the diverse and complex nature of ST due to the diversity of communication, network intelligence, there are many challenges that are present in this sector that should be resolved. Some of these challenges are given below:

7.1.1 Technical Staff

The technical staff is required to run this complex and compatible architecture, however processing of efficient ML, understanding of communications networks and smooth running of these topologies over the network with the better understanding of fast data driven techniques and operating systems. If the staff is technical, then it is easy to handle data managing schemes and machine integration with the software strategies to gain efficient results of ST IN manufacturing sector to meet the security, sustainability of computer related results.

7.1.2 Difficult to Handle Huge Data Analytic and Management Techniques

There are many data analytic techniques to cover the collection of data, sensing of data, and process of future decision on the bases of these data management techniques, the running of different IIOT to run in an effective way by handling and retrieving a large amount of data.

7.1.3 System Integration

The compatibility of different smart technology with machine learning, the new manufacturing system required IPV6 connectivity for smooth running of different interfaces. From the previous decades many technologies and platforms are used to attain the productive result for the manufacturing sector, this is a bit difficult to integrate all the platforms. The technology moved from mechanical manufacturing to man production and then proceeded further to IT (information Technology) and now a day's cyber physical production system is used.

7.1.4 Big Data Analytic Tools

This is difficult to handle a large amount of data and then analysis of that data has occurred, many data analytic tools are used, and the big issue behind these tools is sharing and maintaining that record on different networks.

7.1.5 Robustness and Security Issue

To cover the security issue, cyber and many artificial intelligence tools are used but lack of security is also occurring in different technologies over the network, the trust issue in the use of smart technology arises because all the data between different resources are through use of different communication networks.

7.1.6 Use of Wireless Technologies and Different Protocols

This is not easy to use the different technologies by use of wireless, all the communication between the machine and human is done through the use of networks by using different topologies over the centralized or distributed network is used. There are many communication networks and wireless technologies are used, this is not easy to take decision which communication network is better for the productive manufacturing system. There are many technical issues that are related to communication networks like latency bandwidth and many more factors.

7.1.7 Invention of Specific Operating System

There is a need for specific operating systems that are used in manufacturing sectors.TinyOS and ConTiki are the most commonly used operating systems that meet the requirement of the smart manufacturing techniques.

7.1.8 Supply Chain is Complex

The challenge behind the smart manufacturing system is to handle large supply chains between different stakeholders and technologies that are involved in ST all over the world and there is requirement to remove all the conflicts among the communication between different platforms that are used in the smart Industrial Manufacturer.

7.1.9 Customer Trust Involvement

The product that should evolved from smart manufacturing should be system integrated, reliable and durables techniques are used in this term, there are many wireless technologies including many challenges that are used in IOT and Industrial 4.0.The customer should be awarded of these technologies by making effective software model that are human to machine oriented

8. CONCLUSION

This research paper gives perspective review of smart technologies that produce the energetic and profitable products, involving wireless communication by using different operating systems using reliable and effective system and communication protocols. End-to-end visibility enabled by Blockchain technology refers to the clear and complete oversight of a process or supply chain, spanning from its initiation to its conclusion. This is made possible by leveraging Block chain's unique capabilities, where each stage of a transaction generates a data block stored in a public ledger. These blocks are interconnected via cryptographic keys, encoding information about preceding blocks. Any attempt to alter a link in the chain triggers significant changes in the key, immediately noticeable to all network participants. However, smart technologies are also emerging towards robotic technology to reduce the involvement of human machines and human workload. Digital Twins involve the use of sensor technologies and have many benefits to reduce the cost of the products and enhance the quality of products. This fosters transparency, builds trust among stakeholders, and mitigates risks such as counterfeiting and fraud. Ultimately, embracing Blockchain-driven operations management offers a competitive advantage in today's dynamic market landscape.

REFERENCES

Ayyaswamy, K. (2024b). Enhancing Digital Technology Planning, Leadership, and Management to Transform Education. In Bhatia, M., & Mushtaq, M. T. (Eds.), *Navigating Innovative Technologies and Intelligent Systems in Modern Education* (pp. 1–9). IGI Global., DOI: 10.4018/979-8-3693-5370-7.ch001

Brougham, D., & Haar, J. (2018). Smart Technology, Artificial Intelligence, Robotics, and Algorithms (STARA): Employees perceptions of our future workplace. *Journal of Management & Organization*, 24(2), 239–257. DOI: 10.1017/jmo.2016.55

Gobinath, V., Ayyaswamy, K., & Kathirvel, N. (2024). Information Communication Technology and Intelligent Manufacturing Industries Perspective: An Insight. *Asian Science Bulletin*, 2(1), 36–45. DOI: 10.3923/asb.2024.36.45

Gobinath, V. M., Kathirvel, A., Rajesh Kanna, S. K., & Annamalai, K. (July 2024g) “Chapter 5: Smart Technology in Management Industries: A Useful Perspective”, Artificial Intelligence Applied to Industry 4.0, Wiley Publiser, USA, 1st Edition, pp.93-106, , ISBN 978-1-394-21611-6, DOI eBook ISBN, https://www.wiley.com/en-us/Topics+in+Artificial+Intelligence+Applied+to+Industry+4+0-p-9781394216116DOI: 10.1002/9781394216147.ch5

Hedberg, T. D., Bajaj, M., & Camelio, J. A. (2019). Using Graphs to Link Data Across the Product Lifecycle for Enabling Smart Manufacturing Digital Threads. *ASME Journal of Computing and Information Science in Engineering*, 1, 213–224.

Iansiti, M., & Lakhani, K. R. (2017). The truth about blockchain. *Harvard Business Review*, 95(1), 118–127.

Kang, H. S., Lee, J. Y., & Choi, S.. (2016). “Smart manufacturing: Past research, present findings, and future directions”, *Int. J. of Precis. Eng. and Manuf.-. Green Tech.*, 3, 111–128.

Kathirvel, A. (2024a). Applications of Serverless Computing: Systematic Overview. In Aluvalu, R., & Maheswari, U. (Eds.), *Serverless Computing Concepts, Technology and Architecture* (Vol. 221-233). IGI Global., DOI: 10.4018/979-8-3693-1682-5.ch014

Kathirvel, A. (2024c). Innovation and Industry Application: IoT-Based Robotics Frontier of Automation in Industry Application. In Satishkumar, D., & Sivaraja, M. (Eds.), *Internet of Things and AI for Natural Disaster Management and Prediction* (pp. 83–105). IGI Global., DOI: 10.4018/979-8-3693-4284-8.ch004

Kathirvel, A., & Gobinath, V. M. (2024f) "Chapter 1: A Review on Additive Manufactuing in Industrial", Modern Hybird Machince and Super Finishing Process: Technology and Application, CRC Publiser/Chapman and Hall, New York, 1st Edition, ISBN 1032354291, DOI eBook ISBN 9781032354293, https://books.google.co.in/books/about/Modern_Hybrid_Machining_and_Super_Finish.html?id=I_VW0AEACAAJ&source=kp_book_description&redir_esc=y

Kathirvel, A., Gopinath, V. M., Naren, K., Nithyanand, D., & Nirmaladevi, K. (2024i). Manufacturing Smart Industry Perspective an Overview. *American Journal of Engineering and Applied Sciences*, 17(1), 33–39. DOI: 10.3844/ajeassp.2024.33.39

Kathirvel, A., & Maheswaran, C. P. (2023a) "Chapter 8: Enhanced AI-Based Intrusion Detection and Response System for WSN", Artificial Intelligence for Intrusion Detection Systems, CRC Publiser/Chapman and Hall, New York, 1st Edition, ISBN 978-100096755-5, 978-103238665-2 DOI eBook ISBN 9781003346340, https://www.taylorfrancis.com/chapters/edit/10.1201/9781003346340-8/enhanced-ai-based-intrusion-detection-response-system-wsn-kathirvel-maheswaranDOI: 10.1201/9781003346340-8

. Kathirvel A, Maheswaran C P, Subramaniam M, and Naren A K, (March 2023b.) "Chapter 25: Quantum Computers Based on Distributed Computing Systems for the Next Generation: Overview and Applications", Quantum Computers Based on Distributed Computing Systems for the Next Generation, Handbook of Research on Quantum Computing for Smart Environments, IGI Global Publiser, Hershey, Pennsylvania, USA, ISBN13: 9781668466971 ISBN10: 166846697X EISBN13: 9781668466988, DOI: Quantum Computers Based on Distributed Computing Systems for the Next Generation: Overview and Applications: Computer Science & IT Book Chapter | IGI Global (igi-global.com)DOI: 10.4018/978-1-6684-6697-1.ch025

Kathirvel, A., & Naren, A. K. (2024). Critical Approaches to Data Engineering Systems Innovation and Industry Application Using IoT. In *Critical Approaches to Data Engineering Systems and Analysis*. IGI Global. DOI: 10.4018/979-8-3693-2260-4.ch005

Kathirvel, A., & Naren, A. K. (2024d). Diabetes and Pre-Diabetes Prediction by AI Using Tuned XGB Classifier. In Khang, A. (Ed.), *Medical Robotics and AI-Assisted Diagnostics for a High-Tech Healthcare Industry* (pp. 52–64). IGI Global., DOI: 10.4018/979-8-3693-2105-8.ch004

Kathirvel, A., Naren, K., Nithyanand, D., & Santhoshi, B. (2024h). Overview of 5G Technology: Streamlined Virtual Event Experiences. *Advances of Robotic Technology*, 2(1), 1–8. DOI: 10.23880/art-16000109

Kathirvel, A., Rithik, G., & Naren, A. K. "Chapter 11: Automation of IOT Robotics." In *Predicting Natural Disasters With AI and Machine Learning,* edited by D. Satishkumar and M. Sivaraja, 224-249. ISBN 9798369322802, eBook ISBN 9798369322819, Hershey, PA: IGI Global, 2024e. https://doi.org/DOI: 10.4018/979-8-3693-2280-2.ch011

Kathirvel, A., Subramaniam, M., Navaneethan, S., & Sabarinath, C. "Improved IDR Response System for Sensor Network", Journal of Web Engineering, Vol. 20 1, 53–88, 2021a. Publisher: © River Publishers, 15 Feb 2021, ISSN: 1540-9589 (Print Version). ISSN: 1544-5976 (Online Version), https://doi.org/DOI: 10.13052/jwe1540-9589.2013

Kathirvel, A., Sudha, D., Naveneethan, S., Subramaniam, M., Das, D., & Kirubakaran, S. (2022a). AI Based Mobile Bill Payment System using Biometric Fingerprint. *American Journal of Engineering and Applied Sciences*, 15(1), 23–31. DOI: 10.3844/ajeassp.2022.23.31

Khan, S., Khan, M. I., & Haleem, A. (2020) "Prioritisation of Challenges Towards Development of Smart Manufacturing Using BWM Method", Alam M., Shakil K., Khan S. (eds) Internet of Things (IoT). Springer publisher. DOI: 10.1007/978-3-030-37468-6_22

Khan, W. Z., Rehman, M. H., Zangoti, H. M., Afzal, M. K., Armi, N., & Salah, K. (2020). Industrial internet of things: Recent advances, enabling technologies and open challenges. *Computers & Electrical Engineering*, 81, 1–13. DOI: 10.1016/j.compeleceng.2019.106522

Kusiak, A. (2018). Smart manufacturing. *International Journal of Production Research*, 56(2), 508–517. DOI: 10.1080/00207543.2017.1351644

Kusiak, A. (2019). Fundamentals of smart manufacturing: A multi-thread perspective. *Annual Reviews in Control*, 47, 214–220. DOI: 10.1016/j.arcontrol.2019.02.001

Liao, Y., de Freitas Rocha Loures, E., & Deschamps, F. (2018). Industrial Internet of Things: A Systematic Literature Review and Insights. *IEEE Internet of Things Journal*, 5(6), pp4515–pp4525. DOI: 10.1109/JIOT.2018.2834151

. Yuqian Lu, Chao Liu, Kevin I-Kai Wang, Huiyue Huang, Xun Xu, Digital Twin-driven, (2020), "smart manufacturing: Connotation, reference model, applications and research issues", Robotics and Computer-Integrated Manufacturing, Vol. 61, pp1-14.

Mittal, S., Khan, M. A., Purohit, J. K., Menon, K., Romero, D., & Wuest, T. (2020). A smart manufacturing adoption framework for SMEs. *International Journal of Production Research*, 58(5), 1555–1573. DOI: 10.1080/00207543.2019.1661540

Mittal, S., Khan, M. A., Romero, D., & Wuest, T. (2018). A critical review of smart manufacturing & Industry 4.0 maturity models: Implications for small and medium-sized enterprises (SMEs). *Journal of Manufacturing Systems*, 49, 194–214. DOI: 10.1016/j.jmsy.2018.10.005

Naveneethan, S. Madhan E.S., Kathirvel A. (2022) Identifying and Eliminating the Misbehavior Nodes in the Wireless Sensor Network. In: Reddy V.S., Prasad V.K., Wang J., Reddy K. (eds) Soft Computing and Signal Processing. ICSCSP 2021. Advances in Intelligent Systems and Computing, vol 1413. Springer International Publishing, Springer, Singapore. https://doi.org/DOI: 10.1007/978-981-16-7088-6_36

Pérez, M., Carou, D., Rubio, E. M., & Teti, R. (2020). Current advances in additive manufacturing. *Procedia*, 88, 439–444.

Qi, Q., & Tao, F. (2019). A Smart Manufacturing Service System Based on Edge Computing, Fog Computing, and Cloud Computing. *IEEE Access : Practical Innovations, Open Solutions*, 7, 86769–86777. DOI: 10.1109/ACCESS.2019.2923610

Rojas, R. A., & Rauch, E. (2019). A conceptual framework of enablers for smart manufacturing control. *International Journal of Advanced Manufacturing Technology*, 104(1-4), 517–533. DOI: 10.1007/s00170-019-03854-4

Subramaniam, M., Kathirvel, A., Sabitha, E., & Anwar Basha, H. (2021b) "Modified Firefly Algorithm and Fuzzy C-Mean Clustering Based Semantic Information Retrieval", Journal of Web Engineering, Vol. 20 1, 33–52,. Publisher: © River Publishers, 15 Feb 2021, ISSN: 1540-9589 (Print Version). ISSN: 1544-5976 (Online Version), https://doi.org/DOI: 10.13052/jwe1540-9589.2012

Sudha, D., & Kathirvel, A. (2022a).An Intrusion Detection System to Detect and Mitigating Attacks Using Hidden Markov Model (HMM) Energy Monitoring Technique. Stochastic Modeling an Applications, 26(3) (January – June, Special Issue 2022 Part – 9), 467-476, ISSN: 0972-3641.

Sudha, D., & Kathirvel, A. (2023a) The performance enhancement of Aodv protocol using GETUS. International Journal of Early Childhood Special Education (INT-JECSE), Vol. 15, Issue 2, Pp. 115-125. ISSN: 1308-5581. DOI:DOI: 10.48047/INTJECSE/V15I2.11

Sudha, D., & Kathirvel, A. (2023b)The effect of ETUS in various generic attacks in mobile ad hoc networks to improve the performance of Aodv protocol. International Journal of humanities, law, and social sciences, Kanpur philosophers, Vol. 9 Issue 1, N0.1, 467-476, 2022. ISSN: 2348-8301.

Tao, F., Qi, Q., Liu, A., & Kusiak, A. (2018). Data-driven smart manufacturing. *Journal of Manufacturing Systems*, 48, 157–169. DOI: 10.1016/j.jmsy.2018.01.006

Tao, F., Qi, Q., Liu, A., & Kusiak, A. (2018). Data-driven smart manufacturing. *Journal of Manufacturing Systems*, 48, 157–169. DOI: 10.1016/j.jmsy.2018.01.006

Uhlemann, T. H.-J., Lehmann, C., & Steinhilper, R. (2017). The Digital Twin: Realizing the Cyber-Physical Production System for Industry 4.0. *Procedia CIRP*, 61, 335–340. DOI: 10.1016/j.procir.2016.11.152

Wan, J.. (2016). Software-Defined Industrial Internet of Things in the Context of Industry 4.0. *Journal of Sensors*, 16(20), pp7373–pp7380.

Wang, J., Ma, Y., Zhang, L., Gao, R. X., & Wu, D. (2018). Deep learning for smart manufacturing: Methods and applications. *Journal of Manufacturing Systems*, 48, 144–156. DOI: 10.1016/j.jmsy.2018.01.003

Xu, H., Yu, W., Griffith, D., & Golmie, N. (2018). A Survey on Industrial Internet of Things: A Cyber-Physical Systems Perspective. *IEEE Access : Practical Innovations, Open Solutions*, 6, pp78238–pp78259. DOI: 10.1109/ACCESS.2018.2884906 PMID: 35531371

Yao, X., Zhou, J., Lin, Y., Li, Y., Yu, H., & Liu, Y. (2019). Smart manufacturing based on cyber-physical systems and beyond. *Journal of Intelligent Manufacturing*, 30(8), pp2805–pp2817. DOI: 10.1007/s10845-017-1384-5

Yao, X., Zhou, J., Lin, Y., Li, Y., Yu, H., & Liu, Y. (2019). Smart manufacturing based on cyber-physical systems and beyond. *Journal of Intelligent Manufacturing*, 30(8), pp2805–pp2817. DOI: 10.1007/s10845-017-1384-5

Zhang, X., Xu, X., Liu, A., Lu, Q., Xu, L., & Tao, F. (2019). Blockchain-Based Trust Mechanism for IoT-Based Smart Manufacturing System. *IEEE Transactions on Computational Social Systems*, 6(6), pp1386–pp1394. DOI: 10.1109/TCSS.2019.2918467

Zhu, C., Rodrigues, J. J. P. C., Leung, V. C. M., Shu, L., & Yang, L. T. (2018). Trust-Based Communication for the Industrial Internet of Things. *IEEE Communications Magazine*, 56(2), 16–22. DOI: 10.1109/MCOM.2018.1700592

KEY TERMS AND DEFINITION

Blockchain: A blockchain is a public ledger that is digitally distributed and decentralised via a network.

CCT: fourth generation industry named as 4.0

CPPS: Smart Technology that is proposed for the artificial intelligence, this system is collection of collaborative technologies with this global world

MEMs: Micro-electro-mechanical Sensors, motion sensors and torque sensors

RFID: Cost, temperature, motion, environmental, however most popular sensor that is used in automation called Radio Frequency Identification

RoIS: Robotic Interaction Services define HRI functions and the usage of external building blocks, as well as abstracting the hardware in service robots.

WLAN: Wireless Local Area Network communication standards, such as 802.11a, 802.11b and 802.11g, 802.11n, 802.11ac, and 802.11ad, are included in the 802.11 - Wi-Fi protocols that are used.

Chapter 11
China's Digital Silk Road and Its Implications for Africa's Digital Transformation

Mohamad Zreik
https://orcid.org/0000-0002-6812-6529
Sun Yat-sen University, China

ABSTRACT

This chapter delves into China's strategic Digital Silk Road initiative and its impact on Africa's digital transformation. It looks at the impact that Chinese IT giants have had on the African digital environment and how the continent may learn from China's experience. This chapter acts as a guide through the opportunities and problems presented by topics such as data security and digital sovereignty. In-depth case studies of China's digital projects in Africa show the scope and potential of these efforts. Its goal is to help African stakeholders reap the most benefits from and limit any negative consequences of this relationship, so fostering long-term, equitable progress in the digital sphere for all.

INTRODUCTION

The fast economic and social development of China is reflected in the country's rapidly evolving digital landscape. China has become a digital technology superpower during the past two decades. Chinese IT giants like Alibaba, Tencent, and Huawei have been instrumental in this shift by delivering game-changing digital technologies in a variety of markets, from e-commerce and social media to telecom and AI (Hong

DOI: 10.4018/979-8-3693-3579-6.ch011

& Wei, 2022). The Chinese government has made significant investments in digital infrastructure and policies, which has helped to increase the country's population's access to and understanding of digital technologies. The internet economy now makes a sizable contribution to China's GDP, and the country's digital models are influencing developments in other countries (Qin et al., 2022).

Africa, on the other hand, is only getting started on its path to digital transformation, but it has enormous promise. The continent is enjoying a tech boom, propelled by a younger, more computer-savvy population and the rise of mobile technologies despite difficulties such as low internet penetration, digital illiteracy, and inadequate infrastructure. Mobile money services such as Kenya's M-Pesa demonstrate the continent's digital innovation by making banking services available to millions of previously unbanked people (Burns, 2018). E-commerce, agri-tech, and ed-tech businesses in particular are helping to address regional issues and fuel expansion in those industries. However, there are major discrepancies in the adoption and use of digital technology between countries and demographics throughout Africa. Africa needs to overcome these obstacles and take use of collaborations, investments, and knowledge from digitally sophisticated nations like China if it is to reach its full potential.

China's ambitious plan to increase its worldwide influence through digital connectivity and cooperation, the Digital Silk Road (DSR), was unveiled in 2015. The Digital Silk Road is China's plan to build a global network of digital infrastructure such underwater cables, satellite networks, and e-commerce links as part of the larger Belt and Road Initiative (Keane & Yu, 2019). It also involves spreading the word about cutting-edge innovations like 5G, AI, and cloud computing. The DSR reaffirms China's long-term goal of building an international digital economy in which it plays a dominant role (Kokas, 2022).

The DSR provides an opportunity for Africa to address some of its most serious digital concerns, helping the continent speed its digital transformation. State-backed Chinese tech firms are already heavily invested in establishing Africa's digital infrastructure. This include, but is not limited to, the installation of fiber-optic connections, the growth of telecommunications networks, and the provision of inexpensive cellphones.

Additionally, China supplies essential funding and technical experience through the DSR, which can help bridge Africa's digital gap. It also facilitates the exchange of digital governance models, which can serve as a guide for African countries as they regulate and control the digital ecosystems within their borders (Parepa, 2020). It is beneficial for African countries to engage critically with the Digital Silk Road, exploring how this initiative can be aligned with their interests to promote digital inclusion and realize its full potential.

In-depth analysis of the crucial relationship between China's Digital Silk Road and Africa's growing digital transformation will be presented in this chapter. It will begin by explaining the DSR in depth, including its goals, priority sectors, and connection to the broader Belt and Road Initiative. Later the chapter will examine the impact of Chinese IT behemoths like Huawei, Tencent, and Alibaba on the continent of Africa's digitization efforts. It will analyze how China's involvement in ICT projects in Africa is changing the continent's digital landscape and what that could mean for the continent's future economic and social development via the use of specific case studies.

The chapter will also attempt to learn something from the digital development in China. It will examine whether and how African countries might benefit from adopting the digitization tactics that have driven China's economic success. This chapter will focus heavily on the prospects and difficulties of Africa's digital cooperation with China. Data security, digital sovereignty, and the benefits and drawbacks of Africa's growing digital partnership with China will all be discussed critically. In-depth case studies of individual projects and collaborations will show readers the breadth and depth of China's digital endeavors across Africa. These will help demonstrate the results of these efforts and the scope for future cooperation. The chapter will look at how China's participation in Africa's digital transformation fits into the bigger picture. The chapter will also provide recommendations for how African countries might best position themselves to reap the rewards of this partnership while minimizing its potential drawbacks. The overarching goal is to help African politicians, investors, scholars, and stakeholders better grasp China's role in the continent's digital revolution.

This chapter seeks to explore the critical research question: 'What are the implications of China's Digital Silk Road initiative on Africa's digital transformation, and how can African countries navigate the opportunities and challenges presented by this collaboration?' To address this question, the chapter is structured around three main arguments. First, it examines the transformative impact of Chinese technology giants in shaping Africa's digital landscape, emphasizing the potential benefits and risks associated with their involvement. Second, it analyzes the lessons Africa can learn from China's rapid digital development, considering how these insights can be applied to accelerate Africa's digital growth while maintaining digital sovereignty and security. Third, the chapter delves into the broader geopolitical and economic implications of the Digital Silk Road, discussing how African nations can strategically engage with China to ensure equitable and sustainable digital advancement. By exploring these arguments, the chapter aims to offer a comprehensive and coherent analysis of the complex dynamics at play in the evolving digital partnership between China and Africa.

CHINA'S DIGITAL SILK ROAD: AN OVERVIEW

Definition and Objectives of the Digital Silk Road

The DSR, as an extension of China's Belt and Road Initiative, holds particular relevance for Africa's digital development. However, its significance extends beyond mere infrastructural development. This chapter seeks to explore the nuanced implications of the DSR for Africa, not by stating its importance outright but by constructing an argument based on Africa's unique digital landscape. The chapter examines the DSR's alignment with Africa's digital aspirations, its potential to catalyze technological advancement, and the socio-economic transformations it could engender within the African context. This analysis involves a deep dive into the specific needs, challenges, and opportunities in Africa's digital ecosystem, thereby situating the DSR within a framework that is both contextually relevant and critical to understanding its potential impact on the continent.

China's Digital Silk Road is a large-scale initiative that began in 2015 as part of the broader Belt and Road Initiative with the goal of improving digital infrastructure, fostering e-commerce ties, and increasing scientific and technology collaboration among countries along the Belt and Road. China's goal of building a digitally interconnected world, with China at its heart, is reflected in the DSR (Ly, 2020).

The DSR's primary goals are expansive. The first goal is to increase China's worldwide clout by use of digital infrastructure. China's goal is to establish a globally integrated digital economy through the development and improvement of information and communication technology (ICT) infrastructure in BRI countries (Wang, Zhou, & Wang, 2021).

The DSR allows China to promote and sell its digital technology and standards to international markets. The effort relies heavily on cutting-edge technologies, including 5G networks, AI, and cloud storage. Chinese internet titans like Huawei, Alibaba, and Tencent play an important role in this area by serving as catalysts for initiatives in a variety of nations.

The DSR's ultimate goal is to increase digital collaboration between member nations. The promotion of shared digital governance principles and the dissemination of relevant information are all part of this (Okano-Heijmans & Vosse, 2021). China's DSR is an effort to increase its clout in the international political and economic spheres through way of the Internet.

Connection to the Belt and Road Initiative

The Belt and Road Initiative, China's grandiose program introduced in 2013, includes the Digital Silk Road as a central component. The BRI, which has been likened to the Silk Road of old, is an international development strategy that will see infrastructure built and investments made in approximately 140 countries and international organizations across Asia, Europe, Africa, the Middle East, and the Americas (Zreik, 2022).

As the BRI expanded beyond its initial concentration on physical infrastructure including roads, railroads, ports, and energy projects, it spawned the Digital Silk Road in 2015 (Silin et al., 2018). The Digital Silk Road is a digital extension of the BRI that aims to build networks and promote digital collaboration among member countries.

There is mutual reinforcement between the DSR and BRI. Construction of fiber optic cables, telecommunication networks, and satellite systems are all made easier by the BRI's provision of physical infrastructure (Hillman et al., 2021). On the flip side, the BRI's physical infrastructure investments, like ports and trains, benefit from the DSR's digital connection, which is enabled by innovations like AI, cloud computing, and the Internet of Things.

Identification of the Key Sectors of Focus

The goal of the Digital Silk Road is to improve digital connection and cooperation between countries along the route. Information and communication technology infrastructure development is a key area of attention. Fiber optic cables, telecommunications networks, satellite systems, and smart city development all fall under this category. These systems are essential to the functioning of the global digital economy because they facilitate cross-border data transmission, processing, and storage.

The DSR also places emphasis on the digital technology industry, especially emerging technologies like 5G, AI, the cloud, and big data. Technology and standards in these fields are increasingly being exported to DSR nations by Chinese digital titans like Huawei, Tencent, and Alibaba (Zreik, 2023).

Internet commerce is yet another important industry. China plans to improve the digital economy and trade with its DSR partners by facilitating the expansion of cross-border e-commerce along the DSR using a variety of platforms. For instance, Alibaba's eWTP aims to make international trade more accessible to small and medium-sized businesses (Johnston, 2021).

The DSR also places an emphasis on digital cooperation and governance. Collaboration and information sharing among the countries in the DSR region on policies and best practices could be advantageous. Having successfully governed its own digital economy for so long, China is now in a position to help other countries do the same. China's goal in exporting its digital financial services and platforms to DSR nations is to increase financial inclusion and boost economic growth as a result of China's position as a leader in fintech innovation.

CHINA'S INFLUENCE ON AFRICA'S DIGITAL TRANSFORMATION

Examination of How Chinese Tech Giants are Shaping Africa's Digital Infrastructure

Huawei has been playing a crucial role in developing Africa's ICT infrastructure, particularly in the areas of telecommunications and network services. In contrast, Tencent and Alibaba, while not primarily focused on infrastructure, have significantly influenced the digital landscape through their innovative digital platforms and e-commerce solutions. These companies are in a prime position to aid in Africa's digital transformation thanks to their technological expertise, financial resources, and experience in shaping China's digital economy.

Huawei is a pioneer in this effort. It has played a crucial role in the development of 2G, 3G, and even 4G networks across Africa (Jenkins, 2022). Huawei is currently heavily invested in the construction of 5G networks throughout several African nations. Huawei is helping to alleviate poverty on the African continent by delivering low-cost gear that increases connection and decreases the price of access. In addition, Huawei has helped improve Africa's digital capabilities and security infrastructure by building data centers, cloud services, and Safe City Solutions (Abegunrin & Manyeruke, 2020).

While not as pervasive as Huawei, both Tencent and Alibaba have made their mark. WeChat, developed by Tencent, is a messaging app, social media platform, and digital payment system that has gained popularity in a number of African countries (Matemba & Li, 2018). By bridging the gap between African SMEs and international marketplaces, Alibaba's eWTP is actively promoting e-commerce in Africa. Jack Ma, founder of Alibaba Group, has also started programs like the "Netpreneur Prize," which awards money to African business owners in order to spur digital innovation in the continent (Lee & Shen, 2020).

Case Studies of Chinese Involvement in Major African ICT Projects

Huawei and ZTE, two prominent examples of Chinese technology firms, have participated in numerous large-scale ICT projects in Africa. The West Africa Cable System (WACS) ranks as one of the world's most important infrastructure projects. Over 14,000 kilometers in length, this submarine cable links Southern Africa to Europe. The cable, laid by Huawei Marine, connects 14 countries and vastly improves internet availability and speeds for those locations (Hasler, 2020). Because of its scope and significance, this project is a shining example of Chinese technology companies promoting internet access in Africa.

The goal of the joint venture between Huawei and ZTE in Ethiopia is to improve the country's telecommunications network. The project's approximately $1.6 billion in funding was supposed to go toward establishing and expanding 4G networks in the country's capital, Addis Ababa, as well as 2G and 3G networks elsewhere (Workneh, 2016). Millions more Ethiopians now have access to smartphones and the internet as a result of these dramatic enhancements to the country's infrastructure.

Also, in 2019, Huawei established the first 5G network in South Africa in collaboration with a local telecom business called Rain (Alraih et al., 2022). Through their combined efforts, South Africa has entered a new era of high-speed internet, which has fueled both technological advancement and economic expansion.

China's tech firms have played a critical role in assisting African nations in building "smart cities." In Kenya, for instance, Huawei is working side-by-side with the government on the Konza Technopolis project, a smart city initiative that will, among other things, generate thousands of new employments and greatly boost Kenya's gross domestic product (Lahtinen, 2018). These examples show the scope and depth of Chinese involvement in Africa's ICT industry and how Chinese IT businesses are reshaping the continent's digital landscape.

To critically assess the applicability of China's digital development experience in Africa, it is vital to understand the African context in terms of ICT development and the role of Chinese ICT enterprises on the continent. Over the past decades, Africa has witnessed a significant evolution in its ICT sector, characterized by increased mobile connectivity, internet penetration, and the emergence of local digital innovations. Simultaneously, the presence of Chinese ICT companies like Huawei and ZTE in Africa has grown substantially, influencing various aspects of digital infrastructure and services. However, this involvement is not a recent phenomenon but a progression over two decades, intertwining with the broader narrative of Africa's digital journey.

LESSONS FROM CHINA'S DIGITAL EVOLUTION

Discussion on the Role of Digital Technology in China's Socio-Economic Transformation

Lessons can be learned by Africa and other emerging regions from China's digital transformation. China has seen a dramatic socioeconomic transformation over the past three decades, shifting from an agricultural economy to a worldwide technological powerhouse driven in large part by advances in digital technology.

China's case study highlights the value of a well-developed digital infrastructure. The government of China has put a lot of money into developing its informational infrastructure, including its telecommunications networks and data centers (Chen, Mao, & Qiu, 2018). This funding allowed for a surge in internet users and the establishment of industry leaders such as Alibaba, Tencent, and Huawei, among others.

China serves as an example of the revolutionary potential of digital technologies in many fields. When it comes to shopping, for example, e-commerce in China has been a game-changer. Not only have platforms like Alibaba's Taobao and JD.com made life easier for consumers, but they have also opened up national and international marketplaces to millions of small and medium-sized businesses (Jung, Ugboma, & Liow, 2015). Digital finance, spearheaded by Alipay and WeChat Pay, has similarly expanded access to banking services, boosted consumer spending, and fueled expansion in the economy (Shen, 2021).

In addition, the leadership of the Chinese government in this digital transition has been important. It has created a hospitable atmosphere for digital innovation through its rules and regulations. For instance, in 2015, the government unveiled the "Internet Plus" strategy to foster technological innovation and economic growth by combining the internet with conventional sectors (Xu, Zhao, & Wang, 2021).

China's rapid digital development highlights the value of digital competence and literacy. China's education and training programs have produced a sizable pool of digital talent that has powered the country's information economy. Africa may benefit greatly from these insights into China's digital progress as it begins its own digital transformation journey. African countries can improve their use of digital technology for social and economic development by studying China's experiences in this area.

Analysis of Key Strategies Employed by China in Digitalization and How They Can be Adopted by African Nations

There is much that African countries may learn from China's experience with digital transformation, which has been defined by a strong digital infrastructure, innovation-driven development, and a supportive policy environment.

The creation of an all-encompassing digital infrastructure is central to China's plan for digitalization. The rise of IT companies and the proliferation of digital services in China are made possible by the country's high-speed internet, extensive mobile connectivity, and secure data centers. Countries in Africa could follow suit if they made ICT infrastructure investment a top priority. As a first step, the continent needs to be outfitted with internet that is accessible, affordable, and fast. The partnerships with Huawei and ZTE in China show how effective international cooperation can be in this regard.

China has created an environment that promotes digital innovation, going beyond basic infrastructure. China has been the birthplace and rapid expansion of several ground-breaking digital services, such as e-commerce platforms like Alibaba and mobile payment systems like WeChat Pay (Gervasi, 2016). This pioneering spirit originates from a big population of technically savvy workers, a considerable amount of money spent on R&D, and a thriving start-up culture. For Africa to build similar innovation ecosystems, it may be necessary to incentivize research and development, invest in STEM (Science, Technology, Engineering, and Mathematics) education, and encourage entrepreneurship.

The way China regulates and governs the internet is also important. The Chinese government has taken the initiative to create regulations that steer the digital economy, such as the "Internet Plus" plan, and to regulate vital sectors, such as data security and privacy (Roberts et al., 2021). Creating transparent and future-oriented digital policy is crucial for African countries. Such regulations may lay out standards for e-commerce, online purchases, data protection, and digital taxation. Policymakers and regulators could also benefit from efforts to increase their digital literacy.

Last but not least, a key component of China's digitalization agenda is the combination of digital and traditional sectors. Increased productivity and economic growth have resulted from the use of technologies like AI, IoT, and big data to optimize processes across the manufacturing, agricultural, and service sectors (Wang et al., 2022). Many African nations rely heavily on agriculture as an economic driver, and many of them stand to benefit greatly from the implementation of digital technologies. Manufacturing, retail, and services are just some of the industries that could benefit from digitalization.

OPPORTUNITIES AND CHALLENGES FOR AFRICA

Assessment of How Africa Can Benefit From Its Relationship With China in Its Digital Transformation Journey

There are both huge advantages and formidable difficulties for Africa to work with China on its digital revolution. Africa's digital revolution can be propelled by a strategic cooperation with China, a global leader in digital technology, but it requires cautious management to maximize advantages and avoid hazards.

There are a lot of possibilities because of this connection. China's investment in Africa's digital infrastructure, especially through firms like Huawei and ZTE, is helping to close the digital divide (Zreik, 2021). If infrastructure is upgraded, more people will have access to more knowledge, have better communication, be more creative, and spend more money.

In addition, African countries can learn from China's mistakes and successes in using digital technologies for social and economic growth. Successful e-commerce, digital finance, and smart city implementation in China can serve as a model for Africa. Financial inclusion is a major difficulty in many African countries, but examples like China's extensive adoption of mobile payments show how digital finance may help.

Alibaba and Tencent, two of China's largest internet companies, provide cutting-edge digital solutions that might be adapted to the needs of the African market. These options, ranging from e-commerce platforms to digital health services, can help Africa's digital economy expand (Zhang & Chen, 2019).

In addition, Africa can gain from China's efforts to improve infrastructure and human resources. China's training programs, information exchange, and technical aid can equip people with the digital literacy and advanced technical proficiency they'll need to succeed in today's global economy.

However, there are risks associated with Africa's growing digital link with China. Protecting one's digital independence is a top priority. Concerns regarding data privacy, security, and control are warranted because Chinese corporations are building large parts of Africa's digital infrastructure (Gravett, 2020). To secure data privacy and cybersecurity, African countries will need to upgrade their legal and regulatory frameworks.

Concerns have also been raised that Africa's digital economy may come to rely too heavily on Chinese technology and standards, which might eventually stifle both competition and innovation. Exploring the development of domestic IT industries and considering investments from a variety of global players could complement the welcome of Chinese investments in African nations.

There is still a chance of being left out of the digital world. Despite China's efforts, digital transformation benefits may be unequally distributed, contributing to existing inequities. Policymakers will have to work to ensure that digitization measures are accessible to all, particularly those in rural areas, the economically disadvantaged, women, and the elderly.

Exploration of Potential Risks and Obstacles, Including Issues Around Data Security and Digital Sovereignty

There are a number of hazards and challenges that Africa may face as it expands its digital partnership with China. Data security, digital sovereignty, and the danger of being too reliant on a single foreign company are three of the most prominent.

Concern over the safety of one's data is paramount. The widespread availability of digital services has resulted in an explosion of data. There are growing concerns regarding the security of this information as Chinese firms play an increasingly important role in developing Africa's digital infrastructure (Ren, Zeng, & Gozgor, 2023). For instance, if private information were to get into the wrong hands, it might result in identity theft, fraud, and other serious violations of confidentiality. Digital infrastructure could be used as a tool for political manipulation, which raises concerns about the possibility of state surveillance and control (Telo, 2023).

The concept of digital sovereignty is intrinsically tied to the problem of data protection. Concerns have been raised concerning the privacy and security of Africans' personal information as Chinese corporations take over more and more of the continent's digital infrastructure (Bartholomew, 2020). The concept of "digital sovereignty" refers to a state's freedom to make decisions about its own digital infrastructure without fear of interference from other nations. Foreign control over a large chunk of Africa's digital infrastructure raises concerns about the continent's ability to maintain its independent digital identity (Hummel, 2021).

Another major danger is depending too heavily on a single foreign power, in this case China. While China's assistance in Africa's digital transformation has been significant, it is important for the continent to consider the implications of a heavy reliance on Beijing, as this might influence its own creative and competitive capabilities. The countries of Africa may become more susceptible to changes in China's foreign or economic policy as a result.

Compatibility and standards in technology are another area of issue. Compatibility and interoperability across systems and technologies are essential in today's digital economy, which operates on a global scale. Especially if China's standards are not universally accepted, their presence in Africa could hinder the continent's ability to engage in digital trade and communication with the rest of the globe.

There is also serious worry that this could exacerbate already existing inequities. Digital transformation has the potential to provide many benefits, but if it isn't controlled wisely, it could also lead to growing inequality. A digital divide might widen if, for instance, metropolitan areas reap the benefits of digital investments to a greater extent than rural areas. Also, higher-income groups might be more likely to have access to various forms of digital technology (Luo, Zhu, & Song, 2023).

African countries will need to implement strategic digital policies to combat these threats and challenges. Focusing on data privacy and security, fostering digital independence, encouraging local innovation and competition, forming a broad range of tech alliances, and applying digital technologies in a fair and equitable manner are potential areas of emphasis. The obstacles are significant, but they can be overcome with forethought and strategy, paving the way for a digital transformation in Africa that is safe, independent, and beneficial to all its people.

CASE STUDIES: CHINESE DIGITAL INITIATIVES IN AFRICA

The digital activities of the Chinese government in Africa cover a wide range of projects and partnerships across many industries, all of which have had significant effects on Africa's information society. Let's look at a few of the most important projects and what they could mean for the future.

First, China's financial commitment to Africa's information technology infrastructure has been unprecedented. Huawei and ZTE, two Chinese digital behemoths, have played crucial roles in building and improving communication infrastructure across Africa. In Nigeria, for instance, Huawei has been critical in the development and upkeep of the country's 4G networks, allowing millions of citizens to benefit from ultra-fast internet access (Agbebi, 2019). ZTE collaborated with the government of Angola to build the country's first national fiber optic backbone network, which has greatly improved the country's access to the internet (Gagliardone, 2019). The expansion of the digital economy in Africa is dependent on these efforts closing the digital divide.

Second, Chinese firms are forming strategic alliances with their African counterparts in the e-commerce sector. One particularly noteworthy effort by Alibaba is the eWTP, or electronic World Trade Platform. The platform's primary objective is to improve international trade by giving small and medium-sized enterprises (SMEs) easier digital access to international markets. Alibaba and the government of Rwanda collaborated to establish Africa's first eWTP center. Farmers in Rwanda, for example, have seen increased market access and revenue because to this platform's capacity to connect them directly with Chinese consumers (Johnston, 2021).

Even in the field of digital banking, Chinese firms are at the forefront. Several African countries now provide mobile payment options thanks to partnerships between international corporations like WeChat Pay and Alipay and local providers. WeChat Pay, in partnership with Standard Bank, is a widely utilized mobile payment system in South Africa. People who don't have bank accounts may be able to have access to financial services thanks to these types of collaborations (Matemba & Li, 2018).

The development of "smart cities" in Africa also features heavy Chinese involvement. Huawei is supplying smart city solutions, such as data centers and intelligent security systems, to Kenya's Konza Technopolis (Gordon & Meia, 2022). Tencent is working with the government of Mauritius to improve the island nation's digital infrastructure, with an emphasis on developing a smart city plan (Wright, 2021). These initiatives demonstrate how digital technology may be used to better manage cities and raise living standards.

While these efforts are helpful, they do bring up some serious concerns about data privacy, digital sovereignty, and Africa's growing dependence on Chinese technology. To address these challenges effectively, enhancing legal and regulatory infrastructures could be a valuable step for African countries. They need to counter the power of multinational corporations by investing in and developing homegrown technology industries.

In addition, these alliances need to be planned and executed in a way that promotes equality and longevity. As Africa embraces the digital era, ensuring equitable access to the benefits of digitalization for all residents, including those in rural areas, the economically disadvantaged, women, and other marginalized groups, becomes an important consideration.

CONCLUSION

The story of China's role in Africa's digital development is a complicated one, fraught with both promise and peril. China's significant investments in infrastructure, strategic relationships, and knowledge-sharing efforts are collectively speeding Africa's digital revolution, making China's impact on Africa's digital environment unmistakable. However, a detailed understanding and planned approach is necessary to maximize benefits and reduce dangers in light of the tremendous effect and ramifications of this relationship.

Huawei, ZTE, Tencent, and Alibaba, four of China's largest technology companies, are building Africa's digital infrastructure and providing cutting-edge tech solutions and fresh approaches to doing business online. These initiatives are helping to eliminate the digital divide while also encouraging creativity and facilitating economic growth. Transformative changes are occurring in critical industries including e-commerce, digital finance, and smart cities, frequently inspired by China's own successful digitalization experience.

However, there are also major difficulties posed by the growing digital engagement with China. Data privacy, the possibility of state surveillance, and the loss of control over national digital space are all sources of anxiety for those concerned about data security and digital sovereignty. In addition, African countries could be left exposed to external political and economic developments if they become too dependent on a single foreign player, which could reduce competition, impede local innovation, and ultimately hurt the continent.

In light of these facts, African countries need to take deliberate actions to guarantee the success, safety, and longevity of their digital transformation. Improving data protection, cybersecurity, and digital sovereignty requires, among other things, a more robust legal and regulatory framework. African nations may diversify their digital economies if they encourage innovation and competition in their own technological sectors.

Investing in capacity building to acquire the relevant abilities in preparation for the digital age is another important tactic. Emphasizing both advanced technical skills and the enhancement of digital literacy could be key in ensuring that the benefits of digitization reach a broad spectrum of the population.

Engaging in collaborations with China could be a pathway for African countries to foster sustainable and inclusive economic growth, leveraging these ties to benefit their digital transformation journey. It is crucial that China's initiatives and investments in digitization reach underserved groups like those living in rural areas, people with lower incomes, women, and the elderly.

Future prospects for China and Africa's digital partnership hold both promising opportunities and formidable difficulties. The strategic decisions made by African countries will determine the future of this partnership. Africa may use its digital relationship with China to accelerate its digital transformation if proper planning, proactive governance, and an emphasis on inclusivity and sustainability are implemented. This will result in a more digitally-enabled, equitable, and prosperous future for all Africans.

The research holds significant implications for policymakers, business leaders, and scholars, emphasizing the strategic need to navigate the complex terrain of digital globalization with a critical eye towards sustainable development, digital sovereignty, and equitable progress. For policymakers in Africa, the findings underscore the importance of crafting policies that not only foster digital infrastructure development but also ensure data security, protect digital sovereignty, and promote inclusivity, enabling a digital economy that benefits all sectors of society. Business leaders are encouraged to explore collaborative ventures that leverage Chinese technological advancements while cultivating local innovation ecosystems that can sustain economic growth and competitiveness in the digital age. For scholars, this study highlights the urgent need for interdisciplinary research that addresses the rapidly evolving dynamics of international digital relations, offering insights into the broader geopolitical and economic implications of digital initiatives like the Digital Silk Road. Through a careful consideration of these implications, stakeholders across the spectrum can work towards realizing the full potential of Africa's digital transformation in a manner that is both equitable and sustainable, ensuring that the digital future is inclusive and beneficial for all involved.

The findings from the study of China's Digital Silk Road and its implications for Africa's digital transformation open several avenues for future research. Scholars and policymakers are encouraged to delve into longitudinal studies that track the evolution and long-term impacts of digital infrastructure projects across different African regions, offering insights into sustainability and socio-economic outcomes. Comparative analyses between Africa and other regions engaged with the Digital Silk Road could enrich understanding of global digital geopolitics and economic strategies. Additionally, in-depth investigations into the governance models of digital technology in Africa, focusing on data sovereignty, privacy, and cybersecurity, would provide valuable guidance for policy formulation. Further research could also explore the role of indigenous innovation and digital entrepreneurship in Africa, examining how local ecosystems can thrive alongside international partnerships. These areas promise to shed light on the complex dynamics of digital transformation in Africa, contributing to a more nuanced and comprehensive understanding of its future path.

LIMITATIONS

The exploration of "China's Digital Silk Road and Its Implications for Africa's Digital Transformation" faces inherent constraints tied to the specific focus on certain regions and sectors within Africa, which may not fully encapsulate the varied impacts across the continent's diverse landscapes. Additionally, the analysis leans heavily on publicly available data and literature, encountering obstacles in depth due to the proprietary nature of corporate data and the occasional scarcity of comprehensive data from African governments. This reliance places a limit on conducting a detailed empirical assessment of the Digital Silk Road's broader effects.

Moreover, the rapid pace at which digital technology and international relations evolve presents a challenge to maintaining the relevance of the chapter's findings over time. The dynamic nature of the subject matter means that the insights provided might soon be overtaken by new developments, underscoring the necessity for continuous research in this area.

The interdisciplinary approach, while enriching, also introduces complexities in achieving depth across the multiple disciplines intersected, including international relations, the digital economy, and development studies. This breadth of focus might not allow for an exhaustive exploration of specialized aspects within each discipline.

REFERENCES

Abegunrin, O., & Manyeruke, C. (2020). *China's Power in Africa*. Springer International Publishing. DOI: 10.1007/978-3-030-21994-9

Agbebi, M. (2019). Exploring the human capital development dimensions of Chinese investments in Africa: Opportunities, implications and directions for further research. *Journal of Asian and African Studies*, 54(2), 189–210. DOI: 10.1177/0021909618801381

Alraih, S., Shayea, I., Behjati, M., Nordin, R., Abdullah, N. F., Abu-Samah, A., & Nandi, D. (2022). Revolution or evolution? Technical requirements and considerations towards 6G mobile communications. *Sensors (Basel)*, 22(3), 762. DOI: 10.3390/s22030762 PMID: 35161509

Bartholomew, C. (2020). China and 5G. *Issues in Science and Technology*, 36(2), 50–57.

Burns, S. (2018). M-Pesa and the 'market-led' approach to financial inclusion. *Economic Affairs*, 38(3), 406–421. DOI: 10.1111/ecaf.12321

Chen, Y., Mao, Z., & Qiu, J. L. (2018). *Super-sticky WeChat and Chinese society*. Emerald Publishing Limited. DOI: 10.1108/9781787430914

Gagliardone, I. (2019). *China, Africa, and the Future of the Internet*. Bloomsbury Publishing. DOI: 10.5040/9781350219113

Gervasi, M. (2016). *East-Commerce: China e-commerce and the internet of things*. John Wiley & Sons.

Gordon, D., & Meia, N. (Eds.). (2022). *The Digital Silk Road: China's Technological Rise and the Geopolitics of Cyberspace*. Taylor & Francis.

Gravett, W. (2020). Digital neo-colonialism: The Chinese model of internet sovereignty in Africa. *African Human Rights Law Journal*, 20(1), 125–146.

Hasler, J. C. (2020). *Explaining the International Infrastructure of Global Telecommunications: Why Government Ownership Determines a Country's Position in Interstate Communications Networks* (Doctoral dissertation, The George Washington University).

Hillman, J., Sacks, D., Lew, J. J., & Roughead, G. (2021). *China's Belt and Road: Implications for the United States*. Council on Foreign Relations.

Hong, Y., & Wei, Y. (2022). Towards future politics of the cybersphere: China's temporal-spatial governance of digital transition. *Media International Australia, Incorporating Culture & Policy*, 185(1), 9–25. DOI: 10.1177/1329878X221095593

Hummel, P., Braun, M., Tretter, M., & Dabrock, P. (2021). Data sovereignty: A review. *Big Data & Society*, 8(1), 2053951720982012. DOI: 10.1177/2053951720982012

Jenkins, R. (2022). *How China is reshaping the global economy: Development impacts in Africa and Latin America*. Oxford University Press. DOI: 10.1093/oso/9780192866356.001.0001

Johnston, L. A. (2021). World trade, e-commerce, and COVID-19. *China Review*, 21(2), 65–86.

Jung, J. C., Ugboma, M. A., & Liow, A. K. (2015). Does Alibaba's magic work outside China? *Thunderbird International Business Review*, 57(6), 505–518. DOI: 10.1002/tie.21739

Keane, M., & Yu, H. (2019). A digital empire in the making: China's outbound digital platforms. *International Journal of Communication*, 13, 4624–4641.

Kokas, A. (2022). *Trafficking data: How China is winning the battle for digital sovereignty*. Oxford University Press. DOI: 10.1093/oso/9780197620502.001.0001

Lahtinen, A. (2018). *China's Diplomacy and Economic Activities in Africa: Relations on the move*. Springer. DOI: 10.1007/978-3-319-69353-8

Lee, H. L., & Shen, Z. J. M. (2020). Supply chain and logistics innovations with the Belt and Road Initiative. *Journal of Management Science and Engineering*, 5(2), 77–86. DOI: 10.1016/j.jmse.2020.05.001

Luo, X., Zhu, S., & Song, Z. (2023). Quantifying the Income-Increasing Effect of Digital Agriculture: Take the New Agricultural Tools of Smartphone as an Example. *International Journal of Environmental Research and Public Health*, 20(4), 3127. DOI: 10.3390/ijerph20043127 PMID: 36833825

Ly, B. (2020). Challenge and perspective for digital Silk road. *Cogent Business & Management*, 7(1), 1804180. DOI: 10.1080/23311975.2020.1804180

Matemba, E. D., & Li, G. (2018). Consumers' willingness to adopt and use WeChat wallet: An empirical study in South Africa. *Technology in Society*, 53, 55–68. DOI: 10.1016/j.techsoc.2017.12.001

Okano-Heijmans, M., & Vosse, W. (2021). Promoting open and inclusive connectivity: The case for digital development cooperation. *Research in Globalization*, 3, 100061. DOI: 10.1016/j.resglo.2021.100061

Parepa, L. A. (2020). The Belt and Road Initiative as continuity in Chinese foreign policy. *Journal of Contemporary East Asia Studies*, 9(2), 175–201. DOI: 10.1080/24761028.2020.1848370

Qin, T., Wang, L., Zhou, Y., Guo, L., Jiang, G., & Zhang, L. (2022). Digital technology-and-services-driven sustainable transformation of agriculture: Cases of China and the EU. *Agriculture*, 12(2), 297. DOI: 10.3390/agriculture12020297

Ren, X., Zeng, G., & Gozgor, G. (2023). How does digital finance affect industrial structure upgrading? Evidence from Chinese prefecture-level cities. *Journal of Environmental Management*, 330, 117125. DOI: 10.1016/j.jenvman.2022.117125 PMID: 36603250

Roberts, H., Cowls, J., Morley, J., Taddeo, M., Wang, V., & Floridi, L. (2021). The Chinese approach to artificial intelligence: An analysis of policy, ethics, and regulation. *AI & Society*, 36(1), 59–77. DOI: 10.1007/s00146-020-00992-2

Shen, H. (2021). *Alibaba: Infrastructuring global china*. Routledge. DOI: 10.4324/9781003169055

Silin, Y., Kapustina, L., Trevisan, I., & Drevalev, A. (2018). The silk road economic belt: Balance of interests. *Economic and Political Studies*, 6(3), 293–318. DOI: 10.1080/20954816.2018.1499072

Telo, J. (2023). Smart City Security Threats and Countermeasures in the Context of Emerging Technologies. *International Journal of Intelligent Automation and Computing*, 6(1), 31–45.

Wang, D., Zhou, T., & Wang, M. (2021). Information and communication technology (ICT), digital divide and urbanization: Evidence from Chinese cities. *Technology in Society*, 64, 101516. DOI: 10.1016/j.techsoc.2020.101516

Wang, L., Wu, Y., Huang, Z., & Wang, Y. (2022). How big data drives green economic development: Evidence from China. *Frontiers in Environmental Science*, 10, 1055162. DOI: 10.3389/fenvs.2022.1055162

Workneh, T. W. (2016). Chinese multinationals in the Ethiopian telecommunications sector. *Communication, Culture & Critique*, 9(1), 126–147. DOI: 10.1111/cccr.12129

Wright, C. (2021). China's Digital Colonialism: Espionage and Repression Along the Digital Silk Road. *SAIS Review (Paul H. Nitze School of Advanced International Studies)*, 41(2), 89–113. DOI: 10.1353/sais.2021.0020

Xu, W., Zhao, C., & Wang, X. (2021). The Role of the Internet Technology in the Employment Structural Transformation under Background of "Internet Plus" in China. *Theoretical Economics Letters*, 11(5), 1020–1037. DOI: 10.4236/tel.2021.115065

Zhang, M. L., & Chen, M. S. (2019). *China's digital economy: Opportunities and risks*. International Monetary Fund.

Zreik, M. (2021, September 1). The Potential of a Sino-Lebanese Partnership through the Belt and Road Initiative (BRI). *Contemporary Arab Affairs*, 14(3), 125–145. DOI: 10.1525/caa.2021.14.3.125

Zreik, M. (2022). The Chinese presence in the Arab region: Lebanon at the heart of the Belt and Road Initiative. *International Journal of Business and Systems Research*, 16(5-6), 644–662. DOI: 10.1504/IJBSR.2022.125477

Zreik, M. (2023). Analytical study on Foreign Direct Investment Divestment Inflows and Outflows in Developing Economies: Evidence of China. *Chinese Economy*, 56(6), 415–430. DOI: 10.1080/10971475.2023.2193118

ADDITIONAL READING SOURCES

Al Shaher, S., Yanzhe, M., & Zreik, M. (2023). Navigating the Digital Frontier: Responsible Innovation in China's Digital Silk Road. *Migration Letters : An International Journal of Migration Studies*, 20(S9), 1297–1315.

Ayodele, O. (2021). The digital transformation of diplomacy: Implications for the African Union and continental diplomacy. *South African Journal of International Affairs*, 28(3), 379–401. DOI: 10.1080/10220461.2021.1968944

Eguegu, O. (2022). The Digital Silk Road: Connecting Africa with New Norms of Digital Development. *Asia Policy*, 29(3), 30–39. DOI: 10.1353/asp.2022.0049

ElMassah, S., & Mohieldin, M. (2020). Digital transformation and localizing the sustainable development goals (SDGs). *Ecological Economics*, 169, 106490. DOI: 10.1016/j.ecolecon.2019.106490

Gertzen, W. M., Van der Lingen, E., & Steyn, H. (2022). Goals and benefits of digital transformation projects: Insights into project selection criteria. *Suid-Afrikaanse Tydskrif vir Ekonomiese en Bestuurswetenskappe*, 25(1), 4158. DOI: 10.4102/sajems.v25i1.4158

Ho, C. Y., Narins, T. P., & Sung, W. (2023). Developing information and communication technology with the belt and road initiative and the digital silk road. *Telecommunications Policy*, 47(10), 102672. DOI: 10.1016/j.telpol.2023.102672

Mapingire, K., Smuts, H., & Van der Merwe, A. (2022). Components of a Digital Transformation Strategy: A South African Perspective. In *Proceedings of Sixth International Congress on Information and Communication Technology: ICICT 2021,London,* Volume 2 (pp. 437-447). Springer Singapore. DOI: 10.1007/978-981-16-2380-6_38

Mhlungu, N. S., Chen, J. Y., & Alkema, P. (2019). The underlying factors of a successful organisational digital transformation. *South African Journal of Information Management*, 21(1), 1–10. DOI: 10.4102/sajim.v21i1.995

Ndemo, B., & Weiss, T. (2017). Making sense of Africa's emerging digital transformation and its many futures. *Africa Journal of Management*, 3(3-4), 328–347. DOI: 10.1080/23322373.2017.1400260

Zeleza, P. T., & Okanda, P. M. (2021). Enhancing the digital transformation of African universities. *Journal of Higher Education in Africa/Revue de l'enseignement supérieur en Afrique, 19*(1), 1-28.

KEY TERMS AND DEFINITIONS

Alibaba: A Chinese multinational conglomerate specializing in e-commerce, retail, internet, and technology, known for its online marketplaces like Alibaba.com and the consumer-to-consumer platform Taobao.

Belt and Road Initiative (BRI): A global development strategy adopted by the Chinese government involving infrastructure development and investments in countries in Europe, Asia, Africa, the Middle East, and the Americas.

Data Security: The practice of protecting digital data from unauthorized access, corruption, or theft throughout its lifecycle.

Digital Inclusion: The efforts and policies designed to ensure that all individuals and communities, including the most disadvantaged, have access to and use of Information and Communication Technologies.

Digital Silk Road (DSR): An initiative by China, part of the larger Belt and Road Initiative (BRI), aimed at enhancing global digital connectivity and cooperation. It focuses on building digital infrastructure, like fiber optic cables and satellite networks, and promoting digital commerce and technology sharing among participating countries.

Digital Sovereignty: The concept that states should have the right to manage and control their own digital data, infrastructure, and communication networks without external influence.

Digital Transformation: The integration of digital technology into all areas of a business or a country's economy, fundamentally changing how operations are run and value delivered to customers or citizens.

E-Commerce: The buying and selling of goods or services using the internet, and the transfer of money and data to execute these transactions.

Huawei: A Chinese multinational technology company that provides telecommunications equipment and sells consumer electronics, including smartphones, and is heavily involved in the development of 5G networks globally.

ICT Infrastructure: Information and Communication Technology infrastructure refers to the composite hardware, software, network resources, and services required for the existence, operation, and management of an enterprise IT environment or country's digital services.

Smart Cities: An urban area that uses different types of electronic methods and sensors to collect data, with insights gained from that data used to manage assets, resources, and services efficiently.

Tencent: A Chinese multinational conglomerate holding company, with subsidiaries in various Internet-related services and products, entertainment, AI, and technology both in China and globally.

Chapter 12
Eatertainment: Staging Food and Experiences for Brand Competitiveness in Restaurants

Reka Danhauser
Universidade Católica Portuguesa, Portugal

Anna Carolina Boechat
https://orcid.org/0000-0002-9252-6602
Universidade Europeia, Portugal

Ana Isabel Inácio
https://orcid.org/0000-0002-5004-2634
ISG Business and Economics School, Portugal

Patrícia Tavares
https://orcid.org/0000-0003-2404-342X
Universidade Católica Portuguesa, Portugal

ABSTRACT

The "eatertainment" phenomenon has become widespread as one of the latest trends for many restaurant brands. The main goal of the present work is to examine the relationship between the visual presentation of food and its influence on the multidimensional notion of brand experience. The research was exploratory and three focus group discussions were held. Findings revealed that food presentation enhances various dimensions of brand experience, although its impact varies: it significantly influences sensory brand experiences, mainly the visual and taste perceptions; it is less prominent in affective brand experiences, where other factors play a more significant role; it also shapes behavioral brand experiences, encouraging

DOI: 10.4018/979-8-3693-3579-6.ch012

photography, online sharing, and culinary exploration; and its impact on intellectual brand experiences varies, with some seeing it inspiring and others merely aesthetic. However, plating is just one aspect of a holistic brand experience, necessitating a multifaceted strategy for contemporary diners.

INTRODUCTION

Throughout the years, food has evolved from being a basic human necessity to serving various roles in contemporary society (Civitello, 2008). Currently, restaurants, cafés, and pubs are not only places for consumption. Rather, they have transitioned into multifunctional spaces that facilitate leisure, socialization, and even professional interactions, reflecting the cultural significance of eating out (Astuti & Hanan, 2016; Fieldhouse, 1996). Hence, these developments highlight the notable transformation in dining practices, reinforcing their fundamental role in many aspects of everyday life and society as a whole.

In this context, branding has emerged as a vital success factor due to the high level of competition (McCracken, 2022; Robitaille, 2023). As a result, it is not enough for restaurants to simply serve delicious meals. They must showcase their unique offers and create a distinct identity, and experiences, allowing them to form emotional relationships with their customers. This is where "eatertainment" takes place and space. As a result, a good brand strategy is essential not just for creating long-lasting impressions but also for standing out in a crowded market (Bistrohub, 2020; Kakadiya, 2023; McCracken, 2022; Robitaille, 2023). Today's customers seek businesses that resonate with their emotions, engage their senses, and align with their lifestyles. Hence, they prioritize brands that emphasize meaningful and enjoyable experiences (Bapat & Thanigan, 2016; Hwang et al., 2021; Moreira et al., 2017; Schmitt et al., 2014). Consequently, restaurants should not only provide good quality and price but also offer distinctive experiences that encompass both practical and symbolic values for guests (Ong et al., 2018).

Moreover, delving deeper into the contemporary dining culture, the increased significance of food presentation is undeniable. Plating in restaurants does not only act as a decoration tool but it is also considered to be part of a holistic dining experience, which might enhance the overall enjoyment of a meal, as well as influence guests' food choices and preferences (Cifci et al., 2021; Michel et al., 2014). Additionally, while the appreciation of an aesthetically presented meal is timeless, the rise of the digital age has increased its significance, reflecting the widespread trend of sharing food experiences online (Spence et al., 2016).

Although an increasing body of literature focuses on the art of plating, a significant gap remains in understanding its interplay with various brand constructs. Therefore, the present study aims to help bridge this gap by examining how plating within the context of brunch restaurants can influence the overall brand experience in its various dimensions. The selection of brunch for this study is due to its inherently photogenic appeal, which aligns with the trends of today's social media-centric culture (Future Food Strategy Group, 2018; Hartley, 2021). Describe the general perspective of the chapter. End by specifically stating the objectives of the chapter.

BACKGROUND

Food reflects who we were, who we are, and who we might become. While genetics, environment, and lifestyle influence us, food plays a crucial role in our survival, quality of life, and physical and mental development at any age. It is a basic necessity, shaping not only daily routines but also social, political, economic, and aesthetic aspects across all societies. Thus, the meaning of food extends far beyond its role in meeting nutritional needs (Sobreira et al., 2018; Civitello, 2008).

Today, food is not just about survival; it's a public health issue with significant implications for human well-being. More than a universal need, food is also a habit, a cultural asset, and a reflection of our relationship with nature and the environment. The history of food mirrors human history, offering insights into society and culture. It highlights key transitions—from hunting and subsistence farming to industrialization—marked by imbalances of scarcity and abundance, environmental degradation, and globalized trade. Recently, movements like slow food, organic farming, and local food consumption have emerged alongside a rise in food intolerances and allergies, leading to diets that medicine now views as essential both individually and collectively.

It's not just about what we eat, but how we eat—the cultural practices, routines, and who we share meals with. Taste is shaped by society and culture, leading to conventions that dictate what can be eaten, when, and under what circumstances, such as specific foods, seasons, or diets. The eating customs of each culture show that eating is a dynamic act, influenced by time and place, marking civilization. Unlike other living beings that eat without social or cultural constraints, humans are defined by these conventions.

We live in a globalized, fragmented, and polarized world, driven by media, self-interest, and concerns about health and aesthetics, where food has become a spectacle and a shared media phenomenon. On one hand, movements focus on ecology, the SDGs, and slow food, while on the other, we witness war, famine, and poverty. This isn't a simple duality, but a spectrum of realities reflecting humanity's disjointed presence on the planet. In today's "risk society" (Beck, 1986) and "liquid modernity" (Bauman, 1990), food embodies these paradoxes—overconsumption alongside media dependency. As Elias (1978) argued, food reflects civility and table manners, defining tribes, fostering belonging, and expressing ways of life. It carries a complex cultural symbolism, blending the personal pleasure of eating with the social act of sharing, both governed by community conventions and shared identities.

Sharing food is a fundamental way, humans address basic needs while distinguishing themselves from animals (Koponen & Mustonen, 2020). The concept of "commensality"—eating together at the same table — captures the essence of human sociability. Public eating, however, has evolved into a way to express both individual and collective identity (Crowther, 2018). Through shared consumption, food symbolizes the "we" of social groups and occasionally all of humankind. Thus, food becomes a powerful symbol of the collective self, while also representing individuals, as it embodies personal identity and serves as a metonym for the self (Pilcher, 2012).

"Food porn" thrives on social media, where meals are often photographed before being eaten, almost replacing the traditional act of giving thanks. It has become a social marker, akin to what people buy, wear, drive, or the places they visit. Consumption is now a key part of identity formation (Miles, 1996), and social media provides the ideal platform for this. Food is also deeply tied to ethnic and national identity (Civitello, 2011), turning meals into performances of collective interaction, amplified by social networks. Food symbolism plays a significant role in many religions, with dietary restrictions reflecting spiritual beliefs (Fieldhouse, 1996). For example, Jewish dietary laws (kashrut) define what foods are permissible for consumption.

Food is consumed not only for its nutritional value but also for what it represents in its social and cultural context, becoming a marker of status, social position, and inclusion or exclusion, as Bourdieu (1979) suggests. Ritualized meals, like Christian Christmas dinners, Jewish Passover feasts, or the Muslim observance of Ramadan, embody collective performances that reinforce social order and strengthen group solidarity, as Durkheim theorizes. Cultural cuisines are maintained through specific food practices, including common ingredients, distinctive flavors, and preparation methods, which are tied to both daily and festive traditions. Shared food habits foster a sense of belonging and cultural identity, making them difficult to abandon (Fieldhouse, 1996), underscoring the importance of studying this topic to better understand its societal impact.

Food consumption, once primarily a household activity, has shifted outside the home and evolved into a leisure pursuit (Astuti & Hanan, 2016). Restaurants have become central to social life, with dining out being a popular activity. Sharing meals and drinks in comfortable settings has become one of life's pleasures, and more meals are now consumed away from home than ever before (Walker, 2017). Restaurants, coffee shops, and pubs are not only places to eat but also spaces for socializing, working, relaxing, and conversation (Astuti & Hanan, 2016; Fieldhouse, 1996).

The concept of restaurants has evolved over centuries, emerging in various forms across different regions without a single point of origin (Crowther, 2018; Rawson & Shore, 2019; Sloan & Leith, 2003). Today, restaurants are integral to modern life, but their development from early taverns, cookshops, and inns took a millennium (Akdeniz, 2019). A restaurant is now a place where meals are served at various times, with customers choosing from a diverse menu (Rawson & Shore, 2019; Sloan & Leith, 2003). The term originated from a French soup vendor in the 1700s, with 'restaurant' deriving from 'restorative,' as these establishments initially offered medicinal soups (Cracknell & Nobis, 1989; Sloan & Leith, 2003). The term's evolution to its current meaning reflects the modernization of Western culture, a topic of interest for cultural theorists and historians (Sloan & Leith, 2003).

Restaurants existed in various regions long before the French Revolution, driven by economic and social factors (Akdeniz, 2019; Kiefer, 2002). The first restaurants actually originated in 12th-century China, amid administrative changes and urban growth (Akdeniz, 2019; Kiefer, 2002; Rawson & Shore, 2019).

The modern restaurant is often linked to revolutionary Paris, with A. Boulanger, a soup vendor, considered the first restaurateur, opening in 1765 (Walker, 2021). By the late 1780s, grand restaurants in Paris set the classic model. By the 1820s, restaurants catered to travelers and locals alike, offering a social dining experience (Crowther, 2018). Restaurants became hubs of urban sociability, serving all social classes and becoming more diverse as dining out grew popular among all socio-economic groups, including women (Rawson & Shore, 2019). They symbolize the democratization of luxury and the spread of fashion, playing a role in the civilizing process (Sloan & Leith, 2003).

By reviewing several reports on current and future market trends (Restaurant Industry 2030 by the National Restaurant Association, 2019; 2023 Dining Trends Report by TouchBistro, 2023; Trends and Tactics: How to Navigate the Changing Landscape of the European Restaurant Industry by Merckaert, 2023b), four trends play critical roles in the restaurant industry known today: technology, health-consciousness, sustainability, and unique experiences.

The digital revolution of the food and beverage industry has been a long time in the making, with the pandemic accelerating the process (Merckaert, 2023a). And although the pandemic could have increased takeaway and delivery, it is evident that customers still enjoy the experience of dining in. According to a study conducted by TouchBistro (2023), ordering food home has never been easier; nevertheless, most diners agree that it cannot replace the full experience of going out to restaurants.

Moreover, in recent years, there has been a noticeable increase in customers' interest in health and wellness (Georgiev, n.d; Owner, n.d.). The demand for unprocessed whole foods is increasing as the population develops more chronic illnesses and individuals educate themselves on the advantages of nutritious eating (Owner, n.d.). Furthermore, younger generations are more willing to spend extra for healthy meals that include 'buzzwords' like all-natural, organic, and GMO-free. Vegan food is also becoming a popular trend and many restaurants are catching up with it (Georgiev, n.d.). It is also likely that more customers will have dietary restrictions such as vegetarian, gluten-free, lactose-free or paleo among many others and operators will need to provide more alternatives to accommodate special diets (National Restaurant Organization, 2019; Wasserstrom, 2019). Nonetheless, it is also important to note that offering healthier meal alternatives may attract a more varied clientele and attract guests who may otherwise avoid visiting (Owner, n.d.).

However, trends in the restaurant sector can be about more than just cuisine or service. "Eatertainment", originally started in places like Disney, became widespread as one of the latest trends for many restaurant brands. To create a memorable event, an experience occurs to engage individual customers, when a company intentionally uses services as the stage, and goods as props. Experiences are memorable, as services are intangible and goods are tangible. Experiences are the heart of the entertainment business. Walt Disney has creatively exploited this concept. But today the concept is taking root in businesses far removed from amusement parks. New technologies encourage whole new types of experiences, such as multi-player or interactive games, Internet chat rooms, simulators and virtual reality. In the food industry theme restaurants such as the Hard Rock Café or Planet Hollywood, promise that concept of experience plus entertainment, also known as "eatertainment."

The environment has become one of the most recent concerns causing customer changes in the food and beverage market (Euler, 2022; Georgiev, n.d.; Voicu, 2023a). Environmentally friendly activities have grown in the recent decade and are expected to grow even faster in the upcoming years (National Restaurant Organization, 2019). In order to be more sustainable, restaurants need to consider recycling initiatives, sustainably sourced products, energy-efficient equipment and even eco-friendly designs. Since food deliveries are also a crucial part of today's industry, restaurants must consider environmentally friendly packaging (Burton-Hughes, 2020). In fact, 43% of diners would be willing to pay more to fund restaurants' sustainability measures, and some customers indicated they would not order if sustainable packaging alternatives are not available (Merckaert, 2023a).

Nevertheless, the social aspect of sustainability should also be mentioned. It concerns initiatives like aiding the local community, establishing appropriate relationships with community members, guests, employees, or suppliers, and complying to ethical standards (Božić & Milošević, 2021). 77% of consumers worldwide prefer purchasing goods from socially responsible businesses, and even in the restaurant industry, they are willing to spend 5-10% more due to the brand's devotion to the community. As a result, restaurants should become more socially responsible, benefiting both their surroundings and reputation. This could include sourcing products locally and seasonally to support local farmers and communities; or even supporting social causes through donations for charities or partnerships with organizations which tackle issues such as food insecurity or hunger (Khachatryan, 2023). In addition, nowadays customers also expect restaurants to address issues of equality and diversity, as well as their commitment to create a society that is more inclusive and fair (McCain Foods, 2021).

Brunch Restaurants and Plating

The term 'brunch' combines 'breakfast' and 'lunch' (Martins, 2018), reflecting its mix of the two meals. It typically includes a variety of dishes suitable for both breakfast and lunch (Peters-Desteract, 2018; Ternikar, 2014). By the 1920s, brunch had become a sign of class, and by the 1950s, it symbolized comfort and convenience (Ternikar, 2014).

In recent decades, brunch has evolved into a meal of comfort and casual entertaining, often associated with luxurious dining (Ellison, 2022; Ternikar, 2014). Today, it represents gastronomic luxury and comfort food (Ternikar, 2014). Millennials, who value experiences over material possessions, have driven the brunch revolution, making it popular among them and Gen Z (Future Food Strategy Group, 2018; Mintel, 2022). The creative nature of brunch often makes it irresistible to photograph before eating (Future Food Strategy Group, 2018; Hartley, 2021).

Plating refers to "both the preparation and presentation of food on a plate", with the two primary goals of visual and sensual attraction (Meulen, 2017, p. 236). This presentation involves the combination of various visual attributes of the plateware to create a visually appealing and cohesive presentation (Spence et al., 2014). Relatively little research has been done on the cultural history of plates and plating, compared to the studies of eating. Nevertheless, "we eat first with our eyes." (Apicius in Spence et al., 2016, p. 53; Delwiche, 2012) is a well-known expression, allegedly coined by the Roman gourmand Apicius in the first century AD, has recently been supported by numerous scientific studies demonstrating its relevance in today's world (Gambetti & Han, 2022; Michel et al., 2014; Spence et al., 2016). Visual compositions of plates are becoming increasingly important since findings suggest that presenting food in an aesthetically pleasing way can improve the overall enjoyment of the meal (Cifci et al., 2021; Michel et al., 2014). Eatertainment is aligned with this trend, therefore, plating should be viewed as a component of a complex dining experience, which today plays a critical role in the creative processes in the kitchen and people's food preferences (Cifci et al., 2021).

The growing significance of this trend can be attributed to the rapid increase in digital interfaces and audiovisual media, justifying the massive spread of "food porn." With the proliferation of smartphones, tablets, and other electronic devices, individuals now have constant access to digital screens (Gambetti & Han, 2022; Spence et al., 2016). These technologies have not only increased exposure to digital images of food but also heightened the desire to share meal images on social media (Spence et al., 2016). A large portion of Instagram images are food-related, with approximately 85% of photos featuring food content and 208 million posts with the hashtag 'food' (Gambetti & Han, 2022; Spence et al., 2022).

The phrase 'camera eats first' illustrates how sharing meal pictures on social media has become a fundamental part of dining experiences (Gambetti & Han, 2022; Wachyuni & Yusuf, 2021; Yong et al., 2020). Consequently, people often choose restaurants based on food images shared by other customers. To increase engagement, restaurants also implement image-driven social media strategies, sharing pictures of their dishes on social media and websites (Gambetti & Han, 2022).

Color, shape, and balance are significant characteristics of food that customers value, aiming for a harmonious and appealing composition that is both aesthetically pleasing and nutritionally satisfying (Cifci et al., 2021). The character of the chef is also important, as the design of a chef's plate can reflect their identity and personality (Edwards, 2023). Chefs today have the aura of "rock stars," globally known and celebrated by the media through shows, YouTube videos, and series. Names like Gordon Ramsay, Jamie Oliver, and Anthony Bourdain have become global phenomena, with Michelin Stars akin to the "Oscars" in this field, becoming brands that sell experiences.

The Importance of Branding in the Restaurant Industry

Branding is a vital component of the restaurant industry (McCracken, 2022; Robitaille, 2023). According to Dabeva (2012), the ability to differentiate, emphasize, and highlight the distinctive attributes of products and services is critical in the contemporary landscape. Therefore, operating a restaurant requires more than just serving delicious meals (McCracken, 2022). Without a unique identity, restaurants cannot connect with customers on an emotional level; thus, besides offering excellent food and service, a carefully planned brand strategy becomes critical in creating lasting impressions among customers and distinguishing oneself from competitors in the market (Bistrohub, 2020; Kakadiya, 2023; McCracken, 2022; Robitaille, 2023). As a result, it is widely understood that establishing and maintaining powerful brands is critical for succeeding in the hospitality sector (Gardini, 2014; Kwun, 2012).

Nonetheless, in addition to ensuring long-term success for businesses, restaurant owners may notice a variety of beneficial outcomes that can result from proper branding and brand management (Dabeva, 2012). Restaurant branding aids in developing a unique identity and personality for businesses, to tell their stories, how it wants to be perceived by customers, allowing the business to maintain a consistent and differentiated image (Dobos, 2022). Serving as an informative tool and an indicator of the restaurant's quality and image since it allows effective communication about products and services.

Therefore, customers can learn about products and services, have consistent quality standards, and receive clear and trustworthy information about the brand (Genc, 2010). Restaurants may also build a strong emotional connection with customers, leading to frequent visits and therefore increased customer loyalty, especially when paired with excellent cuisine and a memorable dining experience (Dobos, 202). By building trust and loyalty among diners, restaurants can also profit from better customer recalls and positive word-of-mouth (Cuboh, 2022; Robitaille, 2023). Branding is also closely connected to credibility. Given the sensitivity surrounding food, people are unlikely to dine in a restaurant that does not appear to be trustworthy (Dobos, 2022). Restaurants may successfully decrease perceived risks, such as financial, social, and safety concerns, by building a recognizable brand with which customers are familiar and have positive associations (Dabeva, 2012). Finally, developing a solid brand opens up greater potential for brand extension (Kwun, 2012). Hence, as a result of a successful brand strategy, restaurants may improve revenue performance by introducing new items or services with lower costs (Dabeva, 2012).

When examining the restaurant industry, it's essential to recognize that brands and brand management influence customers through various channels. Dining at a restaurant is a sensory experience, so the brand must consider multiple elements that affect guests throughout their journey, from booking a table to leaving the restaurant. A successful brand strategy encompasses every visible aspect of the restaurant and every component customers interact with, including staff, menus, advertisements, and food presentation (McCracken, 2022). Each element contributes to the brand's development, and the goal is to combine them in the most effective way. Instead of viewing elements as independent entities, restaurant owners should synchronize them to align with the overall business concept (Dabeva, 2012).

Over the years, economic and social landscapes have transformed significantly. At the start of the millennium, experiences became the foundation of a new type of economy and marketing approach (Schmitt et al., 2014). Pine and Gilmore (1998) described this phenomenon as the 'experience economy,' where businesses shift from offering products and services to offering experiences (Cleff et al., 2014). The economic value of society has evolved through various phases: from a commodity-based phase before the Industrial Revolution, to a manufacturing phase during the revolution, and then to the service-focused economy of the 20th century (Pine & Gilmore, 1998; Schmitt et al., 2014). With the advent of the 'experience economy,' the most valuable economic offerings have become enjoyable and informative experiences (Schmitt et al., 2014).

Contemporary customers are not only considered rational but also emotional, who do not base their choices of products and services solely on functionality and benefits (Khan & Fatma, 2017; Moreira et al., 2017). They seek products and marketing initiatives that engage their senses, connect with their emotions, and spark intellectual interest while also matching their lifestyle and, most importantly, delivering experiences (Moreira et al., 2017; Schmitt et al., 2014). They desire more than functionality; they seek enjoyable experiences with their chosen brands (Bapat & Thanigan, 2016; Hwang et al., 2021).

The importance of brand experience can be observed in marketing and brand management literature (Khan & Fatma, 2017). Hence, the need to create distinctive and memorable brand experiences has been highlighted over the last ten years in order to build better connections between brands and customers and improve the overall performance of businesses. The concept of experience has been extensively studied across various contexts by numerous researchers, contributing to the development of key ideas and associations (Pina & Dias, 2021). Holbrook and Hirschman (1982) were pioneers in introducing the concept of 'experience' in marketing, suggesting that consumption is not purely rational but also involves an experiential component, encompassing "fantasies, feelings, and fun" (Pina & Dias, 2021). Pine and Gilmore (1998) further advanced the field by introducing the 'experience economy,' highlighting the personal nature of experiences.

Building on these foundations, Schmitt (1999a, 1999b) explored experience from a multidimensional perspective, enriching the marketing literature with various constructs of 'experience.' Research has examined experience in multiple contexts and stages of the consumer life cycle (Nysveen et al., 2012), leading to constructs such as consumption experience, service experience, shopping experience, product experience, customer experience, and brand experience. Consequently, the overall brand experience comprises several experiential notions, including consumption, service, shopping, and product experiences (Brakus et al., 2009).

Brakus et al. (2009, p. 53) were the first to provide a comprehensive definition of brand experience, describing it "as subjective, internal consumer responses (sensations, feelings, and cognitions) and behavioral responses evoked by brand-related stimuli that are part of a brand's design and identity, packaging, communications, and environments". Ha and Perks (2005) took a cognitive approach to brand experiences, and the term could be described as how well-informed and familiar a consumer is with a specific brand or brand category, leading to better memorability and deeper significance for individuals, potentially fostering greater trust in the brand and enhancing their ability to classify brands by their attributes (Ha & Perks, 2005). Moreover, brand experience is defined as consumers' perceptions of the brand at every interaction with it, from advertising impressions to human encounters. Yet, direct personal connections are the most important aspect of creating an outstanding brand experience (Alloza, 2008).

The elements representing a brand's identity and design, which can act as stimuli, can include the name, slogan, logo, signage, packaging, mascot, the environment of the physical store, or any other communication forms like advertisements, websites, brochures, or even events. Nonetheless, the stimuli could also be in the form of the brand's visual identity, involving shapes, colors, typefaces, or any other design elements (Brakus et al., 2009). Moreover, brand experiences can vary in their level of intensity and strength, as well as in their valence, origin and duration. It was Brakus et al. (2009, p. 53) who provided a more detailed examination of this phenomenon in their study and stated that "in particular, brand experience differs from evaluative, affective, and associative constructs, such as brand attitudes, brand involvement, brand attachment, customer delight, and brand personality." They claimed that as opposed to brand attitudes, which are concerning more general impressions, such as liking or disliking brands, experiences are specific reactions to the interactions with a brand. Also, while brand involvement and attachment necessitate a personal interest in or connection to the brand, brand experiences could occur without these. In contrast to customer delight, which occurs after product use, brand experiences can happen at any interaction, whether expected or unexpected. Finally, unlike brand personality, which attributes human traits to businesses, brand experiences are about genuine reactions to brand-related stimuli (Brakus et al., 2009). Holbrook and Hirschman (1982) were the first to present the notion of experience within this field, highlighting the experimental view of consumption by claiming that it could also involve "various playful leisure activities, sensory pleasures, daydreams, esthetic enjoyment and emotional responses" (Holbrook & Hirschman, 1982, p. 132). Hence, this could already hint at the dimensions of activities, sensations, and emotions (Nysveen et al., 2012). Pine and Gilmore (1999) proposed the four 'e' dimensions of experiences: 'entertainment', 'educational', 'escapist', and 'esthetic' (aesthetic). The four zones also lie along the axes of active versus passive (describes the participation), and immersion versus absorption (describes the connection) (Cleff et al., 2014; Pine & Gilmore 1998).

Schmitt (1999a, 1999b) identified five distinct dimensions of experiences, including 'sense' (sensory), 'think' (cognitive), 'feel' (affective), 'act' (physical) and 'relate' (social). The element of 'sense' uses the five senses to create sensory experiences; the 'think' element refers to the engagement of the mind in order to deliver intellectual, solution-oriented experiences that inspire customers' creativity; when it comes to appealing to customers' feelings and emotions, the element of 'feel' prevalent; the element of 'act' represents physical experiences, lifestyles, and behaviors; and finally, the 'relate' element stands for social experiences that help a person connect to their ideal selves, other individuals, or cultures (Bapat & Thanigan, 2016; Beig & Nika, 2019; Schmitt, 1999b).

It was Brakus et al. (2009) who first empirically identified the brand experience dimensions. It is worth noting that the five types of experiences identified by Schmitt (1999) served as the foundation for their study on brand experiences. Nonetheless, as opposed to Schmitt (1999), Brakus et al. (2009) created a customer-focused approach that saw experiences as unique, internal, and behavioral responses of customers rather than as tactical aspects of experiential marketing. They proposed four underlying brand experience dimensions, which include 'sensory', 'affective', 'behavioral', and 'intellectual' experiences with the help of cognitive science, philosophical studies, as well as experience marketing and management (Brakus et al., 2009; Hwang et al., 2021).

'Sensory experiences' are primarily focusing on customers' visual, auditory, smell, taste, and touch perceptions. Hence, it includes the activation of senses via brand exposure (Huaman-Ramirez & Merunka, 2019), visual elements are seen as key in sensory experiences, since images are easily remembered by customers. Visual brand experiences give consumers an aesthetic feel of the brand as well as its surroundings. However, the importance of tasting, smelling, hearing, and touching varies by the product and services of businesses. Moreover, taste and touch are direct experiences, while smell and sound can be indirectly felt through a brand's atmosphere (Huaman-Ramirez & Merunka, 2019).

'Affective experiences' include all the feelings, emotions and moods sparked by interacting with businesses. In simple terms, it captures the emotional bond between the consumer and the brand (Hwang et al., 2021). This subdimension of experiences is imprinted in the brains of customers and causes either positive (such as love for the brand or enthusiasm for the brand) or negative (such as hatred for the brand or disgust) emotions (Mostafa & Kasamani, 2020).

Moreover, 'behavioral experiences', can encourage certain physical acts, behaviors, and lifestyles after interacting with a brand (Pina & Dias, 2021). Hence, it also entails doing tasks in a more efficient manner and even creating improvements to one's lifestyle (Hwang et al., 2021). In fact, customers are enthusiastic to participate in physical activities in order to create memorable experiences with a brand. Furthermore, the body may be used as a source of well-being, as a representation of emotions, and as a form of symbolic representation (Huaman-Ramirez & Merunka, 2019). It is also worth highlighting that behavioral experiences leave a long-lasting impression on customers and are seen as a crucial component of successful marketing practices (Mostafa & Kasamani, 2020).

Finally, 'intellectual experiences' involve creative and analytical thinking induced by encounters with a brand. It is believed that customers can clearly better recall brands with intellectual experiences and remember memories associated with them. Thus, intellectual experiences should aim to convince and create positive and satisfying memories for customers, as intellectual experience could form people's opinions about a brand (Mostafa & Kasamani, 2020).

Brakus et al. (2009) argued that brand experiences positively influence brand satisfaction, brand loyalty, as well as brand personality. Since positive brand experiences offer value for customers, it could lead to increased satisfaction among customers. Moreover, a favorable experience is one that customers see as different and memorable, one they would want to relive and recommend to others (Moreira et al., 2017). Brand experience influences not just previous satisfaction assessments, but also upcoming customer loyalty (Mostafa & Kasamani, 2020). Therefore, brand experience helps businesses stand out, win customer loyalty, and establish long-term connections with customers (Mostafa & Kasamani, 2020; Ong et al., 2018). Furthermore, brand experience directly increases brand attachment since favorable brand interactions can deepen customers' emotional bonds.

Thus, as brand experience becomes increasingly important in today's world, marketers study its effects and how they relate to other brand elements, which could be beneficial to improve marketing strategies (Bapat & Thanigan, 2016). It is not at all unexpected that today's top global corporations prioritize providing favorable experiences as one of their strategic objectives (Shamim & Butt, 2013). Therefore, to remain relevant in contemporary markets, businesses must focus on every facet of brand experience, as it not only provides increased brand value, but also drives higher sales; and ensures frequent engagement and marketplace dominance. Customers seek tangible benefits and unique experiences when it comes to restaurants. Hence, competition has shifted from only offering good quality and prices to providing distinctive dining experiences (Ong et al., 2018). Within the restaurant sector, customer experience studies often overlook the potential for experiences deriving from sensations, emotions, behaviors, and intellect in the context of branding. However, since distinctive brand experiences will provide a long-lasting competitive advantage in the restaurant sector, it is crucial to consider experiences based on the brand rather than just from the customer's viewpoint (Ong et al., 2018).

Brand experiences for products are generally simpler than those for services, making them easier to manage. Mosley (2007) explains this by highlighting the complexity in service-oriented businesses, which can be viewed from two perspectives: managing the variety of services provided (operational complexity) and handling customer interactions (interpersonal complexity). The former deals with the range and steps of services, while the latter focuses on relationship quality and understanding client needs.

Services differ from products in terms of intangibility (less or no physical presence), inseparability (simultaneous provision and consumption), heterogeneity (variation in service performance), and perishability (cannot be stored for future use) (Mandil, 2016; Nysveen et al., 2012). However, Mandil (2016) notes that services can bridge the gap between products and experiences. As services become more experiential, there is increased interaction with employees, greater personal touch, and more customization options (Mandil, 2016). This aligns with Pine and Gilmore's (1998) observation of a shift from product orientation to services, and eventually to experiences.

Due to their complex nature, services involve multi-level customer interactions, central to the concept of experience. These characteristics highlight the need to focus more on service-based brand experiences, which may be more relevant for understanding brand experiences than product-based ones (Mandil, 2016).

In the restaurant context, diners of today value not only practical, but also symbolic aspects, making the dining experience essential for return visits (Ong et al., 2018). Customers choose restaurants not just to meet their immediate needs, but also to have a delightful brand experience, which could actually be enhanced by several brand clues: "the quality of food, order-handling system, ambience, and behavior and appearance of the staff" (Khan & Fatma, 2017, p. 448-449). As a consequence, a restaurant brand is more likely to provide customers with a favorable brand experience if it has a high-quality cuisine, an effective order-handling system, a relaxing atmosphere, and well-trained employees who exhibit appropriate behavior and appearance. Besides these brand clues, they revealed that diners are more inclined to visit a restaurant for a new brand experience if they find the marketing messages appealing. Hence, marketers should create campaigns that are both meaningful and emotionally engaging. Moreover, customers now rely more on websites for gathering information, so it is important to update these sites as well. Finally, the significance of organizing thematic events (e.g., charity fundraisers) that might elicit pleasant emotions was noted as a way, businesses can strengthen the connection between the brand and customers (Khan & Fatma, 2017).

This study highlights that memorable brand experiences not only boost marketing success but also build trust, customer satisfaction, loyalty, positive brand attitudes, word-of-mouth, and increased brand credibility (Khan & Fatma, 2017). In conclusion, effective branding and brand management are essential in the competitive restaurant industry. Although still emerging, the concept of brand experience has gained significant importance due to changing customer attitudes and preferences. As society enters the 'experience economy,' brands focus more on creating memorable experiences rather than just offering products and services, emphasizing the importance of emotional connections with customers.

Brand experiences are particularly crucial for restaurants due to their multidimensional nature. A restaurant's brand strategy must encompass all visible aspects that customers interact with, including plating, which can be considered a brand-related stimulus. This makes it relevant to study whether aesthetic plating can enhance the overall brand experience of diners. Therefore, given the pivotal role of brand experiences in the restaurant sector, this research explores whether aesthetic plating in restaurants can significantly enhance diners' perceived brand experience.

Adopting a primary research question: How does plating in restaurants influence the overall brand experience of diners? This work breaks this question into four sub questions, by following Brakus et al. (2009) concept of brand experience, addressing the four dimensions of sensory, affective, behavioral, and intellectual experience. Hence the following four research subquestions:

- RQ1A: How does food presentation in restaurants influence the sensory brand experience of diners?
- RQ1B: How does food presentation in restaurants influence the affective brand experience of diners?
- RQ1C: How does food presentation in restaurants influence the behavioral brand experience of diners?
- RQ1D: How does food presentation in restaurants influence the intellectual brand experience of diners?

METHODOLOGY

This exploratory research adopts a social constructivist and interpretivist philosophical worldview, utilizing a qualitative research approach (Creswell, 2017; Kelliher, 2005). It employs a focus group strategy (Babbie, 2011; Creswell, 2017; Hair et al., 2021; Mainardes et al., 2010) to deeply explore the relationship between food plating and diners' perceived brand experience. According to Powell and Single (1996, p. 499), a focus group is "a group of individuals selected and assembled by researchers to discuss and comment on, from personal experience, the topic that is the subject of the research." This approach uses guided, interactive conversations to provide detailed insights into complex experiences and the reasons behind beliefs, behaviors, perceptions, and attitudes (Powell & Single, 1996).

Focus groups allow researchers to understand how and why individuals think about a specific topic, offering insights into their viewpoints and priorities. By fostering direct engagement, this strategy brings researchers closer to the topic, making it easier to understand participants' perspectives (Lane et al., 2001). This method is known for collecting real-life information in social settings, making it more reflective of genuine thoughts and behaviors. Its flexibility allows adaptation to various scenarios and situations. Additionally, this method is effective and trustworthy, contributing to the study's validity. It can provide quick results and is considered cost-effective compared to other methods. Group dynamics often reveal aspects of the issue that might not emerge in one-on-one interviews (Babbie, 2011). Given the goal of this research to address the perceived influence of food plating on diners' brand experience, and considering the social dynamics and relevance of plating, especially on social media, focus groups were deemed an appropriate data collection strategy.

Sampling, Participants Selection and Data Collection Procedures

Focus group participants often come from purposive sampling, where individuals are selected for their relevance to the research rather than their representativeness (Hair et al., 2021; Lane et al., 2001). This method allows for the collection of rich data that significantly reflect the characteristics of the population from which the sample is drawn (Lane et al., 2001). Qualitative exploratory research prioritizes thick descriptions, process identification and mapping, and exploring individual nuances and perceptions about a topic over representativeness. To mitigate the risk of unrepresentative insights from a single small group, multiple focus groups were conducted. Three focus groups were held in Hungary, each consisting of five participants, including both male and female members. Despite their diverse educational and professional backgrounds, all participants were members of Generation Z and from Hungary. Generation Z is particularly important for the hospitality sector due to their high customer lifetime value potential (Babbie, 2011).

Members of Generation Z highly value authentic brand experiences. In fact, they are typically more inclined toward creating memories and engaging in experiences rather than acquiring material possessions (Inphantry, 2017). This generation seeks pleasure in experiences that are not limited to large-scale concerts or travels; they seek enjoyment in experiences that are part of everyday life as well (We Are Collider, 2022). More than half of Generation Z and Millennials have increased their frequency of eating breakfast at foodservice establishments, making brunch an ideal opportunity to meet their demands and help foodservice firms build a loyal client base (Future Food Strategy Group, 2018; Mintel, 2022). Additionally, Generation Z highly values unique, relatable, and photogenic food experiences. They seek out new and appealing foods that not only satisfy their taste but are also suitable for sharing on social media, reflecting their enthusiasm for capturing and sharing these experiences (Campisi, 2022; McLymont, 2022).

A specific restaurant was intentionally chosen as the study's center to examine how plating might influence brand experiences. Participants were invited for brunch at this restaurant before the focus group discussions. This provided them with a practical, real-world experience, ensuring richer and more in-depth conversations during the focus group meetings. Given the geographical focus of the research, the study was conducted in Budapest, Hungary. A well-known brunch restaurant in the city was chosen as the ideal venue, representing contemporary brunch trends in Budapest and providing an appropriate setting to examine the relationship between food plating and brand experiences. Although the restaurant opened in early 2023, it quickly gained popularity in the lively city environment, attracting neighborhood employees, tourists, and local customers from various age groups. The restaurant offers all-day breakfast, allowing customers to enjoy breakfast options throughout the day, which is rare among other brunch spots in Budapest.

Participants Characteristics

Group 1 of participants was composed by 5 individuals with ages from 23 to 25 years, being 3 females and 2 males. Group 2 had also 5 participants from 23 to 24 years old, and 3 females and 1 male. On the other hand, Group 3 had also 5 individuals with the age distribution as Group 1 and 4 females and 1 male. To ensure the privacy and confidentiality of the participants, their identities have been anonymized. Participants were identified by a unique code, comprising the number of the focus group they participated in (either 1, 2, or 3), followed by a specific number assigned to them within that group.

In this study, these young people are all Hungarians from Generation Z, and in the constitution of the three groups, gender issues were ignored in favor of the diversity of backgrounds and occupation that these young people have. Therefore, in terms of results, gender issues did not prove to be relevant. Yet, it is believed that the results could be richer if the background of these young people were different, which is why three groups were created, with no people having the same occupation or background.

Data Collection Instrument: Focus Group Script

The script for the focus group sessions, consisted of five different sections: 1) Introduction to the study; 2) Warm-up questions; 3) General questions about the appreciation of aesthetic food plating; 4) Questions about the overall dining experience in the chosen restaurant; 5) Questions about the brand experience in the chosen restaurant (sensory, affective, behavioral, intellectual).

The introduction provided an overview of the research issue, ensuring participants understood the main topic and objective. Section 2 used warm-up questions as icebreakers, allowing participants to discuss their dining habits. In Section 3, participants were asked about their appreciation of aesthetic food plating, inspired by the Centrality of Visual Product Aesthetics (CVPA) scale (Bloch et al., 2003). Questions were adjusted to food presentation, asking how much they value appealing food, if it enhances their eating experience, and if it influences their choice of meals or locations. They also considered if appealing food presentation would encourage them to recommend or return to the restaurant. Section 4 addressed participants' experiences at the chosen restaurant. They shared their overall thoughts on their recent visit, focusing on the visual appeal of their meals and whether any elements of the presentation stood out. Participants discussed if meal presentation was a defining characteristic of the restaurant. Section 5, inspired by Brakus et al.' s (2009), brand experience framework, explored the four dimensions of guests' experiences. The sensory dimension examined how the dining experience appealed to their senses and if the food presentation enhanced sensory engagement. The affective dimension looked at the emotional connection with the restaurant's brand, discussing feelings and emotions during dining and how food presentation influenced these ties. The behavioral dimension focused on diners' actions, such as taking photos or writing reviews. It meant to assess how the restaurant's food presentation might have influenced guests' activities or engagements. To assess the intellectual dimension of the dining experience, participants were asked if their time at the restaurant sparked their curiosity or offered fresh perspectives on dining. This also aimed to understand if the restaurant's plating style played a role in stimulating such intellectual reflections.

Data Analysis Procedures

After the focus group discussions, which were audio recorded, the sessions were later transcribed to prepare for data analysis. The focus groups were conducted in Hungarian. Thus, to make the data universally accessible the transcriptions were translated into English.

For data analysis, we used thematic analysis to interpret focus group discussions. According to Braun and Clarke (2006), thematic analysis identifies, analyzes, and reports patterns (themes) within data. This technique focuses on identifying common themes from group dynamics and participant interaction, including personal attitudes, ideas, beliefs, and societal conventions (Massey, 2011). Instead of numerical metrics, thematic analysis explores whether the data provides meaningful insights to address the research question (Scharp & Sanders, 2018).

This study follows Braun and Clarke's (2006) six-step framework for thematic analysis:

1. Familiarizing with the data: transcribing and reading through the material.
2. Creating initial codes: identifying interesting parts of the data and grouping related codes.
3. Discovering themes: sorting codes into themes and gathering relevant information.
4. Reviewing themes: evaluating if the themes work with the coded data and the entire dataset, creating a 'map' of the analysis.
5. Naming and defining themes: refining each theme and the overall narrative, giving each theme a clear name and meaning.
6. Writing the narrative: writing the final analysis based on the developed themes and codes.

In this research, we slightly modified the sequence to align with the pre-established focus group script, which already includes clear guidelines and overarching themes. This allowed us to approach step three, 'Discovering themes,' more quickly, using the pre-defined themes as the foundation for step two.

FINDINGS AND DISCUSSION

Eating Out Habits

The focus group discussions began by exploring participants' eating out habits. This theme aimed to show the frequency and occasions for eating out. The data revealed varied habits: some participants eat out regularly, while others reserve it for special occasions. Workplace dining also emerged as a common practice, with many opting for on-site cafeterias or nearby restaurants. Participants 1.2, 1.3, 2.2, 2.3, 2.4, and 2.5 highlighted these trends. For example, Participant 1.2 said, "I eat at least five times a week in restaurants, or even more, as I don't really eat at home." Similarly, Participant 2.4 noted, "I eat out pretty frequently, like once or twice a week with friends. And I also always eat at the restaurant at my workplace."

On the other hand, Participant 2.5 said, "I usually only eat at restaurants on special occasions like family gatherings, birthdays, etc., once or twice a month." Similarly, Participant 3.3 mentioned, "I go to restaurants mostly on occasions, when I celebrate." These two participants view eating out as an occasional activity rather than a regular habit. The occasions for eating out vary, including special events and everyday activities. Many participants eat out to celebrate birthdays or family gatherings, but also see it as part of their social life with partners and friends, or as a leisure activity. For example, Participant 1.1 explained, "I think I eat out once or twice a week with my friends or on a date with my boyfriend, but I also really like to celebrate special events like birthdays in restaurants." This perspective shows that eating out can be both a routine social activity and a way to celebrate special occasions.

Interestingly, Participants 3.1 and 3.2 view dining out as an integral part of their lifestyle rather than just a leisure activity. Participant 3.1 admitted feeling guilty about eating out often, especially during holidays, but enjoys it as a hedonistic pleasure. Participant 3.2 added that dining out is part of their everyday life, not just for special occasions, eating out once a day in some form. Participant 3.5 offered a different perspective, enjoying cooking at home for cost efficiency but seeking unique dining experiences at restaurants for meals they can't prepare themselves. This highlights that for some, eating out is about the unique experience that restaurants offer, beyond just the food.

Factors of Restaurant Selection

The focus group conversation then shifted to factors participants consider important when selecting a restaurant. A recurring theme was the variety and options of meals available on the menu. Participants emphasized the need for a diverse selection to satisfy various preferences. Participant 1.3 mentioned, "The selection of food is the most important, as well as looking at the pictures," and Participant 1.2 added, "I like to check if there's something I would like to eat." Many participants also highlighted the importance of an updated, available online menu, supported by Participants 1.2, 1.3, 1.4, and 2.5.

Price was another consistent factor in the decision-making process. For some, it serves as an initial filter before other criteria. Participant 3.4 stated, "I decide how much I want to pay first, and then choose something I can afford." Additionally, value for money was important to avoid overpriced places. The atmosphere and ambience of a restaurant are crucial factors. It's not just about the feel inside but also the initial perception based on online pictures of the interior and food. Participant 1.4 emphasized the need to see the current menu, prices, food selection, and atmosphere online. Similarly, Participant 2.1 mentioned avoiding places with unappealing atmospheres. Online platforms like Google Maps and TripAdvisor are valuable for evaluating restaurants' location, ambiance, food selection, and customer reviews. Multiple participants, including 1.1, 1.3, 1.4, 1.5, 2.1, 2.2, and 3.1, mentioned using these tools. Participant 1.1 highlighted the importance of reviews, stating she avoids places with at least five bad reviews to avoid wasting money on a bad experience.

Finally, the last factor that was mentioned by multiple participants, were recommendations from family, friends, and acquaintances. This points out to more trusted and reliable sources. For example, Participant 3.2 highlighted it as the most important factor, claiming that "If someone already said that 'Oh, this one's really good!', then I'm more likely to try it, because I feel like I have to experience it too." This viewpoint was also supported by Participants 2.2 and 2.4.

Perceptions and Practices Regarding Brunch

The research on brunch restaurants revealed a universal appreciation for brunch food across all focus groups. Participants expressed their love for typical brunch menus, with Participant 1.4 stating, "brunch food is literally the best," and others echoing similar sentiments.

However, brunch habits vary. Some participants, like 1.1, 1.5, 2.1, 2.2, 2.5, and 3.1, only visit brunch restaurants while on vacation. Participant 1.5 also enjoys brunch on weekends. Despite the love for brunch, logistical issues like work schedules often prevent frequent visits. Participant 3.5 highlighted this by saying, "... it's hard to have a 'slow morning' because of work, but still, I like it very much."

Participant 1.4 mentioned that it's "hard to find occasion when you have a job," highlighting the difficulty of finding time for brunch. Brunch is seen as more than just food; it's a cultural phenomenon. Participants noted that part of the cost is for the experience and atmosphere. Participant 1.1 said brunch restaurants are overpriced due to their cultural appeal, and Participant 1.3 mentioned the luxurious branding of brunch. Participant 1.5 added that brunch is an 'Instagrammable' occasion.

However, participants also criticized the high costs. Many feel the meals are overpriced for what they are. Participant 3.3 pointed out that similar meals could be made at home for much less, citing an English breakfast as an example. The issue of pricing seemed to be particularly about Budapest, as multiple participants noted that brunch spots in the city tend to be overpriced: "although I don't mind spending money on brunch, I would say that brunch places in Budapest are really overpriced" (Participant 1.4). Nonetheless, after talking about the prices of such restaurants, Participant 1.2 added, "I think you kind of buy the experience of these restaurants."

Aesthetic Appreciation of Food Plating

The discussion then focused on the importance of aesthetic food plating in restaurants. All participants agreed that visually appealing plating enhances their dining experience and adds value. Participant 1.1 noted that aesthetic plating influences how one eats, while Participant 2.3 mentioned it signals precision and effort from the restaurant.

Participants also felt that appealing plating can boost their appetite and enjoyment of the meal. For example, Participant 2.1 said it adds to the dining experience, and Participant 1.4 mentioned it improves their appetite and satisfaction. Participant 3.3 emphasized that dining is a multisensory experience, involving sight and smell, and restaurants should consider this. Participant 3.2 agreed with the perspective that restaurants need to pay attention on plating and should be included in the whole experience: "I also agree with this. When someone goes to a restaurant, they also pay for the experience, not just for the food. Therefore, the experience also includes plates, which has been done aesthetically and precisely, that you might not be able to do at home. Otherwise, I could also just eat food at home. So, from restaurants, I expect a nice plate which adds value to the visuals, as I pay the price for it".

Participants noted that the visual aspect of food often influences their choice of restaurant or meal. This is especially true when viewing online images or observing neighboring tables. Participant 3.1 explained that pictures online significantly impact decisions, even unconsciously. Participant 2.4 added that the visual presentation is the first impression of the dish and the restaurant, often seen in social media posts. Participant 3.4 also added, "A lot of times it happened to me, that I was in a restaurant, and I saw a plate there from another guest, and I just thought, I need that dish now." This phenomenon was also mentioned by Participant 1.5, who said "...when you're sitting in a restaurant and you see on the table next to you a plate, and you want to have the same dish immediately just based on the look." In addition, this viewpoint was also supported by Participant 1.2, 1.3, and 1.5. Participants 2.2 and 3.1 highlighted the importance of food presentation for social media. Participant 2.2 said, "I like to post on social media, so nice plating matters." Participant 3.1, a content creator, emphasized, "Aesthetic plates are crucial for my posts on Instagram, TikTok, etc. If they don't look good, I don't post them."

However, while attractive plating can draw customers, it doesn't guarantee they'll return or recommend the restaurant. Participants agreed that taste, quality, service, and ambience are more influential. Participant 2.4 mentioned, "If the overall experience is good, I'd recommend it, but not just for aesthetic plating." Participant 3.4 added, "It's crucial as a first impression, but other factors matter too." Participant 3.2 concluded, "Looks can attract, but revisits and recommendations depend on delicious food, good staff, and ambience." Participants 1.2 and 2.3 mentioned they would recommend a restaurant based on the aesthetic appeal of the plates for those who prioritize looks, like for Instagram. Participant 1.2 said they would recommend a restaurant for its aesthetic appeal if the person values looks. Participant 2.3 added that even if they didn't like it, they would still recommend it for someone who loves aesthetic plating.

Additionally, some participants noted that they could overlook plating if the food is delicious, the portions are good, and the value for money is fair. This was especially true for rustic and authentic Hungarian restaurants. Participant 2.4 mentioned that despite simple presentation, the food's flavor makes them return frequently. However, on top of all of this, one interesting comment came up by Participant 1.3, who expressed that there could be extremes in food plating, especially in fine dining settings, where the visual aspects sometimes overshadow the quantity or even the food itself.

Overall Dining Experience at Berries & Bite Brunch

The focus group conversation continued to focus on the participants' recent experience at Berries & Bite Brunch. Most participants in all three focus groups seemed to agree on Berries & Bite Brunch's minimalistic approach to interior design and thought they could have had potential to make it nicer even with attention on smaller details. While some did not have a problem with the more clear and simple approach, many people would have preferred a more 'cosy' or 'rustic' environment. Participants appreciated the combination of berry and white colors in the interior design. They generally found the food's taste and quality to be positive, describing it as "good," "really good," "delicious," or "really delicious." However, some noted that while the food was good, it wasn't extraordinary.

The visual presentation of the food was praised for its color harmony and appealing plating. Participants liked how the vivid colors of the ingredients contrasted with the pastel tones of the plateware. They also appreciated the precise arrangement and shaping of the ingredients. The plateware itself was noted for its appropriate size, shape, and color, enhancing the overall dining experience. Participants felt that the plating style at Berries & Bite Brunch was similar to other brunch places, making it visually pleasing but not distinctive. This sentiment was shared across all focus groups, with many participants agreeing through non-verbal cues like nodding. While participants appreciated the creative concept of berries in the logo, interior, and menu, they suggested that the restaurant could have done more to emphasize this theme. They recommended incorporating berries or berry colors consistently in each dish to make the restaurant's identity more recognizable. However, some participants cautioned against overdoing this theme.

Brand Experience at Berries & Bite Brunch

The discussion continued to explore the brand experience participants encountered at Berries & Bite Brunch, encompassing the sensory, affective, behavioural, and finally intellectual dimensions of it.

Sensory Dimension

Participants discussed the sensory dimension of their brand experience, focusing on the engagement of their five senses while dining. Visually, most were engaged by the restaurant's interior design, color scheme, and wall art, though some found the wall information overwhelming. Participant 1.2 noted the unified design but found the wall motifs a bit much. Participant 2.4 suggested that staff attire could be more berries-themed to enhance the visual experience. The presentation of food also played a significant role, with Participant 1.2 and Participant 2.5 stating that the plating exceeded their expectations and enhanced their sensory experience. Participants noted that dining at the restaurant stimulated their sense of taste, with many describing the food as delicious. The presentation of the food also enhanced their appetite and taste experience. For example, Participant 2.1 mentioned that seeing the food improved their appetite, while Participant 1.1 said the plating's colors and arrangement made them enjoy the meal more.

Regarding the auditory experience, feedback varied. Some participants enjoyed the background music and the sounds of other diners, finding it added to the atmosphere. However, others found the noise overwhelming or distracting. For instance, Participant 1.2 enjoyed the music and ambient sounds, while Participant 3.3 found the noise too loud. Participant 2.2 wished for more background music. Additionally, some participants noted that the texture of certain foods, like crunchy elements, slightly enhanced their auditory experience, though this was not a universal sentiment. The focus groups were generally disappointed with the restaurant's lack of appealing smells, which are often crucial in food experiences.

Many felt that the aroma of freshly baked goods or coffee could have enhanced the ambience. Participant 2.2 emphasized this by saying, "…for the smell, I really missed it – when I pass by restaurants and they have a good smell, I immediately consider going in, so that could be a huge plus actually." Participant 1.1 agreed, noting the absence of these smells. Additionally, the sense of touch was largely ignored, with Participant 1.1 stating, "Touching is not relevant I think."

Affective Dimension

The discussion then explored the emotional responses and connections participants felt at Berries & Bite Brunch. Most participants felt pleasant and calm, especially when enjoying good food and company. Many expressed excitements when selecting meals, waiting for food, and seeing the dishes: "I got really excited when looking at the menu… just reading the ingredients made me feel hyped" (Participant 1.4).

However, these feelings didn't translate into a deeper emotional connection with the brand. Participant 1.1 said, "I think I didn't really have any kind of deep feelings, I felt pleasant in the restaurant". Some negative emotions were also noted. Participant 1.2 mentioned feeling excited but also experienced negative feelings due to a neutral and grumpy waiter. Participant 1.5 added, "…it can be a bit of a mood killer when the waiter seemed irritated by the guests".

Regarding food presentation, some participants felt that aesthetic plating indicated the restaurant's attention to detail, fostering trust and positive feelings towards the brand. Despite the positive emotions from the aesthetic food presentation, many participants felt that Berries & Bite Brunch's plating was not unique enough to create a deeper emotional connection. Participant 3.2 noted, "I wouldn't say I felt closer to the brand, as this plate can be found in most brunch restaurants." Participant 3.5 agreed, saying, "I didn't feel a deep connection as it was not that special". Participants clarified that their affective experience was influenced by multiple factors, such as taste, ambiance, and staff interaction, rather than just the food presentation. Participant 1.2 mentioned, "I think I got the wow feeling from the plate, but my feelings towards the brand were influenced by various things." Participant 2.4 added, "Although the plating was nice, it didn't really help to build an emotional connection with the brand."

Behavioural Dimension

The dialogue then explored the behavioral dimension of the brand experience at Berries & Bite Brunch. A common behavior was taking photos of meals and sharing them on social media, driven by the aesthetic appeal of the plates. Participant 3.1 noted, "I actually made pictures only from the plates, so that 'behavior' was evoked by them." Most participants said they would share their experiences with family and friends rather than writing reviews. Participant 2.1 mentioned, "I wouldn't write a review, but I would tell my friends and family how the experience was if they ask about it." Participant 3.5 added, "To write a review, a restaurant needs to be either exceptional or really bad." Additionally, the food presentation inspired many participants to consider replicating the dishes at home.

Intellectual Dimension

The focus groups had mixed responses regarding the intellectual engagement at Berries & Bite Brunch. Some participants (1.4, 1.5, 2.1, 3.2) were intellectually stimulated by the menu variety, special ingredients, or educational wall art about berries. Participants 2.2 and 2.5 offered unique perspectives on the restaurant's business model. Participant 2.5 thought starting a business in the breakfast and brunch sector might not be as challenging as assumed: "I had a thought, that it wouldn't be that complicated to start a business in this field." Participant 2.2 found it interesting how a simple meal like breakfast could be transformed into an upscale experience: "From a simple breakfast, you can create something that feels fancy, available for the upper class." Others, like Participants 1.2 and 3.4, were skeptical about a restaurant's ability to evoke intellectual thoughts. Participant 1.2 said, "I enjoyed the food here, but I wouldn't say it particularly sparked any deep thoughts or curiosity in me." Participant 3.4 added, "To be honest, I wouldn't say that any restaurant can really provoke thoughts – maybe creativity in terms of the ingredients and how the food is prepared but that's it."

Participants were divided on whether food presentation could enhance intellectual experience. Some felt it was purely aesthetic and didn't stimulate intellectual curiosity. Participant 1.2 said, "Their plating style, while beautiful, doesn't drive intellectual engagement but enhances the overall experience." Participant 2.1 agreed, stating, "Plating cannot evoke intellectual thoughts." However, Participant 3.4 noted that while Berries & Bite Brunch's presentation might not engage diners intellectually, other formats like fine dining and molecular gastronomy could offer deeper intellectual stimulation.

End-of-Discussion Insights

The focus group discussion highlighted the importance of plating in shaping guests' experiences and perceptions of a restaurant brand. Participants agreed that aesthetic plating is crucial, especially in the social media age, where it can serve as free advertising. However, they also emphasized that visual appeal alone is not enough; taste, service quality, and ambiance are equally important for a holistic brand experience. Participant 2.1 noted that the importance of plating varies by restaurant type. For example, nice plating is expected in a brunch restaurant but might be confusing in a traditional Hungarian restaurant. This suggests that the role of plating depends on the restaurant's style.

Regarding Berries & Bite Brunch, participants had mixed views. While some found the plating to be the most positive aspect of their experience, they agreed that a good restaurant brand cannot rely solely on presentation. Participant 1.2 mentioned that the appealing presentation saved their overall experience, despite the food being just good and the atmosphere and staff not being exceptional. On the other hand, Participant 3.2 suggested that Berries & Bite Brunch needs to offer something more unique in its plating to add bigger value to the dining experience. Although the plating was aesthetically pleasing, she argued that good plating has become a standard expectation, and therefore, for the restaurant to stand out, it needs a bit more special touch: "I think most of the brunch restaurants follow the same patterns, and this restaurant didn't give a plus – so to be engaged with all these dimensions of brand experience, I think a bit more is needed."

Contextual Insights

The focus group findings revealed two distinct categories of eating out preferences: some participants view it as a regular, everyday activity, while others save it for special occasions. Many participants dine out almost daily, supporting Walker's (2017) observation that meals consumed away from home have increased over time. Those who frequently dine out do so as part of their social life, work routines, or lifestyle, validating that dining establishments serve purposes beyond just eating (Astuti & Hanan, 2016; Fieldhouse, 1996). Conversely, a small subgroup only dines out on special occasions with close people, highlighting the symbolic power of food as a means to demonstrate belonging and hospitality (Fieldhouse, 1996). Their experiences confirm literature's focus on 'commensality' and the sense of community that eating embodies (Fieldhouse, 1996; Koponen & Mustonen, 2020; Pilcher, 2012).

Additionally, some participants found restaurants suitable for various purposes, such as casual meetups, work lunches, or special events, highlighting the multifaceted role of restaurants in contemporary life. These results align with studies indicating that food consumption has evolved into a multifaceted social and individual experience (Crowther, 2018; Pilcher, 2012). During the focus group discussions, participants highlighted several key factors influencing their restaurant choices. The decision-making process is multifaceted, involving both practical and experiential factors. Google Maps and online reviews serve as digital reference points, while ambiance, food variety, and price remain important.

A frequently mentioned point was the importance of menu variety, reflecting participants' varied tastes. The need for up-to-date online menus was also emphasized, aligning with literature that shows diners increasingly rely on online resources. In fact, 84% of diners examine the menu online before visiting a restaurant (National Restaurant Association, 2019), highlighting the need for restaurants to maintain an active digital presence. The atmosphere of a restaurant is also vital, as dining out is about the entire experience, not just the food. Interestingly, participants form initial impressions based on online images of a restaurant's interiors, décor, and dishes, indicating that what diners see online heavily influences their decision to visit.

Participants indicated that platforms like Google Maps are crucial in their restaurant decision-making, offering comprehensive evaluations and insights. Peer reviews, both positive and negative, significantly influence their choices, highlighting the importance of restaurants providing outstanding customer experiences. This aligns with the trend of increasing technological integration in the restaurant industry, predicted to be fully digital by 2030 (National Restaurant Association, 2019). Despite the rise of digital platforms, personal recommendations from trusted individuals remain valuable. Price is also a key consideration, with participants seeking value for their expenditure without sacrificing quality.

Interestingly, themes like sustainability and health consciousness, emphasized in recent studies (Euler, 2022; Georgiev, n.d.; Owner, n.d.; Voicu, 2023a), were absent from the focus group discussions. This suggests a potential gap between industry trends and the actual preferences of the participants.

Participants admired brunch for its food selection and overall experience, associating it with vacation, pleasure, and leisure. This aligns with literature describing brunch as a symbol of comfort and convenience (Ternikar, 2014). They also viewed brunch as luxurious and comforting, echoing Ternikar's (2014) characterization of it as gastronomic luxury. Brunch's digital appeal was noted, with participants seeing it as an 'Instagrammable' occasion, supported by sources like Hartley (2021) and the Future Food Strategy Group (2018). However, participants raised concerns about logistics and pricing, particularly in Budapest, where they felt brunch was overpriced. This contrasts with the Future Food Strategy Group's (2018) claim of casual, affordable pricing.

Participants highlighted the importance of aesthetic food plating in enhancing the dining experience. A well-constructed plate was seen as a sign of a restaurant's attention to detail and quality. This aligns with literature emphasizing the role of visual presentation in dining satisfaction (Cifci et al., 2021; Michel et al., 2014). The visual appeal of food influenced participants' choices of restaurants and dishes. They often based their initial impressions on the appearance of dishes and were swayed by beautifully plated food at neighboring tables. This supports research showing that visual cues significantly impact eating habits and food selection (Spence et al., 2014; Campo et al., 2017; Spence, 2017).

Participants noted the use of technology to assess the visual appeal of dishes, highlighting the importance of well-presented food in attracting diners and serving as a marketing tool. This connection between aesthetic plating and social media was emphasized, especially by content creators, aligning with literature on the trend of sharing food experiences online (Gambetti & Han, 2022; Spence et al., 2022). While aesthetics influenced initial impressions and dining choices, they were not the sole determinants of repeat visits or recommendations. The overall dining experience, including taste, food quality, service, and ambiance, was more crucial for customer loyalty. For some, especially in authentic Hungarian restaurants or familiar places, taste could overshadow the need for aesthetic plating, showing that flavor sometimes takes precedence over visual appeal.

One participant raised a concern that extreme aesthetic presentation in fine dining can overshadow practical aspects like quantity and essence of a dish. This aligns with Paaki et al.'s (2019) research, which emphasizes that aesthetics should also serve a practical purpose. Feedback from the dining experience at Berries & Bite Brunch highlighted criticism of its minimalist interior. Participants desired a more 'cosy' or 'rustic' ambiance, feeling the design was inauthentic and overly targeted at younger diners. This suggests that while trendy, the minimalist design may not effectively differentiate the restaurant. Berries & Bite Brunch might benefit from revisiting their interior design to better align with their brand identity and audience.

Feedback on the food at Berries & Bite Brunch was positive regarding taste and quality, but it wasn't seen as unique or outstanding. This aligns with McCracken's (2022) view that success in the restaurant industry requires more than just good food; it needs a unique culinary identity. Participants praised the visual appeal of the food plating, noting the effective use of color and arrangement, which aligns with guidelines in the literature (Campo et al., 2017; Spence & Piqueras-Fiszman, 2014). However, the presentation style was not seen as unique, suggesting that while Berries & Bite Brunch excelled in aesthetics, they did not establish a distinctive brand identity through their plating techniques. The 'Berries' theme at Berries & Bite Brunch was appreciated but could be more consistently integrated, especially in plating. Participants suggested incorporating berries or the color palette in each dish to strengthen brand identity. Dabeva (2012) emphasizes the importance of harmonizing branding elements to express the business concept. Enhancing the theme's expression across all touchpoints might resonate better with customers. Regarding the brand experience, the interior design and color schemes created a visually pleasant atmosphere, though some participants desired more warmth and found certain visual elements overwhelming. Visual cohesion is crucial for a cozy and engaging environment. Suggestions included staff uniforms aligning with the berries theme to enhance visual experience.

The auditory experience received mixed feedback. Balancing background noise is essential, as some diners prefer a lively atmosphere while others seek quiet for conversation. Background music should be carefully chosen to cater to diverse preferences. Loud conversations are often unavoidable in a full restaurant. The olfactory dimension was a missed opportunity at Berries & Bite Brunch. Participants suggested that appealing smells like freshly baked products or coffee could enhance the dining experience, setting expectations and evoking anticipation. The lack of engaging scents might be due to the types of dishes served or preparation methods. Touch was negligible in shaping the overall experience.

Taste, primarily related to the plating, will be discussed in the next section. The affective dimension showed a mix of positive emotions and missed opportunities for deeper connections. Participants felt a general pleasantness, enjoying good food and company, which created a soothing atmosphere. The excitement of meal selection and presentation indicated success in creating moments of heightened emotion. However, these feelings were not deep enough to foster strong emotional bonds, potentially affecting long-term brand loyalty or advocacy. The role of staff significantly influences affective experiences. Neutral or grumpy waiters introduced negative elements, reducing overall positive emotions. Consistent staff training and a friendly atmosphere are crucial, as highlighted by Mosley (2007) and Khan and Fatma (2017), who emphasize the importance of staff behavior in shaping brand experiences.

Regarding the behavioral dimension, participants' engagement varied. They shared their experiences with family and friends but did not write online reviews unless the experience was exceptionally good or poor. This suggests room for improvement to encourage positive reviews. The restaurant's wall art not only provided visual appeal but also motivated some participants to include more berries in their diets. This aligns with literature on behavioral brand experiences, which states that brand interaction can encourage physical acts and lifestyle changes (Bapat & Thanigan, 2016; Beig & Nika, 2019; Hwang et al., 2021; Pina & Dias, 2021).

Not all feedback was positive. Some participants felt their interaction with Berries & Bite Brunch was limited to dining, missing deeper brand engagement. Intellectual experiences varied. Some participants found the menu's diversity, unique ingredients, and informative wall art intellectually engaging. Others discussed how brunch restaurants elevate a simple meal for a wealthier audience, reflecting on the brand's influence. This aligns with Ternikar (2014), who describes brunch as a symbol of gastronomic luxury. However, some participants were skeptical about the intellectual stimulation a restaurant can provide. While culinary creativity was appreciated, it didn't always lead to broader cognitive reflections. This highlights that intellectual ambiance may not resonate with everyone, as what stimulates one diner may not affect another.

Addressing the Research Questions

It is crucial to highlight again that this study was conducted in an exploratory manner, guided by research questions rather than predefined hypotheses. This approach was chosen due to the significant gap in existing literature concerning the relationship between food presentation and brand experience in restaurants. Thus, the subsequent section will address the posed research questions in detail in order to offer foundational insights into this area.

RQ1A: How does food presentation in restaurants influence the sensory brand experience of diners?

At Berries & Bite Brunch, plating significantly shaped the sensory brand experience, impacting visual, auditory, smell, taste, and touch perceptions (Huaman-Ramirez & Merunka, 2019; Mostafa & Kasamani, 2020; Moreira et al., 2017; Pina & Dias, 2021; Shamim & Butt, 2013). Food presentation acted as a key brand touchpoint (Huaman-Ramirez & Merunka, 2019). The visual dimension was the most dominant, with participants appreciating the color contrasts and food arrangement, enhancing the visual experience (Huaman-Ramirez & Merunka, 2019; Hwang et al., 2021; Mostafa & Kasamani, 2020). Plating also influenced taste, as the visual appeal of the food enhanced participants' appetites (Huaman-Ramirez & Merunka, 2019).

Interestingly, plating slightly influenced the auditory experience, with some participants noting that crunchy textures enhanced their auditory dimension. This aligns with literature on the role of sounds in dining experiences (Schifferstein et al., 2020; Spence & Piqueras-Fiszman, 2014; Zampini & Spence, 2004). However, there was a negligible correlation between food plating and the sensory experiences of smell and touch. The underestimation of smell in food experiences is noted in studies, as many people associate smell sensations with taste (Schifferstein et al., 2020; Spence, 2015). The type of dishes at Berries & Bite Brunch, like toasts, may not emit strong aromas, explaining the overlooked role of smell. Touch could be an area for exploration, as textures and temperatures could add engagement (Kravchyk, 2022; Schifferstein et al., 2020; Spence & Piqueras-Fiszman, 2014).

RQ1B: How does food presentation in restaurants influence the affective brand experience of diners?

In the settings of Berries & Bite Brunch, while plating played a role in shaping the affective brand experience – encompassing a variety of feelings, emotions, and moods evoked during interactions with businesses and signifying the emotional bond between consumer and brand (Bapat & Thanigan, 2016; Beig & Nika, 2019; Huaman-Ramirez & Merunka, 2019; Hwang et al., 2021; Mostafa & Kasamani, 2020; Pina & Dias, 2021; Shamim & Butt, 2013) – participants pointed out that it was not the only factor determining it. Several participants viewed Berries & Bite Brunch's aesthetic plating as a direct reflection of the restaurant's precision and dedication to quality. This perception of the restaurant's commitment to detail, as seen by the plating, generated a sense of trust and enhanced more positive sentiments towards the brand. Moreover, as mentioned previously, when the dishes arrived at the tables, participants also mentioned the feeling of excitement evoked by the overall picture of the dish. While the plates were aesthetically pleasing, the food presentation lacked uniqueness, making it hard to form a strong emotional bond with the Berries & Bite Brunch brand. Participants noted that the plating style was common and could be found in many other brunch places. This aligns with Dabeva's (2012) view that differentiation is crucial in today's competitive environment. Without a unique identity, restaurants may struggle to connect emotionally with customers and stand out from competitors (Bistrohub, 2020; Kakadiya, 2023; McCracken, 2022; Robitaille, 2023).

Participants noted that the overall dining experience is crucial. Plating alone doesn't entirely shape their emotional connection to a brand; it's influenced by factors like food flavor, restaurant ambiance, and staff interactions. This multifaceted nature of dining experiences is supported by Ivkov et al. (2016) and Merckaert (2023b). While Berries & Bite Brunch's plating fostered positive sentiments, its lack of uniqueness lessened its impact. The overall emotional bond with the brand was shaped by the entire dining experience, not just the food presentation.

RQ1C: How does food presentation in restaurants influence the behavioural brand experience of diners?

The behavioural brand experience at Berries & Bite Brunch was notably influenced by the restaurant's food presentation. Participants frequently photographed their meals and shared them on social media, particularly Instagram, due to the visual appeal of the dishes. This aligns with the trend of sharing food images online, often described as 'camera eats first' (Gambetti & Han, 2022; Wachyuni & Yusuf, 2021; Yong et al., 2020). Beyond digital sharing, the attractive plating inspired some diners to replicate the dishes at home, highlighting the powerful impact of visually appealing food. One participant even tried a dish outside his comfort zone because it looked appealing, showing how presentation can influence preferences and purchasing decisions (Campo et al., 2017; Spence, 2017). In summary, plating can encourage diners to take and share photos, inspire home cooking, and even try new dishes, significantly shaping their dining behavior and choices.

RQ1D: How does food presentation in restaurants influence the intellectual brand experience of diners?

Participants had varied opinions on whether food presentation at Berries & Bite Brunch stimulated intellectual curiosity. Some saw it as a tool for culinary creativity and exploration, wondering about ingredient pairings and how to replicate or improve them. This reflects both analytical and creative thinking. Others felt the presentation was mainly aesthetic and didn't evoke deeper intellectual engagement. One participant noted that while Berries & Bite Brunch's plating might not achieve significant intellectual engagement, fine dining and molecular gastronomy could stimulate profound intellectual curiosity due to their innovative techniques. Thus, while some diners found the plating intellectually stimulating, others viewed it as merely adding to the sensory experience. The level of intellectual engagement seems to depend on individual perspectives and the broader dining context.

RQ1: How does food presentation in restaurants influence the overall brand experience of diners?

The presentation of food significantly impacts how guests perceive a restaurant brand, especially in the age of social media. Participants noted that a well-plated dish can serve as free advertising, making a strong first impression both in person and online. However, they agreed that taste, service quality, and ambiance are also crucial in shaping the dining experience. The type or theme of a restaurant and its target audience also influence the importance of plating. Effective plating strategies may vary based on these factors. For Berries & Bite Brunch, some participants found the plating to be a highlight, while others saw it as a standard expectation, emphasizing the need for continuous innovation. In summary, while plating plays a significant role, the overall brand experience is shaped by a combination of elements, not just one factor.

CONCLUSION

This research explores how food presentation in brunch restaurants influences brand experience. It highlights that visually appealing dishes enhance dining experiences and reflect a restaurant's quality and dedication. The visual appeal significantly impacts customers' first impressions and choices, aligning with previous studies. In the digital age, diners often use social media to evaluate and share their dining experiences. Attractive food presentation can draw customers and serve as free marketing through user-generated content. Restaurants must meet these visual expectations to maintain a strong online brand image, with platforms like Google Maps playing a crucial role in dining decisions.

The research shows that food presentation influences various dimensions of brand experience, though the extent varies. For sensory brand experience (RQ1A), food presentation mainly affects visual and taste perceptions, with minimal impact on smell and touch. Other factors like interior design, décor, sounds, and aromas also contribute to sensory experiences. Restaurants should focus on appealing plating, thematic design, balanced sounds, and pleasant aromas to enhance the dining atmosphere. For affective brand experience (RQ1B), while food presentation elicits positive emotions, its impact is lessened due to similarities with other brunch places. Factors like staff behavior and overall atmosphere play a more crucial role in influencing emotions. Therefore, a strategy focusing on staff interactions and a unique ambiance is essential for deeper emotional connections and brand loyalty. For RQ1C (behavioural brand experience), food presentation at the restaurant encouraged diners to photograph, share online, and explore culinary options. Décor and wall art also influenced behavior. Restaurants should focus on both food presentation and overall ambiance to boost brand engagement. For RQ1D (intellectual brand experience), food presentation inspired creativity and analytical thinking in some diners, while others saw it as merely aesthetic. Diverse menus and unique ingredients also played a role. However, the restaurant may struggle to provide deep intellectual engagement for all guests. Efforts to create an intellectually stimulating atmosphere are positive but may not resonate with everyone.

While plating captures diners' attention and creates a first impression, it is just one part of the overall brand experience. Modern diners seek a holistic experience that includes the taste of the food, restaurant ambiance, quality of service, and the entire dining journey. Restaurants should adopt a comprehensive strategy addressing all touchpoints to create a memorable brand experience. Additionally, the importance of plating varies by restaurant type, theme, and target audience. Brunch places, for example, often focus on being 'Instagrammable' to appeal to their digital-savvy clientele. For establishments like Berries & Bite Brunch, while plating can be a distinct element for some visitors, others see it as a basic expectation. This highlights the evolving standards in contemporary dining. With good plating becoming a standard, restaurants must innovate to stand out in a competitive market.

In today's dining landscape, where the experience extends beyond food, plating enhances the brand experience. Unique plating techniques can make lasting impressions, especially in the digital age. Different restaurants need varied plating strategies to meet their target audiences' expectations. However, plating is just one element of a restaurant's overall brand experience, which is a multifaceted interplay of various factors.Section title should be "Conclusion," not "Conclusions." Provide discussion of the overall coverage of the chapter and concluding remarks.

REFERENCES

Akdeniz, D. (2019). Tracking the history of restaurants through the art of painting. *Gastroia: Journal of Gastronomy and Travel Research*, 3(2), 346–365. DOI: 10.32958/gastoria.512544

Alloza, A. (2008). Brand engagement and brand experience at BBVA, the transformation of a 150 years old company. *Corporate Reputation Review*, 11(4), 371–379. DOI: 10.1057/crr.2008.31

Astuti, S., & Hanan, H. (2016). The Behaviour of Consumer Society in Consuming Food at Restaurants and Cafes. *Journal of ASIAN Behavioural Studies*, 1(1), 71–79. DOI: 10.21834/jabs.v1i1.173

Babbie, E. R. (2011). *The Basics of Social Research 5th (fifth) Edition.* Wadsworth, Cengage Learning.

Bapat, D., & Thanigan, J. (2016). Exploring Relationship among Brand Experience Dimensions, Brand Evaluation and Brand Loyalty. *Global Business Review*, 17(6), 1357–1372. DOI: 10.1177/0972150916660401

Bauman, Z. (1990). *Liquid Modernity*. Polity Press.

Beck, U. (1986). *Risk Society: Towards a New Modernity*. SAGE Publications.

Beig, F. A., & Nika, F. A. (2019). Brand experience and brand equity. *Vision: The Journal of Business Perspective*, 23(4), 410–417. DOI: 10.1177/0972262919860963

Bistrohub. (2020, September 15). The importance of branding in your restaurant. *Bistrohub*. Retrieved June 10, 2023, from https://bistrohub.co/en/the-importance-of-branding-in-your-restaurant/

Bloch, P., Brunel, F. F., & Arnold, T. J. (2003). Individual differences in the centrality of visual product aesthetics: Concept and measurement. *The Journal of Consumer Research*, 29(4), 551–565. DOI: 10.1086/346250

Božić, A. L., & Milošević, S. (2021). Contemporary Trends in the Restaurant Industry and Gastronomy. *Journal of Hospitality & Tourism Research (Washington, D.C.)*, 45(5), 905–907. DOI: 10.1177/10963480211020559

Brakus, J. J., Schmitt, B. H., & Zarantonello, L. (2009). Brand Experience: What is it? How is it measured? Does it affect loyalty? *Journal of Marketing*, 73(3), 52–68. DOI: 10.1509/jmkg.73.3.052

Braun, V., & Clarke, V. (2006). Using thematic analysis in psychology. *Qualitative Research in Psychology*, 3(2), 77–101. DOI: 10.1191/1478088706qp063oa

Burton-Hughes, L. (2020, February 7). How to Improve Sustainability in Restaurants. *The Hub | High Speed Training*. Retrieved May 1, 2023, from https://www.highspeedtraining.co.uk/hub/improve-sustainability-in-restaurants/

Campisi, V. (2022). *The rise of photogenic food: How social media is changing dining*. Food Journal.

Campo, R., Loporcaro, G., & Baldassarre, F. (2017). The effects of food aesthetics on consumers. Visual stimuli and food marketing. *DIEM: Dubrovnik International Economic Meeting, 3*(1), pp. 553-565.

Cifci, H., Gok, I., Atsiz, O., & Cifci, I. (2021). Insights into the Art of Plating in Gastronomy: A Content Analysis of Master Chefs' Perspectives. *Journal of Culinary Science &Amp. Technology*, •••, 1–26.

Civitello, L. (2008). *Cuisine and Culture: A History of Food and People* (2nd ed.). John Wiley & Sons.

Civitello, L. (2011). *Cuisine and Culture: A History of Food and People* (3rd ed.). John Wiley & Sons.

Cleff, T., Lin, I., & Walter, N. (2014). Can you feel it? - The effect of brand experience on brand equity. *IUP Journal of Brand Management*, 11(2), 7–27.

Cracknell, H. L., & Nobis, G. (1989). *The history of culinary arts*. Butterworth-Heinemann.

Creswell, J. W., & Creswell, J. D. (2017). *Research design: Qualitative, quantitative, and mixed methods approaches* (3rd ed.). Sage publications.

Crowther, G. (2018). *Eating Culture: An Anthropological Guide to Food* (2nd ed.). University of Toronto Press.

Cuboh. (2022, May 27). Restaurant Branding 101: How to Brand Your Restaurant. *Cuboh*. Retrieved June 12, 2023, from https://www.cuboh.com/blog/the-importance-of-branding-in-your-restaurant

Dabeva, T. (2012). Contemporary characteristics of restaurant branding. *Tourism and Hospitality Management*, •••, 254–261.

Delwiche, J. F. (2012). You eat with your eyes first. *Physiology &Amp. Behaviour*, 107(4), 502–504.

Dobos, O. (2022, September 9). *Branding for restaurants: what is it, how to do it, and tips for success*. GloriaFood. Retrieved June 11, 2023, from https://www.restaurant-website-builder.com/branding-for-restaurants

Durkheim, E. (1912). *The elementary forms of the religious life*. Free Press.

Edwards, E. (2023, May 3). Top Chefs Share Their 10 Food Plating and Presentation Tips. *Lightspeed*. Retrieved May 8, 2023, from https://www.lightspeedhq.com/blog/10-food-plating-and-presentation-tips/

Elias, N. (1978). *The Civilizing Process*. Blackwell.

Ellison, R. (2022, August 11). *The Rise of Brunch - A Look at How This Meal Has Grown in Popularity - Fairway Foodservice*. Fairway Foodservice. Retrieved May 2, 2023, from https://www.fairwayfoodservice.com/articles/the-rise-of-brunch-a-look-at-how-this-meal-has-grown-in-popularity/

Euler, B. E. (2022, December). These are the Food Trends 2023 | KTCHNrebel. www.ktchnrebel.com. Retrieved April 30, 2023, from https://www.ktchnrebel.com/foodtrend-report-2023/

Fieldhouse, P. (1996). *Food and Nutrition: Customs and culture*. Springer.

Future Food Strategy Group. (2018, April 25). The Booming Brunch Scene: The most popular day-part for dining out. *Future Food*. Retrieved May 3, 2023, from https://futurefood.com.au/blog/2018/4/25/the-booming-brunch-scene-the-most-popular-day-part-for-dining-out

Gambetti, A., & Han, Q. (2022). Camera eats first: Exploring food aesthetics portrayed on social media using deep learning. *International Journal of Contemporary Hospitality Management*, 34(9), 3300–3331. DOI: 10.1108/IJCHM-09-2021-1206

Gardini, M. (2014). The Challenge of Branding and Brand Management: Perspectives from the Hospitality Industry. In *Tourism and Leisure: Current Issues and Perspectives of Development* (pp. 247–268). Springer Gabler.

Genc, R. (2010). Strategic brand management in hospitality sector: how to manage co-branding in hotels and restaurants. *Acta Universitatis Danubius. Œconomica, 6*(3), pp. 33-46.

George, P. (2021, December 29). The Evolution of Brunch. *Out & About Magazine*. Retrieved May 2, 2023, from https://outandaboutnow.com/the-evolution-of-brunch/

Georgiev, N. (n.d.). *Restaurant Industry Trends: 6 Restaurant Trends in 2022*. Retrieved March 30, 2023, from https://www.bluecart.com/blog/restaurant-industry-trends

Ha, H., & Perks, H. (2005). Effects of consumer perceptions of brand experience on the web: Brand familiarity, satisfaction and brand trust. *Journal of Consumer Behaviour*, 4(6), 438–452. DOI: 10.1002/cb.29

Hair, J. F., Page, M., & Brunsveld, N. (2020). *Essentials of business research methods* (4th ed.). Routledge.

Hartley, A. (2021, August 16). *Understanding the Brunch Craze in Restaurants*. Alto-Hartley. Retrieved May 2, 2023, from https://altohartley.com/understanding-the-brunch-crazein-restaurants/

Holbrook, M. B., & Hirschman, E. C. (1982). The experiential aspects of consumption: Consumer fantasies, feelings, and fun. *The Journal of Consumer Research*, 9(2), 132–140. DOI: 10.1086/208906

Huaman-Ramirez, R., & Merunka, D. (2019). Brand experience effects on brand attachment: The role of brand trust, age, and income. *European Business Review*, 31(5), 1–55. DOI: 10.1108/EBR-02-2017-0039

Hwang, J., Choe, J. Y., Kim, H. M., & Kim, J. (2021). Human baristas and robot baristas: How does brand experience affect brand satisfaction, brand attitude, brand attachment, and brand loyalty? *International Journal of Hospitality Management*, 99(103050), 1–12. DOI: 10.1016/j.ijhm.2021.103050

Inphantry. (2017, October 27). *Generation Z craves experiential marketing. Here's Why*. Retrieved July 9, 2023, from https://www.inphantry.com/generation-z-craves-experiential-marketing-heres-why/

Ivkov, M., Blešić, I., Simat, K., Demirović, D., Bozic, S., & Stefanovic, V. (2016). Innovations in the restaurant industry: An exploratory study. *Ekonomika Poljoprivrede (1979), 63*(4), pp. 1169–1186.

Kakadiya, J. (2023, April 6). The Importance of Branding in Restaurant Menu. *PhotoADKing*. Retrieved June 11, 2023, from https://blog.photoadking.com/importance-of-branding-in-restaurant-menu/

Kelliher, F. (2005). Interpretivism and the Pursuit of Research Legitimisation: An Integrated Approach to Single Case Design. *Electronic Journal of Business Research Methods*, 3(2), 123–132.

Khachatryan, S. (2023, April 13). Restaurants and Social Responsibility: The Future of Sustainability in the Industry. *Orders.co*. Retrieved May 6, 2023, from https://orders.co/blog/restaurants-and-social-responsibility-the-future-of-sustainability-in-the-industry/

Khan, H., Laizet, F., Moulton, J., & Youldon, T. (2020, August 5). Reimagining European restaurants for the next normal. *McKinsey & Company*. Retrieved May 4, 2023, from https://www.mckinsey.com/industries/retail/our-insights/reimagining-european-restaurants-for-the-next-normal

Khan, I., & Fatma, M. (2017). Antecedents and outcomes of brand experience: An empirical study. *Journal of Brand Management*, 24(5), 439–452. DOI: 10.1057/s41262-017-0040-x

Khan, I., & Rahman, Z. (2015). A review and future directions of brand experience research. *International Strategic Management Review*, 3(1–2), 1–14. DOI: 10.1016/j.ism.2015.09.003

Kiefer, N. M. (2002). Economics and the origin of the restaurant. *The Cornell Hotel and Restaurant Administration Quarterly*, 43(4), 58–64. DOI: 10.1016/S0010-8804(02)80042-8

Koponen, S., & Mustonen, P. (2020). Eating alone, or commensality redefined? Solo dining and the aestheticization of eating (out). *Journal of Consumer Culture*, 22(2), 359–377. DOI: 10.1177/1469540520955219

Kravchyk, R. (2022, May 29). Sensory complexity and cross-modality of the perceived taste | The Perfectionist. *The Perfectionist.* Retrieved May 5, 2023, from https://www.theperfectionist.pl/flavours-are-not-what-they-seem-sensory-complexity-and-cross-modality-of-the-perceived-taste/

Kwun, D. J. (2012). Brand Management in the Hospitality Industry. *Journal of Tourism & Hospitality (Los Angeles, Calif.)*, 01(01), 1–2. DOI: 10.4172/2167-0269.1000e104

Lane, P., McKenna, H., Ryan, A., & Fleming, P. (2001). Focus group methodology. *Nurse Researcher*, 8(3), 45–59. DOI: 10.7748/nr2001.04.8.3.45.c6157

Mainardes, E. W., Alves, H., & Raposo, M. (2010). An exploratory research on the stakeholders of a university. *Journal of Management and Strategy*, 1(1). Advance online publication. DOI: 10.5430/jms.v1n1p76

Mandil, K. A. (2016). *Modelling the relationship between brand experience dimensions and the antecedents of happiness within the context of dining services* [PhD dissertation, Brunel University London].

Martens, M. (1999). A philosophy for sensory science. *Food Quality and Preference*, 10(4–5), 233–244. DOI: 10.1016/S0950-3293(99)00024-5

Massey, O. T. (2011). A proposed model for the analysis and interpretation of focus groups in evaluation research. *Evaluation and Program Planning*, 34(1), 21–28. DOI: 10.1016/j.evalprogplan.2010.06.003 PMID: 20655593

McCain Foods. (2021, February 9). Why restaurants should balance value, quality and social responsibility. *Restaurant Business*. Retrieved May 1, 2023, from https://www.restaurantbusinessonline.com/operations/why-restaurants-should-balance-value-quality-social-responsibility

McCracken, R. (2022, July 20). Why restaurant branding is a game changer in hospitality | TableCheck Blog. *TableCheck*. Retrieved June 10, 2023, from https://www.tablecheck.com/en/blog/restaurant-branding-hospitality-game-changer/

McLymont, V. (2022, October 11). What Gen Z wants: A look inside their food preferences. *Hospitality News*. Retrieved August 10, 2023, from https://hospitalitynewsny.com/what-gen-z-wants-a-look-inside-their-food-preferences/

Merckaert, G. (2023a, January 20). 85+ Restaurant Industry Statistics That Every F&B Operator & Manager Needs to Know in 2023. *Apicbase*. Retrieved March 31, 2023, from https://get.apicbase.com/restaurant-industry-statistics-2023/

Merckaert, G. (2023b, March 17). Trends and Tactics: How to Navigate the Changing Landscape of the European Restaurant Industry. *Apicbase*. Retrieved March 30, 2023, from https://get.apicbase.com/restaurant-industry-europe/

Meulen, N. V. D., & Wiesel, J. (2017). Plating Food On the Pictorial Arrangement of Cuisine on the Plate. In *Culinary Turn* (pp. 235–250).

Michel, C., Velasco, C., Fraemohs, P., & Spence, C. (2015). Studying the impact of plating on ratings of the food served in a naturalistic dining context. *Appetite*, 90, 45–50. DOI: 10.1016/j.appet.2015.02.030 PMID: 25728885

Michel, C., Velasco, C., Gatti, E., & Spence, C. (2014). A taste of Kandinsky: Assessing the influence of the artistic visual presentation of food on the dining experience. *Flavour (London)*, 3(1), 1–11. DOI: 10.1186/2044-7248-3-7

Miles, S. (1996). *Consumerism: As a way of life*. Sage Publications.

Mintel. (2022). Restaurant Breakfast and Brunch Trends Market Report – US – 2022. In https://store.mintel.com/report/us-restaurant-breakfast-and-brunch-trends-market-report

Moreira, A. C., Fortes, N., & Santiago, R. (2017). Influence of sensory stimuli on brand experience, brand equity and purchase intention. *Journal of Business Economics and Management*, 18(1), 68–83. DOI: 10.3846/16111699.2016.1252793

Mosley, R. (2007). Customer experience, organisational culture and the employer brand. *Journal of Brand Management*, 15(2), 123–134. DOI: 10.1057/palgrave.bm.2550124

Mostafa, R. B., & Kasamani, T. (2020). Brand experience and brand loyalty: Is it a matter of emotions? *Asia Pacific Journal of Marketing and Logistics*, 33(4), 1033–1051. DOI: 10.1108/APJML-11-2019-0669

National Restaurant Association. (2022). *2022 State of Restaurant Industry*. Retrieved April 3, 2023, from_https://xchange.avixa.org/documents/pdf-report-2022-state-of-restaurant-industry

National Restaurant Organization. (2019). *Restaurant Industry 2030*. Retrieved March 29, 2023, from_https://restaurant.org/research-and-media/research/research-reports/restaurant-industry-2030/

Nysveen, H., Pedersen, P. B., & Skard, S. (2012). Brand experiences in service organizations: Exploring the individual effects of brand experience dimensions. *Journal of Brand Management*, 20(5), 404–423. DOI: 10.1057/bm.2012.31

Ong, C. H., Wei, L. H., & Ramayah, T. (2018). Impact of brand experience on loyalty. *Journal of Hospitality Marketing & Management*, 27(7), 1–20. DOI: 10.1080/19368623.2018.1445055

Owner. (n.d.). The State of The Restaurant Industry In 2023 (Free Report). *Owner*. Retrieved March 26, 2023, from_https://www.owner.com/blog/restaurant-industry

Paakki, M., Aaltojärvi, I., Sandell, M., & Hopia, A. (2019). The importance of the visual aesthetics of colours in food at a workday lunch. *International Journal of Gastronomy and Food Science*, 16(100131), 1–9. DOI: 10.1016/j.ijgfs.2018.12.001

Peters-Desteract, L. (2020). *Innovative business plan for a brunch restaurant in Toulouse, France* [MA dissertation]. ISCTE Business School.

Pilcher, J. M. (2012). *The Oxford Handbook of Food History*. Oxford University Press. DOI: 10.1093/oxfordhb/9780199729937.001.0001

Pina, R., & Dias, Á. (2021). The influence of brand experiences on consumer-based brand equity. *Journal of Brand Management*, 28(2), 99–115. DOI: 10.1057/s41262-020-00215-5

Pine, B. J., & Gilmore, J. H. (1998). Welcome to the experience economy. *Harvard Business Review*, 76(4), 97–105. PMID: 10181589

Powell, R. A., & Single, H. M. (1996). Focus groups. *International Journal for Quality in Health Care : Journal of the International Society for Quality in Health Care*, 8(5), 499–504. DOI: 10.1093/intqhc/8.5.499 PMID: 9117204

Rawson, K., & Shore, E. (2019). *Dining Out: A Global History of Restaurants*. Reaktion Books.

Robitaille, J. (2023, March 13). The Importance of Restaurant Branding. *Casa Media*. Retrieved June 10, 2023, from https://casamedia.com/branding/importance-restaurant-branding/

Scharp, K. M., & Sanders, M. L. (2018). What is a theme? Teaching thematic analysis in qualitative communication research methods. *Communication Teacher*, 33(2), 117–121. DOI: 10.1080/17404622.2018.1536794

Schifferstein, H. N., Kudrowitz, B. M., & Breuer, C. (2020). Food Perception and Aesthetics - Linking Sensory Science to Culinary Practice. *Journal of Culinary Science &Amp. Technology*, 20(4), 293–335.

Schmidt, S. (2016, November). When Service à la Française Met Service à la Russe. *Manuscript Cookbooks Survey*. Retrieved May 7, 2023, from https://www.manuscriptcookbookssurvey.org/when-service-a-la-francaise-met-service-a-la-russe/

Schmitt, B. H. (1999a). *Experiential marketing: how to get customers to sense, feel, think, act, and relate to your company and brands*. The Free Press New York.

Schmitt, B. H. (1999b). Experiential marketing. *Journal of Marketing Management*, 15(1–3), 53–67. DOI: 10.1362/026725799784870496

Schmitt, B. H., Brakus, J. J., & Zarantonello, L. (2014). The current state and future of brand experience. *Journal of Brand Management*, 21(9), 727–733. DOI: 10.1057/bm.2014.34

Shamim, A., & Butt, M. A. (2013). A critical model of brand experience consequences. *Asia Pacific Journal of Marketing and Logistics*, 25(1), 102–117. DOI: 10.1108/13555851311290957

Sloan, D., & Leith, P. (2012). *Culinary Taste*. Routledge. DOI: 10.4324/9780080476957

Sobreira, L. T. P., Mepe, G., & Nardoto, G. B. (2018). Anthropology of Food: An Essay on Food Transition and Transformations in Brazil. *Journal of Food Nutrition and Population Health*, 02(01), 1–8.

Spence, C. (2015). Just how much of what we taste derives from the sense of smell? *Flavour (London)*, 4(1), 30. DOI: 10.1186/s13411-015-0040-2

Spence, C. (2017). *Gastrophysics: The new science of eating.* Penguin UK.

Spence, C. (2021). What's the Story With Blue Steak? On the Unexpected Popularity of Blue Foods. *Frontiers in Psychology*, 12, 1–17. DOI: 10.3389/fpsyg.2021.638703 PMID: 33737898

Spence, C., Motoki, K., & Petit, O. (2022). Factors influencing the visual deliciousness / eye-appeal of food. *Food Quality and Preference*, 102(104672), 1–16. DOI: 10.1016/j.foodqual.2022.104672

Spence, C., Okajima, K., Cheok, A. D., Petit, O., & Michel, C. (2016). Eating with our eyes: From visual hunger to digital satiation. *Brain and Cognition*, 110, 53–63. DOI: 10.1016/j.bandc.2015.08.006 PMID: 26432045

Spence, C., & Piqueras-Fiszman, B. (2014). *The Perfect Meal: The Multisensory Science of Food and Dining*. Wiley-Blackwell. DOI: 10.1002/9781118491003

Spence, C., Piqueras-Fiszman, B., Michel, C., & Deroy, O. (2014). Plating manifesto (II): The art and science of plating. *Flavour (London)*, 3(1), 1–12. DOI: 10.1186/2044-7248-3-4

Spence, C., & Van Doorn, G. (2022). Visual communication via the design of food and beverage packaging. *Cognitive Research: Principles and Implications*, 7(1), 42. DOI: 10.1186/s41235-022-00391-9 PMID: 35551542

Spence, C., Youssef, J., Michel, C., & Woods, A. T. (2019). Assessing the aesthetic oblique effect in painting and plating. *International Journal of Gastronomy and Food Science*, 17(100168), 1–29. DOI: 10.1016/j.ijgfs.2019.100168

Ternikar, F. (2014). *Brunch: A History*. Rowman & Littlefield. DOI: 10.5771/9781442229433

TouchBistro. (2023). *2023 Dining Trends Report.* Retrieved April 1, 2023, from https://www.touchbistro.com/blog/diner-trends-report/

Voicu, L. (2023, March 27). 10 Restaurant Sustainability Practices You Can Implement Immediately. *ORACLE GloriaFood.* Retrieved May 4, 2023, from https://www.gloriafood.com/restaurant-sustainability-practices

Wachyuni, S. S., & Yusuf, L. (2021). Camera eat first: Tourist motivation in sharing food photograph on Instagram. *International Journal of Tourism & Hospitality Reviews*, 8(1), 62–70. DOI: 10.18510/ijthr.2021.815

Walker, J. R. (2017). *Restaurant concepts, management and operations.* Wiley Global Education.

Walker, J. R. (2021). *The Restaurant: From Concept to Operation*. John Wiley & Sons.

Wasserstrom. (2019, August 30). Dietary Restrictions for Restaurants: Understanding Fads, Trends & Lifestyle. *The Official Wasserstrom Blog*. Retrieved May 4, 2023, from https://www.wasserstrom.com/blog/2019/08/08/dietary-restrictions-for-restaurants-understanding-fads-trends-lifestyle/

We Are Collider. (2022, November 22). Millennials fuelled the experience economy; will Gen Z kill it? *We Are Collider*. Retrieved August 10, 2023, from https://wearecollider.com/insights/millennials-fuelled-the-experience-economy-will-gen-z-kill-it/

Yong, J. Y., Tong, E. M. W., & Liu, J. C. J. (2020). When the camera eats first: Exploring how meal-time cell phone photography affects eating behaviours. *Appetite*, 154(104787), 1–7. DOI: 10.1016/j.appet.2020.104787 PMID: 32579971

Zampini, M., & Spence, C. (2004). The role of auditory cues in modulating the perceived crispness and staleness of potato chips. *Journal of Sensory Studies*, 19(5), 347–363. DOI: 10.1111/j.1745-459x.2004.080403.x

Chapter 13 Marketing Communications and Social Responsibilities of Businesses: Strategies and Initiatives

Pratap Chandra Mandal
https://orcid.org/0000-0002-0300-4002
Indian Institute of Management, Shillong, India

ABSTRACT

Companies and businesses have responsibilities towards their customers and the society at large. Such responsibilities should be revealed in their business actions and marketing communications done by them. The objective of the study is to understand the various aspects of socially responsible marketing communications. The methodology adopted is a conceptual analysis of the various aspects of marketing communications done by companies. Companies should understand their roles in socially responsible marketing communications viz. advertising, sales promotion, and personal selling. They should realize their responsibilities towards promotion of social responsibility and should adopt initiatives towards socially responsible marketing communications. Such initiatives will instill a sense of trust and belief in the minds of customers and the society at large. This will in turn allow companies to fulfill their social responsibilities and to achieve business excellence in the long run.

DOI: 10.4018/979-8-3693-3579-6.ch013

1. INTRODUCTION

Companies require understanding customer requirements and preferences to serve them well. Companies should also create value for customers (Zeithaml, Verleye, Hatak, Koller, Zauner, 2020). They should communicate the value created for customers clearly and persuasively. Promotion is not a single tool but rather a mix of several tools. A company will carefully coordinate these promotion elements to engage customers and to build clear, consistent, and compelling messages and content about an organization and its products (Adnan, Khan, & Hayee, 2021).

Companies should communicate about their offerings and value proposition to the target markets in a socially responsible manner (Gao, Zhang, Li, Li, & Shao, 2022). In the era of the internet, digital marketing and social media play important roles in socially responsible marketing communications. These means of communications make it easier for companies to communicate with the target audience. However, companies should not communicate all those aspects which may not be desirable by the target markets and by the society at large (Jin & He, 2018). However, the major aim of companies is to generate profits and revenues for themselves and to achieve growth. In the quest for generating profits and revenues and achieving business growth, companies may sometimes cross the line of ethics while communicating about their offerings (Bhattacharyya, 2023). Companies do not have responsibilities only towards their customers. They also have responsibilities towards the society at large. They should ensure that the marketing communications done by them do not create a negative impact on the society (Nur, Akmaliah, Chairul, & Safira, 2021).

The responsibilities of companies and businesses while communicating about their offerings cannot be overemphasized. It is imperative for companies to be ethical while promoting their offerings to the target markets. The impact of marketing communications on the target markets and on the society in the era of the internet and social media may be substantial. Although the topic is important, few researchers have focused on the social impact of marketing communications and on the social responsibilities of companies while promoting their offerings. The study aims to address this research gap.

The objective of the study is to analyze the various aspects of marketing communications done by companies in the era of the internet and social media. The study focuses more on those aspects for which companies require to be socially responsible. The study also focuses on the various legal and ethical issues in socially responsible marketing communications.

The methodology adopted is a conceptual analysis of the various aspects of marketing communications done by companies. The study focuses on the conceptual analysis of those aspects of marketing communications for which companies are socially responsible. Primary data is not collected and empirical analysis is not done.

The novelty and the contributions of the study lie in the fact that an in-depth discussion about the various aspects of social responsibilities of companies and businesses is done. The social responsibilities of companies should be revealed in the promotions and marketing communications done by them. The roles of digital marketing and social media in socially responsible marketing communications are highlighted. The legal and ethical issues in socially responsible marketing communications, and consumer privacy and consumer privacy bill of rights are discussed. Academicians may study the strategies and initiatives adopted by companies and suggest strategies and initiatives which are effective and which should be adopted by companies to fulfill their social responsibilities. Managers will be sensitized about their responsibilities not only to their customers but also to the society at large. They will be able to incorporate them in the different types of marketing promotion and marketing communications done. They will also be able to adopt initiatives which are beyond their business activities and which support social welfare.

The study is structured as follows:

Companies have different roles in socially responsible marketing communications and these roles are discussed in section 2. The roles of advertising and sales promotion in socially responsible marketing communications are discussed in section 3 while the roles of personal selling are discussed in section 4. The roles of digital marketing and social media in socially responsible communications are discussed in section 5. Section 6 highlights the various legal and ethical issues in socially responsible marketing communications. Section 7 focuses on consumer privacy and consumer privacy bill of rights. Companies have responsibilities towards promotion of social responsibility in their marketing communications and these responsibilities are discussed in section 8. Companies adopt a number of initiatives for socially responsible marketing communications. The initiatives adopted by a number of enlightened companies are discussed in section 9. Section 10 discusses the salient points of the study with sub-sections 10.1 and 10.2 highlighting the theoretical implications and the managerial implications of the study respectively. Section 11 concludes the study with sub-sections 11.1 and 11.2 highlighting the limitations of the study and the avenues of future research respectively.

2. ROLES OF COMPANIES IN SOCIALLY RESPONSIBLE MARKETING COMMUNICATIONS

Companies should be aware of their social responsibilities while communicating about their offerings to the target markets (Singh & Sanjeev, 2022). They should ensure that marketing communications done by them do not create negative influence and do not create harm to the customers and to the society at large (Antoniuk,

Mokiy, Pikh, Bukharina, & Antoniuk, 2021). Marketers may sometimes cross the line of ethics while promoting their offerings. This may result in legal and ethical issues. Companies should ensure that they remain within the realms of ethics while communicating about their offerings.

In shaping its promotion mix, a company must be aware of the many legal and ethical issues surrounding marketing communications. Marketers strive hard to make sure that authentic information reaches their target markets. They work hard to communicate openly and honestly with consumers and retailers. However, marketing communications may result in abuses. Public policy makers have developed a substantial body of laws and regulations to govern advertising, sales promotion, personal selling, and direct marketing (Jackson, 2013). There might be issues regarding advertising, sales promotion, and personal selling (Trihas, Perakakis, Venitourakis, Mastorakis, & Kopanakis, 2013).

3. ADVERTISING AND SALES PROMOTION

Companies cannot communicate false or deceptive advertising which is against law (Mandal, 2019a). Companies may not make false claims through advertising. For example, advertisers cannot suggest that a product cures something when they know that it does not (Ronald, 2008). Companies should also avoid communicating those advertisements which they feel are deceptive, even though no one actually may be deceived (Mandal, 2019a). For example, companies cannot advertise an automobile as getting 32 miles per gallon unless it does so under typical conditions. Also, diet bread cannot be advertised as having fewer calories simply because its slices are thinner (Mandal, 2019a).

Companies must avoid bait-and-switch advertising that attracts buyers under false pretenses (Litwin, 2007). For example, a sewing machine was advertised at $179 by a large retailer. However, when consumers tried to buy the advertised machine, the seller downplayed its features, placed faulty machines on showroom floors, and understated the performance of the machine. It also took other actions so that buyers are forced to switch to a more expensive machine. Such actions are both unethical and illegal (Emamalizadeh, 2015).

Regulations are enforced to monitor the trade promotion activities of companies (Maier, 2021). For example, the Robinson-Patman Act requires that sellers cannot favor certain customers through their use of trade promotions (Mandal, 2021). They must make promotional allowances and services available to all resellers on proportionately equal terms. Similarly, the European Union has introduced EU Unfair Commercial Practices Directives. The directives require that sellers cannot undertake misleading or aggressive practices that disadvantage customers (Maier, 2021).

Companies should aim to avoid the legal pitfalls involved in deceptive advertising and bait-and-switch advertising. Further, companies should use advertising and other forms of marketing communications to encourage and promote socially responsible programs, actions, and ideas. Companies across industries promote a wide range of social and environmental causes and issues related to their brands. For example, Frito-Lay ran television, print, and online advertisements to promote the compostable packaging for its SunChips brand (Evans, Parsons, Jackson, Greenwood, & Ryan, 2020). One print advertisement communicated, "They're 100% compostable. You eat the delicious chips. The earth eats the bag" (Spence, 2012). Frito-Lay also launched a program which challenged consumers to create videos encouraging positive change in the world. The winner of the competition was selected by former U.S. Vice President, Al-Gore. It was aired on national television by Frito-Lay during Earth Week.

Ecover is a Belgium-based manufacturer and distributor of domestic cleaning products (Errichiello & Zschiesche, 2022). Ecover has the mission to "offer efficient and sustainable solutions for the hygienic needs of people today and of future generates", with a vision "to contribute to the process of creating economic, ecological, and social change within our society to build a sustainable future for everyone" (Carter, Jayachandran, & Murdock, 2021). The company slogan of Ecover is *Powered by nature*. A print advertisement of Ecover washing up liquid simply states "protect your future. Begin with the Washing up!" (Jones, 2019).

4. PERSONAL SELLING

Companies should follow the principle and the rules of *fair competition* when employing salespeople (Mandal, 2021). To keep a check on the unethical practices adopted by salespeople, many states have enacted deceptive sales acts that clearly indicate what is not allowed (Mandal, 2021). Salespeople need to follow the basic principles of selling. They may not lie to customers or misled them about product characteristics and functionalities. They cannot exaggerate the advantages of buying the product. The statements made of salespeople should match the claims made in advertisements. This ensures that bait-and-switch practices are avoided (Ahearne, Atefi, Lam, & Pourmasoudi, 2022).

Rules and regulations are enforced to protect consumers who are called on at home or who buy at a location that is not the permanent place of the seller (Willan, 2021). Different rules and regulations apply to those consumers who go to a store in search of a product. Individuals who are called on may be taken by surprise. They may also be especially vulnerable to high-pressure selling techniques. The Federal Trade Commission (FTC) adopted a *three-day cooling-off rule* to provide special protection to those customers who are not actively looking for the products or those who are worried about product performance (Sánchez Abril, Oliva Blázquez, & Martínez Evora., 2018). This rule allows customers who agree in their own homes, workplace, dormitory, or facilities rented by the seller on a temporary basis – such as hotel rooms, convention centers, and restaurants – to buy something costing more than $25, have 72 hours (i.e. three days) in which to cancel a contract or return merchandize and get their money back. The company cannot ask any questions for the products which are returned (Sánchez Abril, Oliva Blázquez, & Martínez Evora., 2018). Similarly, the EU Distance Selling Directive allows a *seven-day cooling-off rule* to provide special protection to customers (Gladys, 2018).

Business-to-business trade involves personal selling. In selling to businesses, salespeople may not offer bribes to purchasing agents or others who can influence a sale (Jia & Wan, 2022). Salespeople are not allowed to obtain or use technical or trade secrets of competitors through bribery or industrial espionage. Also, salespeople must not disparage competitors or competing products by suggesting things that are not true (Jia & Wan, 2022).

5. ROLES OF DIGITAL MARKETING AND SOCIAL MEDIA IN SOCIALLY RESPONSIBLE COMMUNICATIONS

Marketers are able to target and reach a larger audience directly because of the growth and the widespread use of the internet and the social media. Sometimes, such direct communications raise concerns about potential targeting abuses (Luxon, 2019). The internet and the social media allow companies to identify and to target vulnerable segments with questionable products. Companies adopt unfair and unethical means to do deceptive advertising to vulnerable audience. Direct marketing and social media allow companies to reach out to millions of unsuspecting consumers with customized deceptive messages via e-mails. For example, the FBI's Internet Crime Complaint Center alone received more than 31000 complaints against fraudulent and deceptive communications in one single year (Chevers, 2019).

Marketing communication channels like online videos, cinema halls, social media, website games, television integration, radio advertising, and live tours are used extensively by companies to promote awareness among people. Companies employ innovative techniques such as a pile of body bags piled up in front of Philip Morris headquarters. Campaigns also rely on humor, animation, murals at the SXSW music festival, and a Broadway-style song and dance routines to communicate the message (Rhodes et al., 2016).

It is debated by critics and experts whether humor is the right way to communicate about a subject that kills millions of people every year (Huang, Friedman, Lin, & Thrasher, 2018). Young smokers believe that they will not be affected by nicotine-related diseases. So, a pile of body bags may not be an effective way to reach and convince them. It may not also be fair and appropriate to apply this kind of messaging which tobacco companies claim is abusive to them and to their employees. Family members of some individuals who are working for a tobacco company may have emphysema from smoking. Such employees will be in a dilemma related to tobacco advertising. All these issues are debatable and there are no straightforward answers and solutions.

6. LEGAL AND ETHICAL ISSUES IN SOCIALLY RESPONSIBLE MARKETING COMMUNICATIONS

Various legal and ethical issues are related to marketing communications which are not socially responsible. Many companies strive hard to do honest and open communications with consumers and resellers without being deceptive (Mandal, 2019b). However, there are instances where abuses occur intentionally or unintentionally. Detailed laws and regulations are formulated by policy makers, regulatory authorities, and governments to govern advertising, sales promotion, direct marketing, and personal selling (Kees & Andrews, 2019).

Regulatory bodies keep a close watch on the trade promotion activities of companies. The Robinson-Patman Act requires that companies cannot favor certain customers over others in trade promotions (Salant, 2016). Similar promotional allowances and services should be provided by companies to all customers on proportionately equal terms (Salant, 2016).

Companies should abide by the legal regulations. They should avoid deceptive and bait-and-switch advertising. Companies should encourage activities and actions which are socially responsible. They should use advertising and other forms of promotion which showcase such activities and actions. For example, when the explosion and oil spill at the Deepwater Horizon oil rig in the Gulf of Mexico took place in 2010, BP actively took part in the restoration efforts and spent billions of dollars on Gulf Coast (Sharpe, Kaufman, Goldman, Wolkin, & Gribble, 2019). BP initiated a promotion campaign which aimed at restoring tourism along the Gulf Coast. The campaign was named "Voices from the Gulf" and it is organized for three years. The great beaches, fishing, and seafood of the Gulf Coast were focused on in the campaign. The campaign urged tourists to spend a vacation along the coast. Promotions were done on television, social media, and online (Ryan, 2019). Apart from the promotional campaign, BP also arranged an online MyGulf campaign which focused on the daily lives of people who work and live in the Gulf. All such promotional activities and initiatives helped tourism in many Gulf Coast areas surpass pre-spill levels despite a poor economic situation (Ryan, 2019).

7. CONSUMER PRIVACY AND CONSUMER PRIVACY BILL OF RIGHTS

Consumers are concerned and worried about the information shared by them with companies. They fear that the privacy and the security of information might be compromised with. Now-a-days, companies go for more targeted advertising to better cater to the requirements of target markets rather than buying advertisements on websites related to the product being promoted (e.g. a healthy drink on the GNC website). Sustainable entrepreneurs understand and appreciate the importance of consumer privacy and security. This also enhances the authenticity of such entrepreneurs in the minds of consumers (Balmer, Powell, & Greyser, 2011). Women entrepreneurs are more sensitive towards the ethical issues and aim to do ethical marketing communications (Dauhgerty, 2014). Companies understand the requirements and the preferences of customers better by observing and analyzing the online behavior, web browsing activities, and past purchases of individuals (Schmeiser, 2018). It is possible for a retailer like Target to predict and to determine whether any of its female customers is pregnant. Based on such predictions, Target encourages such women or their husbands to visit Target or Target.com and purchase baby-related products. Promotions and coupons are targeted directly for them. Such prediction and estimation allow Target to understand which females are pregnant. However, critics are worried about whether such types of tracking are ethical. They further argue that if the predictions are not correct and if a targeted promotion is sent to the home of an unwed teenage mother whose father was unaware of the pregnancy, such actions may generate unfavorable reactions (Martin, 2019).

Critics emphasize that companies, advertising agencies, and modeling agencies should respect the privacy and the security of the customers. Measures should be taken to protect consumer privacy and security. The Consumer Privacy Bill of Rights was passed to protect consumers against the changing ways in which firms use the data collected from consumers (Mercer, 2020). Consumers also have the option of not to be traced and tracked. The bill allows consumers to exercise the following rights (Mercer, 2020):

1. *Individual Control*: Companies may collect individual data about consumers. However, consumers will have individual control over how the shared data is used.

2. *Transparency*: Privacy practices of companies should be made available to consumers. The privacy practices should be listed down clearly in a lucid manner so that consumers might understand such practices.
3. *Context*: Companies that collect and use personal data will do so in ways consistent with how companies provided the data.
4. *Security:* Companies should ensure that consumer security is not compromised with and consumers will not be at risk.
5. *Access and Accuracy:* Companies should ensure that only accurate personal data of consumers are used. Also, companies should provide consumers with the option of correcting the data that is inaccurate.
6. *Focused Collection:* The collection of consumers' personal data will be limited and restricted by the context of how it was supplied in the first place.
7. *Accountability*: Companies are responsible for establishing systems, training employees, and limiting the access of third parties to consumers' personal data. Companies are accountable and answerable both to consumers and enforcement authorities to ensure that all these issues are taken care of.

It is debatable whether consumers' personal privacy rights are being unjustly invaded by companies that provide them with targeted promotions based on their browsing habits (Liu & Greene, 2020). Experts and critics are also concerned about whether companies ensure privacy rights of consumers while providing them with information which helps consumers to make informed decisions.

8. RESPONSIBILITIES OF COMPANIES TOWARDS PROMOTION OF SOCIAL RESPONSIBILITY

Companies today are aware about their social responsibilities. They incorporate social responsibility in their corporate strategy (Maccarrone & Contri, 2021). Companies employ advertising and marketing to promote socially responsible programs, actions, and ideas. Companies do not focus only on generating revenues and profits for themselves. They recognize that company success is inexorably linked to the welfare of their customers and of the society at large (Shim, Moon, Lee, & Chung, 2021).

Companies across industries support and promote a number of social and environmental issues (Ooi, 2022). For example, over the past many seasons, more than 20 percent of all Super Bowl advertisements have connected brands with a wide range of social and environmental issues (Aaker, 2018). In one season, Microsoft promoted an inspirational advertisement featuring children with physical and cognitive disabilities. These children rose to the top of their games and achieved success with the help of its Xbox Adaptive Controller (Dua, 2019). "The story illustrates

Microsoft's commitment to building accessible technology that levels the playing field and creates opportunity for all of us. When everyone plays, we all win", says Microsoft (Dua, 2019). An Anheuser-Busch Super Bowl advertisement featured the Budweiser Dalmatian. Ears flapped in the breeze. The Budweiser Dalmatian sat atop a wagon pulled by the Budweiser Clydesdales through a field of golden grain on a wind turbine farm, set to Bob Dylan's "Blowin' In the Wind". "Wind never felt better", the advertisement concluded, noting that Budweiser is "now brewed with 100 percent renewable energy. This Bud's is for a better tomorrow" (Dua, 2019).

The above examples indicate that more and more companies are using marketing communications and advertising to take a stand on relevant social and environmental issues. Companies do it not to make money, but because they feel that it is the right thing to do.

9. INITIATIVES ADOPTED BY COMPANIES FOR SOCIALLY RESPONSIBLE COMMUNICATIONS

Companies adopt a number of initiatives to promote marketing communications which are socially responsible (Hayes & Duff, 2022). Companies make social responsibility an integral part of their corporate missions (Kitchen & Tourky, 2022). They employ advertising and marketing to promote socially responsible programs, actions, and ideas. They recognize that the success of companies is not related only to profits but also to the welfare of their customers and the world around them.

Companies in almost every industry support and promote a wide range of social and environmental issues. Chipotle introduced a campaign named *Food with Integrity* which challenges factory farming and promotes organic, local, and family-farmed ingredients (Haarpaintner, 2022). Intel introduced *Girl Rising* campaign which promotes education for girls around the world. Apple introduced *Better* campaign which raises awareness of important environmental issues because Apple feels that "it's our responsibility to make sure that while creating beautiful products, we're also caring for our beautiful planet" (Haarpaintner, 2022).

An all-time classic example is the long-running *Dove Campaign for Real Beauty*. The objective of the campaign was to change stereotypical definitions of beauty with advertisements featuring candid and confident images of real women with all sorts of body types which did not include glamorous actresses and skinny models (Reeves, Chandra, & Kim, 2022). "Our mission is to make more women feel beautiful every day by broadening the definition of beauty," says Dove (Allen, 2022). Over the years, the campaign has evolved with different advertisements, heavily viewed digital videos, and other content. All these have done much to serve the cause of helping women to be comfortable with their own skin. The objective of the campaign is noble motives which are beyond sales and profits. However, the campaign has contributed to the success of Dove. During the first decade of the campaign, the annual revenues of the brand grew from $2.5 billion to more than $4 billion (Fogarty, 2022). The campaign was initiated more than 15 years back. The *Dove Campaign for Real Beauty* campaign is still going strong.

Enlightened companies adopt successful campaigns to communicate and promote social responsibility. They use advertising and marketing to take a stand on relevant social and environmental issues. They do it because they feel that it is the right thing to do. These companies have developed and implemented socially responsible marketing communications strategies and campaigns. The campaigns initiated by some of the companies across diverse industries are highlighted below.

9.1. Google: "Made with Code"

Google launched a marketing and advertising campaign, *Made with Code* to encourage young girls to pursue careers in science and technology. It was a campaign worth $50 million (Johnson, 2018). Google found that 74 percent of middle school girls express interest in science, technology, engineering, and mathematics (STEM). However, after getting promoted to high school, less than one percent of girls plan to major in computer science (Johnson, 2018). Jobs and opportunities in computer science are the most lucrative ones. Such jobs are among the highest paid jobs at present. However, women are underrepresented in the roles that make technology happen.

Google promotes the concept of technology and its importance through advertisements, dedicated digital and social media sites, events, and partnerships with dozens of not-for-profit organizations. The campaign promotes the idea that the things young girls love, from their smartphone apps to fashions to their favorite movies are *made with code*. "Simply put, code is a tool that lets you write your story with technology," says Google (Johnson, 2018).

The *Made with Code* website provides a number of resources to help kids learn coding. The website includes communities for discussing and sharing coding projects and discoveries, and support from coding mentors (Johnson, 2018). *Made with Code* has Twitter, Instagram, and Tumblr pages which are loaded with images, news, and stories that bring coding to life. "Girls start out with a love of science and technology but lose it somewhere along the way," says Google. "Let's help encourage that passion" (Johnson, 2018).

9.2. Walgreens: "Red Nose Day"

Red Nose Day is a global fundraising effort aimed at helping to end child poverty (Wood & Reinert, 2019). Walgreens did not start the initiative. However, Walgreens worked for *Red Nose Day* more than any other organization and made it famous in the United States (Pittman & Sheehan, 2021). Walgreens has a corporate commitment towards child health, education, and well-being. Walgreens became the official and exclusive retailer for *Red Nose Day* in the United States (Wood & Reinert, 2019). *Red Nose Day* is celebrated every year on the fourth Thursday of May. In the weeks leading up to *Red Nose Day*, Walgreens promotes and sells the iconic red noses online and in each of its more than 9000 stores in the United States. It sells the red noses for $1 each, individually or in boxes of 60 or 360. The revenue earned is donated by Walgreens to the *Red Nose Day Fund* (Wood & Reinert, 2019).

Walgreens supports *Red Nose Day* not only by selling red noses. Walgreens has developed a fully integrated marketing communications campaign to support *Red Nose Day* (Pittman & Sheehan, 2021). Walgreens promotes *Red Nose Day* through TV advertisements, online videos, website and social media content, and events and challenges. For example, in the *Red Nose Challenge*, Walgreens invites everyone everywhere to *get seriously silly* on social media or share their *#NosesOn* selfies (Pittman & Sheehan, 2021).

Various other brands support the initiative of *Red Noses Day* in the United States (Wood & Reinert, 2019). However, Walgreens is the only outlet where official *Red Nose Day* noses can be purchased. Walgreens plays an important role in promoting events and in encouraging individuals to donate for the cause. Few years back, Walgreens generated 180 million media impressions and 267000 engagements for *Red Nose Day* (Pittman & Sheehan, 2021). Walgreens raised more than $28 million for *Red Nose Day*. It generated more than $100 million in five years for the cause. One Walgreens executive comments, "By focusing our efforts on initiatives like *Red Nose Day*, we created a unique and powerful opportunity for our customers and the communities we serve to come together to help make an impact on a worthwhile cause like child poverty. By choosing Walgreens, our customers choose to make a difference in the life of a child in need" (Pittman & Sheehan, 2021).

9.3. Gillette: "The Best Men Can Be"

Gillette represents images of men and masculinity for decades. At present, Gillette associates that image to act as an agent for social change (Sternadori & Abitbol, 2022). Gillette responds to the *#metoo* movement with a firm stand. Gillette says, "Turn on the news today and it's easy to believe that men are not at their best. It's time we acknowledge that brands like ours play a role in influencing culture. And as a company that encourages men to be their best, we have a responsibility to make sure we are promoting positive, attainable, inclusive, and healthy versions of what it means to be a man" (Mirzaei, Wilkie, & Siuki, 2022).

Gillette released a short video advertisement with the tagline, *TheBest Man Can Be* in line with its age-old slogan, *The Best a Man Can Get*. The video emphasizes that men should be united to address issues related to *toxic masculinity*, including sexual harassment, sexual discrimination, and bullying (Sternadori & Abitbol, 2022). The initiative urges men to support positive change, to strive to be better, to make us better, and to help each other be better (Sanusi, 2019).

Gillette launched a website, *TheBestMenCanBe.org* to detail the purpose of the campaign and pledges that the brand will "actively challenge the stereotypes and expectations of what it means to be a man everywhere you see Gillette. In the ads we run, the images we publish to social media, the words we choose, and so much more" (Bogen, Williams, Reidy, & Orchowski, 2021). Gillette pledged to donate $1 million to non-profit organizations which offer programs that help men of all ages "achieve their personal best and become role models for the next generation." The brand says, "We've got work to do. And it starts today" (Sternadori & Abitbol, 2022).

Gillette receives appreciation for it's *The Best Men Can Be* campaign. However, Gillette also encounters controversies related to such hot social issues (Bhargava & Bedi, 2022). Within 48 hours of its launch, the video received 20 million views. Gillette received 1.5 million mentions which was more than the mere 10000 mentions during the same period the previous week. However, Gillette also received *thumbs down* responses on YouTube. Critics agreed that there is a need for social change. However, they accused Gillette of polarizing a sensitive issue. They accused that the initiative clubbed all men into one category and it is too preachy. One prominent broadcaster even accused Gillette of contributing to *global assault on masculinity* (Bhargava & Bedi, 2022).

Gillette faced many controversies because of the campaign. However, Gillette supports the campaign and the cause because it expects that *The Best Men Can Be* campaign will spark a different dialogue. A Gillette marketer says, "Successful brands today have to be relevant and engage consumers in topics that matter to them. If we get people to pause, reflect, and to challenge themselves and others to ensure that their actions reflect who they really are, then this campaign will be a success" (Vredenburg, Kapitan, Spry, & Kemper, 2020).

9.4. AT&T: "It Can Wait"

Individuals have the habit of sending text messages from their mobile phones while driving. The issue has worsened and has become serious over the last few years. AT&T, one of the largest voice and data services providers, stepped in with its *It Can Wait* campaign. The campaign requests and urges individuals to put away their phones while driving. AT&T sensitizes individuals about the horrific and potential outcomes of using a phone while driving. This is the heart of the campaign. Such messages are communicated through a slate of television and digital advertisements (Diehl, Marín, & Zreiqat, 2020). AT&T showed a three-minute digital advertisement, "Close to Home". The advertisement shows individuals involved in their daily routine. One young mother runs errands with her daughter securely seat-belted in the back seat. She looks down briefly to check a social media post on her phone, inadvertently crosses the center line, and causes a head-on collision with an oncoming car (Diehl, Marín, & Zreiqat, 2020). It is difficult and painful to watch the advertisement. However, the message in the advertisement is clear: Even a quick glance at your phone while driving cars can have disastrous consequences. "It can wait", admonished the advertisement (Wolfsdorf, Scott, & Herzog, 2019).

It Can Wait campaign tries to make individuals aware not only through advertisements. AT&T also offers a DriveMode app that is enabled automatically when the car is moving. The app silences texts and alerts to notify senders that the user is behind the wheel (Training, 2016). The campaign also promotes the *It Can Wait Pledge* which urges individuals to keep their eyes on the road ahead and not on their mobile phones. More than eight million individuals have taken the pledge (Bergmark, Gliklich, Guo, & Gliklich, 2016). *It Can Wait* campaign shows a virtual reality simulator that puts individuals behind the wheel to experience smartphone-induced crashes. The campaign also arranges celebrity tweets and outreach sessions at schools (Quisenberry, 2015).

The campaign has partnered with more than dozens of companies and media organizations. It has garnered hundreds of millions of impressions (Lisinska, 2019). The campaign was introduced more than ten years back. The campaign seems to be gaining acceptance in the society. One AT&T marketer says, "We do surveys (that show) we are making a difference. About a third of people say our no-texting-while driving message has changed their behavior" (Lisinska, 2019).

9.5. The Rainforest Alliance: "Follow the Frog"

Advertising and promotions allow non-profit organizations to market their social and environmental missions. The Rainforest Alliance is an international non-profit conservation organization which promotes the cause of minimizing harmful human impact on the environment (Giovannucci, Hagen, & Wozniak, 2014). The organization has the Rainforest Alliance Certified logo which can be displayed by companies which are committed to sustainable sourcing from farms and forests. The companies can display the logo on their packages and promotional materials. The logo contains a signature green tree frog symbol. The organization also offers sourcing assistance and training to help businesses improve their sustainability efforts (Asbury & Ball, 2016).

The campaign wanted to boost awareness of pressing ecological challenges such as deforestation, climate change, and water quality. In 2012, the Rainforest Alliance launched an encompassing and clever promotional campaign called *Follow the Frog* (Asbury & Ball, 2016). The campaign contained an award-winning three-minute web video. The video comically illustrated all those aspects on which individuals do not need to focus to save the environment and the planet. The video features a young professional who does crazy and extreme things to save the ecology. He quits his job, leaves his wife and kid, and heads to the heart of the Amazon jungle to join in a crusade against all evil factions that are destroying the environment. The video concludes by assuring that there are easier and simpler ways to save the environment (Asbury & Ball, 2016). It says, "You don't have to go to the ends of the Earth to protect the planet. Just follow the frog" (Giovannucci, Hagen, & Wozniak, 2014). It encourages individuals to buy products that display the Rainforest Alliance Certified seal.

The *Follow the Frog* campaign has gained acceptance among individuals. The campaign encompasses a wide range of marketing content and activities, from print advertisements, retail promotional materials, *Follow the Frog* events, and a *Frog Blog*. The campaign has also created alliance with more than 6000 companies which display and promote the Rainforest Alliance Certified seal. The effectiveness of the campaign is visible. The initial video received more than 5 million YouTube views and 32000 shares. It was also nominated for a Webby award for online video excellence. It won a TED award as the top *Ad Worth Spreading* in the social causes category (Asbury & Ball, 2016). The campaign helped in increasing consumer awareness. It boosted the number of companies committing to sourcing from Rainforest Alliance Certified operations. The campaign was able to motivate major brands such as Lipton, Mars, and McDonald's to join the cause (Asbury & Ball, 2016).

9.6. Always: "Like a Girl"

Always is a feminine hygiene brand from Procter & Gamble (P&G). *Always* has a mission which supports a social cause (Mkamboi, Ndavula, & Mwangi, 2022). Research shows that the confidence of a girl goes down drastically at puberty (Røstvik, 2022). *Always* launched a big-budget advertisement campaign aimed at changing the meaning of doing something *like a girl*. The first advertisement of *Like a Girl* was based on a reality show format. The advertisement urged young women, men, boys, and girls to "show me what it looks like to run like a girl" or "fight like a girl" or "throw like a girl" (Røstvik, 2022). The classic *feminine* stereotypes were portrayed by the women, men, and boys. Young girls took the task positively without showing any signs of any negative connotations. "When did doing something 'like a girl' become an insult?" the advertisement asked. "*Always* wants to change that" (Mkamboi, Ndavula, & Mwangi, 2022). The advertisement showcased inspirational images of girls running, throwing, and kicking with skill and confidence. The advertisement concluded with the statement, "Let's make *#RunLikeAGirl* mean amazing things" (Mkamboi, Ndavula, & Mwangi, 2022).

The *Like a Girl* advertisement ran during Super Bowl XLIX. Reactions of viewers were spectacular. The advertisement secured first or second position in almost every major advertisement-rating poll. It also topped all other Super Bowl advertisements by a wide margin in social media mentions and sharing (Feldheim, 2022). *Like a Girl* also won an Emmy. It was judged as the most outstanding commercial. It also won a Grand Prix award for best PR campaign at the Cannes international advertisement festival. In the first year of its launch, *Like a Girl* secured 85 million views and more than a billion global impressions (Feldheim, 2022).

The *Like a Girl* campaign from *Always* has developed into a full-blown, ongoing educational and support campaign (Um, 2021). The vice president of global feminine care for P&G says, "When you have a message that really addresses such an important and a real issue and it's done in a way that is very consistent with who you are as a brand, I think consumers want to engage with that. We have started an epic battle to show the world that doing things like a girl is downright amazing" (Um, 2021).

10. DISCUSSION

Companies are aware about their responsibilities towards their customers and towards the society at large. They develop marketing communications and do promotion about their offerings in a socially responsible manner. The digital age allows companies to do marketing communications with the target audience better. However, there are various legal and ethical issues surrounding marketing communications. Employees at all levels of the organization must be aware about such issues. Companies must work hard and proactively at communicating openly, honestly, and agreeably with their customers and resellers.

Companies and businesses should understand their roles in socially responsible marketing communications. They should understand and realize that their actions not only influence their customers but also the society at large. Various issues related to social responsibility are involved in marketing communications done through advertising, sales promotion, and personal selling. Companies are aware about the roles of digital marketing and social media in socially responsible marketing communications and should apply such means of communications in an effective manner. Legal and ethical issues are associated with socially responsible marketing communications. Companies should be aware about such issues and should respect consumer privacy in marketing communications. Companies have responsibilities towards promotion of social responsibility. They adopt initiatives for doing socially responsible communications. All these aspects are highlighted and discussed in the study. The study has both theoretical and managerial implications.

10.1. Theoretical Implications

Academicians may study and analyze the various aspects of responsibilities of companies and businesses towards the society and the roles played by marketing communications and promotions in the process. They may study and evaluate the various forms of promotions and marketing communications done by companies. They may analyze the roles of digital marketing, the internet, and social media in socially responsible marketing communications. They may also analyze the various legal and ethical issues and concerns in socially responsible marketing communications. They may analyze the various initiatives in marketing communications taken by companies to fulfill their social responsibilities. They may suggest strategies and initiatives which may be adopted by companies and businesses to fulfill their responsibilities towards the society more effectively.

10.2. Managerial Implications

Companies and businesses will realize and understand their responsibilities towards their customers and the society at large. They will understand their roles in socially responsible marketing communications. They will also realize their social responsibilities while doing promotions through various marketing communication channels like advertising, sales promotion, and personal selling. They should appreciate the roles of digital marketing, the internet, and social media in socially responsible marketing communications. They should also appreciate the various legal and ethical issues and concerns in socially responsible marketing communications. They should ensure consumer privacy in strategies related to marketing communications. They have responsibilities towards promotion of social responsibility and will be able to evaluate the various strategies and initiatives adopted by them for socially responsible marketing communications. Based on the evaluation, they may decide which strategies and initiatives should be retained and which strategies and initiatives should be adopted in future to convince customers about their dedication towards the society and to achieve business excellence.

11. CONCLUSION

Companies and businesses have responsibilities towards their customers and the society at large. These responsibilities should be revealed in the manner in which they communicate about their offerings to the target market. The activities of companies and businesses will assure customers and the society about their responsibilities and commitment towards social and environmental causes.

11.1. Limitations

The study conducted a conceptual analysis of the various aspects of social responsibilities of companies and the various aspects of socially responsible marketing communications. Primary data is not collected and empirical analysis is not done. Also, the study focused on doing a conceptual analysis mainly for the markets in the United States.

11.2. Avenues of Future Research

Companies and businesses should be socially responsible for their actions. Such responsibilities should be revealed in the manner in which they do promotion of their offerings to the target markets and in the initiatives which they adopt for the welfare of their customers and the society at large. Researchers in future may collect primary data related to socially responsible marketing communications done by companies. They may conduct empirical analysis and generate insights which are actionable and implementable. They may conduct research related to the roles of the internet, digital marketing, and social media in socially responsible marketing communications. They may also conduct analysis in markets other than that of the United States and understand the existing initiatives and strategies adopted by companies in socially responsible marketing communications and suggest initiatives and strategies which are more effective and which may be adopted by companies in future.

REFERENCES

Aaker, D. (2018). *Creating signature stories: Strategic messaging that persuades, energizes and inspires*. Morgan James Publishing.

Adnan, M., Khan, A. U., & Hayee, R. (2021). Integrated Marketing Communication and Customer Satisfaction as a function of Customer Loyalty with the Moderating role of Product Innovation Moderation. *Sustainable Business and Society in Emerging Economies*, 3(2), 133–148. DOI: 10.26710/sbsee.v3i2.1910

Ahearne, M., Atefi, Y., Lam, S. K., & Pourmasoudi, M. (2022). The future of buyer–seller interactions: A conceptual framework and research agenda. *Journal of the Academy of Marketing Science*, 50(1), 22–45. DOI: 10.1007/s11747-021-00803-0 PMID: 34608341

Allen, E. (2022). Jess Weiner and the dove real beauty campaign: Selling feminism for profit or social change? *Women Leading Change: Case Studies on Women, Gender, and Feminism*, 6(1), 18–37.

Antoniuk, K., Mokiy, A., Pikh, M., Bukharina, L., & Antoniuk, D. (2021). Ensuring of Consumer Choice Ecologization on the Basis of Consumption Safety: International Experience for Ukraine. In *SHS Web of Conferences* (Vol. 100, p. 05009). EDP Sciences.

Asbury, S., & Ball, R. (2016). *The practical guide to corporate social responsibility: do the right thing*. Routledge. DOI: 10.4324/9781315697659

Balmer, J. M., Powell, S. M., & Greyser, S. A. (2011). Explicating Ethical Corporate Marketing. Insights from the BP Deepwater Horizon Catastrophe: The Ethical Brand That Exploded and Then Imploded. *Journal of Business Ethics*, 102(1), 1–14. DOI: 10.1007/s10551-011-0902-1

Bergmark, R. W., Gliklich, E., Guo, R., & Gliklich, R. E. (2016). Texting while driving: The development and validation of the distracted driving survey and risk score among young adults. *Injury Epidemiology*, 3(1), 1–10. DOI: 10.1186/s40621-016-0073-8 PMID: 27747544

Bhargava, V. R., & Bedi, S. (2022). Brand as promise. *Journal of Business Ethics*, 179(3), 919–936. DOI: 10.1007/s10551-021-04834-z

Bhattacharyya, J. (2023). The structure of sustainability marketing research: A bibliometric review and directions for future research. *Asia-Pacific Journal of Business Administration*, 15(2), 245–286. DOI: 10.1108/APJBA-06-2021-0239

Bogen, K. W., Williams, S. L., Reidy, D. E., & Orchowski, L. M. (2021). We (want to) believe in the best of men: A qualitative analysis of reactions to# Gillette on Twitter. *Psychology of Men & Masculinity*, 22(1), 101–112. DOI: 10.1037/men0000308

Carter, K., Jayachandran, S., & Murdock, M. R. (2021). Building a sustainable shelf: The role of firm sustainability reputation. *Journal of Retailing*, 97(4), 507–522. DOI: 10.1016/j.jretai.2021.03.003

Chevers, D. A. (2019). The Impact of Cybercrime on E-Banking: A Proposed Model. In *CONF-IRM* (p. 11).

Dauhgerty, E. L. (2014). Ethical Dilemmas of Female Agency Owners: Practices in Public Relations and Marketing Communications. *Quarterly Review of Business Disciplines*, 1(2), 173–190.

Diehl, F. S., Marín, L. S., & Zreiqat, M. (2020). An evaluation of AT&T social marketing delivery modes aimed at teen smartphone use while driving. *Journal of Cultural Marketing Strategy*, 5(1), 60–75. DOI: 10.69554/XBTL7959

Dua, T. (2019, February 4). *Here's the Full List of All the Super Bowl Commercials That Ran This Year*, Business Insider, www.businessinsider.com/super-bowl-commercials-2019-list-2019-1 (Accessed on 4th February, 2019)

Emamalizadeh, H. (2015). Bait-and-switch advertising: A conceptual approach. In *Proceedings of the 1983 Academy of Marketing Science (AMS) Annual Conference* (pp. 458-460). Cham: Springer International Publishing. DOI: 10.1007/978-3-319-16937-8_109

Errichiello, O., & Zschiesche, A. (2022). Green Consumption. In *Green Branding: Basics, Success Factors And Instruments For Sustainable Brand And Innovation Management* (pp. 17–57). Springer Fachmedien Wiesbaden. DOI: 10.1007/978-3-658-36060-3_2

Evans, D. M., Parsons, R., Jackson, P., Greenwood, S., & Ryan, A. (2020). Understanding plastic packaging: The co-evolution of materials and society. *Global Environmental Change*, 65, 102166. DOI: 10.1016/j.gloenvcha.2020.102166

Feldheim, D. (2022). *Dare to Lead Like a Girl: How to Survive and Thrive in the Corporate Jungle*. Rowman & Littlefield. DOI: 10.5771/9781538163535

Fogarty, M. J. (2022). Rewired: Utilizing an Integrated Design Campaign to Persuade Generation Z to Rethink Their Hyper-Connected World.

Gao, Q., Zhang, Z., Li, Z., Li, Y., & Shao, X. (2022). Strategic green marketing and cross-border merger and acquisition completion: The role of corporate social responsibility and green patent development. *Journal of Cleaner Production*, 343, 130961. DOI: 10.1016/j.jclepro.2022.130961

Giovannucci, D., von Hagen, O., & Wozniak, J. (2014). Corporate social responsibility and the role of voluntary sustainability standards. In *Voluntary standard systems: A contribution to sustainable development* (pp. 359–384). Springer Berlin Heidelberg. DOI: 10.1007/978-3-642-35716-9_24

Gladys, E. U. (2018). Legal Framework for the Protection of Online Products Consumers in Nigeria. *Nigerian LJ*, 21, 24.

Haarpaintner, É. (2022). What is the relationship between sustainability, corporate social responsibility and quality of life applications at work? *Worldwide Hospitality and Tourism Themes*, 14(3), 236–246. DOI: 10.1108/WHATT-02-2022-0020

Hayes, J. L., & Duff, B. R. (2022). Introducing a special issue on advertising & corporate social responsibility. *International Journal of Advertising*, 41(2), 205–209. DOI: 10.1080/02650487.2022.2056357

Huang, L. L., Friedman, D. B., Lin, F. C., & Thrasher, J. F. (2018). Which types of anti-smoking television advertisements work better in Taiwan? *Health Promotion International*, 33(3), 545–555. PMID: 28011658

Jackson, N. (2013). *Promoting and marketing events: Theory and practice*. Routledge. DOI: 10.4324/9780203583166

Jia, M., & Wan, C. (2022). Why are strangers trusted more during trade fairs? A literature review on the conceptual model of general trust formation. *Journal of Business and Industrial Marketing*, 37(11), 2205–2216. DOI: 10.1108/JBIM-03-2021-0164

Jin, L., & He, Y. (2018). How the frequency and amount of corporate donations affect consumer perception and behavioral responses. *Journal of the Academy of Marketing Science*, 46(6), 1072–1088. DOI: 10.1007/s11747-018-0584-7

Johnson, T. (2018, December 7). Google Event Encourages Girls to Look at Computer Science Careers. *Daily Nonpareil*, www.nonpareilonline.com/news/education/google-eventencourages-girls-to-look-at-computer-science-careers/article_2a08d2dd-b48d-54cf-8c41-c5c91a8ade7e.html (Accessed on 7th December, 2018)

Jones, E. (2019). Rethinking greenwashing: Corporate discourse, unethical practice, and the unmet potential of ethical consumerism. *Sociological Perspectives*, 62(5), 728–754. DOI: 10.1177/0731121419849095

Kees, J., & Andrews, J. C. (2019). Research Issues and Needs at the Intersection of Advertising and Public Policy. *Journal of Advertising*, 48(1), 126–135. DOI: 10.1080/00913367.2019.1599747

Kitchen, P. J., & Tourky, M. E. (2022). Integrated Communication or Integrated Marketing Communication. *Integrated Marketing Communications: A Global Brand-Driven Approach*, 69-87.

Lisinska, A. (2019). Do Initiators Matter in Social Marketing? Understanding Characteristics and Particularities of Social Marketing Campaigns Initiated by Different Agents. *SPNHA Review*, 15(1), 78–104.

Litwin, M. L. (2007). *The ABC's of Strategic Communication: Thousands of Terms, Tips and Techniques*. AuthorHouse.

Luxon, E. M. (2019). Mobilizing Environmental Sentiment Through the Media. *Environmental Politics*, 28(4), 639–662. DOI: 10.1080/09644016.2018.1560743

Maccarrone, P., & Contri, A. M. (2021). Integrating corporate social responsibility into corporate strategy: The role of formal tools. *Sustainability (Basel)*, 13(22), 12551. DOI: 10.3390/su132212551

Maier, A. L. (2021). Political corporate social responsibility in authoritarian contexts. *Journal of International Business Policy*, 4(4), 476–495. DOI: 10.1057/s42214-020-00085-3

Mandal, P. C. (2019a). Social criticisms of marketing: Concerns and initiatives. *International Journal of Society Systems Science*, 11(2), 157–170. DOI: 10.1504/IJSSS.2019.100088

Mandal, P. C. (2019b). Ethical Issues and Concerns in Collection of Marketing Information and Marketing Intelligence: Ethical Issue in Collection of Information. [IJBIR]. *International Journal of Business Intelligence Research*, 10(2), 16–28. DOI: 10.4018/IJBIR.2019070102

Mandal, P. C. (2021). Socially-Responsible Marketing Communications: Concerns, Strategies, and Initiatives. [JMME]. *Journal of Media Management and Entrepreneurship*, 3(1), 1–17. DOI: 10.4018/JMME.290301

Martin, B. A. (2019). The Unregulated Underground Market for Your Data: Providing Adequate Protections for Consumer Privacy in the Modern Era. *Iowa Law Review*, 105, 865.

Mercer, S. T. (2020). The Limitations of European Data Protection As A Model for Global Privacy Regulation. *AJIL Unbound*, 114, 20–25. DOI: 10.1017/aju.2019.83

Mirzaei, A., Wilkie, D. C., & Siuki, H. (2022). Woke brand activism authenticity or the lack of it. *Journal of Business Research*, 139, 1–12. DOI: 10.1016/j.jbusres.2021.09.044

Mkamboi, A., Ndavula, J., & Mwangi, S. W. (2022). Utilizing Social Media Platforms for Online Crisis Management: Focus on Always Pads Crisis of 2019. *African Multidisciplinary Journal of Research*, 7(1), 83–94.

Nur, F., Akmaliah, N., Chairul, R., & Safira, S. (2021). Green purchase intention: The power of success in green marketing promotion. *Management Science Letters*, 11(5), 1607–1620.

Ooi, C. S. (2022). Sustainable tourism and the moral limits of the market: can Asia offer better alternatives. In *Asian tourism sustainability* (pp. 177–197). Springer Nature Singapore. DOI: 10.1007/978-981-16-5264-6_10

Pittman, M., & Sheehan, K. (2021). Brand authenticity and strategic response to crises: Symbolic effects of donation type on purchase intent and digital engagement. *Journal of Current Issues and Research in Advertising*, 42(3), 215–235. DOI: 10.1080/10641734.2020.1734503

Quisenberry, P. N. (2015). Texting and driving: Can it be explained by the general theory of crime? *American Journal of Criminal Justice*, 40(2), 303–316. DOI: 10.1007/s12103-014-9249-3

Reeves, M., Chandra, D., & Kim, K. (2022). Cause Marketing in a Time of Polarization. *Advertising & Society Quarterly, 23*(1).

Rhodes, N., Ralston, R., & Bigsby, E. (2016). Teens' Reactance to Anti-Smoking Public Service Announcements: How Norms Set the Stage. *Journal of Health Communication*, 21(5), 575–582. DOI: 10.1080/10810730.2015.1114052 PMID: 27116415

Ronald, L. W. (2008). *Kleppner's advertising procedure*. Pearson Education India.

Røstvik, C. M. (2022). *Cash Flow: The businesses of menstruation*. UCL Press. DOI: 10.14324/111.9781787355385

Ryan, C. B. (2019). # BPCares: Sinking Oil, Spreadable Détournement. *ISLE: Interdisciplinary Studies in Literature and Environment*, 26(4), 882–900. DOI: 10.1093/isle/isy093

Salant, S. W. (2016). The Strategic Use of Early Bird Discounts for Dealers. *Quantitative Marketing and Economics*, 14(2), 97–127. DOI: 10.1007/s11129-016-9167-4

Sánchez Abril, P., Oliva Blázquez, F., & Martínez Evora, J. (2018). The Right of Withdrawal in Consumer Contracts: A comparative analysis of American and European law. *InDret*, 3, 18–13.

Sanusi, N. (2019). Maskulinitas dalam Iklan Gillette We Believe: What Best Man Can Be. *Media Komunika (Jurnal Komunikasi). Universitas Sangga Buana YPKP*, 4(2), 61–69.

Schmeiser, S. (2018). Online Advertising Networks and Consumer Perceptions of Privacy. *Applied Economics Letters*, 25(11), 776–780. DOI: 10.1080/13504851.2017.1366634

Sharpe, J. D., Kaufman, J. A., Goldman, Z. E., Wolkin, A., & Gribble, M. O. (2019). Determinants of Oil-Spill Cleanup Participation Following the Deepwater Horizon Oil Spill. *Environmental Research*, 170, 472–480. DOI: 10.1016/j.envres.2019.01.009 PMID: 30640081

Shim, J., Moon, J., Lee, W. S., & Chung, N. (2021). The impact of CSR on corporate value of restaurant businesses using triple bottom line theory. *Sustainability (Basel)*, 13(4), 2131. DOI: 10.3390/su13042131

Singh, S. K., & Sanjeev, R. (2022). Antecedents of Green Marketing Initiatives: A Hierarchical Regression Model. [IJSESD]. *International Journal of Social Ecology and Sustainable Development*, 13(1), 1–15. DOI: 10.4018/IJSESD.315310

Spence, C. (2012). Auditory contributions to flavour perception and feeding behaviour. *Physiology & Behavior*, 107(4), 505–515. DOI: 10.1016/j.physbeh.2012.04.022 PMID: 22579696

Sternadori, M., & Abitbol, A. (2022). How Male Consumers Respond to "Enlightened Manvertising" Campaigns: Gender Schema, Hostile Sexism, And Political Orientation Feed Attitudes. *Journal of Advertising Research*, 62(1), 87–101. DOI: 10.2501/JAR-2022-006

Training, R. (2016)... *University Transportation Center to Support DOTD Research Needs.*, 31(1), 1–5.

Trihas, N., Perakakis, E., Venitourakis, M., Mastorakis, G., & Kopanakis, I. (2013). Social media as a marketing tool for tourism destinations: The case of Greek municipalities. *Journal of Marketing Vistas*, 3(2), 38–48.

Um, N. (2021). Antecedents and consequences of attitude toward femvertising. *The Journal of the Korea Contents Association*, 21(1), 66–74.

Vredenburg, J., Kapitan, S., Spry, A., & Kemper, J. A. (2020). Brands taking a stand: Authentic brand activism or woke washing? *Journal of Public Policy & Marketing*, 39(4), 444–460. DOI: 10.1177/0743915620947359

Willan, M. (2021). *The essentials of contemporary marketing*. Bloomsbury Publishing.

Wolfsdorf, A., Scott, A., & Herzog, S. (2019). What happened on July 21st: An investigation of trauma and trigger warnings in the English classroom. *Changing English*, 26(2), 198–215. DOI: 10.1080/1358684X.2018.1552514

Wood, N., & Reinert, C. (2019). NONPROFITS AND FOR PROFITS. WHO BENEFITS? *Interdisciplinary Insights: The Journal of Saint Leo University's College of Education and Social Services*, 1(2), 14–33.

Zeithaml, V. A., Verleye, K., Hatak, I., Koller, M., & Zauner, A. (2020). Three decades of customer value research: Paradigmatic roots and future research avenues. *Journal of Service Research*, 23(4), 409–432. DOI: 10.1177/1094670520948134

Chapter 14
Socially Responsible Marketing Communications and Target Marketing:
Strategies and Initiatives

Pratap Chandra Mandal
https://orcid.org/0000-0002-0300-4002
Indian Institute of Management, Shillong, India

ABSTRACT

Companies and businesses have roles in marketing communications and target marketing done by them. They should ensure that such communications are socially responsible. The objective of the study is to understand the various aspects of socially responsible marketing communications and target marketing, and ethical and legal aspects and concerns related to such issues. The methodology adopted is a conceptual analysis of the various aspects and issues. Enlightened companies adopt initiatives for socially responsible marketing communications and target marketing. The discussions will allow academicians and practitioners to understand the importance of social responsibility in marketing communications and target marketing and the various issues related to such communications. Policy makers may investigate about the ethical and legal concerns related to the issues and may implement rules and regulations which will ensure the triple bottom line – people, profit, and planet.

DOI: 10.4018/979-8-3693-3579-6.ch014

1. INTRODUCTION

Companies will be able to satisfy and delight their customers when they understand customer requirements and preferences well (Demirel, 2022). Companies should try to gain in-depth knowledge about customer requirements and preferences so that they can create customer value (Mandal, 2022). In-depth knowledge about customer value will allow companies to communicate the value clearly and persuasively to their target markets (Foltean & van Bruggen, 2022). Companies should apply promotion tools properly to target their customers. Promotion is done not through a single tool but rather through a mix of several tools (Tian, Zhang, & Mei, 2022). Companies should select and coordinate these promotion elements carefully to engage customers and to build, develop, and communicate clear, consistent, and compelling messages and content about their organizations, products, and brands (Adnan, Khan, & Hayee, 2021).

Companies have responsibilities not only towards their customers and target markets but also towards the society at large (Jin & He, 2018). They should communicate about their offerings, brands, and value proposition to the target markets in a socially responsible manner. Companies should communicate only those aspects of value proposition which are of relevance to the target markets and the society at large (Jin & He, 2018). The main objective of any company is to generate revenues and profits and to achieve growth (Bhattacharyya, 2022). However, in the quest of generating revenues and profits, companies often forget the line of ethics. They sometimes do marketing communications which are not socially responsible and target their audience in a manner which is not acceptable by the society at large (Bhattacharyya, 2022). Companies should realize that they have responsibilities not only towards their customers and target markets but also towards the society at large. Consequently, companies should ensure that marketing communications and target marketing are done in a socially responsible manner. They should also ensure that such promotional activities do not create a negative impact on the society (Nur, Akmaliah, Chairul, & Safira, 2021).

The responsibilities of companies and businesses while communicating about their offerings and while doing target marketing cannot be overemphasized. It is imperative for companies to be socially responsible while communicating about their offerings and while doing target marketing. The impact of marketing communications on the target markets and on the society may be substantial. Although the topic is important, there exist few studies which focus on socially responsible marketing communications and target marketing and on the social impact of such activities. The study aims to address this research gap.

The objective of the study is to analyze the various aspects of marketing communications and target marketing done by companies. The study aims to analyze whether such marketing communications and target marketing are done by companies in a socially responsible manner.

The methodology adopted is a conceptual analysis of the various aspects of marketing communications and target marketing done by companies. The study aims to investigate whether such activities are socially responsible or not. Research studies from reputed academic journals are referred to for conducting the analysis. Primary data is not collected and empirical analysis is not done.

The novelty and the contributions of the study lie in the fact that an in-depth analysis of the various aspects of socially responsible marketing communications and socially responsible target marketing is done. Roles and responsibilities of companies and businesses are discussed. Various strategies and initiatives adopted by companies for ensuring socially responsible marketing communications and target marketing are discussed. Various concerns related to privacy and security of data shared by individuals with companies and concerns related to target marketing of vulnerable segments and targeting in the digital era are discussed. The discussions will be beneficial for both academicians and practitioners. Academicians may analyze the social concerns and may suggest better ways of doing marketing communications and target marketing. Practicing managers should ensure that the marketing communications and target marketing done by companies are socially responsible. They should also address the concerns of individuals regarding the privacy and security of customer data. Overall, the study will allow academicians, practicing managers, and policy makers to understand the various aspects of socially responsible marketing communications and target marketing, privacy and security concerns of individuals, and ethical and legal aspects related to such issues.

The study is structured as follows:

The roles of companies in socially responsible marketing communications are highlighted in section 2. Marketing communications through advertising and sales promotion are highlighted in section 3 while marketing communications through personal selling are highlighted in section 4. The responsibilities of companies in socially responsible marketing communications are discussed in section 5. Companies adopt several initiatives for socially responsible marketing communications. Initiatives adopted by enlightened companies are discussed in section 6 and its sub-sections. The concerns related to socially responsible target marketing are highlighted in section 7 with sub-sections 7.1 and 7.2 discussing about target marketing of vulnerable segments and target marketing in the digital era respectively. The roles of companies in socially responsible target marketing are discussed in section 8. The salient points of the study are discussed in section 9 with sub-sections 9.1 and 9.2 highlighting the theoretical implications and the managerial implications of the study respectively.

Section 10 concludes the study with sub-sections 10.1 and 10.2 highlighting the limitations of the study and the avenues of future research respectively.

2. ROLES OF COMPANIES IN SOCIALLY RESPONSIBLE MARKETING COMMUNICATIONS

Companies should be aware about their social responsibilities while communicating about their offerings and brands to the target markets (Singh & Sanjeev, 2022). They should ensure that the various types of promotions done by them do not create harm to the customers and to the society at large (Antoniuk, Mokiy, Pikh, Bukharina, & Antoniuk, 2021). In the quest of generating revenues and profits for themselves, companies may sometimes cross the line of ethics while promoting their offerings. Such activities may result in legal and ethical issues (Coffie, Anning-Dorson, Hinson, Bosah, & Anani-Bossman, 2022). Consequently, companies should ensure that they remain within the realm of ethics while communicating about their offerings.

Companies must be aware of the many legal and ethical issues surrounding marketing communications while shaping their promotion mix (Coffie et al., 2022). Many companies are aware of their social responsibilities and strive hard to ensure that the communications done with target markets and resellers are open and honest (Letunovska, 2020). However, abuses may occur. To ensure that the marketing communications done by companies are socially responsible, public policy makers may develop and implement a substantial body of laws and regulations to govern advertising, personal selling, sales promotion, and direct marketing (Jackson, 2013). There might be issues when companies do marketing communications through advertising, sales promotion, and personal selling (Trihas, Perakakis, Venitourakis, Mastorakis, & Kopanakis, 2013).

3. ADVERTISING AND SALES PROMOTION

Companies may face issues when they do marketing communications through advertising and sales promotion. Law requires that companies should avoid false or deceptive advertising (Mandal, 2019). Companies should not make claims which they themselves know are not true. For example, advertisers cannot claim that a product cures something when they know that it actually does not. Companies should avoid communicating such advertisements which have the capacity of deceiving individuals even when they actually do not deceive anyone (Ronald, 2008). For example, an automobile cannot be advertised as getting 32 miles per gallon unless it does so under typical conditions. Also, diet bread cannot be advertised as having fewer calories simply because its slices are thinner.

Bait-and-switch advertising is a type of promotion which tries to attract individuals under false pretenses (Litwin, 2007). For example, a sewing machine was advertised by a large retailer at $179. The retailer promised a number of product features and benefits at $179. However, when consumers tried to buy the advertised machine, the seller downplayed its features, placed faulty machines on showroom floors, understated the machine's performance, and took their actions in an attempt to switch buyers to a more expensive machine. Companies should avoid such types of advertising. Such actions are both unethical and illegal (Emamalizadeh, 2015).

Regulations are enforced to keep a close watch on the trade promotion activities of companies (Maier, 2021). For example, Robinson-Patman Act is enforced so that sellers cannot favor a group of customers through their use of trade promotions (Mandal, 2021). Sellers should make promotional allowances and services available to all resellers on proportionately equal terms (Barter, 2019).

Companies should avoid legal pitfalls such as deceptive and bait-and-switch advertising. Moreover, companies should use advertising and other forms of promotion to encourage and to promote socially responsible programs, actions, and ideas (Wagner, 2014). A wide range of social and environmental causes related to their brands are promoted by companies in almost every industry (Wagner, 2014). For example, companies like AT&T, The Rainforest Alliance, and Google have adopted a number of initiatives to promote social and environmental causes related to their brands (Bendell & Huvaj, 2020).

4. PERSONAL SELLING

Salespeople are assigned by companies to convince and to sell products to customers. However, salespeople should follow the rules of *fair competition*. Sales acts are enforced in many states to clearly indicate what is not allowed and which types of behavior are deceptive for salespeople (Mandal, 2021). For example, salespeople may not lie to customers or mislead them about the benefits and advantages of buying a particular product. Salespeople should promote products in a manner so that their statements match advertising claims. This may help companies to avoid bait-and-switch practices (Ahearne, Atefi, Lam, & Pourmasoudi, 2022).

Many consumers may be called on at home by salespeople. Some consumers may buy at a location that is not the seller's permanent place of business. Some consumers may go to a store in search of a product. Individuals who are called on may be surprised and be especially vulnerable to high-pressure selling techniques. The Federal Trade Commission (FTC) has a *three-day cooling-off rule* to give special protection to customers who are not seeking products (Sánchez Abril, Oliva Blázquez, & Martínez Evora, 2018). Customers may agree to buy in their own homes, workplaces, dormitories, or facilities rented by the seller on a temporary basis. Such places may include hotel rooms, convention centers, and restaurants. Customers who agree to buy a product costing more than $25 have 72 hours in which they can cancel or return merchandize and get their money back. No questions can be asked in such cases (Sánchez Abril et al., 2018).

Personal selling may happen in business domains. In selling to businesses, salespeople may not offer bribes to purchasing agents or others who can influence a sale (Jia & Wan, 2021). Salespeople are not allowed to obtain and to use technical trade secrets of competitors through bribery or industrial espionage. Finally, salespeople cannot disparage competitors or competing products by suggesting things that they themselves know are not true (Jia & Wan, 2021).

5. RESPONSIBILITIES OF COMPANIES IN SOCIALLY RESPONSIBLE MARKETING COMMUNICATIONS

Companies realize and understand their responsibilities towards the society and consequently, in doing socially responsible marketing communications (Maccarrone & Contri, 2021). Most companies incorporate social responsibility into their corporate missions. Companies use various promotion tools viz. advertising and communication programs to promote socially responsible communications, actions, and ideas. Companies do not focus only on revenues and profits. Rather, they recognize that company success is inexorably linked to the welfare of customers and the world around them (Shim, Moon, Lee, & Chung, 2021).

Companies from all industries have a focus on sustainability and promote a number of social and environmental issues (Ooi, 2022). For example, over the past many seasons, more than 20 percent of all Super Bowl advertisements have connected brands with a wide range of social and environmental issues (Aaker, 2018). In one instance, Microsoft inspired children with physical and cognitive disabilities by featuring and promoting them in an inspirational advertisement. These children rose to the top of their games and achieved success with the help of its Xbox Adaptive Controller (Dua, 2019). "The story illustrates Microsoft's commitment to building accessible technology that levels the playing field and creates opportunity for all of us. When everyone plays, we all win," says Microsoft (Dua, 2019). The ears of Budweiser Dalmatian flapped in the breeze as shown in an Anheuser-Busch Super Bowl advertisement. The Budweiser Dalmatian sat on the top of a wagon pulled by the Budweiser Clydesdales through a field of golden grain on a wind turbine farm, set to Bob Dylan's "Blowin' In the Wind". The advertisement concluded, "Wind never felt better". The advertisement further noted that Budweiser is "now brewed with 100 percent renewable energy. This Bud's is for a better tomorrow" (Dua, 2019).

The above examples indicate that companies are serious about social and environmental issues and take stand on those issues which are relevant. Companies do it because they feel that they are socially responsible and they feel that it is the right thing to do.

6. INITIATIVES BY COMPANIES FOR SOCIALLY RESPONSIBLE MARKETING COMMUNICATIONS

Companies adopt a number of initiatives to promote socially responsible marketing communications (Hayes & Duff, 2022). Companies incorporate and integrate social responsibility in their corporate missions (Kitchen & Tourky, 2022). They employ advertising and other promotion tools to promote socially responsible programs, ideas, and actions (Hayes & Duff, 2022). Companies focus on triple bottom line – people, profit, and planet – to succeed in the business environment.

Companies in every industry recognize, support, and promote a wide range of social and environmental issues (Camilleri, 2022). Chipotle introduced *Food with Integrity* campaign. The campaign challenges factory farming and promotes organic, local, and family-farmed ingredients (Haarpaintner, 2022). The importance and the necessity of educating girls around the world is promoted by Intel through its *Girls Rising* campaign (Jain, 2018). Girls are usually not encouraged to pursue careers in science and technology. Google encourages young girls to pursue careers in science and technology through its *Made with Code* campaign (Tasmin, Ahmed, & Motahar, 2019). Apple introduced *Better* campaign to raise and build awareness of important and relevant environmental issues. Apple believes that "it's our responsibility to make sure that while creating beautiful products, we're also caring for our beautiful planet" (Halvorsen, 2021).

Companies support and take a stand on relevant social and environmental issues and not only generate revenues and profits. They feel that it is the right thing to do. Companies adopt a number of initiatives and campaigns promoting social responsibility. Some of the campaigns are highlighted.

6.1. Dove: Dove Campaign for Real Beauty

Dove started a campaign to change the perceptions of individuals about beauty. The campaign aimed to change stereotypical definitions of beauty (Allen, 2022). The advertisements featured candid and confident images of real women of all sorts of body types. The women chosen were not glamorous actresses or skinny models.

"Our mission is to make more women feel beautiful every day by broadening the definition of beauty," says Dove (Danthinne, Giorgianni, Austin, & Rodgers, 2022). The constantly evolving *Campaign for Real Beauty* consisted of advertisements, heavily viewed digital videos, and other content. The campaign contributed substantially to lift the self-esteem of women. The campaign had noble motives beyond generating revenues and profits (Akena, 2022). The campaign made Dove a success. During the first decade of the campaign, the annual revenues from the brand grew from $2.5 billion to more than $4 billion (Millimen, 2015).

6.2. AT&T: It Can Wait

The habit of sending and receiving text messages while driving has worsened in recent times (Turnbull, Khanal, & Dakin, 2021). AT&T, one of the largest voice and data services providers in the United States, introduced its *It Can Wait* campaign which urges individuals to put away their phones while driving (Myers, 2021). The campaign tries to sensitize individuals about the horrific potential outcomes of using a phone while driving. Such communications are done through a number of television and digital advertisements. AT&T posted a three-minute digital advertisement, *Close to Home* (Diehl, Marín, & Zreiqat, 2020). The advertisement shows different individuals going about their day. One of those individuals is a young mother who is doing daily routine jobs with her daughter securely seat-belted in the back seat. She looks down briefly to check a social media post on her phone, inadvertently crosses the center line, and causes a head-on collision with an oncoming car. The advertisement is graphic and heart-breaking to watch. However, the message is clear. Even a quick glance at your phone while driving can have disastrous consequences. "It can wait", admonishes the advertisement (Wolfsdorf, Scott, & Herzog, 2019).

The *It Can Wait* campaign is extensive and advertisements are only the beginning of the extensive campaign. AT&T also offers a DriveMode app that kicks in automatically when the car is in motion to silence texts and alerts to notify senders that the user is behind the wheel (Training, 2016). *It Can Wait* pledge is promoted by the campaign. The pledge urges individuals to commit to keeping their eyes on the road and not on their phones. Individuals are motivated by the noble cause of the initiative and almost 8 million individuals took the pledge (Bergmark, Gliklich, Guo, & Gliklich, 2016). There is even an *It Can Wait* virtual reality simulator that puts individuals behind the wheel to experience crashes which are caused by smartphones. Other efforts include celebrity tweets and outreach sessions at schools (Quisenberry, 2015).

The campaign has the support of more than dozens of partnering companies and media organizations. It has garnered hundreds of millions of impressions. The campaign seems to be raising awareness among individuals (Lisinska, 2019). An

AT&T marketer says, "We do surveys (that show) we are making a difference. About a third of people say our no-texting-while-driving message has changed their behavior" (Lisinska, 2019).

6.3. The Rainforest Alliance: Follow the Frog

Non-profit organizations focus on social and environmental issues. Such organizations rely heavily on advertising and promotion to market their social and environmental missions (Bernardino, 2021). The Rainforest Alliance is an international non-profit conversation organization. Since 1987, it is promoting the cause of minimizing harmful human impact on the environment. The Rainforest Alliance is authorized to issue the Rainforest Alliance Certified logo to companies who are committed to sustainable sourcing from farms and forests. The logo contains a signature green tree frog symbol which can be displayed on packages and promotional materials (Asbury & Ball, 2016). The organization provides assistance and training to businesses which wish to improve their sustainability efforts.

It is essential to develop awareness of pressing ecological challenges such as deforestation, climate change, and water quality. In 2012, the Rainforest Alliance launched an encompassing and clever promotional campaign called *Follow the Frog* (Asbury & Ball, 2016). The campaign was promoted with a three-minute web video which comically illustrated what people *don't* have to do to help sustain the environment (Asbury & Ball, 2016). A young professional is featured in the video. The professional is willing to go to any ecological extremes for ensuring environmental sustainability. He quits his job, leaves his wife and young child, and heads to the heart of the Amazon jungle. In the jungle, he joins a crusade against evil factions that are destroying the environment (Asbury & Ball, 2016). The video concludes with an assurance that there is an easier way to save the environment. It says, "You don't have to go to the ends of the Earth to protect the planet. Just follow the frog" (Giovannucci, Hagen, & Wozniak, 2014). This can be done by choosing products that display the Rainforest Alliance Certified seal. The video won many awards for this initiative.

The *Follow the Frog* campaign has become popular. It encompasses a wide range of marketing content and activities, from print advertisements, retail promotional materials, *Follow the Frog* events, and a *Frog Blog* to alliances with nearly 6000 companies. All these companies display and promote the Rainforest Alliance Certified seal (Christodoulatos, 2021). The campaign is effective with the video capturing more than 5 million views and 32000 shares on the YouTube. The video was nominated for a Webby award for online video excellence and also won a TED award as the top *Ad Worth Spreading* in the social causes category (Asbury & Ball, 2016). The campaign helped in increasing consumer awareness about environmental sustainability. It boosted the number of companies committing to sourcing from Rainforest Alliance Certified operations, including major brands such as Lipton, McDonald's, and Mars (Asbury & Ball, 2016).

6.4. Always: Like a Girl

Always is a feminine hygiene brand from P&G. Research shows that the confidence of a girl plummets at puberty (Røstvik, 2022). *Always* launched a big-budget advertisement campaign aimed at changing the meaning of doing something *like a girl*. The first *Like a Girl* advertisement used a reality show format. The advertisement asked young women, men, boys, and girls to "show me what it looks like to run like a girl" or "fight like a girl" or "throw like a girl" (Mkamboi, Ndavula, & Mwangi, 2022). The classic *feminine* stereotypes were portrayed by the women, men, and boys. Young girls, however, attacked the tasks with gusto, devoid of any negative connotations. The advertisement asked, "When did doing something *like a girl* become an insult?". The advertisement closed with "Always wants to change that" (Mkamboi et al., 2022). Inspirational images of girls running, throwing, and kicking with skill and confidence were shown in the advertisement. The advertisement concluded, "Let's make *#RunLikeAGirl* mean amazing things" (Rodrigues, 2016).

The *Like a Girl* advertisement was shown during Super Bowl XLIX. The advertisement received spectacular reactions from viewers (Feldheim, 2022). The advertisement placed first or second in virtually every major ad-rating poll. The advertisement also topped all other Super Bowl advertisements by a wide margin in social media mentions and sharing (Feldheim, 2022). *Like a Girl* won the Emmy for the most outstanding commercial. It also won a Grand Prix award for the best PR campaign at the Cannes international advertisement festival. Advertisements of *Like a Girl* received more than 85 million views and more than a billion global impressions in the first year of launch (Feldheim, 2022).

The effort of *Always* in the form of *Like a Girl* has expanded into a full-blown, ongoing educational and support campaign (Um, 2021). "When you have a message that really addresses such an important and a real issue and it's done in a way that is very consistent with who you are as a brand, I think consumers want to engage with that," says P&G's vice president of global feminine care. "We have started an epic battle to show the world that doing things like a girl is downright amazing" (Um, 2021).

6.5. Google: Made with Code

Google launched its *Made with Code* marketing and advertising campaign to encourage young girls to pursue careers in science and technology. The project is worth $50 million (Johnson, 2018). Google found that 74 percent of middle school girls express interest in science, technology, engineering, and mathematics (STEM). However, by high school, only one percent retain their interest and think of planning to major in computer science (Johnson, 2018). The campaign promotes the idea that the things young girls love, from their smartphone apps to fashions to their favorite movies, are *made with code*. The idea is promoted through advertisements, dedicated digital and social media sites, events, and partnerships with not-for-profit organizations. Google says, "Simply put, code is a tool that lets you write your story with technology. Girls start out with a love of science and technology, but lose it somewhere along the way. Let's help encourage that passion" (Johnson, 2018).

7. CONCERNS IN SOCIALLY RESPONSIBLE TARGET MARKETING

Smart targeting allows companies to improve their focus on the segments that they intend to satisfy the best and most profitably. Smart targeting enables companies to be more efficient and effective (Mandal, 2021). Consumers are benefited from targeting because specific groups of consumers are targeted by companies with offers which are carefully prepared as per their requirements and preferences (Ferrell, Hartline, & Hochstein, 2021). However, target marketing may sometimes result in controversy and concern. A major issue related to target marketing is the targeting of vulnerable or disadvantaged customers with controversial or potentially harmful products (Laczniak & Shultz, 2021).

7.1. Target Marketing of Vulnerable Segments

Companies aim to target vulnerable segments like kids, children, and senior citizens to influence them and to generate higher revenues and profits (Mandal, 2021). Fast-food chains have generated controversary over the years with their attempts to target inner-city minority consumers. Critics accuse such chains of pitching their high-fat, salt-laden fare to low-income, urban residents who are much more likely than suburbanites to be heavy customers (Mokhtar, Javeed, & Khan, 2021). Individuals in poor urban areas are targeted by big banks and mortgage leaders. They are targeted with attractive adjustable-rate home mortgages that they cannot really afford (Mandal, 2021).

Kids and children are considered as vulnerable audience in particular. Companies from a wide range of industries viz. cereals, soft drinks, fast food, toys, and fashion target their marketing efforts towards children (Mandal, 2021). Such companies are criticized because of such activities. Critics worry that enticing premium offers and high-powered advertising appeals will overwhelm the defenses of children. For example, McDonald's has been criticized by various health advocates and parent groups for its initiatives related to its popular *Happy Meals* offers (Singireddy, 2020). The offers feature trinkets and other items tied in with popular children's movies and television shows. Such offers create a powerful and irresistible connection between children and less-healthy eating. McDonald's has responded by putting the *Happy Meals* on a diet. The overall calorie count is reduced by 20 percent and fruits are added to every meal. *Happy Meals* are promoted only with milk, water, and juice. For a two-week span each year, McDonald's replaces the toys in its *Happy Meals* with children's books (Singireddy, 2020).

7.2. Target Marketing in the Digital Era

The digital era has changed the manner in which companies do target marketing (Purnomo, 2021). Children and kids are more vulnerable to targeted marketing messages in the digital era. Parents can easily detect and control the traditional child-directed television and print advertisements which contain fairly obvious pitches (Fatma, 2021). However, marketing in digital media may be subtly embedded within the content and viewed by children on personal, small-screen devices. It is difficult for parents to keep track of such activities of kids (Fatma, 2021). In digital platforms, the lines between educational, entertainment, and commercial content are often blurred. At present, kids and children are the major consumers of online and digital content. One expert advises that kids "shouldn't be entirely left to their own devices" (Carvalho & Eler, 2017).

Concerns about potential targeting abuses increase with the growth of the internet, smartphones, and other carefully targeted direct media (Mandal, 2021). Communications on the internet and mobile phones allow companies to do precise targeting, letting the makers of questionable products or deceptive advertisers zero in on the most vulnerable audiences (Laczniak & Shultz, 2021). Unscrupulous marketers can prepare tailor-made and deceptive messages which can be communicated by email directly to millions of unsuspecting customers. For example, more than 269000 complaints were filed in a year in the Federal Bureau of Investigation's Internet Crime Complaint Center website (Campina & Rodrigues, 2022).

Digital movements of individuals can be tracked by companies with the help of advanced and sophisticated analytical techniques. Tracking and analysis through such techniques allow companies to build detailed and accurate customer profiles containing highly sensitive personal information (Hauser, Flath, & Thiesse, 2021). Companies may use such profiles to hypertarget individual customers with personalized brand messages and offers. However, in such targeting, companies carry the risk of stalking customers and need to fix the line between serving customers better and stalking them (Hauser et al., 2021).

In the digital era with technological developments, smartphones and laptops know about individuals much more than anyone else does. In truth, digital devices may know about individuals much more than the individuals themselves. Smartphones and other digital equipment have become fundamental extensions and integral parts of the lives of human beings. Whatever individuals do – at work, at play, socializing, shopping – phone, tablet, laptop, or desktop of individuals are integrally related to all actions. These devices are carried by individuals everywhere, entertain them, connect them with friends, help in browsing and shopping, and feed individuals with news and information. Such devices can also keep track of intimate voice, text, and email conversations of individuals. Simultaneously, all such information may be shared instantly with marketers. The information may be sensitive and companies have developed sophisticated new ways that border on wizardry to extract intimate insights about customers. Such information is important for marketers and brands because of the richness and depth of the insights gained.

8. ROLES OF COMPANIES IN SOCIALLY RESPONSIBLE TARGET MARKETING

Companies and marketers have important roles and responsibilities in target marketing. Such target marketing should be socially responsible. Critics worry that sometimes the target marketing done by companies is not socially responsible (Mandal, 2021). Marketers are of the opinion that using such personal and sensitive information, if analyzed properly, serves the purposes of both customers and of companies. Customers receive tailored and relevant information and offers from brands that really understand and interest them (Yang, Kim, & Yoo, 2013). However, critics and privacy advocates worry that such intimate information in the hands of unscrupulous marketers could result in more harm than benefit to customers. Critics feel that companies employ *big data* and hypertargeting less as "getting to know customers better to serve them better" and more as *stalking* customers and *profiling* them (Zhang, Qu, Jie, & Liang, 2022). Companies should realize and understand that customers worry about the privacy and security of the data shared. Customers are willing to share some personal information if it means getting better service or deals. However, companies should understand that many customers worry that marketers might go too far.

Companies should understand that, in target marketing, the issue is not really *who* is targeted but rather *how* and for *what*. Controversies arise when marketers attempt to profit at the expense of targeted segments. Companies should not unfairly target vulnerable segments or target them with questionable products or tactics. Companies should understand that socially responsible target marketing calls for segmentation and targeting that serve not just the interests of the company but also the interests of the individuals targeted and of the society at large.

9. DISCUSSION

Companies should be aware about their responsibilities not only towards their customers but also towards the society at large. Companies should do marketing communications and target marketing which are socially responsible. Individuals at all levels of the organization must be aware of the many legal and ethical issues surrounding marketing communications. Companies must work hard and proactively at communicating openly, honestly, and agreeably with their customers, resellers, and the society at large. They should be aware about their roles and responsibilities in target marketing. Companies should be socially responsible when employing marketing communication channels like advertising, sales promotion, and personal selling. They adopt a number of initiatives in marketing communications to fulfill their responsibilities towards building, developing, and maintaining a sustainable environment. There are concerns in socially responsible target marketing. Concerns relate to targeting of vulnerable segments viz. kids, children, and senior citizens and targeting of individuals in the digital era. Companies should be aware about their responsibilities in target marketing. All these aspects are discussed in the study. The study has both theoretical and managerial implications.

9.1. Theoretical Implications

Academicians may study the various aspects of marketing communications and target marketing done by companies. They may investigate whether marketing communications and target marketing done by companies are socially responsible. They may study and analyze the roles and responsibilities of companies in marketing communications and target marketing. They may also study and analyze the various strategies and initiatives adopted by companies. They may analyze the legal and ethical issues related to marketing communications and target marketing done by companies. They may analyze whether companies are trying to satisfy the triple bottom line – people, profit, and planet. Based on the analysis, academicians may suggest corrective actions which will allow companies to do socially responsible marketing communications and target marketing effectively.

9.2. Managerial Implications

Companies and businesses should realize and understand the importance of environmental sustainability. They should understand that the strategies and initiatives adopted by them should be done in a socially responsible manner. Companies should not only think of generating revenues and profits for themselves but should also be sensitive about the welfare of customers and the welfare of the society at large. They should do marketing communications and target marketing which are socially responsible. Companies and practicing managers require understanding their roles and responsibilities in marketing communications and target marketing. In the digital age, individuals are worried and concerned about the privacy and security of data shared by them with companies, more than ever. Companies should understand such concerns, ensure the privacy and security of data shared by individuals, and should assure individuals that the data are in safe hands.

10. CONCLUSION

The study discussed about the various aspects of marketing communications and target marketing done by companies. It is imperative for companies to be socially responsible in the marketing communications and target marketing done by them. They should understand their roles and responsibilities in marketing communications and target marketing. Critics, individuals, and the society at large have concerns about target marketing. Companies should address such concerns. They should adopt such strategies and initiatives which reveal their willingness and sincerity in doing socially responsible marketing communications and target marketing.

10.1. Limitations

The study conducted a conceptual analysis of the various aspects of socially responsible marketing communications and socially responsible target marketing done by companies and businesses. Primary data is not collected and empirical analysis is not done. Also, the analysis is done mainly with a focus on the markets in the United States.

10.2. Avenues of Future Research

Socially responsible marketing communications and socially responsible target marketing are imperative for companies and businesses at present. The various activities of companies and businesses concern not only the customers but also the society at large. Researchers may collect primary data and conduct empirical analysis to suggest strategies and initiatives to be adopted by companies and businesses to ensure environmental sustainability. Individuals are worried and concerned about the privacy and security of the data shared by them with companies. Research should be done to ensure that the shared data is safe and secure. Researchers may conduct empirical analysis for markets other than that of the United States and suggest improvements in the strategies and initiatives adopted by companies. Researchers may suggest a business environment which ensures the triple bottom line – people, profit, and planet. This will ensure that the ecology is sustained.

REFERENCES

Aaker, D. (2018). *Creating signature stories: Strategic messaging that persuades, energizes and inspires*. Morgan James Publishing.

Adnan, M., Khan, A. U., & Hayee, R. (2021). Integrated Marketing Communication and Customer Satisfaction as a function of Customer Loyalty with the Moderating role of Product Innovation Moderation. *Sustainable Business and Society in Emerging Economies*, 3(2), 133–148. DOI: 10.26710/sbsee.v3i2.1910

Ahearne, M., Atefi, Y., Lam, S. K., & Pourmasoudi, M. (2022). The future of buyer–seller interactions: A conceptual framework and research agenda. *Journal of the Academy of Marketing Science*, 50(1), 22–45. DOI: 10.1007/s11747-021-00803-0 PMID: 34608341

Akena, F. A. (2022). The "Miss Curvy Uganda" pageant: Representation, commodification and exploitation of women's bodies. *Social Semiotics*, 32(2), 224–239. DOI: 10.1080/10350330.2020.1743524

Allen, E. (2022). Jess Weiner and the Dove Real Beauty Campaign: Selling Feminism for Profit or Social Change? *Women Leading Change: Case Studies on Women, Gender, and Feminism*, 6(1), 18–37.

Antoniuk, K., Mokiy, A., Pikh, M., Bukharina, L., & Antoniuk, D. (2021). Ensuring of Consumer Choice Ecologization on the Basis of Consumption Safety: International Experience for Ukraine. In *SHS Web of Conferences* (Vol. 100, p. 05009). EDP Sciences.

Asbury, S., & Ball, R. (2016). *The practical guide to corporate social responsibility: do the right thing*. Routledge. DOI: 10.4324/9781315697659

Barter, A. (2019). *Shifting environmental attitudes and influencing pro-environmental behaviour through the process of managing how, when and in what way we communicate* (Doctoral dissertation, Stellenbosch: Stellenbosch University).

Bendell, B. L., & Huvaj, M. N. (2020). Does stakeholder engagement through corporate social and environmental behaviors affect innovation? *Journal of Business Research*, 119, 685–696. DOI: 10.1016/j.jbusres.2018.10.028

Bergmark, R. W., Gliklich, E., Guo, R., & Gliklich, R. E. (2016). Texting while driving: The development and validation of the distracted driving survey and risk score among young adults. *Injury Epidemiology*, 3(1), 1–10. DOI: 10.1186/s40621-016-0073-8 PMID: 27747544

Bernardino, P. (2021). Engaging employees through corporate social responsibility programs: Aligning corporate social responsibility and employee engagement. *Journal of Organizational Psychology*, 21(1), 105–113.

Bhattacharyya, J. (2022). The structure of sustainability marketing research: a bibliometric review and directions for future research. *Asia-Pacific Journal of Business Administration*, (ahead-of-print).

Camilleri, M. A. (2022). Strategic attributions of corporate social responsibility and environmental management: The business case for doing well by doing good! *Sustainable Development (Bradford)*, 30(3), 409–422. DOI: 10.1002/sd.2256

Campina, A., & Rodrigues, C. (2022, February). Cybercrime and the Council of Europe Budapest Convention: prevention, criminalization, and International Cooperation. In *The Book of Full Papers-7th International Zeugma Conference on Scientific Researches* (Vol. 1, No. 1, pp. 112-123). IKSAD.

Carvalho, L. G. D., & Eler, M. M. (2017, April). Security requirements and tests for smart toys. In *International Conference on Enterprise Information Systems* (pp. 291-312). Springer, Cham.

Christodoulatos, M. L. (2021). *Ethical chocolate consumption by millennials in the Netherlands* (Doctoral dissertation).

Coffie, S., Anning-Dorson, T., Hinson, R. E., Bosah, G., & Anani-Bossman, A. (2022). Marketing Communications in Emerging Economies: Conclusions and Recommendations. In *Marketing Communications in Emerging Economies, Volume II* (pp. 277-296). Palgrave Macmillan, Cham.

Danthinne, E. S., Giorgianni, F. E., Austin, S. B., & Rodgers, R. F. (2022). Beauty and the big business: A qualitative investigation of the development and implementation of the CVS Beauty Mark. *Body Image*, 41, 74–83. DOI: 10.1016/j.bodyim.2022.01.012 PMID: 35240392

Demirel, D. (2022). The effect of service quality on customer satisfaction in digital age: Customer satisfaction based examination of digital CRM. *Journal of Business Economics and Management*, 23(3), 507–531. DOI: 10.3846/jbem.2022.15328

Diehl, F. S., Marín, L. S., & Zreiqat, M. (2020). An evaluation of AT&T social marketing delivery modes aimed at teen smartphone use while driving. *Journal of Cultural Marketing Strategy*, 5(1), 60–75. DOI: 10.69554/XBTL7959

Dua, T. (2019, February 4). Here's the Full List of All the Super Bowl Commercials That Ran This Year. *Business Insider*, www.businessinsider.com/super-bowl-commercials-2019-list-2019-1 (Accessed on 4th February, 2019)

Emamalizadeh, H. (2015). Bait-And-Switch Advertising: A Conceptual Approach. In *Proceedings of the 1983 Academy of Marketing Science (AMS) Annual Conference* (pp. 458-460). Springer, Cham. DOI: 10.1007/978-3-319-16937-8_109

Fatma, S. (2021). Use and Misuse of Technology in Marketing: Cases from India. In *Research Anthology on Strategies for Using Social Media as a Service and Tool in Business* (pp. 106-116). IGI Global.

Feldheim, D. (2022). *Dare to Lead Like a Girl: How to Survive and Thrive in the Corporate Jungle*. Rowman & Littlefield. DOI: 10.5771/9781538163535

Ferrell, O. C., Hartline, M., & Hochstein, B. W. (2021). *Marketing strategy*. Cengage Learning.

Foltean, F. S., & van Bruggen, G. H. (2022). Digital Technologies, Marketing Agility, and Marketing Management Support Systems: How to Remain Competitive in Changing Markets. In *Organizational Innovation in the Digital Age* (pp. 1–38). Springer. DOI: 10.1007/978-3-030-98183-9_1

Giovannucci, D., Hagen, O. V., & Wozniak, J. (2014). Corporate social responsibility and the role of voluntary sustainability standards. In *Voluntary Standard Systems* (pp. 359–384). Springer. DOI: 10.1007/978-3-642-35716-9_24

Haarpaintner, É. (2022). What is the relationship between sustainability, corporate social responsibility and quality of life applications at work?. *Worldwide Hospitality and Tourism Themes*, (ahead-of-print).

Halvorsen, S. V. (2021). *Assessing and explaining the proficiency of social media marketing in small ministries: a comparative case study* (Master's thesis).

Hauser, M., Flath, C. M., & Thiesse, F. (2021). Catch me if you scan: Data-driven prescriptive modeling for smart store environments. *European Journal of Operational Research*, 294(3), 860–873. DOI: 10.1016/j.ejor.2020.12.047

Hayes, J. L., & Duff, B. R. (2022). Introducing a special issue on advertising & corporate social responsibility. *International Journal of Advertising*, 41(2), 1–5. DOI: 10.1080/02650487.2022.2056357

Jackson, N. (2013). *Promoting and marketing events: Theory and practice*. Routledge. DOI: 10.4324/9780203583166

Jain, D. (2018). *The journey of a southern feminist*. SAGE Publishing India. DOI: 10.4135/9789353280710

Jia, M., & Wan, C. (2021). Why are strangers trusted more during trade fairs? A literature review on the conceptual model of general trust formation. *Journal of Business and Industrial Marketing*.

Jin, L., & He, Y. (2018). How the frequency and amount of corporate donations affect consumer perception and behavioral responses. *Journal of the Academy of Marketing Science*, 46(6), 1072–1088. DOI: 10.1007/s11747-018-0584-7

Johnson, T. (2018, December 7). Google Event Encourages Girls to Look at Computer Science Careers. *Daily Nonpareil*, www.nonpareilonline.com/news/education/google-eventencourages-girls-to-look-at-computer-science-careers/article_2a08d2dd-b48d-54cf-8c41-c5c91a8ade7e.html (Accessed on 7th December, 2018)

Kitchen, P. J., & Tourky, M. E. (2022). Integrated Communication or Integrated Marketing Communication. In *Integrated Marketing Communications* (pp. 69–87). Palgrave Macmillan. DOI: 10.1007/978-3-030-76416-6_5

Laczniak, G., & Shultz, C. (2021). Toward a doctrine of socially responsible marketing (SRM): A macro and normative-ethical perspective. *Journal of Macromarketing*, 41(2), 201–231. DOI: 10.1177/0276146720963682

Letunovska, N. Y. (2020). *Challenges for the health components of a region in the context of globalization* (Doctoral dissertation, Sumy State University).

Lisinska, A. (2019). Do Initiators Matter in Social Marketing? Understanding Characteristics and Particularities of Social Marketing Campaigns Initiated by Different Agents. *SPNHA Review*, 15(1), 7.

Litwin, M. L. (2007). *The ABCs of strategic communication: Thousands of terms, tips and techniques*. AuthorHouse.

Maccarrone, P., & Contri, A. M. (2021). Integrating Corporate Social Responsibility into Corporate Strategy: The Role of Formal Tools. *Sustainability (Basel)*, 13(22), 12551. DOI: 10.3390/su132212551

Maier, A. L. (2021). Political corporate social responsibility in authoritarian contexts. *Journal of International Business Policy*, 4(4), 476–495. DOI: 10.1057/s42214-020-00085-3

Mandal, P. C. (2019). Social criticisms of marketing: Concerns and initiatives. *International Journal of Society Systems Science*, 11(2), 157–170. DOI: 10.1504/IJSSS.2019.100088

Mandal, P. C. (2021). Socially-Responsible Marketing Communications: Concerns, Strategies, and Initiatives. [JMME]. *Journal of Media Management and Entrepreneurship*, 3(1), 1–17. DOI: 10.4018/JMME.290301

Mandal, P. C. (2022). Marketing Information and Marketing Intelligence: Generation of Customer Insights. [IJTD]. *International Journal of Technology Diffusion*, 13(1), 1–14. DOI: 10.4018/IJTD.300745

Millimen, S. K. (2015). *All made-up: the hyperfeminization of fat women*. The University of Toledo.

Mkamboi, A., Ndavula, J., & Mwangi, S. W. (2022). Utilizing Social Media Platforms for Online Crisis Management: Focus on Always Pads Crisis of 2019. *African Multidisciplinary Journal of Research*, 7(1), 83–94.

Mokhtar, S. S. M., Javeed, A., & Khan, M. Y. (2021). *Halal Marketing: Concept and Strategies (UUM Press)*. UUM Press.

Myers, C. (2021). *The Effects of Pedestrian Gestures on Motorist Yielding at Two-Lane Crosswalks* (Doctoral dissertation, University of Kansas).

Nur, F., Akmaliah, N., Chairul, R., & Safira, S. (2021). Green purchase intention: The power of success in green marketing promotion. *Management Science Letters*, 11(5), 1607–1620.

Ooi, C. S. (2022). Sustainable Tourism and the Moral Limits of the Market: Can Asia Offer Better Alternatives. In *Asian Tourism Sustainability* (pp. 177–197). Springer. DOI: 10.1007/978-981-16-5264-6_10

Purnomo, H. (2021). Software trends in improving local tourism product marketing in Indonesia: A systematic review. *International Journal of Business. Economics and Management*, 4(1), 77–84.

Quisenberry, P. N. (2015). Texting and driving: Can it be explained by the general theory of crime? *American Journal of Criminal Justice*, 40(2), 303–316. DOI: 10.1007/s12103-014-9249-3

Rodrigues, R. A. (2016). *Femvertising: empowering women through the hashtag? A comparative analysis of consumers' reaction to feminist advertising on twitter* (Doctoral dissertation, Universidade de Lisboa (Portugal)).

Ronald, L. W. (2008). *Kleppner's advertising procedure*. Pearson Education India.

Røstvik, C. M. (2022). *Cash Flow: The businesses of menstruation*. UCL Press. DOI: 10.14324/111.9781787355385

Sánchez Abril, P., Oliva Blázquez, F., & Martínez Evora, J. (2018). The Right of Withdrawal in Consumer Contracts: A comparative analysis of American and European law. *InDret*, 3, 18–13.

Shim, J., Moon, J., Lee, W. S., & Chung, N. (2021). The impact of CSR on corporate value of restaurant businesses using triple bottom line theory. *Sustainability (Basel)*, 13(4), 2131. DOI: 10.3390/su13042131

Singh, S. K., & Sanjeev, R. (2022). Antecedents of Green Marketing Initiatives: A Hierarchical Regression Model. [IJSESD]. *International Journal of Social Ecology and Sustainable Development*, 13(1), 1–15. DOI: 10.4018/IJSESD.315310

Singireddy, M. (2020). Mcdonald's: Global Marketing. *International Journal of Health and Economic Development*, 6(2), 16–27.

Tasmin, M., Ahmed, N., & Motahar, T. (2019, December). Gender Disparity in Computer Science Education in Bangladesh: A Study of Women's Participation in Computer Science. In *2019 IEEE International Conference on Engineering, Technology and Education (TALE)* (pp. 1-7). IEEE. DOI: 10.1109/TALE48000.2019.9225981

Tian, T., Zhang, Y., & Mei, Y. (2022). Intelligent analysis of precision marketing of green agricultural products based on big data and GIS. *Earth Science Informatics*, 15(3), 1–12. DOI: 10.1007/s12145-022-00789-5

Training, R. (2016)... *University Transportation Center to Support DOTD Research Needs*, 31(1), 1–5.

Trihas, N., Perakakis, E., Venitourakis, M., Mastorakis, G., & Kopanakis, I. (2013). Social media as a marketing tool for tourism destinations: The case of Greek municipalities. *Journal of Marketing Vistas*, 3(2), 38–48.

Turnbull, P. R., Khanal, S., & Dakin, S. C. (2021). The effect of cellphone position on driving and gaze behaviour. *Scientific Reports*, 11(1), 1–10. DOI: 10.1038/s41598-021-87120-2 PMID: 33833300

Um, N. (2021). Antecedents and consequences of attitude toward femvertising. *The Journal of the Korea Contents Association*, 21(1), 66–74.

Wagner, D. A. (2014). *The Marketing of Global Warming: A Repeated Measures Examination of the Effects of Cognitive Dissonance, Endorsement, and Information on Beliefs in a Social Cause* (Doctoral dissertation, Trident University International).

Wolfsdorf, A., Scott, A., & Herzog, S. (2019). What happened on July 21st: An investigation of trauma and trigger warnings in the English classroom. *Changing English*, 26(2), 198–215. DOI: 10.1080/1358684X.2018.1552514

Yang, B., Kim, Y., & Yoo, C. (2013). The integrated mobile advertising model: The effects of technology-and emotion-based evaluations. *Journal of Business Research*, 66(9), 1345–1352. DOI: 10.1016/j.jbusres.2012.02.035

Zhang, L., Qu, J., Jie, J., & Liang, N. (2022). Big Data in the Internet Industry and Other Emerging Industries. In *China's Publishing Industry in the Era of Big Data* (pp. 27–45). Springer. DOI: 10.1007/978-981-19-0428-8_2

Chapter 15
Looking Back to 1850 in 2025: Historascan to Digitize Historical Journals

Bruno Frutuoso Costa
https://orcid.org/0000-0003-3023-8960
ISCTE, University Institute of Lisbon, Portugal

Bruno Contreiras Mateus
https://orcid.org/0009-0007-0458-8947
IADE, European University of Lisbon, Portugal & ISCTE, University Institute of Lisbon, Portugal

Hugo José Pinto
Inovaworks Command and Control S.A., Portugal

Mohammad Reza Tabrizi
Inovaworks Command and Control S.A., Portugal

ABSTRACT

This chapter analyses current technologies and the challenges involved in extracting and classifying articles and news headlines from historical journals, as well as converting images to text format. The work to develop a tool focused on digitising historical journals was carried out by a multidisciplinary team of experts in media studies, artificial intelligence, image processing, and cultural heritage preservation. The data used derives from two historic Portuguese journals, Diário de Notícias and Jornal de Notícias, which were created in the mid-19th century. This project is based on a mixture of heuristics, computer vision, pattern recognition, and other artificial intelligence and machine learning techniques. The main challenges included

DOI: 10.4018/979-8-3693-3579-6.ch015

the variability in the design of historical journals, preserving the quality of images over time, and continuously improving image processing and OCR techniques to adapt to different styles and periods of newspapers.

INTRODUCTION

The Fourth Industrial Revolution and digital transformation are expressions used to designate the paradigm in which cutting-edge technologies are used in automation and data exchange to help increase the quality, efficiency, and productivity of companies' production processes (Gilchrist, 2016; Skilton & Hovsepian, 2018). However, Industry 4.0 is not limited to the digitization of equipment, procedures, data collection, and IoT networks (Barros et al., 2023a, 2023b). Based on the principles of interconnectivity, information transparency, decentralized decisions, and technical assistance (Hermann et al., 2016), technological developments have made scalable cloud and service solutions available, such as software as a service (SaaS), platform as a service (PaaS), and hybrid cloud. Their combination with cyber-physical systems, algorithms (which integrate implicit business logic), and cybersecurity mechanisms in a mixed infrastructure allocated to public and private datacentres allows for the development of integrated information management solutions. In a complementary sense, artificial intelligence provides these infrastructures with the ability to learn and produce precise, complex, and sophisticated forecasts that support decision-making and the definition of new business directions (Iansiti & Lakhani, 2020; Skilton & Hovsepian, 2018).

The false belief that Industry 4.0 is synonymous with complex and expensive tools only available to multinational companies has prevented many companies from investing in the digitization of key operations in their value chain and in research and development (R&D) capabilities for value-added products and services to respond to market challenges and opportunities (Gilchrist, 2016). In the creative industries, innovation emerged in the UK at the end of the 20th century, encompassing sectors that combine creativity and communication technologies. These include news media, performing arts, visual arts, music, film and video, design, advertising, and leisure software, to name a few (Granado et al., 2020). The main objective was to create economic and social value for companies. At the beginning of the 21st century, innovation was closely linked to the emergence of the Internet (Bonixe, 2020).

Based on the assumption that digital transformation demands speed, breadth, and depth of change from companies (Schwab, 2017; Takeuchi & Nonoka, 2008/2004), the automotive, technological, and biological industries took on the capacity to innovate and lead this paradigm shift early on (Coelho, 2016). In journalism, companies have been forced to migrate to digital platforms at the risk of jeopardizing their revenues, causing various challenges in terms of routines, languages, ethics, and training for professionals (Bonixe, 2020). Innovation in the media industry varies significantly between countries and companies, being influenced by financial factors, organizational culture, and government regulation (Costa & Mateus, in press; de-Lima-Santos & Ceron, 2021; Meier et al., 2022; Pérez-Seijo & Vicente, 2022). As a rule, the adaptation of the business model of newspaper companies occurs slowly and with few forms of innovation (Cardoso et al., 2019; Crespo et al., 2018).

Among the most common innovations are the use of digital platforms and social networks (Meier et al., 2022), the creation of new multimedia formats (Bonixe, 2020; Gehlen & Sousa, 2018; Granado et al., 2020; Silva & Granado, 2021; Vicente & Pérez-Seijo, 2022), the development of immersive apps (Fante, 2018; García-Avilés et al., 2019), the implementation of data analysis tools (de-Lima-Santos & Ceron, 2021), and the adoption of digital business models (Cardoso et al., 2019). In the area of artificial intelligence, data analysis to identify trends and patterns, automatic fact-checking, and the creation of subtitles and automatic transcripts for videos are little used in the news industry (de-Lima-Santos & Ceron, 2021), except for countries such as the USA and China. In these two countries, artificial intelligence has been used to automate repetitive tasks, such as report writing, and to personalize relevant content for news websites. These actions are supported by large data sets that reflect users' interests and behaviours in the digital environment and with companies (de-Lima-Santos & Ceron, 2021).

Innovation and its integration are a complex process that requires a combination of resources, collaborative leadership, and a predisposed organizational culture (García-Avilés et al., 2019; Meier et al., 2022; Pérez-Seijo & Vicente, 2022). Many organizations have limited resources to invest in expensive innovations, which can hinder the adoption of new technologies and infrastructures (García-Avilés et al., 2019; Meier et al., 2022; Pérez-Seijo & Vicente, 2022). Journalists and other professionals may be more resistant to using new technologies because they feel they detract from their core mission of reporting events or because of the perception that they could threaten their work, but also because they lack the technical skills to implement them effectively (Costa & Mateus, in press; García-Avilés et al., 2019; Meier et al., 2022; Pérez-Seijo & Vicente, 2022). Meier et al. (2022) point out that sustainable business models are scarce. Several innovations have proven not to be economically viable in the long term, which can make their adoption difficult and companies fearful of them.

In this area, the automatic digitization of historical newspapers (and other similar publications) has been the subject of research due to the enormous amount of information they contain, and the challenges associated with the variability in the presentation of documents over time, such as the configuration and pagination of the support, as well as the language used, which have evolved over relevant periods (Kumpulainen & Late, 2022; Lemmers et al., 2023; Rhyno, 2017; Ringel, 2023).

The aim of this chapter is to analyse current technologies and the challenges involved in extracting and classifying articles and news headlines from historical newspapers and converting historical newspaper images to text format, in order to present the work carried out to develop a tool focused on digitizing historical newspapers: The HistoraScan.

BACKGROUND

The automatic digitization of historical journals represents a growing field of research of great importance for the preservation and accessibility of cultural heritage (Allen & Hall, 2010; Powell & Paynter, 2009; Skelbye & Dannells, 2021). This area of study focuses on the development of technologies and methods to convert physical collections of old journals into digital formats, thus enabling easier consultation and more efficient preservation of the documents (Kettunen et al., 2020; Koistinen et al., 2017). The aim is not only to create digital copies but also to extract textual and visual information in an accurate and useful way, facilitating research and public access to these valuable historical sources (Ehrmannm et al., 2022).

In recent years, several approaches have been proposed to address the technical challenges of automatically digitising historical journals. Work such as that by Rhyno (2017) highlights the use of advanced optical character recognition (OCR) techniques to deal with the typographic variability and imperfections of old documents. In addition, projects such as the Library of Congress' "Newspaper Navigator"[1] use convolutional neural networks (CNNs) to improve the accuracy of extracting images and text from digitised journals, highlighting the importance of deep learning techniques in this area.

The integration of artificial intelligence (AI) tools is beginning to be studied because of its potential to revolutionise the digitization of historical journals (Spina, 2023; Teel, 2024; Ferro, et al., 2023), with particular attention to OCR error correction and document segmentation (Dhali, 2024). Interdisciplinary collaboration between computer scientists, historians and archivists has been fundamental to progress in this area (Thomas et al., 2024). Collaborative projects such as 'READ'[2] (Recognition and Enrichment of Archival Documents) exemplify how the application of pattern recognition and AI technologies can be enhanced by historical and archival expertise to develop more robust solutions adapted to the specific needs of historical archives. These initiatives not only improve the accuracy of digitisation, but also promote the democratisation of access to historical heritage, allowing a wider audience to explore and value these rich sources of knowledge.

Automatic Digitization of Historical Newspapers

The automatic digitization of historical newspapers refers to the automated process of converting physical historical newspapers into a digital format readable by computers (Allen et al., 2008). This process is essential for preserving and making accessible the information contained in old newspapers, which are valuable resources for historical and cultural research (Bunout et al., 2023). It facilitates access to and exploration of vast collections of historical newspapers, allowing researchers to conduct analyses and studies in a more efficient and comprehensive manner. Additionally, automatic digitization helps preserve the information contained in these documents, protecting it from potential physical deterioration or loss over time

The automatic digitization of historical journals has several applications that cover multiple areas of study and interest. As outlined in Table 1, these applications include research in journalism and media studies, where academics can analyse digitised journals to study the evolution of journalistic practices, media representations, and communication trends over time. In the field of historical research, historians and researchers have the possibility of accessing digitised journals to study past events, trends, and cultural phenomena, allowing specific searches by keywords, topics, or dates. In the area of genealogy, individuals can use digitised journals to trace family history, finding information such as birth announcements, obituaries, wedding notices, and other family events. Cultural heritage institutions such as libraries and archives can use digitization to preserve and make accessible their collections, ensuring that valuable cultural materials are protected for future generations and shared with a wider audience. Furthermore, in the field of education, teachers and students can use digitised historical journals as educational resources, analysing primary sources, examining historical perspectives, and participating in activities that promote critical thinking and research skills.

Table 1. Applications of automatic digitization of historical newspapers

Areas	Examples
Journalism and media studies	Media scholars can analyse digitized newspapers to study the evolution of journalism practices, media representations, and communication trends over time. They can explore changes in reporting styles, editorial policies, and societal attitudes reflected in historical news coverage.
Historical research	Scholars and historians can access digitized historical newspapers to study past events, trends, and a cultural phenomenon. They can search for specific keywords, topics, or dates to gather information and insights.
Genealogy	Individuals interested in tracing their family history can use digitized newspapers to find information about their ancestors, such as birth announcements, obituaries, marriage notices, and other family-related events.

continued on following page

Table 1. Continued

Areas	Examples
Cultural heritage preservation	Libraries, archives, and cultural institutions can digitize historical newspapers to preserve and make accessible their collections. This ensures that valuable cultural heritage materials are safeguarded for future generations and can be shared with a wider audience.
Education	Teachers and students can use digitized historical newspapers as educational resources. They can analyse primary sources, examine historical perspectives, and engage in activities that promote critical thinking and research skills.

Source: Own elaboration

The process involves the use of technologies such as Optical Character Recognition (OCR), image processing, artificial intelligence, and machine learning (Kettunen et al., 2022; Fizaine et al., 2024; Valente et al., 2023). They allow to perform tasks such as converting images of newspaper pages into editable digital text, segmenting content (headlines, articles, images, among others), improving image quality, and organizing digitized data into a suitable format.

The automatic digitization of historical newspapers has been a subject of research for years due to the vast amount of information they contain, and the challenges associated with the variability in document presentation over time (Kumpulainen & Late, 2022; Lemmers et al. 2023; Rhyno, 2017; Ringel, 2023). The challenges are essentially associated with document structure analysis, segmentation and labelling, image processing and OCR, Convolutional Neural Networks (CNN) and Natural Language Processing (NLP), data storage and format, and integration of headlines and article.

A document image is made up of a variety of physical regions. Examples: blocks of text, lines, words, figures, tables, and backgrounds. According to Namboodiri and Jain (2007), the process of analysing the structure and layout of a document aims to break down a given document image into its regions and components, understanding their functional roles and relationships. If we consider that in historical journals, the main elements are sentences in columns, headings, titles, and captions, as well as images and author names, the main challenge is to assign functional labels to these elements (Girdhar et al., 2024).

Segmenting historical journals into different classes (headlines, articles, images, and advertising, among others) involves appropriate image segmentation and labelling techniques. This requires using computer vision to improve image quality and segmentation for enhanced OCR processing. The last one is used to convert text images into editable digital text (Palfray et al., 2012).

CNN is a multi-layer neural network method that learns the hierarchical characteristics of data and is commonly used for image classification tasks in this context (Wang & Gang, 2018). In recent years, CNN has evolved rapidly in the design and calculation of NLP. In turn, NLP can be used to improve OCR accuracy and help extract semantic information from journalistic texts. It allows computers to work on word segmentation (WB), information extraction (IE), relation extraction (RE), named entity recognition (NER), part-of-speech tagging (POS), word sense disambiguation (WSD), text-to-speech (TTS), machine translation (MT), among others (Wang & Gang, 2018).

As the amount of material to be processed is vast and breaks down into various levels of detail, the metadata needs to be reliable to facilitate automatic text processing. It is responsible for guiding OCR, linguistic processing and search (Allen & Schalow, 1999). When storing and formatting data, it is important to organise the digitised data in an appropriate format, dividing the data set into training, test and validation sets and adjusting the model's hyperparameters to obtain the best results. In the case of metadata, these provide reference points that determine the performance of programmes. However, it can also later be used by end users as a guide for searching and interacting with collections of historical journals.

Grouping headlines with their respective articles is a crucial step in facilitating access to information and improving the usability of digitised data (Liebl & Burghard, 2020; Schultze et al., 2024). The quality of the machine-readable text on a journal page can be improved when individual blocks of text are identified as such before OCR is carried out (Klijn, 2008). Segmentation techniques are generally a semi-automatic process, so manual checking of the results is often necessary (Broersma & Harbers, 2018). For example, articles whose text appears on different pages of the journal may need to be followed up on manually. However, this is a time-consuming task, so the solution is to improve the consistency of the journal's layout. The more predictable the layout, the better the automated segmentation (Broersma & Harbers, 2018; Klijn, 2008).

All over the world, libraries offer free access to digitised historical journals through user interfaces (Pestana, 2021). In the initial phase, only features such as search and filter options were made available. As users increasingly desire more options and more advanced tools, another challenge emerges, which concerns the development of simple, usable interfaces for different user groups (Pfanzelter et al., 2021).

In this area, companies specialized in the digitization of historical documents can be direct competitors. These companies offer services to convert physical documents into digital formats, including historical newspapers (Ehrmann et al., 2019). Companies focused on OCR and image processing may be indirect competitors. They provide solutions to convert image text into digital text and may have features that address the specific challenges of digitizing historical newspapers (Rhyno, 2017). On the other hand, some software providers offering document management solutions may be competitors, as some of them may have features related to the organization and classification of historical data (Black, 2023; Schlukbier, 2008).

MAIN FOCUS OF THE CHAPTER

Sample

Given the difficulties in obtaining a set of public data that would provide relevant information and allow the development of the HistoraScan tool to digitise historic journals, we decided, after several searches and attempts, to use a set of data provided by Diário de Notícias (Figure 1) and Jornal de Notícias. These journals are considered as reference presses in the sense that their target audience is leading public opinion, and their content focuses on national and international politics, economics, and culture (Costa & Antunes, 2024; Figueiras, 2005). To this end, they essentially use an analytical and explanatory journalistic angle on the subject, standing out from the rest for their ability to be an agenda-setter (Figueiras, 2005). This means that "their highlights also become the priorities of the public agenda (themes shared by the community), of the other media, and also of the capacity to condition the political agenda, imposing behaviour on it" (Figueiras, 2005, p. 4).

A newspaper or journal is a periodical publication that focuses on the dissemination of news, reports, opinions, and announcements of general or specific interest, usually on a daily or weekly basis. Newspapers are aimed at a wide audience and aim to inform readers about recent events, current affairs, and developments in various areas, including politics, economics, sports, and entertainment. Typical examples include The New York Times and The Guardian. Their accessibility and comprehensive coverage of topics make newspapers a primary source of daily news for the public (Franklin, 2016; McNair, 2017). Newspapers are produced in short time cycles, with tight deadlines to ensure that the news is current, and involve a team of editors, reporters, and photographers working to cover day-to-day events. In addition, journals generally have a wide and diverse target audience (Sterling, 2009).

Despite several attempts to publish cheap newspapers in Portugal, it was with the arrival of the Industrial Revolution that Diário de Notícias (DN) was founded in 1864 by journalist and businessman Eduardo Coelho and his partner and printer Tomás Quintino (Sousa, 2018, 2021). At the time, the doctrinaire, combative, and partisan press characterised the Portuguese journalistic field. This was categorised by a high degree of rotation between the main parties in government. The opposition newspapers fought the government party journals, and vice versa. In this way, DN's industrial model presented itself as politically independent, giving rise to a newsroom with several dozen professionals, in which the reporter became increasingly important. Writing techniques quickly became oriented towards the pursuit of factuality, the separation between information and opinion, and the objectivity, as can be seen in this passage: "It was a journal different from the other Portuguese newspapers of the time, in content (news), in style (clear, concise, precise, and simple), in form, namely in appearance (four-column layout, not two or even one, as was customary)" (Sousa, 2018, p. 171).

Figure 1. Historical image from the Portuguese newspaper "Diário de Notícias"

Diario de Noticias

Source: Own material

Jornal de Notícias (JN) was founded in 1888 by José Diogo Arroio, Manuel Vaz de Miranda, and Aníbal da Costa Morais. Unlike DN, the founding members belonged to the Regeneration Party (Helena, 2017). In this way, the monarchist journal often argued against the government's progressivism, publishing national and international news with a page of commercials. In 1907, it was the journal with the largest circulation in the north of the country, but in that year, it abandoned its monarchist editorial line and adopted a republican slant in defence of the interests of the north. Although with the inclusion of a somewhat sensationalist narrative. The changes implemented enshrined a popular press style that was accessible to a wider audience than the one traditionally made up of political, social, and intellectual elites, the journal's audience par excellence (Helena, 2017).

Currently, Portuguese journalism faces significant challenges, such as the pressure on journalists to produce content quickly. Lack of resources is the main barrier to innovation in the industry (Cardoso et al., 2019). Crespo et al. (2018) show that journalistic companies in Portugal are slowly adapting to new practices. They favour the use of different available narratives and tend to edit with new publishing tools, these being the main forms of innovation. It is considered that the serious economic crisis that journalism has faced in the last decade has reduced teams that don't have the time or technical skills to develop and test new forms of innovation (Bonixe, 2020; Fante, 2018; Miranda et al., 2021; Silva & Granado, 2021; Vicente & Pérez-Seijo, 2022).

Methods

The DN and JN dataset provides data from the last century to the present day. These images contain information about headlines, articles, advertisements, among others. To understand the data, we were going to work with, an exploratory analysis was carried out in the first phase.

The work was carried out by a multidisciplinary team of experts in media studies, artificial intelligence, image processing, and cultural heritage preservation. This project involved a series of new techniques to adapt and evolve the state of the art in retrieving data from newspaper pages published since 1850, based on a mixture of heuristics, computer vision, pattern recognition, and other artificial intelligence and machine learning techniques.

As Figure 2 illustrates, a sequence of page scanning, image segmentation, natural language processing, and heuristics was used to reconstruct first a single article segment, then a complete article, and finally a page and complete relationships between pages. Work to refine these techniques with pilot tests in an iterative process is also discussed.

Figure 2. Steps used in the development of HistoraScan

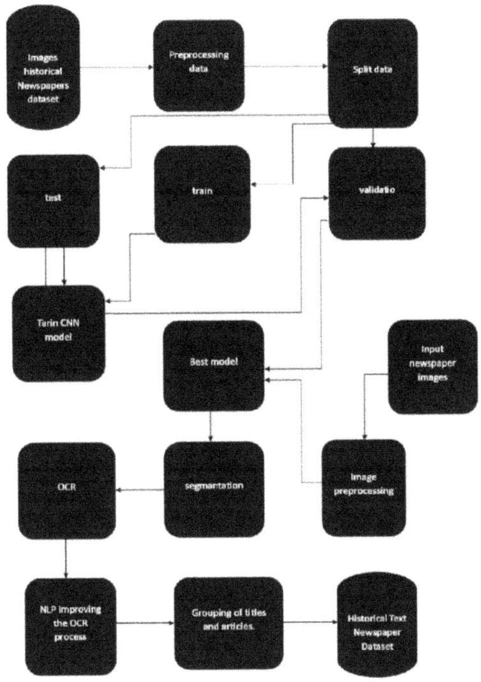

Source: Own elaboration

Models Developed for Segmentation

In our tool, we utilized CNNs for segmenting pages of historical newspapers into distinct sections such as articles, titles, page headers, images, and advertisements (Figure 3 and Figure 4). By leveraging CNNs, which are particularly adept at processing visual data, we aimed to accurately delineate the various components of newspaper pages.

Through extensive training and optimization of the CNN models, we achieved precise segmentation results, enabling efficient extraction of content from digitized newspaper pages. This segmentation process not only facilitates the organization of newspaper data but also enhances the accessibility and usability of historical archives for researchers and enthusiasts.

Additionally, by automating the segmentation task with CNNs, we streamlined the digitization process, saving time and resources compared to manual segmentation methods. The results demonstrate the effectiveness of CNN-based approaches in segmenting complex document layouts and paves the way for further advancements in historical newspaper digitization and analysis. As we can see in Figures 3 and 4, the segmentation of the newspaper image occurred with the identification of several filters that allowed, in this case, only the title and article to be selected.

Figure 3. Segmentation result of articles

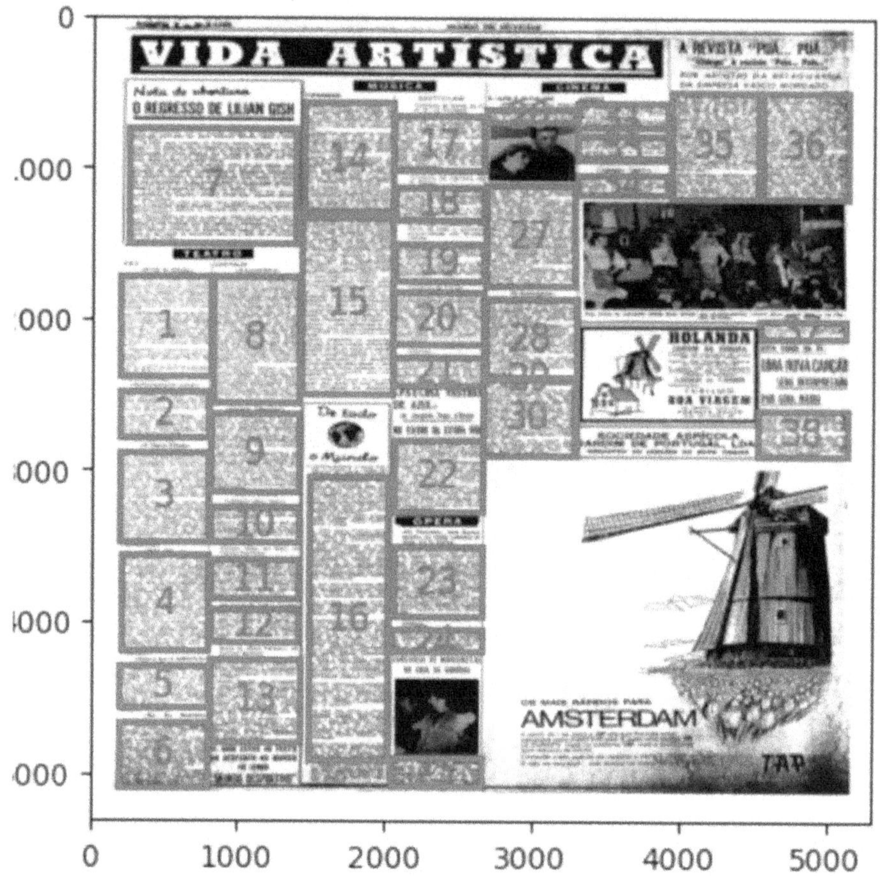

Source: Own elaboration

Figure 4. Segmentation result of titles

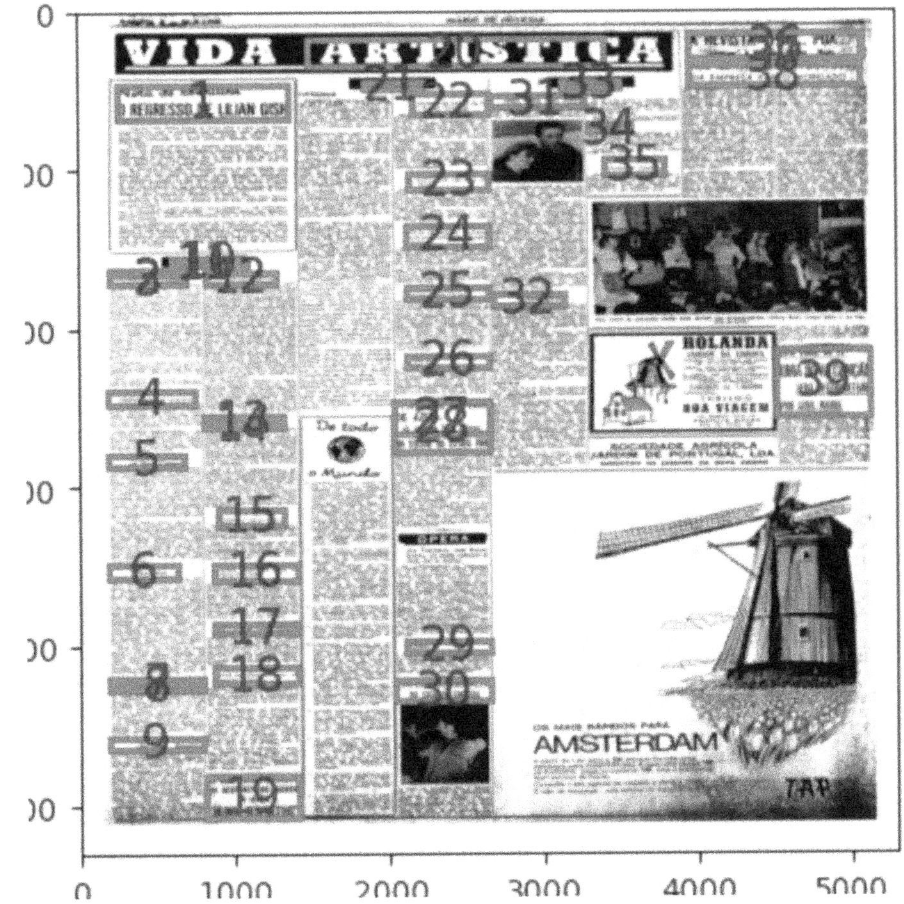

Source: Own elaboration

OCR

After segmenting images, we utilize image segmentation algorithms using computer vision techniques to enhance image qualities, focusing on specific regions such as titles, articles, headers, and images. This segmentation process involves identifying and isolating different components within the image to improve accuracy and readability (Figure 5 and Figure 6). Subsequently, OCR is employed to convert these segmented images into text format. OCR algorithms analyse the content of each segmented region, recognizing characters and patterns to generate a textual representation of the image content.

This involves preprocessing steps to enhance image clarity and remove noise, followed by the application of OCR algorithms to extract text from the segmented regions. Finally, the extracted text is processed and refined using natural language processing techniques to improve accuracy and coherence, ensuring that the converted text retains the original meaning and context from the segmented images. Overall, this process integrates image segmentation, OCR, and natural language processing to effectively convert segmented images into readable text data.

Figure 5. OCR process (Phase 1)

Source: Own elaboration

Figure 6. OCR process (Phase 2)

```
****************************************************
Max variable (10 palabras):  CRISE GRAVE
NO PARTIDO
TRABALHISTA
(Continuado da'l,° pagina)
Wilson persiste

Variable #8 CRISE GRAVE
NO PARTIDO
TRABALHISTA
(Continuado da'l,° pagina)
Wilson persiste
```

Source: Own elaboration

Both figures showcase the extracted text from historical documents, enabling digitization and subsequent manipulation of the text for analysis, search, or any other necessary purpose.

NLP Models

After the OCR process, some outputs may exhibit errors in their original spelling or word separation, particularly due to the complexity of historical newspaper images, which often feature faded text and irregular layouts. To address this challenge and enhance the accuracy of our tool, we employ NLP networks to improve the results in this regard.

Correlating Headlines and Articles

We propose employing logic and various algorithms to mitigate discrepancies between newspaper titles and articles, aiming for a more precise understanding of their relationship. This involves conducting an in-depth analysis of the structure and semantic content of each title and its corresponding article (Figure 7 and Figure 8). NLP algorithms can be employed to discern linguistic patterns and common themes between titles and texts. Additionally, data mining techniques may unveil significant associations based on keyword frequency and content similarity. The utilization of machine learning algorithms, such as classification and clustering, could aid in organizing titles and articles into coherent thematic groups. This approach promises to deepen comprehension of the interplay between newspaper elements, thereby facilitating access to pertinent information for diverse research and analytical purposes.

Figure 7. Correlating headlines and articles

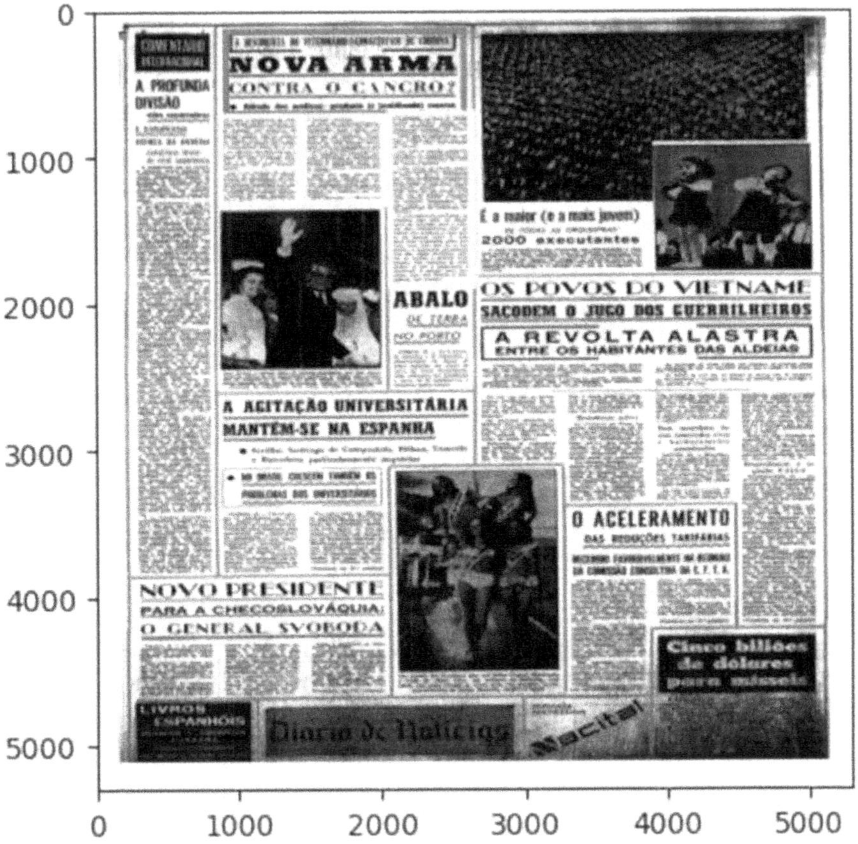

Source: Own elaboration

Figure 8. Correlating headlines and articles

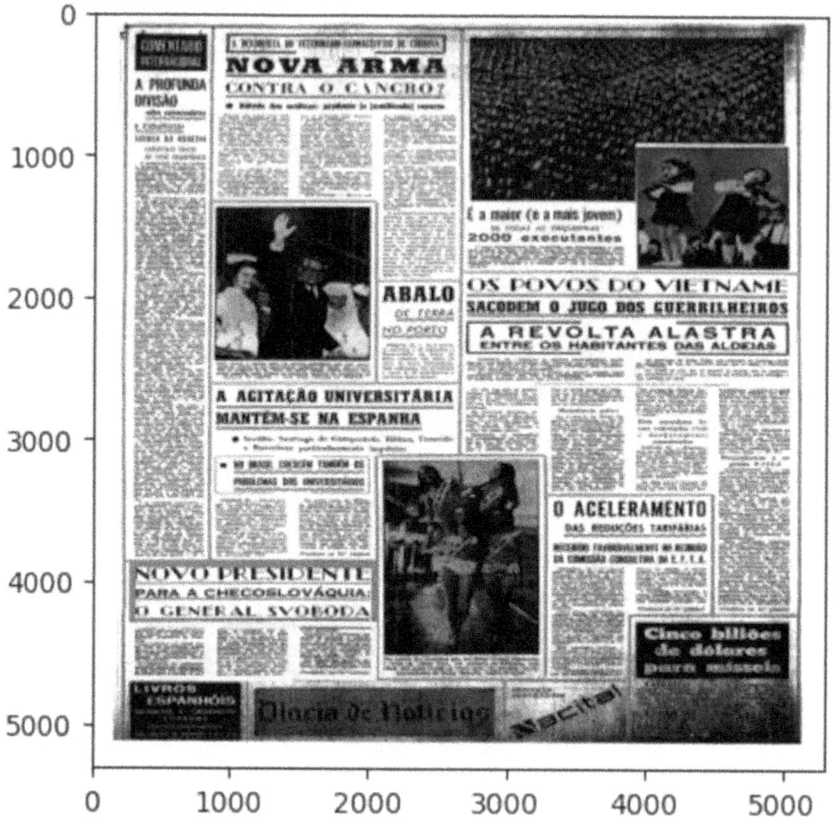

Source: Own elaboration

To finalize the process, we save the extracted text from both the headlines and articles into separate text files in TXT format. This allows for easy access and further analysis of the textual content. This step ensures that the valuable information extracted through the OCR process is preserved and can be utilized for various purposes such as text mining, sentiment analysis, and information retrieval. Developing this idea further, saving the extracted text in TXT format enables researchers and analysts to conduct in-depth textual analysis and explore patterns, themes, and insights within the historical newspaper data. By having the headlines and articles stored in separate files, it facilitates efficient data management and enables targeted analysis of specific sections of the newspapers. Additionally, the TXT format is widely supported and can be easily imported into various text processing tools and platforms for further exploration and research purposes.

DISCUSSION

The proposed methodology effectively addressed the challenges associated with historical newspaper digitization and text extraction. By employing a combination of computer vision techniques, OCR processes, and NLP models, we successfully segmented images, extracted textual content, and correlated headlines with articles. The integration of advanced technologies such as neural networks and natural language processing significantly improved the accuracy and efficiency of the data processing pipeline.

Furthermore, the developed models and algorithms showcased promising results in accurately extracting and correlating textual information from historical newspapers. The ability to store the extracted text in TXT format facilitates further analysis and research, enabling scholars to delve deeper into the historical content and uncover valuable insights.

Innovations Brought by Historascan to the Market

Our tool specifically focuses on the unique challenges associated with digitizing historical newspapers, which sets it apart from generic document digitization solutions. By utilizing advanced techniques such as CNN and NLP, HistoraScan can offer greater accuracy and efficiency in extracting and classifying information from historical newspapers.

Given that the design of historical newspapers may vary over time, HistoraScan may include specific features to adapt to these changes and ensure accurate and consistent digitization. On the other hand, this tool provides an easy-to-use interface and efficient processes for digitization and data organization.

CONCLUSION

As previously demonstrated, HistoraScan focuses on employing neural networks for the segmentation of images from historical newspapers, utilizing various computer vision techniques to enhance the quality of the segmented images. Additionally, we employed an OCR process to convert these images into text format, subsequently leveraging NLP techniques to refine the accuracy of the OCR output.

The main challenges included the variability in the design of historic newspapers, preserving the quality of images over time, and continually improving image processing and OCR techniques to adapt to different newspaper styles and periods. In this sense, HistoraScan has the potential to be the most innovative platform on the market for digitizing, classifying, and searching for information in historical newspaper collections. This tool is aimed at researchers, historians, and history enthusiasts around the world.

The digitization of historical newspapers can offer numerous advantages to contemporary print journals, promoting a rich exchange between the past and the present. Firstly, making digitised archives available allows contemporary journalists to access a wide range of historical references, enriching the contextualization and depth of current stories. This can result in more informed and comprehensive journalism, where reports not only cover present events, but also draw parallels and lessons from the past, offering a more complete perspective to readers. In addition, analysing historical journalistic practices can inspire innovations in editorial approaches and storytelling, taking advantage of techniques that were effective in other times and adapting them to contemporary needs and expectations.

Another significant added value is the potential revitalization of public interest and reader loyalty. By integrating historical content with current news, print journals can attract a wider audience, including historians, researchers, educators, and history enthusiasts. Subscription programmes or dedicated history sections within journals can increase the value perceived by readers, encouraging deeper engagement with the published content. Additionally, digitization allows journals to tell stories about their own evolution, creating an institutional narrative that strengthens the newspaper's identity and brand. This not only preserves institutional memory, but also celebrates the journal's role in society over time, establishing a stronger and more emotional connection with its audience.

Overall, this study contributes to the field of digital humanities by providing a comprehensive framework for digitizing and analysing historical newspapers. The insights gained from this research have the potential to enhance historical research, facilitate cultural heritage preservation, and stimulate interdisciplinary collaborations in the digital humanities domain.

ACKNOWLEDGMENT

This research was supported by the Portuguese Foundation for Science and Technology (FCT – Fundação para a Ciência e a Tecnologia) [2023.04877.BD].

REFERENCES

Allen, R. B., & Hall, C. (2010). Automated Processing of Digitized Historical Newspapers beyond the Article Level: Sections and Regular Features. In Chowdhury, G., Koo, C., & Hunter, J. (Eds.), *The Role of Digital Libraries in a Time of Global Change*. Springer. DOI: 10.1007/978-3-642-13654-2_11

Allen, R. B., & Schalow, J. (1999). Metadata and data structures for the historical newspaper digital library. In *Proceedings of the eighth international conference on Information and knowledge management (CIKM '99)* (pp. 147–153). Association for Computing Machinery. DOI: 10.1145/319950.319971

Allen, R. B., Waldstein, I., & Zhu, W. (2008). Automated processing of digitized historical newspapers: Identification of segments and genres. In Buchanan, G., Masoodian, M., & Cunningham, S. J. (Eds.), *Digital libraries: Universal and ubiquitous access to information* (pp. 379–386). Springer. DOI: 10.1007/978-3-540-89533-6_49

Barros, C., Costa, B. F., & Ramos, D. (2023a). *Guia de boas práticas para a indústria 4.0* [Best practices guide for industry 4.0]. UA Editora.

Barros, C., Costa, B. F., & Ramos, D. (2023b). *Industry 4.0: Best practices for the digital transition*. UA Editora – Universidade de Aveiro.

Black, J. W. (2023). Creating specialized corpora from digitized historical newspaper archives: An iterative bootstrapping approach. *Digital Scholarship in the Humanities*, 38(2), 779–797. DOI: 10.1093/llc/fqac079

Bonixe, L. (2020). Jornalismo radiofónico e inovação: Uma análise à cobertura de acontecimentos mediáticos [Radio journalism and innovation: An analysis of coverage of media events]. *Media & Jornalismo*, 20(36), 153–169. DOI: 10.14195/2183-5462_36_8

Broersma, M., & Harbers, F. (2018). Exploring machine learning to study the long-term transformation of news: Digital newspaper archives, journalism history, and algorithmic transparency. *Digital Journalism (Abingdon, England)*, 6(9), 1150–1164. DOI: 10.1080/21670811.2018.1513337

Bunout, E., Ehrmann, M., & Clavert, F. (2023). *Digitised newspapers: A new eldorado for historians? Reflections on tools, methods and epistemology*. De Gruyter Oldenbourg.

Cardoso, G., Baldi, V., Crespo, M., Pinto-Martinho, A., Pais, P. C., Paisana, M., & Couraceiro, P. (2019). *O que devem saber os jornalistas? Práticas e formação em Portugal* [What should journalists know? Practices and training in Portugal]. OberCom. [REMOVED HYPERLINK FIELD]

Coelho, P. M. (2016). *Rumo à Indústria 4.0* [Towards Industry 4.0] [Unpublished master dissertation]. University of Coimbra, Coimbra, Portugal.

Costa, B. F., & Antunes, E. (2024). Dinâmicas sociais no Facebook: Análise de comentários em conteúdos jornalísticos sobre a suposta tentativa de ataque terrorista à FCUL [Social dynamics on Facebook: Analysis of comments on journalistic content about the alleged attempted terrorist attack on FCUL]. *Observatorio (OBS*), 18*(1), 126-150.

Costa, B. F., & Mateus, B. C. (in press). Innovating in journalism with newsgames: An exploratory study in Portugal. In Bojić, L., Žikić, S., Matthes, J., & Trilling, D. (Eds.), *Navigating the digital age: An in-depth exploration into the intersection of modern technologies and societal transformation.* Institute for Philosophy and Social Theory.

Crespo, M., Foà, C., & Pinto-Martinho, A. (2018). Como o jornalismo lida com a inovação: Um estudo de caso das melhores práticas em Portugal [How journalism deals with innovation: A case study of best practices in Portugal]. *Estudos de Jornalismo*, 9, 75–102.

de-Lima-Santos, M.-F., & Ceron, W. (2021). Artificial Intelligence in news media: Current perceptions and future outlook. *Journalism and Media*, 3(1), 13–26. DOI: 10.3390/journalmedia3010002

Dhali, M. A. (2024). *Artificial Intelligence in Historical Document Analysis: Pattern recognition and machine learning techniques in the study of ancient manuscripts with a focus on the Dead Sea Scrolls* [Unpublished doctoral dissertation]. University of Groningen, Groningen, Netherlands.

Ehrmann, M., & Bunout, E., & Düring. (2019, August). *Historical newspaper user interfaces: A review* [Paper presentation]. *85th IFLA General Conference and Assembly (IFLA)*, Athens, Greece.

Ehrmannm, M., Düring, M., Neudecker, C., & Doucet, A. (2022). Computational Approaches to Digitised Historical Newspapers. *Dagstuhl Reports*, 12(7), 112–129.

Fante, A. (2018). Jornalismo appificado: Arquiteturas da notícia [Appified journalism: News architectures]. *Estudos de Jornalismo*, 9, 112–124.

Ferro, S., Pelillo, M., & Traviglia, A. (2023). AI-assisted digitalisation of historical documents. In Int (pp. 557–562). Arch. Photogramm. Remote Sens. Spatial Inf. Sci., XLVIII-M-2.

Figueiras, R. (2005). Os comentadores na imprensa de referência portuguesa: 1980-1999 [Commentators in the Portuguese reference press: 1980-1999]. In *Anais Intercom - XXVIII Congresso Brasileiro de Ciências da Comunicação* (pp. 1-15). Rio de Janeiro.

Fizaine, F. C., Bard, P., Paindavoine, M., Robin, C., Bouyé, E., Lefèvre, R., & Vinter, A. (2024). Historical text line segmentation using deep learning algorithms: Mask-RCNN against U-Net Networks. *Journal of Imaging*, 10(3), 1–18. DOI: 10.3390/jimaging10030065 PMID: 38535145

Franklin, B. (2016). *The Future of Journalism: In an Age of Digital Media and Economic Uncertainty*. Routledge.

García-Avilés, J. A., Carvajal-Prieto, M., Arias, F., & De Lara-González, A. (2019). How journalists innovate in the newsroom. Proposing a model of the diffusion of innovations in media outlets. *The Journal of Media Innovations*, 5(1), 1–16. DOI: 10.5617/jomi.v5i1.3968

Gehlen, M. A., & Sousa, J. P. (2018). Jornalismo de dados em Portugal: Um estudo exploratório sobre práticas jornalísticas especializadas [Data journalism in Portugal: An exploratory study on specialized journalistic practices]. *Estudos de Jornalismo*, 9, 125–139.

Gilchrist, A. (2016). *Industry 4.0: The Industrial Internet of Things*. Apress. DOI: 10.1007/978-1-4842-2047-4

Girdhar, N., Coustaty, M., & Doucet, A. (2024). Digitizing history: Transitioning historical paper documents to digital content for information retrieval and mining: A comprehensive survey. *IEEE Transactions on Computational Social Systems*, 11(5), 6151–6180. DOI: 10.1109/TCSS.2024.3378419

Granado, A., Silva, D. S., & Vicente, P. N. (2020). Inovação nos media e nas indústrias criativas limítrofes – uma introdução [Innovation in the media and neighboring creative industries – an introduction]. *Media & Jornalismo*, 20(36), 5–9.

Helena, L. (2017). A evolução editorial do Jornal de Notícias e a inclusão de elementos de jornalismo popular [The editorial evolution of Jornal de Notícias and the inclusion of elements of popular journalism]. *Revista Portuguesa de História da Comunicação*, 1, 1–11.

Hermann, M., Pentek, T., & Otto, B. (2016). Design Principles for Industrie 4.0 Scenarios. In *Proceedings of 49th Hawaii International Conference on System Sciences (HICSS)* (pp. 3928-3937). USA. DOI: 10.1109/HICSS.2016.488

Iansiti, M., & Lakhani, K. R. (2020). *Competing in the age of AI: Strategy and leadership when algorithms and networks run the world.* Harvard Business Review Press.

Kettunen, K., Keskustalo, H., Kumpulainen, S., Pääkkönen, T., & Rautiainen, J. (2022, February). *OCR quality affects perceived usefulness of historical newspaper clippings: A user study* [Paper presentation]. *18th Italian Research Conference on Digital Libraries*, Padova, Italy.

Kettunen, K., Koistinen, M., & Kervinen, J. (2020). Ground Truth OCR Sample Data of Finnish Historical Newspapers and Journals in Data Improvement Validation of a re-OCRing Process. *LIBER Quarterly*, 30(1), 1–20. DOI: 10.18352/lq.10322

Klijn, E. (2008). The current state-of-art in newspaper digitization: A Market Perspective. *D-Lib Magazine : the Magazine of the Digital Library Forum*, 14(1/2). Advance online publication. DOI: 10.1045/january2008-klijn

Koistinen, M., Kettunen, K., & Paakkonen, T. (2017). Improving Optical Character Recognition of Finnish Historical Newspapers with a Combination of Fraktur & Antiqua Models and Image Preprocessing. In *Proceedings of the 21st Nordic Conference of Computational Linguistics* (pp. 277–283). Sweden.

Kumpulainen, S., & Late, E. (2022). Struggling with digitized historical newspapers: Contextual barriers to information interaction in history research activities. *Journal of the Association for Information Science and Technology*, 73(7), 1012–1024. DOI: 10.1002/asi.24608

Lemmers, F., Ott, M., & Hermans, S. (2023). Printed mass media and automatic digitisation: The case of Belgian illustrated magazines from the interbellum. *In Monte Artium*, 15(15), 1–31. DOI: 10.4000/ima.427

Liebl, B., & Burghard, M. (2020, November). *From historical newspapers to machine-readable data: The origami ocr pipeline* [Paper presentation]. *Workshop on Computational Humanities Research*, Amsterdam, Netherlands.

McNair, B. (2017). *An Introduction to Political Communication*. Routledge. DOI: 10.4324/9781315750293

Meier, K., Schützeneder, J., García-Avilés, J. A., Valero-Pastor, J. M., Kaltenbrunner, A., Lugschitz, R., Porlezza, C., Ferri, G., Wyss, V., & Saner, M. (2022). Examining the most relevant journalism innovations: A comparative analysis of five European countries from 2010 to 2020. *Journalism and Media*, 3(4), 698–714. DOI: 10.3390/journalmedia3040046

Miranda, J., Fidalgo, J., & Martins, P. (2021). Jornalistas em tempo de pandemia: Novas rotinas profissionais, novos desafios éticos [Journalists in times of pandemic: New professional routines, new ethical challenges]. *Comunicação e Sociedade*, 39, 287–307. DOI: 10.17231/comsoc.39(2021).3176

Namboodiri, A. M., & Jain, A. K. (2007). Document structure and layout analysis. In Chaudhuri, B. B. (Ed.), *Digital document processing: Advances in pattern recognition* (pp. 29–48). Springer. DOI: 10.1007/978-1-84628-726-8_2

Palfray, T., Hebert, D., Nicolas, S., Tranouez, P., & Paquet, T. (2012). Logical segmentation for article extraction in digitized old newspapers. In *Proceedings of the 2012 ACM symposium on Document engineering (DocEng '12)* (pp. 129–132). USA. DOI: 10.1145/2361354.2361383

Pérez-Seijo, S., & Vicente, P. N. (2022). After the hype: How hi-tech is reshaping journalism. In Vázquez-Herrero, J., Silva-Rodríguez, A., Negreira-Rey, M. C., Toural-Bran, C., & López-García, X. (Eds.), *Total Journalism. Studies in Big Data* (pp. 41–52). Springer. DOI: 10.1007/978-3-030-88028-6_4

Pestana, O. (2021). O acesso aos jornais históricos: Considerações sobre o desenvolvimento de coleções digitalizadas [Access to historical newspapers: Considerations for developing digitized collections]. *Media & Jornalismo*, 21(39), 161–174. DOI: 10.14195/2183-5462_39_8

Pfanzelter, E., Oberbichler, S., Marjanen, J., Langlais, P., & Hechl, S. (2021). Digital interfaces of historical newspapers: Opportunities, restrictions and recommendations. *Journal of Data Mining & Digital Humanities*, 1-26.

Powell, T., & Paynter, G. (2009). Going Grey? Comparing the OCR Accuracy Levels of Bitonal and Greyscale Images. *D-Lib Magazine : the Magazine of the Digital Library Forum*, 15(3/4).

Rhyno, A. (2017, April). *Historical newspaper digitization on a shoestring* [Paper presentation]. 2017 IFLA International News Media Conference. Landsbókasafn Íslands – Háskólabókasafn, Reykjavík, Iceland.

Ringel, S. (2023). Digitizing the paper of record: Archiving digital newspapers at the New York Times. *Journalism, 24*(2), 245-261.

Schlukbier, G. (2008). Digitization of old newspapers: Software developments to enable successful conversion. In Walravens, H. (Ed.), *Newspapers collection management: Printed and digital challenges* (pp. 143–146). K. G. Saur. DOI: 10.1515/9783598441011.143

Schultze, C., Kerkfeld, N., Kuebart, K., Weber, P., Wolter, M., & Selgert, F. (2024). Reading yesterday's news. Layout recognition by segmentation of historical newspaper pages.

Schwab, K. (2017). *The fourth industrial revolution*. Crown Publishing Group.

Silva, D. S., & Granado, A. (2021). *Cobertura jornalística dos números da Covid-19: Casos de inovação em Portugal* [Journalistic coverage of Covid-19 numbers: Cases of innovation in Portugal]. Obi.Media.

Skelbye, M. B., & Dannells, D. (2021). OCR Processing of Swedish Historical Newspapers Using Deep Hybrid CNN–LSTM Networks. In *Proceedings of Recent Advances in Natural Language Processing* (pp. 190–198). Online.

Skilton, M., & Hovsepian, F. (2018). *The 4th Industrial Revolution: Responding to the impact of artificial intelligence on business*. Palgrave Macmillan. DOI: 10.1007/978-3-319-62479-2

Sousa, J. P. (2018). Eduardo Coelho e a fundação do Diário de Notícias. [Eduardo Coelho and the founding of Diário de Notícias] In Sousa, J. P. (Ed.), *Notícias em Portugal: estudos sobre a imprensa informativa (século XVI-XX)* [News in Portugal: studies on the informative press (16th-20th century)]. (pp. 163–192). Colecção Livros ICNOVA.

Sousa, J. P. (2021). *Portugal. Pequena história de um grande jornalismo I. Da manufatura à indústria* [Portugal. Small story of great journalism I. From manufacturing to industry]. Livros ICNOVA.

Spina, S. (2023). Artificial Intelligence in archival and historical scholarship workflow: HTS and ChatGPT. *Umanistica Digitale*, 16, 125–140.

Sterling, C. H. (2009). *Encyclopedia of Journalism*. Sage Publications. DOI: 10.4135/9781412972048

Takeuchi, H., & Nonoka, I. (2008). *Gestão do conhecimento* [Knowledge management] (A Thorell, Transl.). Bookman. (First published in 2004).

Teel, Z. (2024). Artificial Intelligence's Role in Digitally Preserving Historic Archives. *Preservation. Digital Technology & Culture*, 53(1), 29–33.

Thomas, A., Gaizauskas, R., & Lu, H. (2024, May). *Leveraging LLMs for Post-OCR Correction of Historical Newspapers* [Paper presentation]. Lingotto Conference Centre, Torino, Italy.

Valente, J., António, J., Mora, C., & Jardim, S. (2023). Developments in image processing using deep learning and reinforcement learning. *Journal of Imaging*, 10(9), 1–22. DOI: 10.3390/jimaging9100207 PMID: 37888314

Vicente, P. N., & Pérez-Seijo, S. (2022). Spatial audio and immersive journalism: Production, narrative design, and sense of presence. *El Profesional de la Información*, 31(5), 1–14. DOI: 10.3145/epi.2022.sep.19

Wang, W., & Gang, J. (2018). Application of Convolutional Neural Network in Natural Language Processing. In *2018 International Conference on Information Systems and Computer Aided Education (ICISCAE)* (pp. 64-70). Jilin Finance Building Hotel. DOI: 10.1109/ICISCAE.2018.8666928

KEY TERMS AND DEFINITIONS

Automatic Digitization of Historical Journals: An automated process of converting physical historical journals into a digital format readable by computers.

Convolutional Neural Networks (CNN): A multi-layer neural network method that learns the hierarchical characteristics of data and is normally used for image classification tasks.

Fourth Industrial Revolution: Expression used to designate the paradigm in which cutting-edge technologies are used in automation and data exchange to help increase the quality, efficiency, and productivity of companies' production processes.

Innovation: Creation of economic and social value for companies.

Journal: A periodical publication dedicated to the dissemination of news, reports, opinions, and announcements of general or specific interest, usually on a daily or weekly basis.

Natural Language Processing (NLP): A sub-field of artificial intelligence that focuses on the interaction between computers and human language. The aim is to enable computers to understand, interpret, process, and generate natural language in a way that is valuable and meaningful.

Optical Character Recognition (OCR): Technology that makes it possible to convert different types of documents, such as scanned paper documents, PDF files, or images captured by a digital camera, into editable and searchable data.

ENDNOTES

1 https://labs.loc.gov/work/experiments/newspaper-navigator/

2 https://www.transkribus.org/

Chapter 16
Digital Communication and the Modern India

Subir Sinha
https://orcid.org/0000-0001-7578-9809
Dum Dum Motijheel College, India

ABSTRACT

Communication is the process of sending and receiving information. It is the vehicle through which human being are developing and maintaining human civilization. Digitization bring revolution in the process of communication and developed a new communication system called digital communication. However in India, the significance of digital communication is enormous and push the nation towards the modernity. The article entitled 'Digital Communication and the Modern India' displayed how digital communication functioned in several fields likes information dissemination, social connectivity, propagation of education, health communication and digital marketing and supporting in the formation of a technologically advance and modern India.

INTRODUCTION

Communication is the linear process of sending and receiving information. It is essential and one of the vital needs of all human beings. The process of communication gain momentum with the advent of digitalization which transformed it into a more advance and a technologically developed communication system called digital communication. Digital communication is the transmission of information or data through digital network or signal. It is the communication in digital form or the electronic exchange of information. In the opening chapter of book 'Principles of Digital Communication' Robert G. Gallager mentioned the digital communication

DOI: 10.4018/979-8-3693-3579-6.ch016

system as "communication system that use such a digital sequence as an interface between the source and the channel input (and similarly between the channel output and final destination." - (Gallager, 2008). Digital form of communication is more effective and highly efficient in compare to analogue system. In contrast to analogue communication system, digital communication is more complex but it mainly controls the problem of noise in the system. Sunil Bhooshan in the opening chapter of his book entitled 'Fundamentals of Analogue and Digital Communication System' mentioned "Digital communication, through complex hardware design, can virtually overcome the problem of noise cluttering the signal." – (Bhooshan, 2022). Digital communication is modern and technologically more advance in nature. It reveals the properties of the synchronous and asynchronous forms of communication. Digital communication shows higher interactive property that enables real time interaction, instant messaging, and wide conferencing. Along with these digital communication shows the capacity where message can be store and forward, and one can read according to his or her own time and preferences. One can easily communicate in real time is a synchronous way of communication whereas one can communicate according to his or her own time or information can be stored or modified according to the necessary which reflects the property of asynchronous mode of communication system. The communication through digital mean shows a wider domain and has also the power to overcome geographical boundaries. One can easily communicate from any part of the world to the other in real time within a very short span of time.

However the roles of digital communication in our society are enormous and significant. It has the capacity to disseminate information, instruction, data, etc across the globe which directly promoting the concept of globalisation. Digital communication is helping to connect the entire globe and turning the society into an advance and unified one. Electronic mail services, digital websites and social networking sites are playing a major role to promote the digital communication system across the globe. They are performing numerous functions and helping in the development of a modern and an advance society. Email service providing the facilities of sending and receiving digital mail across the globe with in a very short span of time. Websites or web portals are highly valuable for the development of the society, public can easily access information, data or message from this websites and can use them according to their necessity. This information makes the public aware and well informed. Social networking sites like Facebook, Whatsapp, Messenger, Google meet, Telegram, etc are also highly valuable and beneficial for the information dissemination, social connectivity, propagation and promotion of educational institutes, health communication, digital or online marketing, etc. Digital marketing websites and online marketing portal and applications like Flipkart, Amazon, Myntra, etc are highly valuable for e-commerce. The roles of digital communication in our society are extra ordinary. Digitalization and digital communication extended its arms, prevail in all sectors and continuously proving beneficial in numerous way.

However in India, digitalization and the digital communication system does not remain behind. It plays several significant roles since its arrival. Digital communication arrived in India with the advent of telecommunication and computer system and gain momentum with the arrival of mobile phones and mobile communication system. Honourable Prime Minister Shri. Narendra Modi in the month of July of 2015 launched an ambitious flagship programme of the Government of India called 'Digital India' with a vision to transform India into a digitally empowered society and knowledge economy. Three key visions of the programme 'Digital India' are digital infrastructure as a core utility to every citizen, governance and service on demand and digital empowerment of citizens. The flagship programme strengthens the root of digital communication in India and accelerated India toward a new dimension. Later in 2018 National Digital Communication Policy was formed and presented several policies related to digital communication and the digitalization of Indian. In the preamble of the National Digital Communications Policy 2018 it was mentioned "Digital India is already unfolding. India's digital profile and footprint is one of the fastest growing in the world. With over a billion mobile phones and digital identities and half a billion internet users, India's mobile data consumption is already the highest in the world."- (National Digital Communication Policy, 2018).

Digital communication is used in India in multi dimensional ways and it is deriving as a key factor behind the national progress. It is proving itself as a blessing of technology and is now becoming as a backbone of the Indian society. It is getting used in various fields and performing various significant roles in the society. In India, most of the department of government as well as non-governmental organisation are using it as a modern way to reach the society as well as the globe. Through digital communication they are disseminating various valuable messages, information, data, which making the people aware and well informed. Digital communication is a way through which India projecting itself in global platform and transforming the society into a modern and an advance one. In India it shows a wide domain and performing several vital functions in a significant way. Information dissemination and instruction propagation, connection among the societal members or the social connectivity, promotion and the propagation of education, support in health communication and health literacy, promoting the digital marketing and the online shopping, etc are among its vital functions.

OBJECTIVES OF THE RESEARCH

Digital communication is playing an active role in the age of digitalisation. In India, digitalization and digital communication is also highly active and performing various significant roles in various sectors. It provides a wide scope to the researchers to conduct research on several areas of digital communication and its significance. The article entitled 'Digital Communication and the Modern India' provides a thorough analyse and will try to give the explanation while focusing on the following objectives:

- To explore the significances of digital communication in the field of information dissemination, social connectivity, promotion and propagation of education, health literacy and digital marketing.
- To find out how digital communication transforming India into a modern and an advance society.

METHODOLOGY OF THE RESEARCH

The article entitled 'Digital Communication and the Modern India' is based on a thorough analysis of various case studies, contents and documents related to digitalization, digital communication and its significance in various fields related to India. It analyses several articles and chapter published in journals and book, along with several statements and contents published and projected in various websites and web portals of Government of India, various educational institutes, and private organisations.

DISCUSSION

The late nineteenth and twentieth century has shown the rise and growth of digitalization and digital communication and the twenty first century showing a momentum in it. It dominantly prevails in every sector across the globe. However in India, digital communication also extended its arms with the help of internet and digital media. The growth of digitalization pushes the government, non government agencies and private organisations to create several official websites, maintain social networking sites and to establish digital communication. Official websites along with social networks play a key role in pushing India towards the establishment of digital communication networks. However the development of the flagship programme called Digital India and the rise of Jio network throughout the nation provides momentum in the process of digitalisation and makes digital communication as a vital form of communication across the nation. In India every sector are now using digital communication and it is also performing various functions in a significant ways. The role of digital communication in information dissemination, social connectivity, promotion and propagation of education, health communication and in digital marketing system are significant. This few major functions of digital communication clearly try to demonstrate how India using advance communication system and developing a modern and an advance society.

Digital Communication and the Information Dissemination

In India, digital communication helps to disseminate various information and data and support in the modernization of various fields. Digital communication is done with the help of digital media which successfully take an active part in the information dissemination process. The wide use of internet, personal computers, laptops, smart mobile phones, tablets, etc provided a wide digital communication platform to the users. Digital communication successfully penetrates deep into the heart of the society and disseminates information among a wide number of audiences across the nation. It shows immense variety of advantages and provides a new platform to the audiences. John R. Barry, Edward A. Lee, and David G. Messerschmitt in the opening part of their book 'Digital Communication' mentioned about the digital transmission of information as "Digital transmission of information has sufficiently overwhelming advantages that it increasingly dominates communication systems, and certainly all new designs." – (Barry, et.al, 2004)

In India, there are numerous news portals and online news channels which disseminate news and blogs in a digital mode. They communicate with their audience through digital means. News portal and online news channels are highly informative digital platform. They are actually acting as an access point of news and information for the viewers. A news portal is a specially designed website that brings news and information from diverse sources in a uniform way. Usually, each information source gets its dedicated area on the page for displaying information; often, the user can configure which ones to display whereas the online news channels provide diverse news from across the nation as well as from the entire globe. In India almost every news media house launched their own news channel and portal for the dissemination of various news and information.

Electronic mail service and social networking sites are also playing a significant role in information dissemination process. The Electronic mail or the Email can easily transmit information or the message within a very short span of time not only throughout the nation but also across the globe. It is promoting information sharing process and supporting the nation with modern postal system. In India there are several email providers such as Gmail, Yahoo mail, Outlook.com, Zoho mail, etc. Government of India is also concern with email system and recognised it as one of the primary mode of communication which helps to empower e-governance. Under the Ministry of Communication and Information Technology, Department of Electronic and Information Technology of the Government of India issued email policy entitled 'Email Policy of the Government of India'. In the opening of the Introduction it was mentioned "The Government use e-mail as a major mode of communication."- (E-mail Policy of Government of India, 2014). The main focus of the policy is to provide a secure email service to the government official. The policy tries to mandates government officials to use government email service for official correspondence.

Social networking site like Facebook, Messenger, Whatsapp, Telegram, Youtube, etc also served a major role in the information dissemination process across the nation. Through social media individual user or any news portal can easily disseminate news or information among the society. Social networking sites like Facebook, Whatsapp, Messenger, Telegram help us to deliver message or information while on the other side social networking sites like Facebook also help us to connect various news portals and blogs from across the globe just by clicking the 'Like' button. Public can choose and like the news portal according to their choice. The news portals in reverse are making the users well informed by providing various valuable news, data and information. Youtube is another important video sharing social networking sites that disseminates various formal and informal information, data and messages across the globe in digital video format. These social networking sites and applications are serving as a vital source of information which turning the society into a knowledge based well informed society.

Digital Communication and the Social Connectivity

Digital communication plays a significant role in social connectivity since the early days of its arrival. Digital communication with the help of various forms of digital media helps to maintain social connectivity not only within India but also across the globe. Social connectivity and communication through digital form is highly beneficial and cost effective. Users can maintain social connectivity and communication with a very minimal cost and can easily communicate globally while overcoming the geographical boundaries. Communication in digital form has the ability to disseminate message, data or information in synchronous a well as in asynchronous way. Email or electronic mail services show wonderfully how digital communication helps to maintain global connectivity. It is supporting in social connectivity process and promoting advance postal system. Email service via internet helps to transmit valuable message from sender to receiver in a convenience and cost effective way. Through email one can send a message to a multiple receivers. It is acting as a highly advance global postal system. Along with email service various social networking sites like Facebook, Messenger, Whatsapp, Telegram, etc also show the significance of digital communication in maintaining the social connectivity. The role of digital communication and digital media in social connectivity are significant. In India communication through digital mean are now getting highly prioritize and a major sections of the Indian citizen are now accustom with the digital communication.

The recent record evolved as a case study which revealed that India reached as the highest Facebook users among all the nations. It is a wonderful example that shows how India using digital communication and social networking sites to maintain social connectivity. The Facebook users in India gradually rise and become highest among the nations of the entire globe according to survey records. According to a report of Statista there are approximately 367 million Facebook users are present in India which make India as a highest Facebook user in the globe. It was mentioned in the report: "There are nearly 367 million Facebook users in India alone, making it the leading country in terms of Facebook audience size."- (Statista, 2024). It suggests that major sections of Indian citizen are now using the social networking site Facebook and maintaining the social connectivity and communication in digital mode.

Digital Communication and the Propagation of Education

Digital communication is not only used for the dissemination of information or to maintenance of social connectivity, it is also used for the promotion and propagation of education. Students and scholars are highly favouring digital communication platforms for their education purposes. Through digital communication platforms scholars and educationist can communicate in synchronous and asynchronous way and can share various texts, articles, video clips along with photographs and graphical representations. It can also be used for webinars, e- conferences and virtual meetings.

In India, the case study of the virtual class of the Indra Gandhi National Open University signifies the role of digitalization and digital communication in propagation and promotion of education. Indra Gandhi National Open University (IGNOU), established by an Act of Parliament in 1985, has continuously striven to build an inclusive knowledge society through inclusive education. It has tried to increase the gross enrolment ratio (GER) by offering high-quality teaching through the open and distance learning (ODL) model. Recently, through the use of digital media and internet facilities, Indra Gandhi National Open University has started to propagate virtual classes. The university developed e-learning platform to deliver an online program. The platform provides a complete online experience from registration to certification. The system is digitalized and digital communication a plays a key role in delivering the virtual lectures.

Along with the case study of Indra Gandhi National Open University several social networking sites like Facebook, Youtube, Whatsapp, Google meet, etc are also playing a key role in propagation and promotion of education. Social networking sites like Facebook, which operate digitally, support the students and educationist in various ways. They can communicate in real time and can discuss on various educational subjects for cooperation and for the better outcome of the education. They can even share educational information, data and various notes on this social networking platform. Facebook is also providing the facilities to create groups where various educationist, scholars and students can enter as a member and can post numerous data and information. Along with these the user of Facebook can select and like various web portals from where they receive various data and information related to formal and informal education. Social Media for them is serving as a platform where they can cooperate and share information and knowledge.

YouTube is another video sharing social networking sites that provides various digital video clips that related to formal and informal education. It teaches the viewer with proper demonstration and illustration. Like the other interactive video sharing sites Youtube also provides the students and educationists a scope to select the video according to their necessity and choice; they can pause and rewind the video clip if they get necessary. It is totally user friendly and tries to provide full satisfaction to the user through its videos.

The digital communication through the social networking site WhatsApp gain huge popularity recently. The social networking site Whatsapp is also promoting education and it development wonderfully. Through the use of Whatsapp students, scholars and educationist can easily transfer any information, data, video or photographs related to education to their classmates or any educationist within a very short span of time. Students and educationist can create educational group where all the members of the group can communicate and can share data and information on the relevant topic but in Whatsapp only Group Admin have the capacity to select group members. WhatsApp attract most of the students and educationist because of its easy functioning and video calling facilities.

Social networking site Google meet is taking an active part in promoting digital communication and education system. It is an online meeting site that is widely used by the teachers, students, scholar and various educationists for various educational purposes. They are widely used for the webinar, online conference, meetings and sometime for taking virtual classes. In India during the period of pandemic various schools, colleges and universities conducted several webinars and virtual classes through the use of Google meet for the benefit of the students and educational development.

Digital Communication and the Health Communication

The roles of digital communication in health sectors are enormous and significant. The vast domains of digital media are helping several Indian health organisations to communicate with the patients, doctors, medical staffs and general public and to propagate various messages on several diseases, health awareness, hygiene, medical facilities, etc to a large number of receivers. The health departments of Government, Municipal Corporation and Non Governmental Organisations are now using digital websites and web portals along with various forms of social media and mass media for the propagations of information related to health, hygiene and sanitation, various diseases and its prevention, etc to the general masses.

Website and the web portal are now become the most valuable and authentic source of information, from where one can collect information or data according to their own preferences and necessity. During the pandemic of COVID-19 the case study of the websites of Health and Family Welfare Department, Government of India was significant. The department regularly updated various data and information related to COVID-19 and its awareness through their web portal. The websites of Ministry of Health and Family Welfare Department of Government of India highlight all the details related to COVID-19 in India. The website provides toll free helpline number and revealed the percentage and figure of 'Active Cases', 'Discharge' and 'Death', 'Latest Update', 'COVID-19 State wise Status' and 'Resources' along with 'FAQ'. Through the use of web portal the department easily disseminate the entire information and messages of social awareness on COVID-19 across the globe. The ministry used wonderfully the digital media platform for the welfare of the nation.

The social networking sites also show the power to maintain digital communication and deliver information about health among the mass within a very short span of time. The health organisations are using the social media as its impacts are much higher than the traditional technique and it can reach easily to the grass root level of our society with a cost effective manner. The message structure and the presentation of the social media attract a large number of masses and the message easily penetrates in the mind of the receivers. It creates a significant effective in health communication. Several health organisations and health departments have open their account and pages on social site like Facebook, Twitter, etc. The organizations disseminate information related to health and health awareness through these pages of social networking sites. During the pandemic of COVID-19 the social media play a significant role in giving information and instructions related to COVID-19 which are highly valuable and create awareness among the public. However digital communication with the help of digital media is acting as an important tool of health communication.

Digital Communication and the Digital Marketing

Digitalisation and the digital communication help to promote digital marketing in a significant way. Digital marketing is one of the most vital and an advance form of marketing system of modern era. It is the form of marketing where the marketers and the business organisations use digital technologies for the marketing communication and to promote their products, goods and service. The term digital marketing mainly refers to the promotional activities or advertising propagated through the digital medium such as social media, email, mobile apps, various website and web portal, etc via computers, laptops or mobile phones. Digital marketing can also be described as a unique modern marketing technique where business organisations or the manufacturing industries display and sell their product through the use of digital technologies. The digital marketing can go beyond the traditional marketing system and can easily pursue more consumers than the traditional one. In the pre digital era, the marketing technique is totally traditional and manual in nature but digitalisation has changed the concept of marketing system. The process of digital marketing is fully turns digital in nature where the sellers or the marketers completely use digital technology and internet facilities to promote marketing of brands, products, goods and service and to communicate with the consumers.

Digital medium and the digital communication system play a significant role in the process of digital marketing. It has changed the concept of marketing and proved success in the promotion of products, goods and services, formation of brand and in the selling of the products. The uses of digital communication in marketing make the business more easy and comfortable from various dimensions. The main purpose of digital communication is to educate the consumers and helps to make them understand what the marketers and business organisations are offering. It also helps in the sustainability of the business organisations and the marketers as it was reflected that the effective communication through digital modes pursuing and motivating the consumers to use their products and services. The process helps to enhance the productivity and growth rate of business organisations. In the age of digitalisation, effective digital communication acts as a backbone of business and marketing system. Digital communication make the digital marketing more efficient and successful and helps in formation of brands, elimination of hoax and support to get feedback from the consumers. Along with sustainability of the business organisations and educate the consumers, digital communication also helps to maintain a cordial relationship among business organisation, marketers and the society. The roles of digital communication in digital marketing proved significant and it transforming the marketing system into a modern and advance one.

The increasing rate in the online shopping has try to justify how marketing system improving with the help of digital technologies. Through digital communication sellers can easily communicate with the consumers and can promote marketing through digital means. Consumers can easily watch the product through digital photographs; judge the quality of the products through product descriptions, consumers can easily communicate with the seller for further enquiries about the product. The total process is digital in nature and provide an advance option to the sellers as well as to the consumers in contrast to the traditional marketing system. The online shopping organisations or the e-commerce platforms like Flipkart, Amazon, Myntra, Jio mart along with Indian retail departmental store like Reliance Smart Bazar are wonderful showing their digital marketing strategies. They are totally focused on consumer's needs and preferences, put advertisements on several websites, send mails directly to the consumers, generate various SMS to the consumers, provide them various rebate and offer on various products, available the facilities of online orders and free deliveries, chat and talk with the consumers directly through digital means. This direct communication with the consumers through digital means easily pursue them and motivate them to buy the products. The marketers and the organisations are taking various advantages from digital communication and successfully promoting the business. However it can be mentioned that the roles of digital communication in digital marketing and e-commerce are significant and enormous.

CONCLUSION

In India, digital communication is performing various essential roles in various fields. It is proving itself as an advance form of communication which is essential to develop a modern society. With the help of internet facilities and technological innovation, India arises as a digitally empowered nation that uses digital communication for various purposes. It successfully supporting in information dissemination and awareness creation, social connectivity, promotion and propagation of education, health communication, digital marketing, etc and helping to transform India into a modern and a technologically advance society. Its role and functions are wide and numerous and too support in the gain of nation economy.

Digital communications technologies are widely getting used across the nation for information dissemination and awareness creation. It successfully disseminating information and making people well informed and aware. Digital platforms like websites, email service and social networking sites are widely used in India for information dissemination. Several departments and organisations of the government of India, municipal corporations, non-governmental organisations and private sectors are using various websites and digital platforms to disseminate information and to create awareness among the public. Social media networks are also widely used in India for the dissemination of information and awareness creation and proving success from numerous ways. Digital communication is highly preferred in India for social connectivity. In India, most of the people are using social network, create their own profile, connects with individual, family, friends, relatives, and colleague and maintain social connectivity through digital means. Electronic mail service, along with social networking sites like Facebook, Whatsapp, Messenger, Telegram, Google meet, Zoom, etc are evolved as vital digital platforms that help to maintain social connectivity not only within the nation but also across the globe. The recent statistical data shows the record of Facebook users across the globe in which India took a leading position which signifies social networking sites and digital communication playing a key role in social connectivity across the India.

In India, digital communication remains successful in promotion and propagation of education. Several official educational websites along with social networking sites Facebook, Whatsapp, Google meet, Google class room, Zoom are widely use for various educational purposes. Educational websites of educational department of the Government of India, educational directorate, UGC, schools, colleges and universities, etc provides valuable information and data. Google class room are used for delivering lecture and demonstration, Google meet and Zoom is used for webinars and e-conferences whereas Facebook and Whatsapp are highly used as a tool for social connection and communication among the students, scholars and teachers. The role of digital communication in health and digital marketing are also significant. In case of health communication, government health departments and various hospitals are using their digital websites to propagate various information and awareness among the public. Digital communication technologies are also used by doctors to communicate with their colleagues, patients and staffs. In case of digital marketing it was came to light that several industries and business organisations are using widely the digital technologies to communicate effectively and to develop an advance marketing system for maintain sustainability of the business. However, the significances of digital communication in India cannot be ignore, India using the digital communication technologies widely and wisely for developing a highly advance and modern society and to give its citizen benefits from various dimensions. Digital communication arrived in India as a blessing of technology and it is used by the entire nation for the betterment and welfare of the society.

REFERENCES

Barry, J. R., Lee, E. A., & Messerschmitt, D. G. (2004). *Digital Communication* (3rd ed., Vol. 1). Springer Science and Business Media LLC. DOI: 10.1007/978-1-4615-0227-2

Bhooshan, S. (2022). *Fundamentals of Analogue and Digital Communication System*. Springer. DOI: 10.1007/978-981-16-4277-7

Department of Electronic and Information Technology of the Government of India. Email Service. (2015). 'EMAIL policy of Government of India. Retrieved from https://www.meity.gov.in/writereaddata/files/E-mail_policy_of_Government_of_India_3.pdf

Dixon. J.S., Statista (2024). Leading Countries based on Facebook audience size as of January 2024 Retrieved from https://www.statista.com/statistics/268136/top-15-countries-based-on-number-of-facebook-users/

Gallger, R. G. (2008). Principles of Digital Communication, Cambridge University Press IGNOU.Preamble; Retrieved from https://www.ignou.ac.in/ignou/aboutignou/profile/2

Ministry of Electronic and information Technology, Government of India. Email Service. Retrieved from https://www.meity.gov.in/secure-email-service-government-india

Ministry of Health and Family Welfare, Government of India. Retrieved from https://www.mohfw.gov.in/

National Digital Communication Policy. Preamble (2018). 'National Digital Communication Policy 2018'. Retrieved from https://www.meity.gov.in/writereaddata/files/National_Digital_Communications_Policy%E2%80%932018.pdf

NHP INDIA -National Health Portal. Retrieved from https://www.nhp.gov.in/

KEY TERMS AND DEFINITIONS

Communication: It is the process of transmission of message from sender to receiver or in other word communication is the process of sending and receiving information. In the process of communication there are mainly six basic elements – sender, message, channel, receiver, noise and feedback.

Digitalization: Digitalization is the process of converting or transforming information, data or any system from analog to digital format.

Digital Communication: Digital communication is the linear exchange of information or data through digital network or signal. It is the communication in digital form.

Digital Education: Imparting education through digital means.

Digital Marketing: It is the form of marketing where the marketers and the business organisations use digital technologies for the marketing communication and to promote their products, goods and service.

Health Communication: It is the communication based on health and health issues. It mainly related with health issue, health awareness, various disease and its prevention, etc.

Social Connectivity: State or the capability of social connection.

Social Media: Media that connects the society through digital means.

Chapter 17
Availability of Academic Resources Presents Opportunities and Challenges

Marcel Ohanga Odhiambo
Mangosuthu University of Technology, South Africa

ABSTRACT

Availability and accessibility of academic resource materials has evolved over time, from textbooks to currently being available online. Tuition or teaching has followed the same trend from classroom-based to online as well. This is in addition to not only the academic resources being provided by various entities but in different formats as well. These phenomena present challenges and opportunities. Cloud computing and Information Communication Technology (ICT) has contributed to this phenomenon. Availability and accessibility of academic resources by various entities and in varying formats has resulted in communication dynamics. This paper aims to review the dynamics due to availability and accessibility of academic resources and concludes that the opportunities presented far outweigh the challenges. Institutions and individuals should therefore leverage their operations on availability and accessibility of academic resources to offer a better service to their clients.

1.0 INTRODUCTION

The availability and accessibility of academic resource materials has evolved over time. Academic resources are no longer confined to the textbooks or academic institutions but are now freely and readily available from several sources and in

DOI: 10.4018/979-8-3693-3579-6.ch017

varying formats. In addition, there is a shift/move whereby academic resources and teaching is not only remote/online but, are remotely hosted.

This has been brought about by the Information Communication Technology (ICT), which is the merging of computing activities such as document processing with communication. Enabling one to prepare a document and communicate or post the document in one session. In addition, the ownership of academic resources and the use of the same are becoming very much separate. Hence, the supplier of computing resources is not necessarily the user at the same time. This has resulted in various entities offering academic resources that are accessed over the Internet by any user from anyplace and anytime. This phenomenon is based on cloud computing.

This chapter reviewed the dynamics due to availability and accessibility of academic resources and concluded that the opportunities presented far outweigh the challenges. Institutions and individuals should therefore leverage their operations on availability and accessibility of academic resources to offer better services to their clients.

2.0 BACKGROUND

Academic resources are currently available from various sources and in varying formats. This has resulted in various entities offering academic resources that are accessed over the Internet by any user from anyplace and anytime. This phenomenon is based on cloud computing. The classroom or teacher/lecturer is no longer the only custodian/source of academic resources. This presents both opportunities and challenges to academics/students.

The teacher/lecturer is faced with students who have access to sources of academic resources. The teacher/lecturer has the task of guiding the students in selecting appropriate and relevant academic resources since not all materials have been vetted by credible authorities and found to be suitable.

Cloud computing platform offers academic resources.

3.0 CLOUD COMPUTING

Cloud computing is a phrase used to describe a variety of computing concepts that involve a large number of computers connected through a real-time communication network such as the Internet. In science, cloud computing is synonym for distributed computing over a network and supports the ability to run a program or application on many connected computers at the same time without knowing the physical location of the computers.

Cloud computing is a computing term metaphor that evolved in the late 2000s, based on utility and consumption of computer resources. Cloud computing involves deploying groups of remote servers and software networks that allow centralized data storage and online access to computer services or resources. Clouds can be classified as public, private or hybrid (Public cloud), (Private cloud).

Cloud computing (Hybrid, 2012), (Paul P.K, Ritam Chatterjee, Sreeramana Aithal) relies on sharing of resources to achieve coherence and economies of scale, similar to a utility (like the electricity grid) over a network (Private cloud). At the foundation of cloud computing is the broader concept of converged infrastructure and shared services as shown in Figure 1.

Cloud computing, or in simpler shorthand just "the cloud", also focuses on maximizing the effectiveness of the shared resources. Cloud resources are usually not only shared by multiple users but are also dynamically reallocated per demand. This is implemented by computer systems allocating resources to users on demand. For example, a cloud computer facility that serves European users during European business hours with a specific application (e.g., email) may reallocate the same resources to serve North American users during North America's business hours with a different application (e.g., a web server). This approach should maximize the use of computing power thus reducing environmental damage as well since less power, air conditioning, rack space, etc. are required for a variety of functions. With cloud computing, multiple users can access a single server to retrieve and update their data without purchasing licenses for different applications.

The term "moving to cloud" also refers to an organization moving away from a traditional CAPEX model (buy the dedicated hardware and depreciate it over a period of time) to the OPEX model (use a shared cloud infrastructure and pay as one uses it).

Cloud computing aims to provide computing resources in the cyber space or Internet in the form of virtual servers available over the Internet. Cloud computing relieves small and medium enterprise the headaches associated with installation and maintenance of computing resources, cut down on IT expenditure (personnel, infrastructure) and users only pay for the services/resources consumed.

Cloud Computing is one of the emerging components of Information Technology and the most promising virtualization technique powered by the Internet and similar technologies. And due to its wider acceptance and advantages (Figure 2), it becomes mandatory in modern days organizations.

Figure 1. Basic structure of a cloud computing model

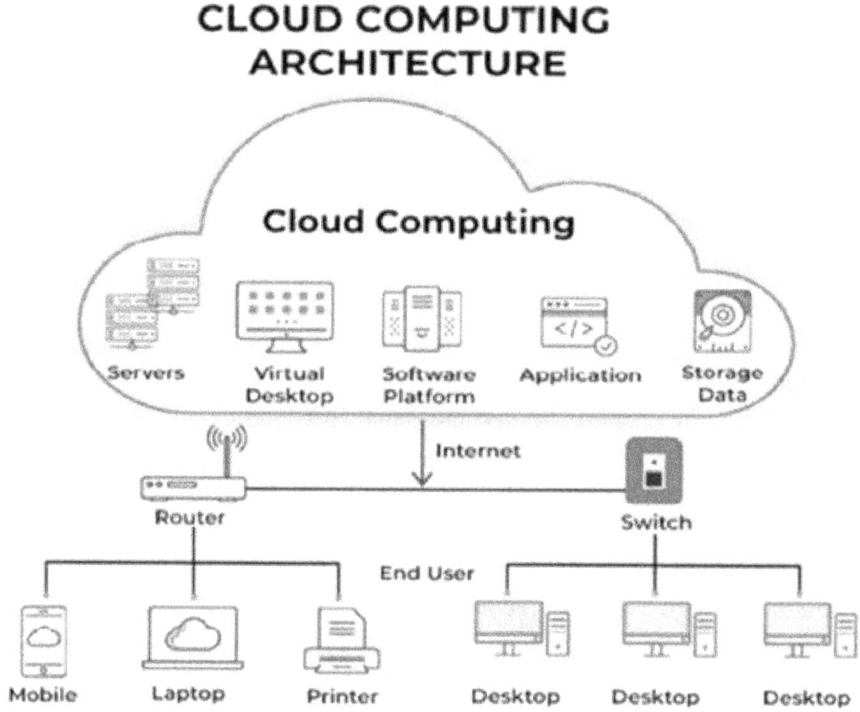

Figure 2. Most valued advantages of cloud computing

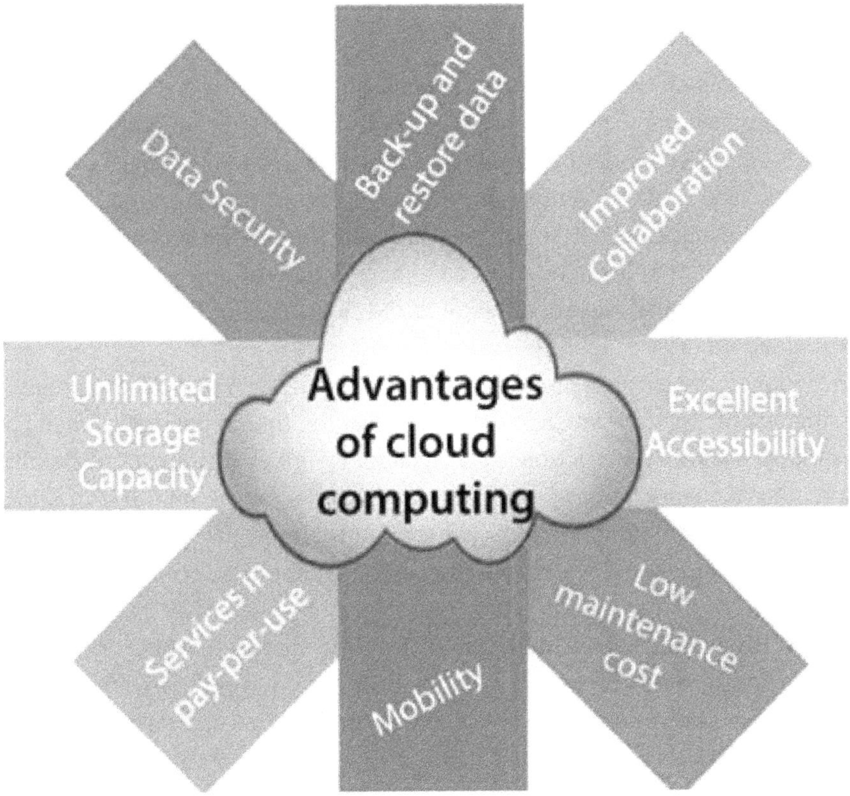

Despite the advantages of cloud computing, social, legal, ownership, responsibility, availability, performance, compliance, cost, regulatory and many other issues amongst others pose challenges. Cloud computing has become a reality and promises to deliver computing services to various sectors of the business community. Nevertheless, security remains a major concern in cloud computing (Paul P.K, Ritam Chatterjee, Sreeramana Aithal), (Cloud computing security), (Gartner).

Cloud security is a collection of procedures and technology designed to address external and internal threats to business security. Organizations need cloud security as they move toward their digital transformation strategy and incorporate cloud-based tools and services as part of their infrastructure (The Biggest Cloud Computing Security Risks Is Impossible to Eliminate).

The terms digital transformation and cloud migration have been used regularly in enterprise settings over recent years. While both phrases can mean different things to different organizations, each is driven by a common denominator: the need for change.

As enterprises embrace these concepts and move toward optimizing their operational approach, new challenges arise when balancing productivity levels and security. While more modern technologies help organizations advance capabilities outside the confines of on-premises infrastructure, transitioning primarily to cloud-based environments can have several implications if not done securely.

Striking the right balance requires an understanding of how modern-day enterprises can benefit from the use of interconnected cloud technologies while deploying the best cloud security practices.

The "cloud" or, more specifically, "cloud computing" refers to the process of accessing resources, software, and databases over the Internet and outside the confines of local hardware restrictions. This technology gives organizations flexibility when scaling their operations by offloading a portion and/or majority of their infrastructure management to third-party hosting providers.

The most common and widely adopted cloud computing services are:

IaaS (Infrastructure-as-a-Service): A hybrid approach, where organizations can manage some of their data and applications on-premises while relying on cloud providers to manage servers, hardware, networking, virtualization, and storage needs.

PaaS (Platform-as-a-Service): Gives organizations the ability to streamline their application development and delivery by providing a custom application framework that automatically manages operating systems, software updates, storage and supporting infrastructure in the cloud.

SaaS (Software-as-a-Service): Cloud-based software hosted online and typically available on a subscription basis. Third-party providers manage all potential technical issues, such as data, middleware, servers and storage, minimizing IT resource expenditures and streamlining maintenance and support functions.

In modern-day enterprises, there has been a growing transition to cloud-based environments and IaaS, Paas or SaaS computing models. The dynamic nature of infrastructure management, especially in scaling applications and services can bring a number of challenges to enterprises when adequately resourcing their departments. These as-a-service models give organizations the ability to offload many of the time-consuming IT-related tasks.

As companies continue to migrate to the cloud, understanding the security requirements for keeping data safe has become critical. While third-party cloud computing providers may take on the management of this infrastructure, the responsibility of data asset security and accountability doesn't necessarily shift along with it.

By default, most cloud providers follow best security practices and take active steps to protect the integrity of their servers. However, organizations need to make their own considerations when protecting data, applications and workloads running on the cloud.

Security threats have become more advanced as the digital landscape continues to evolve. These threats explicitly target cloud computing providers due to an organization's overall lack of visibility in data access and movement. Without taking active steps to improve their cloud security, organizations can face significant governance and compliance risks when managing client information, regardless of where it is stored.

Cloud security should be an important topic of discussion regardless of the size of your enterprise. Cloud infrastructure supports nearly all aspects of modern computing in all industries and across multiple verticals.

However, successful cloud adoption is dependent on putting in place adequate countermeasures to defend against modern-day cyberattacks. Regardless of whether your organization operates in a public, private, or hybrid cloud environment, cloud security solutions and best practices are a necessity when ensuring business continuity.

3.1 Some Cloud Security Challenges

Lack of visibility: It's easy to lose track of how your data is being accessed and by whom, since many cloud services are accessed outside of corporate networks and through third parties.

Multitenancy: Public cloud environments house multiple client infrastructures under the same umbrella, so it's possible your hosted services can get compromised by malicious attackers as collateral damage when targeting other businesses.

Access management and shadow IT: While enterprises may be able to successfully manage and restrict access points across on-premises systems, administering these same levels of restrictions can be challenging in cloud environments. This can be dangerous for organizations that don't deploy bring-your-own device (BYOD) policies and allow unfiltered access to cloud services from any device or geolocation.

Compliance: Regulatory compliance management is oftentimes a source of confusion for enterprises using public or hybrid cloud deployments. Overall accountability for data privacy and security still rests with the enterprise and heavy reliance on third-party solutions to manage this component can lead to costly compliance issues.

Misconfigurations: Misconfigured assets accounted for 86% of breached records in 2019, making the inadvertent insider a key issue for cloud computing environments. Misconfigurations can include leaving default administrative passwords in place or not creating appropriate privacy settings.

4.0 CLOUD COMPUTING APPLICATIONS

Cloud computing is applied in various areas. This section gives a review of some application areas of cloud computing (An overview of cloud security).

4.1 Cloud Computing in Education

Cloud computing application in education (Maaruf Ali, Mahdi H. Miraz), what is the purpose and what are the benefits? With educational organizations worldwide moving to blended or fully remote E-learning due to the COVID-19 pandemic, the challenge of providing secure access and sufficient resources to support E-learning, cloud solutions became even more important. Corporations and educational bodies alike need to host their training materials in easily scalable and protected environments to ensure the cost-effective and efficient availability, accessibility and delivery of learning resources to the clients.

Using cloud computing in E-learning is a great choice as it enables the educators to use cloud scalability, flexibility and security to support the main paradigm of E-learning, which includes training, accessibility from anywhere, at any time, in formats supported by a variety of devices. Running a cloud-based E-learning platform is one of the best decisions an educational body or corporate L&D department can make, and this article will showcase the benefits of delivering E-learning using cloud computing, along with the challenges that arise and the best practices for overcoming them.

Due to flexible data storage and management mechanisms, as well as robust security features, the cloud is the most widely used type of hosting in the EdTech industry. It enables educators to upload vast quantities of E-learning content in various types and formats (videos. texts, presentations, quizzes, questionnaires, etc.), host online classes and interact with their remote learners at the least possible cost.

4.1.1 Cloud Computing and Advantages in Education

Cloud Computing is an emerging component of information technology within network sub-technology. Cloud computing is the virtualization of information technology, infrastructure& services (Gennady. 2011), (Assante et al), 2016), (Elgelany et al 2017), (Ercan, T. 2010), (Etro, F. 2015), (González-Martínez et al 2015). Using cloud computing, hardware, software, systems, platform, database, network can be shared using strong Internet connection. As far as cloud computing is concerned, it has a lot of application areas and among them, education is important one (Paul P.K, Ritam Chatterjee, Sreeramana Aithal), (The Biggest Cloud Computing Security Risks Is Impossible to Eliminate), (Wease, G., Boateng et al 2018). In education, cloud computing is applicable in the following categories:

- Traditional teaching learning process.
- Teaching pedagogy.
- Research activities.
- Office management.
- Educational planning and administration and some other areas as depicted in Figure 3.

Figure 3. Most valued advantages of cloud computing in respect of education

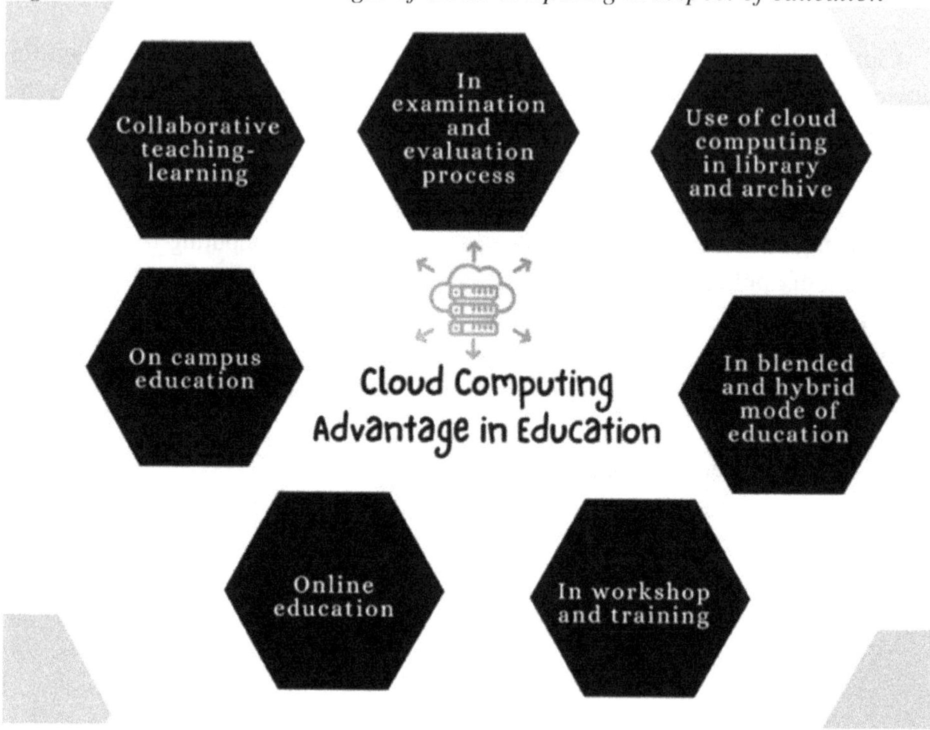

There are several programs at present offered using cloud computing and most are frequently useful in educational premises as employed in the educational field such as:

- Google Slides and Apple Pages
- Apple Keynote and Microsoft Word.
- Microsoft PowerPoint and OneDrive.
- Dropbox.
- Google drive and Docs.
- Capture to cloud.
- Apple iCloud.
- Amazon Cloud Drive.
- Memopal.
- SugarSync.

Regarding advantages, following are important for better teaching learning process using cloud (Figure 3).

Elasticity - Computing gives elasticity to the learners and educational institutes.
Scalability - It offers flexibility and wideness in teaching learning process.
Speed - Cloud computing offers speedy content delivery to all its objects and learners.
Cost management - The initial cost of implementing cloud is minimal but requires a huge amount of funds for technology, purchasing, infrastructure and software. But after implementation, the resources may be utilised in lifelong process.
Remote accessibility - Using cloud computing, anyone can get educational content, instructions and other materials even in a remote and long-distance place.
Time management - Cloud computing gives the flexibility in time management.

4.2 The Application Areas of Cloud Computing in Education

Cloud computing means the virtualization and it is the application of SaaS, PaaS, IaaS. Cloud computing applications in education includes educational management, teaching learning process in traditional teaching learning and also in modern days' educational system. The following are main application areas of cloud computing in education.

(i) **Collaborative teaching-learning** - Cloud computing offers a collaborative teaching learning process as it offers virtualization in educational system. Using cloud computing, any institute may collaborate with other institutes. The faculties, the resources, infrastructure can be shared using this type of collaboration.
(ii) **On campus education** - In on-campus education, that is face to face education or traditional mode of education, cloud computing may be used for utilization of existing resources, sharing of library books and documents, need based online education and meetings, etc.
(iii) **Online education** - Cloud computing is applicable in online and virtual education systems in contemporary scenarios. Today, different educational programmes like certificate-based programmes, diploma, bachelor's degree programmes, master's degree, and doctoral degree programmes are being offered in online mode in real time manner.
(iv) **In workshop and training** - Workshop and training programmes are important in many educational subjects and it is an integral part of overall development like in engineering, computing, health, science, social work. Workshops and trainings are must for complete development. Using cloud computing, workshop and training may be organised in online and virtual mode for saving money, energy and also for instant communication.

(v) **In blended and hybrid mode of education** - Cloud computing is responsible for developing a new age educational system. The combination of traditional mode of education and online mode of education developed a new mode of education which is called blended mode or hybrid mode of education, and which has become popular in developed and developing countries. Many subjects are not suitable in online mode like medicine, engineering, etc, such subjects may be offered in blended or hybrid mode of education. Uses of cloud computing in such educational systems are important and increasing gradually.

(vi) **Use of cloud computing in library and archive** - Applications of cloud computing and virtualization technique basically impact in traditional library management system including document management, barcode and RFID activities. Additionally, in digital library design, development and management, cloud computing is important. In E-learning, online resources and library play vital role and here digital library also plays an important role. Example National Digital Library of India (A project of MoE, Government of India and implemented by IIT Kharagpur), British Library, etc.

(vii) **In examination and evaluation process** - As far as examination and evaluation is concerned, cloud computing is applicable in different areas of examination process. For example, accessing E-answer script using cloud platform, soft-copy based examination online examination live viva-voce etc.

(viii) **Cloud Computing & Education**: The Issues - Cloud computing is very much effective and fruitful (as depicted in Figure 4) in modernizing educational systems and worldwide this trend is growing but still, there are some concerns and issues not only in developing countries but also under-developed countries.

(ix) **Technological integration** - Cloud computing is focused on different kind of applications in the educational sector and there are enormous changes in each application software and systems. Here, the issue of compatibility and technological integration sometimes are considered as most vital.

(x) **Cost of hardware and technology** - Hardware is major factor in implementing cloud-based system because cloud-based hardware are costly when we need to implement private cloud or sometimes hybrid cloud. Apart from hardware, software is also an important aspect in cloud based technological implementation. Cloud based education is purely SAAS, PAAS, and IAAS based.

Figure 4. Core effectiveness of cloud computing and virtualization

(Source: Quitagroup)

4.3 Internet and Bandwidth

Cloud based any system is purely Internet dependent and as far as the educational cloud is concerned, it also requires proper, timely, robust, speedy Internet with proper bandwidth. That is why if in a place Internet is not proper or poor, then implementing cloud-based system becomes difficult.

4.3.1 Remote Accessibility

In developing countries like India, many people reside in rural and remote areas. Therefore, if any candidate seeks cloud based or online education, it may be difficult for them to accommodate.

4.3.2 Security and Infrastructure

Cloud based education system, sometimes purely digital and that is connected with a threat to the individual system by the attackers. Thus, network security, web security, database security all these must be composed together for a better result.

4.3.3 Skill Sets

Cloud base education requires both sided skills and knowledge for its better operation. Many teachers may not be aware about technology and at the same time learners may have lack of skill sets to operate to develop technology.

4.3.4 Summary

Cloud Computing based services are increasing worldwide and its increasing rate is more than 25%. And gradually a large number of students are pursuing online degrees, programs and training through online mode and means. Large number of students now following recorded or offline videos in order to gain knowledge by electronics means and, in this juncture, it is worthy to include the role of cloud computing in online systems. Computer hardware, systems and other technologies become effective with cloud-based services. Mobile learning, distance learning, and web-based collaborative learning become important and effective in modern educational systems, and in all such cases cloud and allied technologies are effective. Apart from developed countries, now many developing countries also putting effort in modernizing the education sector including teaching, learning, and educational administration.

4.4 Cloud Computing in the Academic Environment

A number of organisations such as Amazon, Microsoft, Google, Pamoja, Vodafone Egypt, etc. offer cloud computing environment and infrastructure and academic institutions are already making use of these infrastructure. Cloud computing has extended to the academic space in which tuition content is offered online, enabling students to obtain qualifications online. Some of the educational content provided on the online platforms includes free content also referred to as MOOCS.

The online delivery of tuition is more beneficial for the provision of open distance learning (ODL). The ODL pedagogy thrives on information technology, as it makes the interaction instant and the content accessible anytime, anywhere. This online delivery therefore requires extensive technology infrastructure to support the model and make the content available.

Cloud computing is offering valuable opportunities to learners to access educational materials online. A number of establishments such as google, Microsoft, etc offer study materials online. The infrastructural cost of offering the same can be very daunting to these organizations depending on their financial capability. Cloud computing technology in which the computing resources are held by a different organization enables the users to use the resources and only pay for time and the resources used, offers a more affordable access to learning opportunities. In addition, to making the same educational resources accessible to many learners.

4.4.1 Benefits of Cloud Computing in Academic Environment

Cloud computing offers the benefits of lowering infrastructural cost by residing the educational resources in the cloud. The institution offering the qualification does not necessarily have to meet the infrastructural cost. This could be a major saving in terms of operational as capital expenditures.

Cloud computing enables academic institutions to offer online learning. The online learning makes it possible for many learners at different locations to be taught at the same time. In addition, cloud computing supports availability and accessibility of learning resources from anywhere and at any time.

For the learners, cloud computing makes it possible to not only have access to one source of learning materials but, several sources offered by different entities in the same field. Thus, the learners have access to the same learning resources available from a variety of providers offering the learning resources. Thus, the learners are not restricted to one learning resources provider but are exposed to many learning resources providers.

4.4.2 Challenges of Cloud Computing in Academic Environment

Nevertheless, this diversity of learning resources and the different digitalisation formats of the learning resources present a dilemma to the learners in that it becomes the learner's responsibility to sift through the materials to pick what is relevant for their circumstances. This can at times be a big challenge to the learners especially those who are not Internet savvy. The result could be an overwhelming information overload for the learners.

In addition, the learning resources are offered by different institutions in different formats and styles. Coupled with this is the proliferation of devices used to access the learning resources.

The question for the learners thus, becomes: which sources, formats, styles, devices, etc. is most suitable for learning.

The same challenges experienced by the learners are also experienced by the educators. As for educators, it's made much worse by the fact that most of the educators are not trained to prepare learning materials for delivery Internet delivery, leave alone the ever-changing devices that are available in the market and learners can use to have access to the learning materials. It thus, becomes a question: are educators well equipped/trained to develop learning materials for Internet delivery. This is in addition to the development of the learning materials in various formats that can easily be accessed by the plethora of devices that are readily available to the learners.

The educators are working in an environment in which learning resources are freely available and accessible by the learners as well. The challenge is in guiding both learners and educators is to sift through the available learning resource materials and to select relevant and appropriate materials. The digitalisation of learning resources has resulted in communication dynamics. The impacts of communication dynamics are both negative and positive. Institutions and individuals must adopt or institute ways and means of managing the impacts and to lessen the negative impacts while reaping the benefits of the positive impacts.

5.0 CLOUD COMPUTING IN RESEARCH

The cloud may be a good fit for researchers who are looking to easily share data from multiple locations or to gain access to public datasets. Cloud processing and sharing can provide additional collaboration opportunities for data that is already findable, accessible, interoperable, and reusable (FAIR). Using one cloud repository, researchers working a project gain timely access to work-in-progress and additions or amendments is not only visible but also accessible to all researchers. Cloud computing thus, fast-tracks the processes of a project which involves multiple researchers.

6.0 THE APPLICATION OF CLOUD COMPUTING IN GOVERNMENT MANAGEMENT

Nowadays, the realization of government functions has become more and more inseparable from the network. At the same time, the public also expects the government to provide more and more convenient and fast related services. Interactive functions can also advance with the times. At the same time, the government wants to make public services to be available more efficiently. In the context of big data and cloud computing, the openness and sharing of government-owned data has an extremely important impact on the society. In the process of government management, the traditional management way already can't adapt to the needs of a large number of data exchange, but now the government management leveraging slightly to the research of innovative technology still lags behind, the rise of cloud computing provides a new way for the government management innovation. This article aims to introduce cloud computing to the government management for research purposes, to provide a new way for the government management reform (Kumar, D et al 2017), (Kumar, Det al, 2017).

The current mainstream view of government management innovation is that it refers to a series of new changes in administrative functions, administrative modes, administrative styles, government policies and regulations, and administrative systems caused by changes in administrative environment and tasks. Connotation of the government management innovation is a major factor in cultivating and sets up the government management mode innovation, refers to the government in the process of providing social management and public service, efforts to create a peer in decision science, administration according to law, power and responsibility, strengthening service, pay attention to performance, effective supervision as the main characteristics of the friendly administrative interface, make the public satisfied. With the rapid development of information technology and the continuous advancement of the transformation of government functions, China's e-government construction has also entered a new stage of vigorous development.

By comparing the concept and technology of cloud computing e-government and traditional e-government, we know that cloud computing will bring more infinite possibilities to government management. The development of government management based on cloud computing has brought new opportunities for the transformation of traditional government management, and also provided dynamic support and technical support for the innovation of government management. The improvement of management mode is one of the main ways to improve government management, e-government is an innovation shortcut of government management mode, and the innovation of e-government management mode is to use modern information technology to further improve work efficiency and improve the ability to serve the public.

7.0 CLOUD COMPUTING SECURITY CHALLENGES

Select the right Key Management as a Service to mitigate cloud data security challenges. Stay compliant and retain control over your cloud data irrespective of where it resides, Figure 5 presents some of the identified security issues. Fortanix KMS/HSM. Fortanix EDP. Types: Enterprise Key Management, Tokenization (Hua Zhang).

Cloud security allows one to consolidate protection of cloud-based networks for streamlined, continuous monitoring and analysis of numerous devices, endpoints, and systems. It also enables you to centrally manage software updates and policies from one place and even implement and action disaster recovery plans (Gartner Report), (Free Cloud Computing).

Figure 5. Why is cloud computing security crucial for your business (Zhang)

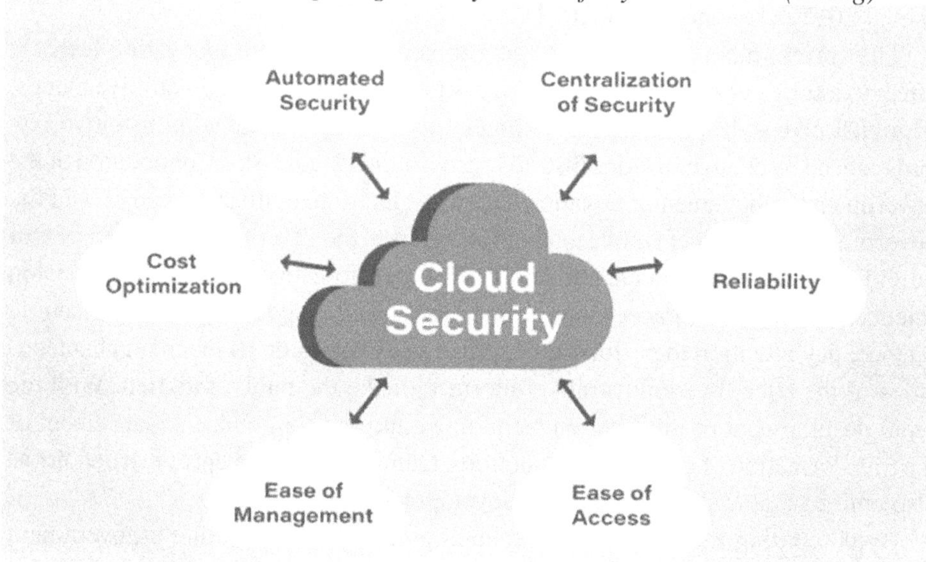

8.0 ARTIFICIAL INTELLIGENCE IN EDUCATION

Artificial Intelligence (AI) has the potential to address some of the biggest challenges in education today, innovate teaching and learning practices, and accelerate progress towards SDG 4. However, rapid technological developments inevitably bring multiple risks and challenges, which have so far outpaced policy debates and regulatory frameworks. UNESCO is committed to supporting Member States to harness the potential of AI technologies for achieving the Education 2030 Agenda, while ensuring that its application in educational contexts is guided by the core principles of inclusion and equity.

UNESCO's mandate calls inherently for a human-centred approach to AI. It aims to shift the conversation to include AI's role in addressing current inequalities regarding access to knowledge, research and the diversity of cultural expressions and to ensure AI does not widen the technological divides within and between countries. The promise of "AI for all" must be that everyone can take advantage of the technological revolution underway and access its fruits, notably in terms of innovation and knowledge.

Furthermore, UNESCO has developed within the framework of the Beijing Consensus a publication aimed at fostering the readiness of education policymakers in artificial intelligence. This publication, Artificial Intelligence and Education:

Guidance for Policymakers, will be of interest to practitioners and professionals in the policymaking and education communities. It aims to generate a shared understanding of the opportunities and challenges that AI offers for education, as well as its implications for the core competencies needed in the AI era (Artificial intelligence in education).

International Forum on artificial intelligence and education through its projects, UNESCO affirms that the deployment of AI technologies in education should be purposed to enhance human capacities and to protect human rights for effective human-machine collaboration in life, learning and work and for sustainable development is depicted in Figure 6. Together with partners, international organizations, and the key values that UNESCO holds as pillars of their mandate, UNESCO hopes to strengthen their leading role in AI in education, as a global laboratory of ideas, standard setter, policy advisor and capacity builder.

Figure 6. International forum on AI and education

'We need to renew this commitment as we move towards an era in which artificial intelligence – a convergence of emerging technologies – is transforming every aspect of our lives (…),' said Ms Stefania Giannini, UNESCO Assistant Director-General for Education at the International Conference on Artificial Intelligence and Education held in Beijing in May 2019. 'We need to steer this revolution in the right direction, to improve livelihoods, to reduce inequalities and promote a fair and inclusive globalization''(Artificial intelligence in education).

Artificial Intelligence (AI) has been transforming various industries, including education. However, implementing AI in education also poses some challenges that need to be addressed. At the same time, there are significant opportunities that AI can bring to education. This article discusses the key challenges and opportunities of AI in education.

8.1 Key Challenges of AI in Education

One of the significant challenges of AI in education is to ensure that it does not replace teachers. AI should be used as a tool to support teachers and not as a substitute for them. The role of teachers is crucial in shaping the minds and lives of students, and AI should not take that away from them. Therefore, it is essential to use AI in conjunction with teachers to enhance their teaching capabilities.

Another challenge is the lack of data privacy and security in AI systems. AI systems generate and collect vast amounts of data, including personal information about students and teachers. This data must be protected from cyber-attacks and breaches. Therefore, it is necessary to have robust data privacy policies and security measures in place to protect this information.

The last significant challenge of AI in education is the digital divide. Not all students have access to technology and the internet, which limits their ability to benefit from AI tools. Therefore, it is essential to ensure that AI tools are accessible to all students, regardless of their socio-economic background.

8.2 Opportunities for AI in Education

AI can revolutionise education by providing personalized learning experiences. AI algorithms can analyse data about students' learning styles, behaviours, and interests to create customised learning paths. This enables educators to offer personalised learning experiences that meet the needs of individual students.

Another opportunity of AI in education is to improve student engagement. AI systems can use gamification techniques to make learning more enjoyable and interactive. This can help students stay engaged and motivated throughout their learning journey.

Lastly, AI can help educators identify learning gaps and provide targeted interventions. AI algorithms can analyse student data to identify areas where students are struggling. Educators can then use this information to provide targeted interventions to support students' learning and help them overcome any challenges.

In conclusion, AI has the potential to transform education significantly. However, to fully realise its benefits, we must address the key challenges of AI in education. By doing so, we can create a future where AI enhances education and provides students with personalised and engaging learning experiences.

9.0 CONCLUSION

In cloud environment, there is no one particular application which suits all requirements. It thus, becomes imperative that academic institutions should make informed choices of what best suits their operational needs for their clients (staff and students).

Cloud computing is a reality that cannot be ignored. An academic institution ignoring cloud computing does so at their own peril. Despite the issues such as security, information integrity, legal, social, etc, associated with cloud computing, cloud computing offers a cost-effective means (without worrying about the infrastructure) for academic institutions to not only offer great learning opportunities for the learners but, also a means to diversify their qualification offering based on the fact that the learning materials are not only readily available in different formats, styles, etc, thus, providing the learners with realistic choices and a conducive learning environment in which to explore the knowledge boundaries.

The ready availability and accessibility of academic resource materials not only from varying entities but in various formats as well presents opportunities and challenges in varying applications areas.

This chapter was aimed at the review of the opportunities and challenges presented by availability and accessibility of academic resource materials. While it was not possible to include all expected opportunities and challenges. From the reviewed samples, it can be concluded that the opportunities far outweigh the challenges.

The opportunities and challenges results in communication dynamics. Individuals and institutions should come up with ways and means to address the communication dynamics for them to leverage on the opportunities to offer improved and better services to the clients.

Cloud computing technology reviewed in this chapter is identified as one of the technologies to support availability, accessibility of academic resource materials and online tuition. Cloud computing has opportunities and challenges. Nevertheless, the opportunities outweigh the challenges. Thus, entities aiming at minimising capital expenditures should leverage on cloud computing to offer efficient services.

With good planning, the communication dynamics can be well managed to lessen the disruptions and adverse effects of the same and consequently reap the benefits of communication dynamics.

REFERENCES

An overview of cloud security https://www.ibm.com/com/topics/cloud-security

Artificial intelligence in education https://www.unesco.org/en/digital-education/artificial-intelligence

Assante, D., Castro, M., Hamburg, I., & Martin, S. (2016). The use of cloud computing in SMEs. *Procedia Computer Science*, 83, 1207–1212. DOI: 10.1016/j.procs.2016.04.250

Cloud computing security, https://en.wikipedia.org/wiki/

Elgelany, A., & Alghabban, W. G. (2017). Cloud computing: Empirical studies in higher education a literature review. *International Journal of Advanced Computer Science and Applications*, 8(10). Advance online publication. DOI: 10.14569/IJACSA.2017.081017

Ercan, T. (2010). Effective use of cloud computing in educational institutions. *Procedia: Social and Behavioral Sciences*, 2(2), 938–942. DOI: 10.1016/j.sbspro.2010.03.130

Etro, F. (2015). The economics of cloud computing. In Cloud Technology: Concepts, Methodologies, Tools, and Applications (pp. 2135-2148). IGI Global. Gartner: Seven Could-Computing Security Risks, https://www.infoworld.comp/print/368853

Free Cloud Computing with AWS - Secure Cloud Services. https://aws.amazon.com

Tarnavsky, G. A., & Vorozhtsov, E. V. (2011). Cloud Computing in Science and Engineering and the "SciShop. ru" Computer Simulation Center. *Engineering, Technology &. Applied Scientific Research*, 1(6), 133–138.

Giannini, S. Think piece by UNESCO Assistant Director-General for Education, Generative Artificial Intelligence in education: Think piece by Stefania Giannini, https://www.unesco.org/en/articles/generative-artificial-intelligence-education-think-piece-stefania-giannini?hub=32618

González-Martínez, J. A., Bote-Lorenzo, M. L., Gómez-Sánchez, E., & Cano-Parra, R. (2015). Cloud computing and education: A state-of-the-art survey. *Computers & Education*, 80, 132–151. DOI: 10.1016/j.compedu.2014.08.017

Marot, T. B. (2020). *Space, the final legal frontier: a look at legal impediments to the increasing commercialisation of outer space* (Doctoral dissertation, University of the Witwatersrand, Johannesburg, South Africa).

Kumar, D., Baranwal, G., Raza, Z., & Vidyarthi, D. P. (2017). A systematic study of double auction mechanisms in cloud computing. *Journal of Systems and Software*, 125, 234–255. DOI: 10.1016/j.jss.2016.12.009

Kumar, D., Samalia, H. V., & Verma, P. (2017). Factors influencing cloud computing adoption by small and medium-sized enterprises (SMEs). In *India*. Pacific Asia Journal of the Association for Information Systems. Maaruf Ali, Mahdi H. Miraz. Cloud Computing Applications.

Paul, P. K., ... Cloud Computing and its Impact in Education, Teaching and Research-A Scientific Review, https://www.researchgate.net/publication/371834367_Cloud_Computing_and_its_Impact_in_Education_Teaching_and_Research-A_Scientific_Review https://www.researchgate.net/publication/255749774_Cloud_Computing_Applications

Private cloud http://http.networkcomputing.com/private-cloud-tech-center

Public cloud, http://www.networkcomputing.com/public-cloud-tech-center

Wease, G., Boateng, K., Yu, C. J., Chan, L., & Barham, H. (2018). Technology assessment: cloud service adoption decision. *Infrastructure and technology management: Contributions from the energy, healthcare and transportation sectors*, 447-471.

Zhang, H. (2020, February). The application of cloud computing in government management. []. IOP Publishing.]. *IOP Conference Series. Materials Science and Engineering*, 750(1), 012166.

Compilation of References

Aaker, D. (2018). *Creating signature stories: Strategic messaging that persuades, energizes and inspires*. Morgan James Publishing.

Abd Halim, M., Foozy, C. F. M., Rahmi, I., & Mustapha, A. (2018). A review of live survey application: SurveyMonkey and SurveyGizmo. *JOIV: International Journal on Informatics Visualization, 2*(4-2), 309-312.

Abdalla, A., & et al. (2022). "Social Media as a Stage: A Behind the Scenes Analysis of Performative Activism, "Cancel Culture," and Effective Allyship". McMaster Undergraduate Journal of Social Psychology, Vol. 1, n.º1, pp. 83-122.

Abdullah, M. (2023). Analyzing augmented reality (AR) and virtual reality (VR) recent development in education. *Social Sciences & Humanities Open*, 8(1), 100532. Advance online publication. DOI: 10.1016/j.ssaho.2023.100532

Abegunrin, O., & Manyeruke, C. (2020). *China's Power in Africa*. Springer International Publishing. DOI: 10.1007/978-3-030-21994-9

Abidin, C. (2020). Mapping Internet Celebrity on TikTok: Exploring Attention Economies and Visibility Labours. *Cultural Science Journal*, 12(1), 77–103. DOI: 10.5334/csci.140

Adelfio, M., Serrano-Estrada, L., Martí-Ciriquián, P., Kain, J. H., & Stenberg, J. (2020). Social Activity in Gothenburg's Intermediate City: Mapping Third Places through Social Media Data. *Applied Spatial Analysis and Policy*, 13(4), 985–1017. Advance online publication. DOI: 10.1007/s12061-020-09338-3

Adetayo, A. J., & Williams-Ilemobola, O. (2021). Librarians' generation and social media adoption in selected academic libraries in Southwestern, Nigeria. *Library Philosophy and Practice (e-Journal), 4984*. https://digitalcommons.unl.edu/libphilprac/4984

Adetayo, A. J. (2022). Harnessing Coopetition for the Survival of Nigerian Public Libraries: Role of Smart Technologies. *The International Information & Library Review*, •••, 1–15. DOI: 10.1080/10572317.2022.2070400

Adetayo, A. J., Komolafe, R. S., & Olalere, J. O. (2022). Virtual Learning and Library Usage during COVID-19 Pandemic: The Adeleke University Experience. *Journal of Library & Information Services in Distance Learning*, 16(1), 1–21. DOI: 10.1080/1533290X.2022.2066744

Adnan, M., Khan, A. U., & Hayee, R. (2021). Integrated Marketing Communication and Customer Satisfaction as a function of Customer Loyalty with the Moderating role of Product Innovation Moderation. *Sustainable Business and Society in Emerging Economies*, 3(2), 133–148. DOI: 10.26710/sbsee.v3i2.1910

Agbebi, M. (2019). Exploring the human capital development dimensions of Chinese investments in Africa: Opportunities, implications and directions for further research. *Journal of Asian and African Studies*, 54(2), 189–210. DOI: 10.1177/0021909618801381

Agina-Obu, R., & Sawyerr-George, O. E. (2023). Evaluation of Users' Satisfaction of Information Resources in University Libraries in Nigeria: A Case Study. *Indian Journal of Information Sources and Services*, 13(1), 1–5. DOI: 10.51983/ijiss-2023.13.1.3378

Ahearne, M., Atefi, Y., Lam, S. K., & Pourmasoudi, M. (2022). The future of buyer–seller interactions: A conceptual framework and research agenda. *Journal of the Academy of Marketing Science*, 50(1), 22–45. DOI: 10.1007/s11747-021-00803-0 PMID: 34608341

Ahmad, H., Islam, M. Z., Ali, R., Haider, A., & Kim, H. S. (2021). Intelligent stretch optimization in information centric Networking-Based tactile internet applications. *Applied Sciences (Basel, Switzerland)*, 11(16), 7351. DOI: 10.3390/app11167351

Ahmad, M. (2023). Connecting People and Places: How Citizen Diplomacy and VGI Are Strengthening Disaster Response and Community Development. In Kavoğlu, S., & Köksoy, E. (Eds.), *Global Perspectives on the Emerging Trends in Public Diplomacy* (pp. 195–226). IGI Global., DOI: 10.4018/978-1-6684-9161-4.ch009

Ahmad, M. (2024). From Crisis to Opportunity: How VGI Is Transforming Disaster Response and Recovery Efforts. In Siyal, S. (Ed.), *Using Crises and Disasters as Opportunities for Innovation and Improvement* (pp. 215–232). IGI Global., DOI: 10.4018/978-1-6684-9522-3.ch013

Ahmed, I., Jeon, G., & Piccialli, F. (2021). From artificial intelligence to explainable artificial intelligence in industry 4.0: A survey on what, how, and where. *IEEE Transactions on Industrial Informatics*, 18(8), 5031–5042. DOI: 10.1109/TII.2022.3146552

Airhihenbuwa, C. O. (1989). Perspectives on AIDS in Africa: strategies for prevention and control. *AIDS Education and Prevention*. Airhihenbuwa, C. O. (1989). Perspectives on AIDS in Africa: strategies for prevention and control. *AIDS Education and Prevention*.

Airhihenbuwa, C. O., Iwelunmor, J., Munodawafa, D., Ford, C. L., Oni, T., Agyemang, C., Mota, C., Ikuomola, O. B., Simbayi, L., Fallah, M. P., Qian, Z., Makinwa, B., Niang, C., & Okosun, I. (2020). Peer reviewed: Culture matters in communicating the global response to COVID-19. *Preventing Chronic Disease*, 17, 200245. https://www.ncbi.nlm.nih.gov/pmc/articles/PMC7367065/. DOI: 10.5888/pcd17.200245

Akbari, J., Salamzadeh, Y., & Farzad, F. S. (2021). Collaborative Governance, a Tool for Enhancing Compliance with COVID-19 Prevention Guidelines. A Study in Iranian Context.

Akdeniz, D. (2019). Tracking the history of restaurants through the art of painting. *Gastroia: Journal of Gastronomy and Travel Research*, 3(2), 346–365. DOI: 10.32958/gastoria.512544

Akena, F. A. (2022). The "Miss Curvy Uganda" pageant: Representation, commodification and exploitation of women's bodies. *Social Semiotics*, 32(2), 224–239. DOI: 10.1080/10350330.2020.1743524

Ala, A., Edge, C., Zumla, A., & Shafi, S. (2021). Specific COVID-19 messaging targeting ethnic minority communities. *EClinicalMedicine*, 35, 100862. https://www.thelancet.com/journals/eclinm/article/PIIS2589-5370(21)00142-5/fulltext. DOI: 10.1016/j.eclinm.2021.100862 PMID: 33997747

Alalwan, N., Cheng, L., Al-Samarraie, H., Yousef, R., Alzahrani, A. I., & Sarsam, S. M. (2020). Challenges and prospects of virtual reality and augmented reality utilization among primary school teachers: A developing country perspective. *Studies in Educational Evaluation*, 66, 100876. DOI: 10.1016/j.stueduc.2020.100876

Al-Amri, S., Hamid, S., Noor, N. F. M., & Gani, A. (2023). A framework for designing interactive mobile training course content using augmented reality. *Multimedia Tools and Applications*, 82(20), 30491–30541. Advance online publication. DOI: 10.1007/s11042-023-14561-4 PMID: 36855614

Al, Ansi., A, M., & Al-Ansi, A. (2020). Future of education post covid-19 pandemic: Reviewing changes in learning environments and latest trends. *Solid State Technology*, 63(6), 201584–201600.

Al-Ansi, A. (2022). Investigating characteristics of learning environments during the COVID-19 pandemic: A systematic review. *Canadian Journal of Learning and Technology*, 48(1). Advance online publication. DOI: 10.21432/cjlt28051

Al-Ansi, A. M., & Fatmawati, I. (2023). Integration of ICT in higher education during covid-19 pandemic: A case study. *International Journal of Learning and Change*, 15(2).

Al-Ansi, A. M., Garad, A., & Al-Ansi, A. (2021). ICT-based learning during Covid-19 outbreak: Advantages, opportunities and challenges. *Gagasan Pendidikan Indonesia*, 2(1), 10–26. DOI: 10.30870/gpi.v2i1.10176

Al-Azawi, R. (2018). Embedding augmented and virtual reality in educational learning method: Present and future. In In2018 9th international conference on Information and communication systems (ICICS) (pp. 218–222). IEEE.

Alberti, P. M., Castaneda, M. J., Castrucci, B. C., & Harrison, L. M. (2020). Engaging With Communities—Lessons (Re) Learned From COVID-19. *PCD Staff, 17*, E65. https://www.cdc.gov/pcd/collections/pdf/US-Public-Health-Response-to-COVID-19.pdf#page=7

Alexander, J., Barton, J., & Goeser, C. (n.d.). Transforming the art museum experience: gallery one. Museumsandtheweb.Com. Arya, A., Hartwick, P., Graham, S., & Nowlan, N. (n.d.). Virtual Space as a Learning Environment: Two Case Studies. http://www.avaya.com

Ali, W. (2020). Online and remote learning in higher education institutes: A necessity in light of COVID-19 pandemic. *Higher Education Studies*, 10(3), 16–25. DOI: 10.5539/hes.v10n3p16

Alizadehsalehi, S., Hadavi, A., & Huang, J. C. (2021). Assessment of AEC students' performance using BIM-into-VR. *Applied Sciences (Basel, Switzerland)*, 11(7), 3225. DOI: 10.3390/app11073225

Allen, E. (2022). Jess Weiner and the dove real beauty campaign: Selling feminism for profit or social change? *Women Leading Change: Case Studies on Women, Gender, and Feminism*, 6(1), 18–37.

Allen, E. (2022). Jess Weiner and the Dove Real Beauty Campaign: Selling Feminism for Profit or Social Change? *Women Leading Change: Case Studies on Women, Gender, and Feminism*, 6(1), 18–37.

Allen, R. B., & Hall, C. (2010). Automated Processing of Digitized Historical Newspapers beyond the Article Level: Sections and Regular Features. In Chowdhury, G., Koo, C., & Hunter, J. (Eds.), *The Role of Digital Libraries in a Time of Global Change*. Springer. DOI: 10.1007/978-3-642-13654-2_11

Allen, R. B., & Schalow, J. (1999). Metadata and data structures for the historical newspaper digital library. In *Proceedings of the eighth international conference on Information and knowledge management (CIKM '99)* (pp. 147–153). Association for Computing Machinery. DOI: 10.1145/319950.319971

Allen, R. B., Waldstein, I., & Zhu, W. (2008). Automated processing of digitized historical newspapers: Identification of segments and genres. In Buchanan, G., Masoodian, M., & Cunningham, S. J. (Eds.), *Digital libraries: Universal and ubiquitous access to information* (pp. 379–386). Springer. DOI: 10.1007/978-3-540-89533-6_49

Alloza, A. (2008). Brand engagement and brand experience at BBVA, the transformation of a 150 years old company. *Corporate Reputation Review*, 11(4), 371–379. DOI: 10.1057/crr.2008.31

Alraih, S., Shayea, I., Behjati, M., Nordin, R., Abdullah, N. F., Abu-Samah, A., & Nandi, D. (2022). Revolution or evolution? Technical requirements and considerations towards 6G mobile communications. *Sensors (Basel)*, 22(3), 762. DOI: 10.3390/s22030762 PMID: 35161509

Alsan, M., & Wanamaker, M. (2018). Tuskegee and the health of black men. *The Quarterly Journal of Economics*, 133(1), 407–455. DOI: 10.1093/qje/qjx029 PMID: 30505005

Amirkhanyan, A., & Meinel, C. (2016). Analysis of the value of public geotagged data from twitter from the perspective of providing situational awareness. *Lecture Notes in Computer Science (Including Subseries Lecture Notes in Artificial Intelligence and Lecture Notes in Bioinformatics), 9844 LNCS*. DOI: 10.1007/978-3-319-45234-0_48

Amnesty International. (29 de maio de 2023). "Who We Are". Available: https://www.amnesty.org/en/who-we-are/

Amnistia Internacional (s.d.). "Como fazemos a mudança acontecer?" Available: https://www.amnistia.pt/como-atuamos/ (last access 30/05/2023).

An overview of cloud security https://www.ibm.com/com/topics/cloud-security

Anderson, K. E. (2020). Getting acquainted with social networks and apps: It is time to talk about TikTok. *Library Hi Tech News*, 37(4), 7–12. DOI: 10.1108/LHTN-01-2020-0001

Anderson-Tarver, C. (2015). Crisis Mapping the 2010 Earthquake in OpenStreetMap Haiti. In *ProQuest Dissertations and Theses*. http://search.proquest.com/docview/1690497435?accountid=8330\nhttp://jn8sf5hk5v.search.serialssolutions.com/?ctx_ver=Z39.88-2004&ctx_enc=info:ofi/enc:UTF-8&rfr_id=info:sid/ProQuest+Dissertations+%26+Theses+A%26I&rft_val_fmt=info:ofi/fmt:kev:mtx:dissertati

ANNALS. (2021, July).. . *AAPSS*, 696, 223. DOI: 10.1177/00027162211050007

Antoniou, V., Morley, J., & Haklay, M. (2010). *Web 2.0 geotagged photos: Assessing the spatial dimension of the phenomenon*. Geomatica.

Antoniuk, K., Mokiy, A., Pikh, M., Bukharina, L., & Antoniuk, D. (2021). Ensuring of Consumer Choice Ecologization on the Basis of Consumption Safety: International Experience for Ukraine. In *SHS Web of Conferences* (Vol. 100, p. 05009). EDP Sciences.

APAV. (s.d.). "Quem Somos". Available: https://apav.pt/apav_v3/index.php/pt/aapav/quem-somos (last access 11/06/2023).

Apuke, O. D. (2017). Quantitative research methods: A synopsis approach. *Kuwait Chapter of Arabian Journal of Business and Management Review*, 33(5471), 1–8.

Arkio. (2024). Arkio. Arkio. https://www.arkio.is/

Artificial intelligence in education https://www.unesco.org/en/digital-education/artificial-intelligence

Asbury, S., & Ball, R. (2016). *The practical guide to corporate social responsibility: do the right thing*. Routledge. DOI: 10.4324/9781315697659

Assante, D., Castro, M., Hamburg, I., & Martin, S. (2016). The use of cloud computing in SMEs. *Procedia Computer Science*, 83, 1207–1212. DOI: 10.1016/j.procs.2016.04.250

Association of College and University Educators (ACUE). (2023). Fostering a culture of belonging: Four insights you need to know. https://acue.org/blog/fostering-a-culture-of-belonging-four-insights-you-need-to-know/

Astuti, S., & Hanan, H. (2016). The Behaviour of Consumer Society in Consuming Food at Restaurants and Cafes. *Journal of ASIAN Behavioural Studies*, 1(1), 71–79. DOI: 10.21834/jabs.v1i1.173

Aukstakalnis, S. (2016). *Practical augmented reality: A guide to the technologies, applications, and human factors for AR and VR*. New York Addison-Wesley Professional.

Ayoub, A., & Pulijala, Y. (2019). The application of virtual reality and augmented reality in Oral & Maxillofacial Surgery. *BMC Oral Health*, 19(1), 1–8. DOI: 10.1186/s12903-019-0937-8 PMID: 31703708

Ayyaswamy, K. (2024b). Enhancing Digital Technology Planning, Leadership, and Management to Transform Education. In Bhatia, M., & Mushtaq, M. T. (Eds.), *Navigating Innovative Technologies and Intelligent Systems in Modern Education* (pp. 1–9). IGI Global., DOI: 10.4018/979-8-3693-5370-7.ch001

Babbie, E. R. (2011). *The Basics of Social Research 5th (fifth) Edition*. Wadsworth, Cengage Learning.

Bachmann, M., Subramaniam, A., Born, J., & Weibel, D. (2023). Virtual reality public speaking training: Effectiveness and user technology acceptance. *Frontiers in Virtual Reality*, 4, 1242544. Advance online publication. DOI: 10.3389/frvir.2023.1242544

Balmer, J. M., Powell, S. M., & Greyser, S. A. (2011). Explicating Ethical Corporate Marketing. Insights from the BP Deepwater Horizon Catastrophe: The Ethical Brand That Exploded and Then Imploded. *Journal of Business Ethics*, 102(1), 1–14. DOI: 10.1007/s10551-011-0902-1

Bandura, A. (1969). Social-learning theory of identificatory processes. In D. A. Goslin (Ed.), *Handbook of socialization theory and research* (pp. *213*- 262. Rand McNally & Company. Bandura, A. (1971). *Social learning theory*. General Learning Press.

Bandura, A. (1972). Modeling theory: Some traditions, trends, and disputes. In Parke, R. D. (Ed.), *Recent trends in social learning theory* (pp. 35–61). Academic Press. DOI: 10.1016/B978-0-12-545050-8.50008-0

Bandura, A. (1986). *Social foundations of thought and action: A social cognitive theory*. Prentice-Hall, Inc.

Bandura, A., & Walters, R. H. (1977). *Social learning theory* (Vol. 1). Prentice Hall.

Bangay, S., & McKenzie, S. (2022). A Classification Schema for Designing Augmented Reality Experiences. *Informing Science and IT Education Conference*. DOI: 10.28945/4942

Ban, H., & Ning, J. (2021). *Online English teaching based on artificial intelligence internet technology embedded system* (Vol. 2021). Mobile Information Systems.

Bansal, G., Rajgopal, K., Chamola, V., Xiong, Z., & Niyato, D. (2022). Healthcare in Metaverse: A Survey on Current Metaverse Applications in Healthcare. *IEEE Access : Practical Innovations, Open Solutions*, 10, 1–1. DOI: 10.1109/ACCESS.2022.3219845

Bapat, D., & Thanigan, J. (2016). Exploring Relationship among Brand Experience Dimensions, Brand Evaluation and Brand Loyalty. *Global Business Review*, 17(6), 1357–1372. DOI: 10.1177/0972150916660401

Barros, C., Costa, B. F., & Ramos, D. (2023a). *Guia de boas práticas para a indústria 4.0* [Best practices guide for industry 4.0]. UA Editora.

Barros, C., Costa, B. F., & Ramos, D. (2023b). *Industry 4.0: Best practices for the digital transition.* UA Editora – Universidade de Aveiro.

Barry, J. R., Lee, E. A., & Messerschmitt, D. G. (2004). *Digital Communication* (3rd ed., Vol. 1). Springer Science and Business Media LLC. DOI: 10.1007/978-1-4615-0227-2

Barter, A. (2019). *Shifting environmental attitudes and influencing pro-environmental behaviour through the process of managing how, when and in what way we communicate* (Doctoral dissertation, Stellenbosch: Stellenbosch University).

Bartholomew, C. (2020). China and 5G. *Issues in Science and Technology*, 36(2), 50–57.

Basch, C. H., Hillyer, G. C., & Jaime, C. (2020). COVID-19 on TikTok: Harnessing an emerging social media platform to convey important public health messages. *International Journal of Adolescent Medicine and Health.* Advance online publication. DOI: 10.1515/IJAMH-2020-0111/MACHINEREADABLECITATION/RIS PMID: 32776899

Basch, C. H., Mohlman, J., Fera, J., Tang, H., Pellicane, A., & Basch, C. E. (2021). Community mitigation of COVID-19 and portrayal of testing on tiktok: Descriptive study. *JMIR Public Health and Surveillance*, 7(6), e29528. https://publichealth.jmir.org/2021/6/e29528. DOI: 10.2196/29528 PMID: 34081591

Basith, A., & Syafi'i, A. (2020). Instagram as Media in Teaching Writing Recount Text for Senior High School Students. *SALEE: Study of Applied Linguistics and English Education*, 1(01), 11–20. DOI: 10.35961/salee.v1i01.66

Bauman, Z. (1990). *Liquid Modernity.* Polity Press.

Baxter, T., Shenoy, S., Lee, H. S., Griffith, T., Rivas-Baxter, A., & Park, S. (2023). Unequal outcomes: The effects of the COVID-19 pandemic on mental health and well-being among Hispanic/Latinos with varying degrees of 'Belonging'. *The International Journal of Social Psychiatry*, 69(4), 853–864. DOI: 10.1177/00207640221140285 PMID: 36573293

Bazzi, S., Fiszbein, M., & Gebresilasse, M. (2020). Rugged Individualism and Collective (In)action During the COVID-19 Pandemic (No. w27776; p. w27776). National Bureau of Economic Research. https://doi.org/DOI: 10.3386/w27776

Beck, D. (2019). Augmented and virtual reality in education: Immersive learning research. *Journal of Educational Computing Research*, 57(7), 1619–1625. DOI: 10.1177/0735633119854035

Beck, U. (1986). *Risk Society: Towards a New Modernity*. SAGE Publications.

Beig, F. A., & Nika, F. A. (2019). Brand experience and brand equity. *Vision: The Journal of Business Perspective*, 23(4), 410–417. DOI: 10.1177/0972262919860963

Bekalu, M. A., Dhawan, D., McCloud, R., Pinnamaneni, R., & Viswanath, K. (2021). Adherence to COVID-19 mitigation measures among American adults: The need for consistent and unified messaging. *Health Education Research*, 36(2), 178–191. DOI: 10.1093/her/cyab002 PMID: 33496331

Bélanger, F., & Crossler, R. E. (2011). Privacy in the digital age: A review of information privacy research in information systems. *Management Information Systems Quarterly*, 35(4), 1017–1041. DOI: 10.2307/41409971

Belkhir, L., & Elmeligi, A. (2018). Assessing ICT global emissions footprint: Trends to 2040 & recommendations. *Journal of Cleaner Production*, 177, 448–463. DOI: 10.1016/j.jclepro.2017.12.239

Bendell, B. L., & Huvaj, M. N. (2020). Does stakeholder engagement through corporate social and environmental behaviors affect innovation? *Journal of Business Research*, 119, 685–696. DOI: 10.1016/j.jbusres.2018.10.028

Bennett, B., Freckelton, I. R., & Wolf, G. (2023). *COVID-19, Law and Regulation: Rights, Freedoms, and Obligations in a Pandemic*. Oxford University Press. DOI: 10.1093/oso/9780192896742.001.0001

Bergmark, R. W., Gliklich, E., Guo, R., & Gliklich, R. E. (2016). Texting while driving: The development and validation of the distracted driving survey and risk score among young adults. *Injury Epidemiology*, 3(1), 1–10. DOI: 10.1186/s40621-016-0073-8 PMID: 27747544

Bernacki, M. L., Greene, J. A., & Crompton, H. (2020). Mobile technology, learning, and achievement: Advances in understanding and measuring the role of mobile technology in education. *Contemporary Educational Psychology*, 60, 101827. DOI: 10.1016/j.cedpsych.2019.101827

Bernardino, P. (2021). Engaging employees through corporate social responsibility programs: Aligning corporate social responsibility and employee engagement. *Journal of Organizational Psychology*, 21(1), 105–113.

Beveridge, C. (2022, January 12). *How to Make YouTube Shorts: Everything You Need to Know*. Hoostsuite. https://blog.hootsuite.com/youtube-shorts/

Bhargava, V. R., & Bedi, S. (2022). Brand as promise. *Journal of Business Ethics*, 179(3), 919–936. DOI: 10.1007/s10551-021-04834-z

Bhattacharyya, J. (2022). The structure of sustainability marketing research: a bibliometric review and directions for future research. *Asia-Pacific Journal of Business Administration*, (ahead-of-print).

Bhattacharyya, J. (2023). The structure of sustainability marketing research: A bibliometric review and directions for future research. *Asia-Pacific Journal of Business Administration*, 15(2), 245–286. DOI: 10.1108/APJBA-06-2021-0239

Bhooshan, S. (2022). *Fundamentals of Analogue and Digital Communication System*. Springer. DOI: 10.1007/978-981-16-4277-7

Biddlestone, M., Green, R., & Douglas, K. M. (2020). Cultural orientation, power, belief in conspiracy theories, and intentions to reduce the spread of COVID-19. *British Journal of Social Psychology*, 59(3), 663–673. DOI: 10.1111/bjso.12397 PMID: 32592420

Billinghurst, M., Clark, A., & Lee, G. (2015). others,“. *A survey of augmented reality,” Foundations and Trends R in Human–Computer Interaction, 8*(2-3), 73-272.

Binkley, M., Erstad, O., Herman, J., Raizen, S., Ripley, M., Miller-Ricci, M., & Rumble, M. (2012). Defining twenty-first century skills. *Assessment and teaching of 21st century skills*, 17-66.

Bin, Y., & Mandal, D. (2019). English teaching practice based on artificial intelligence technology. *Journal of Intelligent & Fuzzy Systems*, 37(3), 3381–3391. DOI: 10.3233/JIFS-179141

Bistrohub. (2020, September 15). The importance of branding in your restaurant. *Bistrohub*. Retrieved June 10, 2023, from https://bistrohub.co/en/the-importance-of-branding-in-your-restaurant/

Biswas, P., Orero, P., Swaminathan, M., Krishnaswamy, K., & Robinson, P. (2021). Adaptive accessible AR/VR systems. In InExtended abstracts of the 2021 CHI Conference on human Factors in computing systems (pp. 1–7).

Black, J. W. (2023). Creating specialized corpora from digitized historical newspaper archives: An iterative bootstrapping approach. *Digital Scholarship in the Humanities*, 38(2), 779–797. DOI: 10.1093/llc/fqac079

Bloch, P., Brunel, F. F., & Arnold, T. J. (2003). Individual differences in the centrality of visual product aesthetics: Concept and measurement. *The Journal of Consumer Research*, 29(4), 551–565. DOI: 10.1086/346250

Bogen, K. W., Williams, S. L., Reidy, D. E., & Orchowski, L. M. (2021). We (want to) believe in the best of men: A qualitative analysis of reactions to# Gillette on Twitter. *Psychology of Men & Masculinity*, 22(1), 101–112. DOI: 10.1037/men0000308

Bonixe, L. (2020). Jornalismo radiofónico e inovação: Uma análise à cobertura de acontecimentos mediáticos [Radio journalism and innovation: An analysis of coverage of media events]. *Media & Jornalismo*, 20(36), 153–169. DOI: 10.14195/2183-5462_36_8

Boyles, B. (2017). Virtual reality and augmented reality in education. West Point, Ny: Center For Teaching Excellence, United States Military Academy.

Božić, A. L., & Milošević, S. (2021). Contemporary Trends in the Restaurant Industry and Gastronomy. *Journal of Hospitality & Tourism Research (Washington, D.C.)*, 45(5), 905–907. DOI: 10.1177/10963480211020559

Brakus, J. J., Schmitt, B. H., & Zarantonello, L. (2009). Brand Experience: What is it? How is it measured? Does it affect loyalty? *Journal of Marketing*, 73(3), 52–68. DOI: 10.1509/jmkg.73.3.052

Brammer, S., Branicki, L., & Linnenluecke, M. K. (2020). COVID-19, societalization, and the future of business in society. *The Academy of Management Perspectives*, 34(4), 493–507. DOI: 10.5465/amp.2019.0053

Braun, V., & Clarke, V. (2006). Using thematic analysis in psychology. *Qualitative Research in Psychology*, 3(2), 77–101. DOI: 10.1191/1478088706qp063oa

Briefing, E. February 9, 2022. https://www.mckinsey.com/business-functions/risk-and-resilience/our-insights/COVID-19-implications-for-business

Broersma, M., & Harbers, F. (2018). Exploring machine learning to study the long-term transformation of news: Digital newspaper archives, journalism history, and algorithmic transparency. *Digital Journalism (Abingdon, England)*, 6(9), 1150–1164. DOI: 10.1080/21670811.2018.1513337

Brougham, D., & Haar, J. (2018). Smart Technology, Artificial Intelligence, Robotics, and Algorithms (STARA): Employees perceptions of our future workplace. *Journal of Management & Organization*, 24(2), 239–257. DOI: 10.1017/jmo.2016.55

Bukar, U. A., Jabar, M. A., Sidi, F., Nor, R. N. H. B., Abdullah, S., & Othman, M. (2020). Crisis informatics in the context of social media crisis communication: Theoretical models, taxonomy, and open issues. *IEEE Access : Practical Innovations, Open Solutions*, 8, 185842–185869. Advance online publication. DOI: 10.1109/ACCESS.2020.3030184

Bunout, E., Ehrmann, M., & Clavert, F. (2023). *Digitised newspapers: A new eldorado for historians? Reflections on tools, methods and epistemology*. De Gruyter Oldenbourg.

Burdea, G. C., & Coiffet, P. (2024). *Virtual Reality Technology* (3rd ed.). John Wiley & Sons Inc. DOI: 10.1002/9781119512608

Burnham, B., & Walden, B. (1997). Interactions in distance education: A report from the other side. In *Proceedings of the Adult Education Research Conference*. Oklahoma State University. http:// www.edst.educ.ubc.ca/aerc/1997/97burnham.html

Burns, S. (2018). M-Pesa and the 'market-led'approach to financial inclusion. *Economic Affairs*, 38(3), 406–421. DOI: 10.1111/ecaf.12321

Burruss, G. W., Jaynes, C. M., Moule, R. K.Jr, & Fairchild, R. E. (2021). Modeling Individual Defiance of COVID-19 Pandemic Mitigation Strategies: Insights From the Expanded Model of Deterrence and Protection Motivation Theory. *Criminal Justice and Behavior*, 48(9), 00938548211010315. DOI: 10.1177/00938548211010315

Burton-Hughes, L. (2020, February 7). How to Improve Sustainability in Restaurants. *The Hub | High Speed Training*. Retrieved May 1, 2023, from https://www.highspeedtraining.co.uk/hub/improve-sustainability-in-restaurants/

Butt, K., Arif, M., & Rafiq, M. (2023). Assessing University Students' Satisfaction with Web-Based Library Services Based on SERVQUAL Model: A Case of a Pakistani University. *Libri*, 73(2), 153–165. DOI: 10.1515/libri-2022-0050

Camilleri, M. A. (2022). Strategic attributions of corporate social responsibility and environmental management: The business case for doing well by doing good! *Sustainable Development (Bradford)*, 30(3), 409–422. DOI: 10.1002/sd.2256

Campina, A., & Rodrigues, C. (2022, February). Cybercrime and the Council of Europe Budapest Convention: prevention, criminalization, and International Co-operation. In *The Book of Full Papers-7th International Zeugma Conference on Scientific Researches* (Vol. 1, No. 1, pp. 112-123). IKSAD.

Campisi, V. (2022). *The rise of photogenic food: How social media is changing dining*. Food Journal.

Campo, R., Loporcaro, G., & Baldassarre, F. (2017). The effects of food aesthetics on consumers. Visual stimuli and food marketing. *DIEM: Dubrovnik International Economic Meeting, 3*(1), pp. 553-565.

Campos, R., Pereira, I., & Simões, J. A. (2016). "Ativismo Digital em Portugal: Um Estudo Exploratório". Sociologia, Problemas e Práticas, n.º82, pp. 27 - 47. DOI:. DOI: 10.7458/SPP2016826977

Cardoso, G., Baldi, V., Crespo, M., Pinto-Martinho, A., Pais, P. C., Paisana, M., & Couraceiro, P. (2019). *O que devem saber os jornalistas? Práticas e formação em Portugal* [What should journalists know? Practices and training in Portugal]. OberCom. [REMOVED HYPERLINK FIELD]

Carmigniani, J., & Furht, B. (2011). Augmented reality: an overview. *Handbook of augmented reality, 10*(2), 3-46.

Carter, K., Jayachandran, S., & Murdock, M. R. (2021). Building a sustainable shelf: The role of firm sustainability reputation. *Journal of Retailing*, 97(4), 507–522. DOI: 10.1016/j.jretai.2021.03.003

Carvalho, L. G. D., & Eler, M. M. (2017, April). Security requirements and tests for smart toys. In *International Conference on Enterprise Information Systems* (pp. 291-312). Springer, Cham.

Castañeda, E. (2020). Analysing Contemporary Social Movements. In Tilly, C., Castañeda, E., & Wood, L. (Eds.), *Social Movements* (Vol. 1768-2018, pp. 167–176). Routledge.

Castells, M. (2010). *The Rise of the Network* (2. ª ed., Vol. 1). Wiley-Blackwell.

Castells, M., & Cardoso, G. (2005). *A Sociedade em Rede: Do Conhecimento à Acção Política*. Imprensa Nacional - Casa da Moeda.

Cénat, J. M., Broussard, C., Darius, W. P., Onesi, O., Auguste, E., El Aouame, A. M., Ukwu, G., Khodabocus, S. N., Labelle, P. R., & Dalexis, R. D. (2023). Social mobilization, education, and prevention of the Ebola virus disease: A scoping review. In *Preventive Medicine* (Vol. 166). DOI: 10.1016/j.ypmed.2022.107328

Chamola, V., Hassija, V., Gupta, V., & Guizani, M. (2020). A Comprehensive Review of the COVID-19 Pandemic and the Role of IoT, Drones, AI, Blockchain, and 5G in Managing its Impact. *IEEE Access : Practical Innovations, Open Solutions*, 8, 90225–90265. DOI: 10.1109/ACCESS.2020.2992341

Chapple, C. (2020, April). *TikTok Crosses 2 Billion Downloads After Best Quarter For Any App Ever*. Sensor Tower. https://sensortower.com/blog/tiktok-downloads-2-billion

Chen, P., Liu, X., Cheng, W., & Huang, R. (2017). A review of using augmented reality in education from 2011 to 2016. Innovations in smart learning, 13–18.

Chen, Y. L., Hsu, C. C., & Lin, C. Y. (2021). Engaging language learners through the use of virtual reality and a robot. In Society for Information Technology & Teacher Education International Conference Association for the Advancement of Computing in Education (AACE)., pp. 906–909.

Chen, C. Y. (2022). Immersive virtual reality to train preservice teachers in managing students' challenging behaviours: A pilot study. *British Journal of Educational Technology*, 53(4), 998–1024. DOI: 10.1111/bjet.13181

Chen, C., Frey, C. B., & Presidente, G. (2021). Culture and contagion: Individualism and compliance with COVID-19 policy. *Journal of Economic Behavior & Organization*, 190, 191–200. DOI: 10.1016/j.jebo.2021.07.026 PMID: 34566218

Cheng, K. H., & Tsai, C. C. (2019). A case study of immersive virtual field trips in an elementary classroom: Students' learning experience and teacher-student interaction behaviors. *Computers & Education*, 140, 103600. DOI: 10.1016/j.compedu.2019.103600

Chen, Y., Mao, Z., & Qiu, J. L. (2018). *Super-sticky WeChat and Chinese society*. Emerald Publishing Limited. DOI: 10.1108/9781787430914

Chen, Z., He, Q., Mao, Z., Chung, H. M., & Maharjan, S. (2019, May 17). A study on the characteristics of douyin short videos and implications for edge caching. *ACM International Conference Proceeding Series*. DOI: 10.1145/3321408.3323082

Chevers, D. A. (2019). The Impact of Cybercrime on E-Banking: A Proposed Model. In *CONF-IRM* (p. 11).

Childs, E., Mohammad, F., Stevens, L., Burbelo, H., Awoke, A., Rewkowski, N., & Manocha, D. (2021). An overview of enhancing distance learning through augmented and virtual reality technologies. arXiv preprint arXiv:2101.11000.

Chon, M., & Park, H. (2020). Social Media Activism in the Digital Age: Testing an Integrative Model of Activism on Contentious Issues. *Journalism & Mass Communication Quarterly*, 97(1), 72–97. DOI: 10.1177/1077699019835896

Christodoulatos, M. L. (2021). *Ethical chocolate consumption by millennials in the Netherlands* (Doctoral dissertation).

Cieri, R. L., Turner, M. L., Carney, R. M., Falkingham, P. L., Kirk, A. M., Wang, T., Jensen, B., Novotny, J., Tveite, J., Gatesy, S. M., Laidlaw, D. H., Kaplan, H., Moorman, A. F. M., Howell, M., Engel, B., Cruz, C., Smith, A., Gerichs, W., Lian, Y., & Farmer, C. G. (2021). Virtual and augmented reality: New tools for visualizing, analyzing, and communicating complex morphology. *Journal of Morphology*, 282(12), 1785–1800. DOI: 10.1002/jmor.21421 PMID: 34689352

Cifci, H., Gok, I., Atsiz, O., & Cifci, I. (2021). Insights into the Art of Plating in Gastronomy: A Content Analysis of Master Chefs' Perspectives. *Journal of Culinary Science &Amp. Technology*, •••, 1–26.

Civitello, L. (2008). *Cuisine and Culture: A History of Food and People* (2nd ed.). John Wiley & Sons.

Cleff, T., Lin, I., & Walter, N. (2014). Can you feel it? - The effect of brand experience on brand equity. *IUP Journal of Brand Management*, 11(2), 7–27.

Cloud computing security, https://en.wikipedia.org/wiki/

Coelho, P. M. (2016). *Rumo à Indústria 4.0* [Towards Industry 4.0] [Unpublished master dissertation]. University of Coimbra, Coimbra, Portugal.

Coffie, S., Anning-Dorson, T., Hinson, R. E., Bosah, G., & Anani-Bossman, A. (2022). Marketing Communications in Emerging Economies: Conclusions and Recommendations. In *Marketing Communications in Emerging Economies, Volume II* (pp. 277-296). Palgrave Macmillan, Cham.

Cortez, P., Moro, S., Rita, P., King, D., & Hall, J. (2018). Insights from a text mining survey on expert systems research from2000 to 2016. *Expert Systems: International Journal of Knowledge Engineering and Neural Networks*, 35(3), e12280. DOI: 10.1111/exsy.12280

Costa, B. F., & Antunes, E. (2024). Dinâmicas sociais no Facebook: Análise de comentários em conteúdos jornalísticos sobre a suposta tentativa de ataque terrorista à FCUL [Social dynamics on Facebook: Analysis of comments on journalistic content about the alleged attempted terrorist attack on FCUL]. *Observatorio (OBS*), 18*(1), 126-150.

Costa, B. F., & Mateus, B. C. (in press). Innovating in journalism with newsgames: An exploratory study in Portugal. In Bojić, L., Žikić, S., Matthes, J., & Trilling, D. (Eds.), *Navigating the digital age: An in-depth exploration into the intersection of modern technologies and societal transformation*. Institute for Philosophy and Social Theory.

Cracknell, H. L., & Nobis, G. (1989). *The history of culinary arts*. Butterworth-Heinemann.

Crawford, C. M. (2001). Developing webs of significance through communications: Appropriate interactive activities for distributed learning environments. *Campus-Wide Information Systems*, 18(2), 68–72. DOI: 10.1108/10650740110386675

Crawford, C. M. (2003). Emerging learning environments: Enhancing the online community. *Academic Exchange Quarterly*, 7(4), 131–135.

Crawford, C. M., Andrews, S., & Wallace, J. K. Y. (2021). Co-creative collegial communities of instructional engagement. [IJHIoT]. *International Journal of Hyperconnectivity and the Internet of Things*, 5(2), 38–56. DOI: 10.4018/IJHIoT.2021070103

Crawford, C. M., Andrews, S., & Young Wallace, J. K. (2022). Implicit cognitive vulnerability through nudges, boosts, and bounces. *International Journal of Hyperconnectivity and the Internet of Things*, 6(1), 1–14. DOI: 10.4018/IJHIoT.285588

Crespo, M., Foà, C., & Pinto-Martinho, A. (2018). Como o jornalismo lida com a inovação: Um estudo de caso das melhores práticas em Portugal [How journalism deals with innovation: A case study of best practices in Portugal]. *Estudos de Jornalismo*, 9, 75–102.

Creswell, J. W., & Creswell, J. D. (2017). *Research design: Qualitative, quantitative, and mixed methods approaches* (3rd ed.). Sage publications.

Criollo-C, S., Altamirano-Suarez, E., Jaramillo-Villacís, L., Vidal-Pacheco, K., Guerrero-Arias, A., & Luján-Mora, S. (2022). Sustainable teaching and learning through a mobile application: A case study. *Sustainability (Basel)*, 14(11), 6663. DOI: 10.3390/su14116663

Criollo-C, S., Guerrero-Arias, A., Jaramillo-Alcázar, Á., & Luján-Mora, S. (2021). Mobile learning technologies for education: Benefits and pending issues. *Applied Sciences (Basel, Switzerland)*, 11(9), 4111. DOI: 10.3390/app11094111

Crisp, J. (2024). The Syrian Emergency and Its Impact on the Evolution of Global Refugee Policy. In *Urban Displacement*. DOI: 10.2307/jj.9891587.15

Crowther, G. (2018). *Eating Culture: An Anthropological Guide to Food* (2nd ed.). University of Toronto Press.

Cuboh. (2022, May 27). Restaurant Branding 101: How to Brand Your Restaurant. *Cuboh.* Retrieved June 12, 2023, from https://www.cuboh.com/blog/the-importance-of-branding-in-your-restaurant

Cuellar, N. G., Aquino, E., Dawson, M. A., Garcia-Dia, M. J., Im, E. O., Jurado, L. F. M., Lee, Y. S., Littlejohn, S., Tom-Orme, L., & Toney, D. A. (2020). Culturally congruent health care of COVID-19 in minorities in the United States: A clinical practice paper from the National Coalition of Ethnic Minority Nurse Associations. *Journal of Transcultural Nursing*, 31(5), 434–443. DOI: 10.1177/1043659620941578 PMID: 32794439

Cuofano, G. (2023). "What Is Relationship Marketing and Why It Matters in Business". https://fourweekmba.com/what-is-relationship-marketing/

Dabeva, T. (2012). Contemporary characteristics of restaurant branding. *Tourism and Hospitality Management*, •••, 254–261.

Dai, C., & Ke, F. (2022). Educational applications of artificial intelligence in simulation-based learning: A systematic mapping review. *Computers and Education: Artificial Intelligence*, 3, 1–17. DOI: 10.1016/j.caeai.2022.100087

Daniels, H. (Ed.). (1996). *An introduction to Vygotsky*. Routledge.

Danthinne, E. S., Giorgianni, F. E., Austin, S. B., & Rodgers, R. F. (2022). Beauty and the big business: A qualitative investigation of the development and implementation of the CVS Beauty Mark. *Body Image*, 41, 74–83. DOI: 10.1016/j.bodyim.2022.01.012 PMID: 35240392

Dauhgerty, E. L. (2014). Ethical Dilemmas of Female Agency Owners: Practices in Public Relations and Marketing Communications. *Quarterly Review of Business Disciplines*, 1(2), 173–190.

de Oliveira, M. E., & Corr^ea, C. G. (2020). Virtual reality and augmented reality applications in agriculture: A literature review. In 2020 22nd symposium on virtual and augmented reality (svr) (pp. 1–9). IEEE. DOI: 10.1109/SVR51698.2020.00017

de-Lima-Santos, M.-F., & Ceron, W. (2021). Artificial Intelligence in news media: Current perceptions and future outlook. *Journalism and Media*, 3(1), 13–26. DOI: 10.3390/journalmedia3010002

Delwiche, J. F. (2012). You eat with your eyes first. *Physiology &Amp. Behaviour*, 107(4), 502–504.

Demirel, D. (2022). The effect of service quality on customer satisfaction in digital age: Customer satisfaction based examination of digital CRM. *Journal of Business Economics and Management*, 23(3), 507–531. DOI: 10.3846/jbem.2022.15328

Demirtas-Madran, H. A. (2021). Individual differences in accepting restrictions and compliance with recommended preventive behaviors for COVID-19: A discussion based on the key approaches and current research on fear appeals. *Frontiers in Psychology*, 12, 2084. https://www.frontiersin.org/articles/10.3389/fpsyg.2021.558437/pdf. DOI: 10.3389/fpsyg.2021.558437

Department of Electronic and Information Technology of the Government of India. Email Service. (2015). ‘EMAIL policy of Government of India. Retrieved from https://www.meity.gov.in/writereaddata/files/E-mail_policy_of_Government_of_India_3.pdf

Dhali, M. A. (2024). *Artificial Intelligence in Historical Document Analysis: Pattern recognition and machine learning techniques in the study of ancient manuscripts with a focus on the Dead Sea Scrolls* [Unpublished doctoral dissertation]. University of Groningen, Groningen, Netherlands.

Diehl, F. S., Marín, L. S., & Zreiqat, M. (2020). An evaluation of AT&T social marketing delivery modes aimed at teen smartphone use while driving. *Journal of Cultural Marketing Strategy*, 5(1), 60–75. DOI: 10.69554/XBTL7959

Dieter Schmalstieg, & Hollerer, T. (2016). Augmented Reality: Principles and Practices. Addison-Wesley Professional.

Dillard, J. P., Tian, X., Cruz, S. M., Smith, R. A., & Shen, L. (2023). Persuasive messages, social norms, and reactance: A study of masking behavior during a COVID-19 campus health campaign. *Health Communication*, 38(7), 1338–1348. DOI: 10.1080/10410236.2021.2007579 PMID: 34879752

Divekar, R. R., Drozdal, J., & Chabot, S. (2021). Foreign language acquisition via artificial intelligence and extended reality: Design and evaluation. *Computer Assisted Language Learning*, , 1–29.

Dixon. J.S., Statista (2024). Leading Countries based on Facebook audience size as of January 2024 Retrieved from https://www.statista.com/statistics/268136/top-15-countries-based-on-number-of-facebook-users/

Dobos, O. (2022, September 9). *Branding for restaurants: what is it, how to do it, and tips for success*. GloriaFood. Retrieved June 11, 2023, from https://www.restaurant-website-builder.com/branding-for-restaurants

Donthu, N., & Gustafsson, A. (2020). Effects of COVID-19 on business and research. *Journal of Business Research*, 117, 284–289. DOI: 10.1016/j.jbusres.2020.06.008 PMID: 32536736

Dorfman, P., Javidan, M., Hanges, P., Dastmalchian, A., & House, R. (2012). GLOBE: A twenty year journey into the intriguing world of culture and leadership. *Journal of World Business*, 47(4), 504–518. DOI: 10.1016/j.jwb.2012.01.004

Du, R., Turner, E., Dzitsiuk, M., Prasso, L., Duarte, I., Dourgarian, J., . . . Kim, D. (2020). DepthLab: Real-time 3D interaction with depth maps for mobile augmented reality. In InProceedings of the 33rd annual ACM Symposium on user interface Software and technology (pp. 829–843).

Dua, T. (2019, February 4). *Here's the Full List of All the Super Bowl Commercials That Ran This Year*, Business Insider, www.businessinsider.com/super-bowl-commercials-2019-list-2019-1 (Accessed on 4th February, 2019)

Dua, T. (2019, February 4). Here's the Full List of All the Super Bowl Commercials That Ran This Year. *Business Insider*, www.businessinsider.com/super-bowl-commercials-2019-list-2019-1 (Accessed on 4th February, 2019)

Durkheim, E. (1912). *The elementary forms of the religious life*. Free Press.

Eckert, M., Volmerg, J. S., & Friedrich, C. M. (2019). Augmented reality in medicine: Systematic and bibliographic review. *JMIR mHealth and uHealth*, 7(4), e10967. DOI: 10.2196/10967 PMID: 31025950

Edwards, E. (2023, May 3). Top Chefs Share Their 10 Food Plating and Presentation Tips. *Lightspeed*. Retrieved May 8, 2023, from https://www.lightspeedhq.com/blog/10-food-plating-and-presentation-tips/

Ehrmann, M., & Bunout, E., & Düring. (2019, August). *Historical newspaper user interfaces: A review* [Paper presentation]. *85th IFLA General Conference and Assembly (IFLA)*, Athens, Greece.

Ehrmannm, M., Düring, M., Neudecker, C., & Doucet, A. (2022). Computational Approaches to Digitised Historical Newspapers. *Dagstuhl Reports*, 12(7), 112–129.

Elamroussi, A. (2021) Omicron surge is 'unlike anything we've ever seen,' expert says. *CNN*https://www.cnn.com/2021/12/30/health/us-coronavirus-thursday/index.html

Elgelany, A., & Alghabban, W. G. (2017). Cloud computing: Empirical studies in higher education a literature review. *International Journal of Advanced Computer Science and Applications*, 8(10). Advance online publication. DOI: 10.14569/IJACSA.2017.081017

Elias, N. (1978). *The Civilizing Process*. Blackwell.

Elliott, C., Rose, M., & van Arnhem, J.-P. (2018). *Augmented and virtual reality in libraries*. Rowman and Littlefield.

Ellison, R. (2022, August 11). *The Rise of Brunch - A Look at How This Meal Has Grown in Popularity - Fairway Foodservice*. Fairway Foodservice. Retrieved May 2, 2023, from https://www.fairwayfoodservice.com/articles/the-rise-of-brunch-a-look-at-how-this-meal-has-grown-in-popularity/

Elmqaddem, N. (2019). Augmented reality and virtual reality in education. Myth or reality? *International Journal of Emerging Technologies in Learning*, 14(3), 234. DOI: 10.3991/ijet.v14i03.9289

Emamalizadeh, H. (2015). Bait-and-switch advertising: A conceptual approach. In *Proceedings of the 1983 Academy of Marketing Science (AMS) Annual Conference* (pp. 458-460). Cham: Springer International Publishing. DOI: 10.1007/978-3-319-16937-8_109

Ercan, T. (2010). Effective use of cloud computing in educational institutions. *Procedia: Social and Behavioral Sciences*, 2(2), 938–942. DOI: 10.1016/j.sbspro.2010.03.130

Errichiello, O., & Zschiesche, A. (2022). Green Consumption. In *Green Branding: Basics, Success Factors And Instruments For Sustainable Brand And Innovation Management* (pp. 17–57). Springer Fachmedien Wiesbaden. DOI: 10.1007/978-3-658-36060-3_2

Escamilla-Fajardo, P., Alguacil, M., & López-Carril, S. (2021). Incorporating TikTok in higher education: Pedagogical perspectives from a corporal expression sport sciences course. *Journal of Hospitality, Leisure, Sport and Tourism Education*, 28, 100302. DOI: 10.1016/j.jhlste.2021.100302

Essel, H. B., Vlachopoulos, D., & Tachie-Menson, A. (2021). The relationship between the nomophobic levels of higher education students in Ghana and academic achievement. *PLoS One*, 16(6), e0252880. DOI: 10.1371/journal.pone.0252880 PMID: 34133434

Etro, F. (2015). The economics of cloud computing. In Cloud Technology: Concepts, Methodologies, Tools, and Applications (pp. 2135-2148). IGI Global. Gartner: Seven Could-Computing Security Risks, https://www.infoworld.comp/print/368853

Euler, B. E. (2022, December). These are the Food Trends 2023 | KTCHNrebel. www.ktchnrebel.com. Retrieved April 30, 2023, from https://www.ktchnrebel.com/foodtrend-report-2023/

Evans, D. M., Parsons, R., Jackson, P., Greenwood, S., & Ryan, A. (2020). Understanding plastic packaging: The co-evolution of materials and society. *Global Environmental Change*, 65, 102166. DOI: 10.1016/j.gloenvcha.2020.102166

Experience Doja Cat's Concert in VR. (2024, January 9). Meta. https://about.fb.com/news/2024/01/doja-cat-vr-8concert

Fan, C., Wu, F., & Mostafavi, A. (2020). A Hybrid Machine Learning Pipeline for Automated Mapping of Events and Locations from Social Media in Disasters. *IEEE Access : Practical Innovations, Open Solutions*, 8, 10478–10490. Advance online publication. DOI: 10.1109/ACCESS.2020.2965550

Fante, A. (2018). Jornalismo appificado: Arquiteturas da notícia [Appified journalism: News architectures]. *Estudos de Jornalismo*, 9, 112–124.

Fan, Y., Li, X., Zhang, L., Wan, S., Zhang, L., & Zhou, F. (2022). SARS-CoV-2 Omicron variant: Recent progress and future perspectives. *Signal Transduction and Targeted Therapy*, 7(1), 1–11. DOI: 10.1038/s41392-022-00997-x PMID: 35484110

Farber, S. A. (2008). US scientists' role in the eugenics movement (1907–1939): A contemporary biologist's perspective. *Zebrafish*, 5(4), 243–245. DOI: 10.1089/zeb.2008.0576 PMID: 19133822

Fatma, S. (2021). Use and Misuse of Technology in Marketing: Cases from India. In *Research Anthology on Strategies for Using Social Media as a Service and Tool in Business* (pp. 106-116). IGI Global.

Fauzi, M. A. (2023). Social media in disaster management: review of the literature and future trends through bibliometric analysis. In *Natural Hazards* (Vol. 118, Issue 2). DOI: 10.1007/s11069-023-06079-7

Feldheim, D. (2022). *Dare to Lead Like a Girl: How to Survive and Thrive in the Corporate Jungle*. Rowman & Littlefield. DOI: 10.5771/9781538163535

Ferrell, O. C., Hartline, M., & Hochstein, B. W. (2021). *Marketing strategy*. Cengage Learning.

Ferro, S., Pelillo, M., & Traviglia, A. (2023). AI-assisted digitalisation of historical documents. In Int (pp. 557–562). Arch. Photogramm. Remote Sens. Spatial Inf. Sci., XLVIII-M-2.

Fidalgo, C. G., Yan, Y., Cho, H., Sousa, M., Lindlbauer, D., & Jorge, J. (2023). A Survey on ijuRemote Assistance and Training in Mixed Reality Environments. 29(5), 2291–2303. .DOI: 10.1109/TVCG.2023.3247081

Fieldhouse, P. (1996). *Food and Nutrition: Customs and culture*. Springer.

Figueiras, R. (2005). Os comentadores na imprensa de referência portuguesa: 1980-1999 [Commentators in the Portuguese reference press: 1980-1999]. In *Anais Intercom - XXVIII Congresso Brasileiro de Ciências da Comunicação* (pp. 1-15). Rio de Janeiro.

Figueroa, J. F., Wadhera, R. K., Lee, D., Yeh, R. W., & Sommers, B. D. (2020). Community-level factors associated with racial and ethnic disparities in COVID-19 rates in Massachusetts. *Health Affairs*, 39(11), 1984–1992. DOI: 10.1377/hlthaff.2020.01040 PMID: 32853056

Firestone, S. (2020). Importance of Communication During a Crisis. In *Biblical Principles of Crisis Leadership*. DOI: 10.1007/978-3-030-44955-1_7

Fizaine, F. C., Bard, P., Paindavoine, M., Robin, C., Bouyé, E., Lefèvre, R., & Vinter, A. (2024). Historical text line segmentation using deep learning algorithms: Mask-RCNN against U-Net Networks. *Journal of Imaging*, 10(3), 1–18. DOI: 10.3390/jimaging10030065 PMID: 38535145

Flavián, C., Ibáñez-Sánchez, S., & Orús, C. (2019, July 1). The impact of virtual, augmented, and mixed reality technologies on the customer experience. *Journal of Business Research*, 100, 547–560. DOI: 10.1016/j.jbusres.2018.10.050

Fogarty, M. J. (2022). Rewired: Utilizing an Integrated Design Campaign to Persuade Generation Z to Rethink Their Hyper-Connected World.

Foltean, F. S., & van Bruggen, G. H. (2022). Digital Technologies, Marketing Agility, and Marketing Management Support Systems: How to Remain Competitive in Changing Markets. In *Organizational Innovation in the Digital Age* (pp. 1–38). Springer. DOI: 10.1007/978-3-030-98183-9_1

Fraga-Lamas, P., Fernandez-Carames, T. M., Blanco-Novoa, O., & Vilar-Montesinos, M. A. (2018). A Review on Industrial Augmented Reality Systems for the Industry 4.0 Shipyard. *IEEE Access : Practical Innovations, Open Solutions*, 6, 13358–13375. DOI: 10.1109/ACCESS.2018.2808326

Franklin, B. (2016). *The Future of Journalism: In an Age of Digital Media and Economic Uncertainty*. Routledge.

Free Cloud Computing with AWS - Secure Cloud Services. https://aws.amazon.com

Frey, C. B., Chen, C., & Presidente, G. (2020). Democracy, Culture, and Contagion. *Political Regimes and Countries Responsiveness to COVID*, 19, 20.

Future Food Strategy Group. (2018, April 25). The Booming Brunch Scene: The most popular day-part for dining out. *Future Food*. Retrieved May 3, 2023, from https://futurefood.com.au/blog/2018/4/25/the-booming-brunch-scene-the-most-popular-day-part-for-dining-out

Fu, Y., Hu, Y., & Sundstedt, V. (2022). A systematic literature review of virtual, augmented, and mixed reality game applications in healthcare. [HEALTH]. *ACM Transactions on Computing for Healthcare*, 3(2), 1–27. DOI: 10.1145/3472303

Gagliardone, I. (2019). *China, Africa, and the Future of the Internet*. Bloomsbury Publishing. DOI: 10.5040/9781350219113

Galdas, P. (2017). Revisiting bias in qualitative research: Reflections on its relationship with funding and impact. *International Journal of Qualitative Methods*, 16(1), 1609406917748992. DOI: 10.1177/1609406917748992

Gallger, R. G. (2008). Principles of Digital Communication, Cambridge University Press IGNOU.Preamble; Retrieved from https://www.ignou.ac.in/ignou/aboutignou/profile/2

Gambetti, A., & Han, Q. (2022). Camera eats first: Exploring food aesthetics portrayed on social media using deep learning. *International Journal of Contemporary Hospitality Management*, 34(9), 3300–3331. DOI: 10.1108/IJCHM-09-2021-1206

Gandedkar, N. H., Wong, M. T., & Darendeliler, M. A. (2021). Role of virtual reality (VR), augmented reality (AR) and artificial intelligence (AI) in tertiary education and research of orthodontics: An insight. [WB Saunders]. *Seminars in Orthodontics*, 27(2), 69–77. DOI: 10.1053/j.sodo.2021.05.003

Gao, Q., Zhang, Z., Li, Z., Li, Y., & Shao, X. (2022). Strategic green marketing and cross-border merger and acquisition completion: The role of corporate social responsibility and green patent development. *Journal of Cleaner Production*, 343, 130961. DOI: 10.1016/j.jclepro.2022.130961

Garad, A., Al-Ansi, A. M., & Qamari, I. N. (2021). The role of e-learning infrastructure and cognitive competence in distance learning effectiveness during the covid-19 pandemic. *Journal Cakrawala Pendidikan*, 40(1), 81–91. DOI: 10.21831/cp.v40i1.33474

García-Avilés, J. A., Carvajal-Prieto, M., Arias, F., & De Lara-González, A. (2019). How journalists innovate in the newsroom. Proposing a model of the diffusion of innovations in media outlets. *The Journal of Media Innovations*, 5(1), 1–16. DOI: 10.5617/jomi.v5i1.3968

Gardiner, S., Chen, J., Abreu Novais, M., Dupré, K., & Castley, J. G. (2023). Analyzing and Leveraging Social Media Disaster Communication of Natural Hazards: Community Sentiment and Messaging Regarding the Australian 2019/20 Bushfires. *Societies (Basel, Switzerland)*, 13(6), 138. Advance online publication. DOI: 10.3390/soc13060138

Gardini, M. (2014). The Challenge of Branding and Brand Management: Perspectives from the Hospitality Industry. In *Tourism and Leisure: Current Issues and Perspectives of Development* (pp. 247–268). Springer Gabler.

Gargrish, S., Mantri, A., & Kaur, D. P. (2020). Augmented reality-based learning environment to enhance teaching-learning experience in geometry education. *Procedia Computer Science*, 172, 1039–1046. DOI: 10.1016/j.procs.2020.05.152

Garzon, ´. J., Pavon, ´. J., & Baldiris, S. (2019). Systematic review and meta-analysis of augmented reality in educational settings. *Virtual Reality (Waltham Cross)*, 23(4), 447–459. DOI: 10.1007/s10055-019-00379-9

Ge, R., & Hsiao, T.-C. (2020, May 1). A Summary of Virtual Reality, Augmented Reality and Mixed Reality Technologies in Film and Television Creative Industries. IEEE Xplore. .DOI: 10.1109/ECBIOS50299.2020.9203607

Gehlen, M. A., & Sousa, J. P. (2018). Jornalismo de dados em Portugal: Um estudo exploratório sobre práticas jornalísticas especializadas [Data journalism in Portugal: An exploratory study on specialized journalistic practices]. *Estudos de Jornalismo*, 9, 125–139.

Genc, R. (2010). Strategic brand management in hospitality sector: how to manage co-branding in hotels and restaurants. *Acta Universitatis Danubius. Œconomica, 6*(3), pp. 33-46.

George, P. (2021, December 29). The Evolution of Brunch. *Out & About Magazine*. Retrieved May 2, 2023, from https://outandaboutnow.com/the-evolution-of-brunch/

Georgiev, N. (n.d.). *Restaurant Industry Trends: 6 Restaurant Trends in 2022*. Retrieved March 30, 2023, from https://www.bluecart.com/blog/restaurant-industry-trends

Gerbaudo, P. (2017). From Cyber-Autonomism to Cyber-Populism: An Ideological History of Digital Activism. *TripleC*, 15(2), 477–489. DOI: 10.31269/triplec.v15i2.773

Germani, A., Buratta, L., Delvecchio, E., & Mazzeschi, C. (2020). Emerging Adults and COVID-19: The Role of Individualism-Collectivism on Perceived Risks and Psychological Maladjustment. *International Journal of Environmental Research and Public Health*, 17(10), 3497. DOI: 10.3390/ijerph17103497 PMID: 32429536

Geroimenko, V. (2020). *Augmented Reality in Education. Springer Series on Cultural Computing*. Springer International Publishing. DOI: 10.1007/978-3-030-42156-4

Gerry, S., Koschmann, T., & Suthers, D. (2006). Computer-Supported Collaborative Learning. In *Cambridge handbook of the learning science* (pp. 409–426). CambridgeUniversity Press.

Gervasi, M. (2016). *East-Commerce: China e-commerce and the internet of things*. John Wiley & Sons.

Giannini, S. Think piece by UNESCO Assistant Director-General for Education, Generative Artificial Intelligence in education: Think piece by Stefania Giannini, https://www.unesco.org/en/articles/generative-artificial-intelligence-education-think-piece-stefania-giannini?hub=32618

Gilchrist, A. (2016). *Industry 4.0: The Industrial Internet of Things*. Apress. DOI: 10.1007/978-1-4842-2047-4

Giovannucci, D., von Hagen, O., & Wozniak, J. (2014). Corporate social responsibility and the role of voluntary sustainability standards. In *Voluntary standard systems: A contribution to sustainable development* (pp. 359–384). Springer Berlin Heidelberg. DOI: 10.1007/978-3-642-35716-9_24

Girdhar, N., Coustaty, M., & Doucet, A. (2024). Digitizing history: Transitioning historical paper documents to digital content for information retrieval and mining: A comprehensive survey. *IEEE Transactions on Computational Social Systems*, 11(5), 6151–6180. DOI: 10.1109/TCSS.2024.3378419

Gladys, E. U. (2018). Legal Framework for the Protection of Online Products Consumers in Nigeria. *Nigerian LJ*, 21, 24.

Glenn, C. L. (2015). Activism or "Slacktivism?": Digital Media and Organizing for Social Change. *Communication Teacher*, 29(2), 81–85. DOI: 10.1080/17404622.2014.1003310

Gobinath, V. M., Kathirvel, A., Rajesh Kanna, S. K., & Annamalai, K. (July 2024g) "Chapter 5: Smart Technology in Management Industries: A Useful Perspective", Artificial Intelligence Applied to Industry 4.0, Wiley Publiser, USA, 1st Edition, pp.93-106, , ISBN 978-1-394-21611-6, DOI eBook ISBN, https://www.wiley.com/en-us/Topics+in+Artificial+Intelligence+Applied+to+Industry+4+0-p-9781394216116DOI: 10.1002/9781394216147.ch5

Gobinath, V., Ayyaswamy, K., & Kathirvel, N. (2024). Information Communication Technology and Intelligent Manufacturing Industries Perspective: An Insight. *Asian Science Bulletin*, 2(1), 36–45. DOI: 10.3923/asb.2024.36.45

Goldstein, I., Koijen, R. S., & Mueller, H. M. (2021). COVID-19 and its impact on financial markets and the real economy. *Review of Financial Studies*, 34(11), 5135–5148. DOI: 10.1093/rfs/hhab085

González-Martínez, J. A., Bote-Lorenzo, M. L., Gómez-Sánchez, E., & Cano-Parra, R. (2015). Cloud computing and education: A state-of-the-art survey. *Computers & Education*, 80, 132–151. DOI: 10.1016/j.compedu.2014.08.017

Goodchild, M. F. (2007). Citizens as sensors: The world of volunteered geography. In *GeoJournal*. DOI: 10.1007/s10708-007-9111-y

Gordon, D., & Meia, N. (Eds.). (2022). *The Digital Silk Road: China's Technological Rise and the Geopolitics of Cyberspace*. Taylor & Francis.

Granado, A., Silva, D. S., & Vicente, P. N. (2020). Inovação nos media e nas indústrias criativas limítrofes – uma introdução [Innovation in the media and neighboring creative industries – an introduction]. *Media & Jornalismo*, 20(36), 5–9.

Gravett, W. (2020). Digital neo-colonialism: The Chinese model of internet sovereignty in Africa. *African Human Rights Law Journal*, 20(1), 125–146.

Grubert, J., Langlotz, T., Zollmann, S., & Regenbrecht, H. (2017). Towards Pervasive Augmented Reality: Context-Awareness in Augmented Reality. *IEEE Transactions on Visualization and Computer Graphics*, 23(6), 1706–1724. DOI: 10.1109/TVCG.2016.2543720 PMID: 27008668

Guan, C., Mou, J., & Jiang, Z. (2020). Artificial intelligence innovation in education: A twenty-year data-driven historical analysis. *International Journal of Innovation Studies*, 4(4), 134–147. DOI: 10.1016/j.ijis.2020.09.001

Gudoniene, D., & Rutkauskiene, D. (2019). Virtual and augmented reality in education. *Baltic Journal of Modern Computing*, 7(2), 293–300. DOI: 10.22364/bjmc.2019.7.2.07

Guerra Rodriguez, Y., & Hudson, B. L. (2020). [Impact on the Hispanic Community: How Understanding Culture Can Improve Outcomes.]. *COVID*, •••, 19s.

Gunter, U., & Önder, I. (2021). An Exploratory Analysis of Geotagged Photos From Instagram for Residents of and Visitors to Vienna. *Journal of Hospitality & Tourism Research (Washington, D.C.)*, 45(2), 373–398. Advance online publication. DOI: 10.1177/1096348020963689

Guo, X., Guo, Y., & Liu, Y. (2021). The development of extended reality in education: Inspiration from the research literature. *Sustainability (Basel)*, 13(24), 13776. DOI: 10.3390/su132413776

Haarpaintner, É. (2022). What is the relationship between sustainability, corporate social responsibility and quality of life applications at work?. *Worldwide Hospitality and Tourism Themes*, (ahead-of-print).

Haarpaintner, É. (2022). What is the relationship between sustainability, corporate social responsibility and quality of life applications at work? *Worldwide Hospitality and Tourism Themes*, 14(3), 236–246. DOI: 10.1108/WHATT-02-2022-0020

Ha, H., & Perks, H. (2005). Effects of consumer perceptions of brand experience on the web: Brand familiarity, satisfaction and brand trust. *Journal of Consumer Behaviour*, 4(6), 438–452. DOI: 10.1002/cb.29

Hair, J. F., Page, M., & Brunsveld, N. (2020). *Essentials of business research methods* (4th ed.). Routledge.

Halvorsen, S. V. (2021). *Assessing and explaining the proficiency of social media marketing in small ministries: a comparative case study* (Master's thesis).

Hamilton, T., II. E. C., Hamilton, T., Welsh, C., II, E. C., & Welsh, C. (2017, February 20). Glitched: A VR Talk Show. IMDb. https://www.imdb.com/title/tt10318060/

Han, Y., Miao, Y., Sha, H., & Liu, Y. (2024). Exploring Influences of Appearance and Voice Realism of Virtual Humans on Communication Effectiveness in Social Virtual Reality. .DOI: 10.1109/VRW62533.2024.00157

Hantono, B. S., Nugroho, L. E., & Santosa, P. I. (2018). Meta-review of augmented reality in education. In In2018 10th international conference on information technology and electrical engineering (ICITEE) (pp. 312–315). IEEE. DOI: 10.1109/ICITEED.2018.8534888

Hartley, A. (2021, August 16). *Understanding the Brunch Craze in Restaurants.* Alto-Hartley. Retrieved May 2, 2023, from_https://altohartley.com/understanding-the-brunch-crazein-restaurants/

Hasler, J. C. (2020). *Explaining the International Infrastructure of Global Telecommunications: Why Government Ownership Determines a Country's Position in Interstate Communications Networks* (Doctoral dissertation, The George Washington University).

Haupt, B. (2021). The Use of Crisis Communication Strategies in Emergency Management. *Journal of Homeland Security and Emergency Management*, 18(2), 125–150. Advance online publication. DOI: 10.1515/jhsem-2020-0039

Hauser, M., Flath, C. M., & Thiesse, F. (2021). Catch me if you scan: Data-driven prescriptive modeling for smart store environments. *European Journal of Operational Research*, 294(3), 860–873. DOI: 10.1016/j.ejor.2020.12.047

Haworth, B., Bruce, E., & Middleton, P. (2012). Emerging Technologies for Risk Reduction : Assessing The Potential Use of Social Media and VGI for Increasing Community Engagement. *Australian Journal of Emergency Management*, 30(3), 36–41.

Hayes, J. L., & Duff, B. R. (2022). Introducing a special issue on advertising & corporate social responsibility. *International Journal of Advertising*, 41(2), 205–209. DOI: 10.1080/02650487.2022.2056357

Hayward, N. (2018). "A Short History of Activism". Available: https://www.womenshistory.org/resources/general/short-history-activism

Hedberg, T. D., Bajaj, M., & Camelio, J. A. (2019). Using Graphs to Link Data Across the Product Lifecycle for Enabling Smart Manufacturing Digital Threads. *ASME Journal of Computing and Information Science in Engineering*, 1, 213–224.

Helena, L. (2017). A evolução editorial do Jornal de Notícias e a inclusão de elementos de jornalismo popular [The editorial evolution of Jornal de Notícias and the inclusion of elements of popular journalism]. *Revista Portuguesa de História da Comunicação*, 1, 1–11.

Hermann, M., Pentek, T., & Otto, B. (2016). Design Principles for Industrie 4.0 Scenarios. In *Proceedings of 49th Hawaii International Conference on System Sciences (HICSS)* (pp. 3928-3937). USA. DOI: 10.1109/HICSS.2016.488

He, W., Cai, D., Geng, G., & Klug, D. (2023). Factors influencing wearing face mask in public during COVID-19 outbreak: A qualitative study. *Disaster Medicine and Public Health Preparedness*, 17, e141. DOI: 10.1017/dmp.2022.52 PMID: 35241205

Hillman, D., Willis, D. J., & Gunawardena, C. (1994). Learner-interface interaction in distance education: An extension of contemporary models and strategies for practitioners. *American Journal of Distance Education*, 8(2), 30–42. DOI: 10.1080/08923649409526853

Hillman, J., Sacks, D., Lew, J. J., & Roughead, G. (2021). *China's Belt and Road: Implications for the United States*. Council on Foreign Relations.

Hills, S., & Eraso, Y. (2021). Factors associated with non-adherence to social distancing rules during the COVID-19 pandemic: A logistic regression analysis. *BMC Public Health*, 21(1), 352. DOI: 10.1186/s12889-021-10379-7 PMID: 33581734

Hilts, P. J. "Experiments On Children Are Reviewed." *The New York Times*, The New York Times, (15 Apr. 1998,) https://www.nytimes.com/1998/04/15/nyregion/experiments-on-children-are-reviewed.html

Holbrook, M. B., & Hirschman, E. C. (1982). The experiential aspects of consumption: Consumer fantasies, feelings, and fun. *The Journal of Consumer Research*, 9(2), 132–140. DOI: 10.1086/208906

Holtgrave, D. R., Barranco, M. A., Tesoriero, J. M., Blog, D. S., & Rosenberg, E. S. (2020). Assessing racial and ethnic disparities using a COVID-19 outcomes continuum for New York State. *Annals of Epidemiology*, 48, 9–14. DOI: 10.1016/j.annepidem.2020.06.010 PMID: 32723697

Hong, Y., & Wei, Y. (2022). Towards future politics of the cybersphere: China's temporal-spatial governance of digital transition. *Media International Australia, Incorporating Culture & Policy*, 185(1), 9–25. DOI: 10.1177/1329878X221095593

HOTOSM. (2023a). *HOT'S RESPONSE TO THE TURKEY & SYRIA EARTHQUAKE.* https://www.hotosm.org/projects/join-the-turkey-and-syria-earthquake-response/

HOTOSM. (2023b). *What We Do | Humanitarian OpenStreetMap Team.* https://www.hotosm.org/what-we-do

Huaman-Ramirez, R., & Merunka, D. (2019). Brand experience effects on brand attachment: The role of brand trust, age, and income. *European Business Review*, 31(5), 1–55. DOI: 10.1108/EBR-02-2017-0039

Huang, S. (2019). Augmented reality and virtual reality: The power of AR and VR for business: M. Claudia tom dieck, timothy jung. Switzerland: Springer, 2019, 335 pp.119.99, ISBN 978-3-030-06245-3.

Huang, J., Saleh, S., & Liu, Y. (2021). A review on artificial intelligence in education. *Academic Journal of Interdisciplinary Studies*, 10(3), 206–206. DOI: 10.36941/ajis-2021-0077

Huang, K. T., Ball, C., Francis, J., Ratan, R., Boumis, J., & Fordham, J. (2019). Augmented versus virtual reality in education: An exploratory study examining science knowledge retention when using augmented reality/virtual reality mobile applications. *Cyberpsychology, Behavior, and Social Networking*, 22(2), 105–110. DOI: 10.1089/cyber.2018.0150 PMID: 30657334

Huang, L. L., Friedman, D. B., Lin, F. C., & Thrasher, J. F. (2018). Which types of anti-smoking television advertisements work better in Taiwan? *Health Promotion International*, 33(3), 545–555. PMID: 28011658

Hummel, P., Braun, M., Tretter, M., & Dabrock, P. (2021). Data sovereignty: A review. *Big Data & Society*, 8(1), 2053951720982012. DOI: 10.1177/2053951720982012

Huynh, T. L. D. (2020). Does culture matter social distancing under the COVID-19 pandemic? *Safety Science*, 130, 104872. https://www.ncbi.nlm.nih.gov/pmc/articles/PMC7284251/. DOI: 10.1016/j.ssci.2020.104872 PMID: 32550745

Hwang, J., Choe, J. Y., Kim, H. M., & Kim, J. (2021). Human baristas and robot baristas: How does brand experience affect brand satisfaction, brand attitude, brand attachment, and brand loyalty? *International Journal of Hospitality Management*, 99(103050), 1–12. DOI: 10.1016/j.ijhm.2021.103050

Iansiti, M., & Lakhani, K. R. (2017). The truth about blockchain. *Harvard Business Review*, 95(1), 118–127.

Iansiti, M., & Lakhani, K. R. (2020). *Competing in the age of AI: Strategy and leadership when algorithms and networks run the world*. Harvard Business Review Press.

Ikuomola, O. B., & Airhihenbuwa, C. O. An Application of the PEN-3 Model. https://www.globaldiasporanews.com/an-application-of-the-pen-3-model/

ILGA. (s.d.). "Porque existimos". Available: https://ilga-portugal.pt/associacao/porqueexistimos/ (last access 14/06/2023).

IMEDTRS. (2023). IMEDTRS. https://imedtrs.com/

Imottesjo, H., & Kain, J. H. (2022). The Urban CoCreation Lab—an integrated platform for remote and simultaneous collaborative urban planning and design through web-based desktop 3D modeling, head-mounted virtual reality and mobile augmented reality: Prototyping a minimum viable product and developing specifications for a minimum marketable product. *Applied Sciences (Basel, Switzerland)*, 12(2), 797. DOI: 10.3390/app12020797

Inphantry. (2017, October 27). *Generation Z craves experiential marketing. Here's Why*. Retrieved July 9, 2023, from https://www.inphantry.com/generation-z-craves-experiential-marketing-heres-why/

Insights, H. (n.d.). Country comparison between China and the United States. Retrieved from https://www.hofstede-insights.com/country-comparison/china,the-usa/

Insights, H. 2021. Compare countries. Retrieved from https://www.hofstede-insights.com/product/compare-countries/

Instagram. (2020, August 5). *Introducing Instagram Reels*. https://about.instagram.com/blog/announcements/introducing-instagram-reels-announcement

Instagram. (2021, January 5). *The Dos and Dont's for Reels*. https://www.instagram.com/p/CJrNqV6h3RK/

Issa, E., Tsou, M. H., Nara, A., & Spitzberg, B. (2017). Understanding the spatio-temporal characteristics of Twitter data with geotagged and non-geotagged content: Two case studies with the topic of flu and Ted (movie). *Annals of GIS*, 23(3), 219–235. Advance online publication. DOI: 10.1080/19475683.2017.1343257

Ivkov, M., Blešić, I., Simat, K., Demirović, D., Bozic, S., & Stefanovic, V. (2016). Innovations in the restaurant industry: An exploratory study. *Ekonomika Poljoprivrede (1979), 63*(4), pp. 1169–1186.

Izzy, S., Tahir, Z., Cote, D. J., Al Jarrah, A., Roberts, M. B., Turbett, S., Kadar, A., & Smirnakis, S.

Jackson, N. (2013). *Promoting and marketing events: Theory and practice*. Routledge. DOI: 10.4324/9780203583166

Jacob, R. J. K., Girouard, A., Hirshfield, L. M., Horn, M. S., Shaer, O., Solovey, E. T., & Zigelbaum, J. (2008). Reality-based interaction. *Proceedings of the SIGCHI Conference on Human Factors in Computing Systems*, 201–210. DOI: 10.1145/1357054.1357089

Jain, D. (2018). *The journey of a southern feminist*. SAGE Publishing India. DOI: 10.4135/9789353280710

Järvelä, S., & Renninger, K. (2014). *Designing for learning: Interest, motivation, and engagement*. London.

Jauhiainen, J. S. (2021). Entrepreneurship and Innovation Events during the COVID-19 Pandemic: The User Preferences of VirBELA Virtual 3D Platform at the SHIFT Event Organized in Finland. *Sustainability (Basel)*, 13(7), 3802. DOI: 10.3390/su13073802

Javornik, A. (2016). Augmented reality: Research agenda for studying the impact of its media characteristics on consumer behaviour. *Journal of Retailing and Consumer Services*, 30, 252–261. DOI: 10.1016/j.jretconser.2016.02.004

Jenkins, R. (2022). *How China is reshaping the global economy: Development impacts in Africa and Latin America*. Oxford University Press. DOI: 10.1093/oso/9780192866356.001.0001

Jia, M., & Wan, C. (2022). Why are strangers trusted more during trade fairs? A literature review on the conceptual model of general trust formation. *Journal of Business and Industrial Marketing*, 37(11), 2205–2216. DOI: 10.1108/JBIM-03-2021-0164

Jiang, S., Wei, Q., & Zhang, L. (2020). Impacts of Cultural Difference on the Transmission of COVID-19Individualism vs. Collectivism (SSRN Scholarly Paper ID 3646229). Social Science Research Network. https://doi.org/DOI: 10.2139/ssrn.3646229

Jin, L., & He, Y. (2018). How the frequency and amount of corporate donations affect consumer perception and behavioral responses. *Journal of the Academy of Marketing Science*, 46(6), 1072–1088. DOI: 10.1007/s11747-018-0584-7

Johnson, T. (2018, December 7). Google Event Encourages Girls to Look at Computer Science Careers. *Daily Nonpareil*, www.nonpareilonline.com/news/education/google-eventencourages-girls-to-look-at-computer-science-careers/article_2a08d2dd-b48d-54cf-8c41-c5c91a8ade7e.html (Accessed on 7th December, 2018)

Johnson, W. L., Vilhjálmsson, H. H., & Marsella, S. (2005). Serious games for language learning: how much game, how much AI? In AIED, vol. 125, no. 1, pp. 306–313.

Johnson, W. L., & Valente, A. (2009). Tactical language and culture training systems: Using AI to teach foreign languages and cultures. *AI Magazine*, 30(2), 72–72. DOI: 10.1609/aimag.v30i2.2240

Johnston, L. A. (2021). World trade, e-commerce, and COVID-19. *China Review*, 21(2), 65–86.

John, Y. (2023). Do Electronic Services Improve Students' Satisfaction among Public Sector Organizations? The Case of Tanzanian Higher Learning Institutions. *Perspectives on Global Development and Technology*, 22(1–2), 27–43. DOI: 10.1163/15691497-12341650

Jones, C., & Barajas, J. (2021). 'It feels like freedom,' 8 people describe getting their COVID-19 vaccine. *PBS*. https://www.pbs.org/newshour/health/it-feels-like-freedom-8-people-describe-getting-their-covid-19-vaccine

Jones, E. (2019). Rethinking greenwashing: Corporate discourse, unethical practice, and the unmet potential of ethical consumerism. *Sociological Perspectives*, 62(5), 728–754. DOI: 10.1177/0731121419849095

Jordan, T. (2001). Mapping Hacktivism: Mass Virtual Direct Action (MVDA), Individual Virtual Direct Action (IVDA) And Cyber-wars. *Computer Fraud & Security*, 2001(4), 8–11. DOI: 10.1016/S1361-3723(01)00416-X

Jordan, T. (2002). *Activism! Direct Action, Hacktivism and the Future of Society*. Reaktion Books.

Jovancic, N. (2021) 5 Types of Bias in Research and How to Make Your Surveys Bias-Free.

Joyce, M. (12 de abril de 2011). "The 7 activist uses of digital tech: the case of popular resistance in Egypt". Available: https://www.nonviolent-conflict.org/the-7-activist-uses-of-digital-tech-the-case-ofpopular-resistance-in-egypt/ (last access: 17/02/2023).

Jung, J. C., Ugboma, M. A., & Liow, A. K. (2015). Does Alibaba's magic work outside China? *Thunderbird International Business Review*, 57(6), 505–518. DOI: 10.1002/tie.21739

Kakadiya, J. (2023, April 6). The Importance of Branding in Restaurant Menu. *PhotoADKing*. Retrieved June 11, 2023, from https://blog.photoadking.com/importance-of-branding-in-restaurant-menu/

Kalam, A., Hossain, M. A., Jahan, N., & Kim, M. (2021). New Normal and Business Sustainability in the Age of Global pandemic. *Asia-Pacific Journal of Business, 12*(1), 71-86. https://www.koreascience.or.kr/article/JAKO202113150643397.pdf

Kaliyadan, F., & Kulkarni, V. (2019). Types of variables, descriptive statistics, and sample size. *Indian Dermatology Online Journal*, 10(1), 82. DOI: 10.4103/idoj.IDOJ_468_18 PMID: 30775310

Kaminska, ´. D., Sapinski, ´. T., Wiak, S., Tikk, T., Haamer, R. E., Avots, E., Helmi, A., Ozcinar, C., & Anbarjafari, G. (2019). Virtual reality and its applications in education: Survey. *Information (Basel)*, 10(10), 318. DOI: 10.3390/info10100318

Kang, H. S., Lee, J. Y., & Choi, S.. (2016). "Smart manufacturing: Past research, present findings, and future directions", *Int. J. of Precis. Eng. and Manuf.-. Green Tech.*, 3, 111–128.

Kasowski, J., Johnson, B. A., Neydavood, R., Akkaraju, A., & Beyeler, M. (2023). A systematic review of extended reality (XR) for understanding and augmenting vision loss. *Journal of Vision (Charlottesville, Va.)*, 23(5), 5. DOI: 10.1167/jov.23.5.5 PMID: 37140911

Kathirvel, A., & Gobinath, V. M. (2024f) "Chapter 1: A Review on Additive Manufactuing in Industrial", Modern Hybird Machince and Super Finishing Process: Technology and Application, CRC Publiser/Chapman and Hall, New York, 1st Edition, ISBN 1032354291, DOI eBook ISBN 9781032354293, https://books.google.co.in/books/about/Modern_Hybrid_Machining_and_Super_Finish.html?id=I_VW0AEACAAJ&source=kp_book_description&redir_esc=y

Kathirvel, A., & Maheswaran, C. P. (2023a) “Chapter 8: Enhanced AI-Based Intrusion Detection and Response System for WSN”, Artificial Intelligence for Intrusion Detection Systems, CRC Publiser/Chapman and Hall, New York, 1st Edition, ISBN 978-100096755-5, 978-103238665-2 DOI eBook ISBN 9781003346340, https://www.taylorfrancis.com/chapters/edit/10.1201/9781003346340-8/enhanced-ai-based-intrusion-detection-response-system-wsn-kathirvel-maheswaranDOI: 10.1201/9781003346340-8

Kathirvel, A., Rithik, G., & Naren, A. K. “Chapter 11: Automation of IOT Robotics.” In *Predicting Natural Disasters With AI and Machine Learning,* edited by D. Satishkumar and M. Sivaraja, 224-249. ISBN 9798369322802, eBook ISBN 9798369322819, Hershey, PA: IGI Global, 2024e. https://doi.org/DOI: 10.4018/979-8-3693-2280-2.ch011

Kathirvel, A., Subramaniam, M., Navaneethan, S., & Sabarinath, C. “Improved IDR Response System for Sensor Network”, Journal of Web Engineering, Vol. 20 1, 53–88, 2021a. Publisher: © River Publishers, 15 Feb 2021, ISSN: 1540-9589 (Print Version). ISSN: 1544-5976 (Online Version), https://doi.org/DOI: 10.13052/jwe1540-9589.2013

Kathirvel, A. (2024a). Applications of Serverless Computing: Systematic Overview. In Aluvalu, R., & Maheswari, U. (Eds.), *Serverless Computing Concepts, Technology and Architecture* (Vol. 221-233). IGI Global., DOI: 10.4018/979-8-3693-1682-5.ch014

Kathirvel, A. (2024c). Innovation and Industry Application: IoT-Based Robotics Frontier of Automation in Industry Application. In Satishkumar, D., & Sivaraja, M. (Eds.), *Internet of Things and AI for Natural Disaster Management and Prediction* (pp. 83–105). IGI Global., DOI: 10.4018/979-8-3693-4284-8.ch004

Kathirvel, A., Gopinath, V. M., Naren, K., Nithyanand, D., & Nirmaladevi, K. (2024i). Manufacturing Smart Industry Perspective an Overview. *American Journal of Engineering and Applied Sciences*, 17(1), 33–39. DOI: 10.3844/ajeassp.2024.33.39

Kathirvel, A., & Naren, A. K. (2024). Critical Approaches to Data Engineering Systems Innovation and Industry Application Using IoT. In *Critical Approaches to Data Engineering Systems and Analysis*. IGI Global. DOI: 10.4018/979-8-3693-2260-4.ch005

Kathirvel, A., & Naren, A. K. (2024d). Diabetes and Pre-Diabetes Prediction by AI Using Tuned XGB Classifier. In Khang, A. (Ed.), *Medical Robotics and AI-Assisted Diagnostics for a High-Tech Healthcare Industry* (pp. 52–64). IGI Global., DOI: 10.4018/979-8-3693-2105-8.ch004

Kathirvel, A., Naren, K., Nithyanand, D., & Santhoshi, B. (2024h). Overview of 5G Technology: Streamlined Virtual Event Experiences. *Advances of Robotic Technology*, 2(1), 1–8. DOI: 10.23880/art-16000109

Kathirvel, A., Sudha, D., Naveneethan, S., Subramaniam, M., Das, D., & Kirubakaran, S. (2022a). AI Based Mobile Bill Payment System using Biometric Fingerprint. *American Journal of Engineering and Applied Sciences*, 15(1), 23–31. DOI: 10.3844/ajeassp.2022.23.31

Keane, M., & Yu, H. (2019). A digital empire in the making: China's outbound digital platforms. *International Journal of Communication*, 13, 4624–4641.

Kees, J., & Andrews, J. C. (2019). Research Issues and Needs at the Intersection of Advertising and Public Policy. *Journal of Advertising*, 48(1), 126–135. DOI: 10.1080/00913367.2019.1599747

Ke, F., Pachman, M., & Dai, Z. (2020). Investigating educational affordances of virtual reality for simulation-based teaching Wireless Communications and Mobile Computing training with graduate teaching assistants. *Journal of Computing in Higher Education*, 32(3), 607–627. DOI: 10.1007/s12528-020-09249-9

Kelliher, F. (2005). Interpretivism and the Pursuit of Research Legitimisation: An Integrated Approach to Single Case Design. *Electronic Journal of Business Research Methods*, 3(2), 123–132.

Kenwright, B. (2023). Impact of XR on Mental Health: Are we Playing with Fire? Cornell University. https://doi.org//arxiv.2304.01648.DOI: 10.48550

Kettunen, K., Keskustalo, H., Kumpulainen, S., Pääkkönen, T., & Rautiainen, J. (2022, February). *OCR quality affects perceived usefulness of historical newspaper clippings: A user study* [Paper presentation]. *18th Italian Research Conference on Digital Libraries*, Padova, Italy.

Kettunen, K., Koistinen, M., & Kervinen, J. (2020). Ground Truth OCR Sample Data of Finnish Historical Newspapers and Journals in Data Improvement Validation of a re-OCRing Process. *LIBER Quarterly*, 30(1), 1–20. DOI: 10.18352/lq.10322

Khachatryan, S. (2023, April 13). Restaurants and Social Responsibility: The Future of Sustainability in the Industry. *Orders.co*. Retrieved May 6, 2023, from_https://orders.co/blog/restaurants-and-social-responsibility-the-future-of-sustainability-in-the-industry/

Khan, H., Laizet, F., Moulton, J., & Youldon, T. (2020, August 5). Reimagining European restaurants for the next normal. *McKinsey & Company*. Retrieved May 4, 2023, from https://www.mckinsey.com/industries/retail/our-insights/reimagining-european-restaurants-for-the-next-normal

Khan, S., Khan, M. I., & Haleem, A. (2020) "Prioritisation of Challenges Towards Development of Smart Manufacturing Using BWM Method", Alam M., Shakil K., Khan S. (eds) Internet of Things (IoT). Springer publisher. DOI: 10.1007/978-3-030-37468-6_22

Khan, I., & Fatma, M. (2017). Antecedents and outcomes of brand experience: An empirical study. *Journal of Brand Management*, 24(5), 439–452. DOI: 10.1057/s41262-017-0040-x

Khan, I., & Rahman, Z. (2015). A review and future directions of brand experience research. *International Strategic Management Review*, 3(1–2), 1–14. DOI: 10.1016/j.ism.2015.09.003

Khan, M. L. (2017). Social media engagement: What motivates user participation and consumption on YouTube? *Computers in Human Behavior*, 66, 236–247. DOI: 10.1016/j.chb.2016.09.024

Khan, W. Z., Rehman, M. H., Zangoti, H. M., Afzal, M. K., Armi, N., & Salah, K. (2020). Industrial internet of things: Recent advances, enabling technologies and open challenges. *Computers & Electrical Engineering*, 81, 1–13. DOI: 10.1016/j.compeleceng.2019.106522

Kiefer, N. M. (2002). Economics and the origin of the restaurant. *The Cornell Hotel and Restaurant Administration Quarterly*, 43(4), 58–64. DOI: 10.1016/S0010-8804(02)80042-8

Kiesberg, P. (2015). A structured approach to defense simulation train- ing. *ERCIM News*, •••, 28–29.

Kilteni, K., Groten, R., & Slater, M. (2012). The Sense of Embodiment in virtual reality. In Presence: Teleoperators and Virtual Environments (Vol. 21, Issue 4, pp. 373–387). MIT Press Journals. DOI: 10.1162/PRES_a_00124

Kim, C., Yoon, H. C., Kim, D. H., & Do, Y. R. (2018). Spectroscopic influence of virtual reality and augmented reality display devices on the human nonvisual characteristics and melatonin suppression response. *IEEE Photonics Journal*, 10(4), 1–11. DOI: 10.1109/JPHOT.2021.3107852

Kim, N. Y., Cha, Y., & Kim, H. S. (2019). Future English learning: Chatbots and artificial intelligence. *Multimedia-Assisted Language Learning*, 22(3), 32–53.

King, D. L. (2018, November 1). *Making Library Videos*. American Libraries Magazine. https://americanlibrariesmagazine.org/2018/11/01/making-library-videos/

King, D. L. (2022). *TikTok And Your Library - It Might Be Time!* Princh Library Blog. https://princh.com/blog-tiktok-and-your-library/#.YoUCrcPMLIU

Kitanov, S., Petrov, I., & Janevski, T. (2021). 6G MOBILE NETWORKS: RESEARCH TRENDS, CHALLENGES AND POTENTIAL SOLUTIONS. *Journal of Electrical Engineering and Information Technologies*, 6(2), 67–77. DOI: 10.51466/JEEIT2162186067k

Kitchen, P. J., & Tourky, M. E. (2022). Integrated Communication or Integrated Marketing Communication. *Integrated Marketing Communications: A Global Brand-Driven Approach*, 69-87.

Kitchen, P. J., & Tourky, M. E. (2022). Integrated Communication or Integrated Marketing Communication. In *Integrated Marketing Communications* (pp. 69–87). Palgrave Macmillan. DOI: 10.1007/978-3-030-76416-6_5

Klijn, E. (2008). The current state-of-art in newspaper digitization: A Market Perspective. *D-Lib Magazine : the Magazine of the Digital Library Forum*, 14(1/2). Advance online publication. DOI: 10.1045/january2008-klijn

Klimova, B., & Polakova, P. (2020). Students' perceptions of an EFL vocabulary learning mobile application. *Education Sciences*, 10(2), 37. DOI: 10.3390/educsci10020037

Koistinen, M., Kettunen, K., & Paakkonen, T. (2017). Improving Optical Character Recognition of Finnish Historical Newspapers with a Combination of Fraktur & Antiqua Models and Image Preprocessing. In *Proceedings of the 21st Nordic Conference of Computational Linguistics* (pp. 277–283). Sweden.

Kokas, A. (2022). *Trafficking data: How China is winning the battle for digital sovereignty*. Oxford University Press. DOI: 10.1093/oso/9780197620502.001.0001

Kong, D. (2018, January 5). *2017 Short Video Industry Report*. http://www.199it.com/archives/672181.html

Koponen, S., & Mustonen, P. (2020). Eating alone, or commensality redefined? Solo dining and the aestheticization of eating (out). *Journal of Consumer Culture*, 22(2), 359–377. DOI: 10.1177/1469540520955219

Kose, ¨. H., & Güner-Yildiz, N. (2021). Augmented reality (AR) as a learning material in special needs education. *Education and Information Technologies*, 26(2), 1921–1936. DOI: 10.1007/s10639-020-10326-w

Krauß, V., Berkholz, J., Recki, L., & Boden, A. (2023). Beyond Well-Intentioned: An HCI Students' Ethical Assessment of Their Own XR Designs. .DOI: 10.1109/ISMAR59233.2023.00020

Kravchyk, R. (2022, May 29). Sensory complexity and cross-modality of the perceived taste | The Perfectionist. *The Perfectionist.* Retrieved May 5, 2023, from https://www.theperfectionist.pl/flavours-are-not-what-they-seem-sensory-complexity-and-cross-modality-of-the-perceived-taste/

Kricorian, K., & Turner, K. (2021). COVID-19 Vaccine Acceptance and Beliefs among Black and Hispanic Americans. *PLoS One*, 16(8), e0256122. DOI: 10.1371/journal.pone.0256122 PMID: 34428216

Krüger, J. M., Buchholz, A., & Bodemer, D. (2019). Augmented reality in education: Three unique characteristics from a user's perspective. In Proc. 27th Int. Conf. on Comput. in Educ, 412–422.

Kumar, P. P., Thallapalli, R., Akshay, R., Sai, K. S., Sai, K. S., & Srujan, G. S. (2022). Stateof-the-Art: Implementation of augmented reality and virtual reality with the integration of 5G in the classroom. InAIP Conference Proceedings, 2418, 20069 (AIP Publishing LLC).

Kumar, D., Baranwal, G., Raza, Z., & Vidyarthi, D. P. (2017). A systematic study of double auction mechanisms in cloud computing. *Journal of Systems and Software*, 125, 234–255. DOI: 10.1016/j.jss.2016.12.009

Kumar, D., Samalia, H. V., & Verma, P. (2017). Factors influencing cloud computing adoption by small and medium-sized enterprises (SMEs). In *India*. Pacific Asia Journal of the Association for Information Systems. Maaruf Ali, Mahdi H. Miraz. Cloud Computing Applications.

Kumpulainen, S., & Late, E. (2022). Struggling with digitized historical newspapers: Contextual barriers to information interaction in history research activities. *Journal of the Association for Information Science and Technology*, 73(7), 1012–1024. DOI: 10.1002/asi.24608

Kusiak, A. (2018). Smart manufacturing. *International Journal of Production Research*, 56(2), 508–517. DOI: 10.1080/00207543.2017.1351644

Kusiak, A. (2019). Fundamentals of smart manufacturing: A multi-thread perspective. *Annual Reviews in Control*, 47, 214–220. DOI: 10.1016/j.arcontrol.2019.02.001

Kutlaca, M., & Radke, H. R. M. (2022). Towards an understanding of performative allyship: Definition, antecedents and consequences. *Social and Personality Psychology Compass*, •••, 1–12. DOI: 10.1111/spc3.12724

Kwun, D. J. (2012). Brand Management in the Hospitality Industry. *Journal of Tourism & Hospitality (Los Angeles, Calif.)*, 01(01), 1–2. DOI: 10.4172/2167-0269.1000e104

Lacoche, J., Pallamin, N., Boggini, T., & Royan, J. (2017). Collaborators awareness for user cohabitation in co-located collaborative virtual environments. *Proceedings of the 23rd ACM Symposium on Virtual Reality Software and Technology*, 1–9. DOI: 10.1145/3139131.3139142

Laczniak, G., & Shultz, C. (2021). Toward a doctrine of socially responsible marketing (SRM): A macro and normative-ethical perspective. *Journal of Macromarketing*, 41(2), 201–231. DOI: 10.1177/0276146720963682

Lagman, J. D. N., Vergara, R. J. D., & Sarmiento, P. J. D. (2021). Culture as a contributory factor in combatting the COVID-19 pandemic. *Journal of Public Health (Oxford, England)*, 43(2), e289–e290. https://academic.oup.com/jpubhealth/article/43/2/e289/6102664?login=true. DOI: 10.1093/pubmed/fdaa280 PMID: 33454771

Lahtinen, A. (2018). *China's Diplomacy and Economic Activities in Africa: Relations on the move*. Springer. DOI: 10.1007/978-3-319-69353-8

Lai, N. Y. G., Wong, K. H., Yu, L. J., & Kang, H. S. (2020). Virtual reality (VR) in engineering education and training: A bibliometric analysis. In InProceedings of the 2nd world symposium on software engineering (pp. 161–165).

Lane, P., McKenna, H., Ryan, A., & Fleming, P. (2001). Focus group methodology. *Nurse Researcher*, 8(3), 45–59. DOI: 10.7748/nr2001.04.8.3.45.c6157

Lavrentieva, O., Arkhypov, I., Kuchma, O., & Uchitel, A. (2020). Use of simulators together with virtual and augmented reality in the system of welders' vocational training: Past, present, and future.

Lazer, D. M., Baum, M. A., Benkler, Y., Berinsky, A. J., Greenhill, K. M., Menczer, F., & Rothschild, D. (2018). The science of fake news. *Science*, 359(6380), 1094–1096. DOI: 10.1126/science.aao2998 PMID: 29590025

Lead Quizzes. (July 4, 2021). https://www.leadquizzes.com/blog/types-of-bias-in-research/

Leander, K. M., & Burriss, S. K. (2020). Critical literacy for a posthuman world: When people read, and become, with machines. *British Journal of Educational Technology*, 51(4), 1262–1276. DOI: 10.1111/bjet.12924

LearnDash Collaborator. (2020). *How is AI being used in e-learning in 2020?* https://www.learndash.com/how-is-ai-being-used-in-e-learning-in-2020/

Lee, B., Sedlmair, M., & Schmalstieg, D. (2023). Design Patterns for Situated Visualization in Augmented Reality. *IEEE Transactions on Visualization and Computer Graphics*, •••, 1–12. DOI: 10.1109/TVCG.2023.3327398 PMID: 37883275

Lee, H. L., & Shen, Z. J. M. (2020). Supply chain and logistics innovations with the Belt and Road Initiative. *Journal of Management Science and Engineering*, 5(2), 77–86. DOI: 10.1016/j.jmse.2020.05.001

Lemmers, F., Ott, M., & Hermans, S. (2023). Printed mass media and automatic digitisation: The case of Belgian illustrated magazines from the interbellum. *In Monte Artium*, 15(15), 1–31. DOI: 10.4000/ima.427

Leonhardt, J. M., & Pezzuti, T. (2021). EXPRESS: Vaccination Acceptance across Cultures: The Roles of Collectivism, Empathy, and Homophily. Journal of International Marketing, 1069031X211073179.

Lerman, R. (2020, August 5). *Instagram Reels, Facebook's new TikTok competitor, is released amid political scrutiny over the Chinese social media platform*. The Washington Post. https://www.washingtonpost.com/technology/2020/08/05/tiktok-instagram-reels-facebook/

Letunovska, N. Y. (2020). *Challenges for the health components of a region in the context of globalization* (Doctoral dissertation, Sumy State University).

Lewis, R. B. (1999). SIMSTAT with wordstat: A comprehensive statistical package with a content analysis module. *Field Methods*, 11(2), 166–179. DOI: 10.1177/1525822X9901100207

Li, Y., Ch'ng, E., Cai, S., & See, S. (2018). Multiuser Interaction with Hybrid VR and AR for Cultural Heritage Objects. 2018 3rd Digital Heritage International Congress (DigitalHERITAGE) Held Jointly with 2018 24th International Conference on Virtual Systems & Multimedia (VSMM 2018), 1–8. DOI: 10.1109/DigitalHeritage.2018.8810126

Liang, X., Liu, F., Wang, L., Zheng, B., & Sun, Y. (2023). Internet of Cultural Things: Current research, challenges and opportunities. *Computers, Materials & Continua*, 74(1), 469–488. DOI: 10.32604/cmc.2023.029641

Liao, Y., de Freitas Rocha Loures, E., & Deschamps, F. (2018). Industrial Internet of Things: A Systematic Literature Review and Insights. *IEEE Internet of Things Journal*, 5(6), pp4515–pp4525. DOI: 10.1109/JIOT.2018.2834151

Li, B. J., Peña, J., & Jung, Y. (2023). Editorial: VR/AR and wellbeing: The use of immersive technologies in promoting health outcomes. *Frontiers in Virtual Reality*, 3, 1119919. Advance online publication. DOI: 10.3389/frvir.2022.1119919

Liebl, B., & Burghard, M. (2020, November). *From historical newspapers to machine-readable data: The origami ocr pipeline* [Paper presentation]. *Workshop on Computational Humanities Research*, Amsterdam, Netherlands.

Liguori, E. W., & Pittz, T. G. (2020). Strategies for small business: Surviving and thriving in the era of COVID-19. *Journal of the International Council for Small Business*, 1(2), 106–110. DOI: 10.1080/26437015.2020.1779538

Li, L., Bensi, M., & Baecher, G. (2023). Exploring the potential of social media crowdsourcing for post-earthquake damage assessment. *International Journal of Disaster Risk Reduction*, 98, 104062. Advance online publication. DOI: 10.1016/j.ijdrr.2023.104062

Lisinska, A. (2019). Do Initiators Matter in Social Marketing? Understanding Characteristics and Particularities of Social Marketing Campaigns Initiated by Different Agents. *SPNHA Review*, 15(1), 78–104.

Litwin, M. L. (2007). *The ABC's of Strategic Communication: Thousands of Terms, Tips and Techniques*. AuthorHouse.

Litwin, M. L. (2007). *The ABCs of strategic communication: Thousands of terms, tips and techniques*. AuthorHouse.

Liu, N., Zhang, F., Wei, C., Jia, Y., Shang, Z., Sun, L., Wu, L., Sun, Z., Zhou, Y., Wang, Y., & Liu, W. (2020). Prevalence and predictors of PTSS during COVID-19 outbreak in China hardest-hit areas: Gender differences matter. *Psychiatry Research*, 112921, 112921. https://www.ncbi.nlm.nih.gov/pmc/articles/PMC7102622/. DOI: 10.1016/j.psychres.2020.112921 PMID: 32240896

Liu, Z., Li, Q., Chen, X., Wu, C., Ishihara, S., Li, J., & Ji, Y. (2020). Point Cloud Video Streaming in 5G Systems and Beyond: Challenges and Solutions. *IEEE Network*. https://doi.org/DOI: 10.36227/techrxiv.13138940.v1

Li, X., Yi, W., Chi, H. L., Wang, X., & Chan, A. P. (2018). A critical review of virtual and augmented reality (VR/AR) applications in construction safety. *Automation in Construction*, 86, 150–162. DOI: 10.1016/j.autcon.2017.11.003

Lopez, L.III, Hart, L. H.III, & Katz, M. H. (2021). Racial and Ethnic Health Disparities Related to COVID-19. *Journal of the American Medical Association*, 325(8), 719–720. DOI: 10.1001/jama.2020.26443 PMID: 33480972

Lorenz, T. (2019). *TikTok Clubs in High School? It Was a Thing*. The New York Times. https://www.nytimes.com/2019/10/19/style/high-school-tiktok-clubs.html

Lotfollahzadeh, A., Rastgoo, L., Shirinzadeh, I., Kharghani Moghadam, S. M., & Ebrahimi, H. (2021). Investigating the compliance of COVID-19 protocols in the workplaces of Ardabil, Iran. *Work*, (Preprint), 1. 1 Jan. 2021: 1031 – 1037.7.

Luo, C., Chen, A., Cui, B., & Liao, W. (2021). Exploring public perceptions of the COVID-19 vaccine online from a cultural perspective: Semantic network analysis of two social media platforms in the United States and China. *Telematics and Informatics*, 65, 101712. https://www.ncbi.nlm.nih.gov/pmc/articles/pmc8429027/. DOI: 10.1016/j.tele.2021.101712 PMID: 34887618

Luo, M., & Cheng, L. (2020). Exploration of interactive foreign language teaching mode based on artificial intelligence. *International Conference on Computer Vision, Image and Deep Learning (CVIDL)*, pp. 285–290. DOI: 10.1109/CVIDL51233.2020.00-84

Luo, X., Zhu, S., & Song, Z. (2023). Quantifying the Income-Increasing Effect of Digital Agriculture: Take the New Agricultural Tools of Smartphone as an Example. *International Journal of Environmental Research and Public Health*, 20(4), 3127. DOI: 10.3390/ijerph20043127 PMID: 36833825

Luxon, E. M. (2019). Mobilizing Environmental Sentiment Through the Media. *Environmental Politics*, 28(4), 639–662. DOI: 10.1080/09644016.2018.1560743

Ly, B. (2020). Challenge and perspective for digital Silk road. *Cogent Business & Management*, 7(1), 1804180. DOI: 10.1080/23311975.2020.1804180

M, Akbari., N, Ha., & S, Kok. (2022). A systematic review of AR/VR in operations and supply chain management: maturity, current trends and future directions. Journal of Global Operations and Strategic Sourcing.

M., Feske, S. K., Zafonte, R., Fishman, J. A., & El Khoury, J. (2020). Characteristics and outcomes of Hispanic and Latino patients with COVID-19 in comparison with other ethnic and racial groups. *Open Forum Infectious Diseases*, 7(10), ofaa401. DOI: 10.1093/ofid/ofaa401

Ma, L. (2021). An immersive context teaching method for college English based on artificial intelligence and machine learning in virtual reality technology. Mobile Information Systems, vol. 2021, Article ID 2637439, 7 pages. DOI: 10.1155/2021/2637439

Maaravi, Y., Levy, A., Gur, T., Confino, D., & Segal, S. (2021). "The Tragedy of the Commons": How Individualism and Collectivism Affected the Spread of the COVID-19 Pandemic. *Frontiers in Public Health*, 9, 627559. Advance online publication. DOI: 10.3389/fpubh.2021.627559 PMID: 33643992

Maccarrone, P., & Contri, A. M. (2021). Integrating corporate social responsibility into corporate strategy: The role of formal tools. *Sustainability (Basel)*, 13(22), 12551. DOI: 10.3390/su132212551

Maier, A. L. (2021). Political corporate social responsibility in authoritarian contexts. *Journal of International Business Policy*, 4(4), 476–495. DOI: 10.1057/s42214-020-00085-3

Mainardes, E. W., Alves, H., & Raposo, M. (2010). An exploratory research on the stakeholders of a university. *Journal of Management and Strategy*, 1(1). Advance online publication. DOI: 10.5430/jms.v1n1p76

Ma, L., Wang, Y., & Kim, J. (2023). How health organizations communicate about COVID-19 on social media: A comparative content analysis. *Journal of Communication in Healthcare*, 16(1), 83–92. Advance online publication. DOI: 10.1080/17538068.2022.2103334 PMID: 36919810

Malecki, K. M., Keating, J. A., & Safdar, N. (2021). Crisis communication and public perception of COVID-19 risk in the era of social media. *Clinical Infectious Diseases*, 72(4), 697–702. https://www.ncbi.nlm.nih.gov/pmc/articles/PMC7337650/ . DOI: 10.1093/cid/ciaa758 PMID: 32544242

Mandal, P. C. (2019a). Social criticisms of marketing: Concerns and initiatives. *International Journal of Society Systems Science*, 11(2), 157–170. DOI: 10.1504/IJSSS.2019.100088

Mandal, P. C. (2019b). Ethical Issues and Concerns in Collection of Marketing Information and Marketing Intelligence: Ethical Issue in Collection of Information. [IJBIR]. *International Journal of Business Intelligence Research*, 10(2), 16–28. DOI: 10.4018/IJBIR.2019070102

Mandal, P. C. (2021). Socially-Responsible Marketing Communications: Concerns, Strategies, and Initiatives. [JMME]. *Journal of Media Management and Entrepreneurship*, 3(1), 1–17. DOI: 10.4018/JMME.290301

Mandal, P. C. (2022). Marketing Information and Marketing Intelligence: Generation of Customer Insights. [IJTD]. *International Journal of Technology Diffusion*, 13(1), 1–14. DOI: 10.4018/IJTD.300745

Mandil, K. A. (2016). *Modelling the relationship between brand experience dimensions and the antecedents of happiness within the context of dining services* [PhD dissertation, Brunel University London].

Markowitz, D., & Bailenson, J. (2019). *Virtual Reality and Communication*. Communication., DOI: 10.1093/obo/9780199756841-0222

Marot, T. B. (2020). *Space, the final legal frontier: a look at legal impediments to the increasing commercialisation of outer space* (Doctoral dissertation, University of the Witwatersrand, Johannesburg, South Africa).

Martens, M. (1999). A philosophy for sensory science. *Food Quality and Preference*, 10(4–5), 233–244. DOI: 10.1016/S0950-3293(99)00024-5

Martin, B. A. (2019). The Unregulated Underground Market for Your Data: Providing Adequate Protections for Consumer Privacy in the Modern Era. *Iowa Law Review*, 105, 865.

Martin, C. M. (2012). One-minute video: Marketing your library to faculty. *RSR. Reference Services Review*, 40(4), 589–600. DOI: 10.1108/00907321211277387

Massey, O. T. (2011). A proposed model for the analysis and interpretation of focus groups in evaluation research. *Evaluation and Program Planning*, 34(1), 21–28. DOI: 10.1016/j.evalprogplan.2010.06.003 PMID: 20655593

Matemba, E. D., & Li, G. (2018). Consumers' willingness to adopt and use WeChat wallet: An empirical study in South Africa. *Technology in Society*, 53, 55–68. DOI: 10.1016/j.techsoc.2017.12.001

Maunder, R. E. (2018). Students' peer relationships and their contribution to university adjustment: The need to belong in the university community. *Journal of Further and Higher Education*, 42(6), 756–768. DOI: 10.1080/0309877X.2017.1311996

McCafferty, D. (2011). Activism Vs. Slacktivism. *Communications of the ACM*, 54(12), 17–19. DOI: 10.1145/2043174.2043182

McCain Foods. (2021, February 9). Why restaurants should balance value, quality and social responsibility. *Restaurant Business*. Retrieved May 1, 2023, from https://www.restaurantbusinessonline.com/operations/why-restaurants-should-balance-value-quality-social-responsibility

McClaughlin, E., Vilar-Lluch, S., Parnell, T., Knight, D., Nichele, E., Adolphs, S., Clos, J., & Schiazza, G. (2023). The reception of public health messages during the COVID-19 pandemic. *Applied Corpus Linguistics*, 3(1), 100037. DOI: 10.1016/j.acorp.2022.100037 PMID: 37521321

McCracken, R. (2022, July 20). Why restaurant branding is a game changer in hospitality | TableCheck Blog. *TableCheck*. Retrieved June 10, 2023, from https://www.tablecheck.com/en/blog/restaurant-branding-hospitality-game-changer/

McLaughlin, K. (2022). *COVID-19: Implications for business February 9, 2022*. McKinsey.

McLeod, S. (2011). *Albert Bandura's social learning theory*. Simply Psychology.

McLymont, V. (2022, October 11). What Gen Z wants: A look inside their food preferences. *Hospitality News*. Retrieved August 10, 2023, from https://hospitalitynewsny.com/what-gen-z-wants-a-look-inside-their-food-preferences/

McNair, B. (2017). *An Introduction to Political Communication*. Routledge. DOI: 10.4324/9781315750293

McRoberts, S., Ma, H., Hall, A., & Yarosh, S. (2017). Share first, save later: Performance of self through Snapchat stories. *Conference on Human Factors in Computing Systems - Proceedings, 2017-May*, 6902–6911. DOI: 10.1145/3025453.3025771

Meier, K., Schützeneder, J., García-Avilés, J. A., Valero-Pastor, J. M., Kaltenbrunner, A., Lugschitz, R., Porlezza, C., Ferri, G., Wyss, V., & Saner, M. (2022). Examining the most relevant journalism innovations: A comparative analysis of five European countries from 2010 to 2020. *Journalism and Media*, 3(4), 698–714. DOI: 10.3390/journalmedia3040046

Meng, K. S., & Leung, L. (2021). Factors influencing TikTok engagement behaviors in China: An examination of gratifications sought, narcissism, and the Big Five personality traits. *Telecommunications Policy*, 45(7), 102172. DOI: 10.1016/j.telpol.2021.102172

Mercer, S. T. (2020). The Limitations of European Data Protection As A Model for Global Privacy Regulation. *AJIL Unbound*, 114, 20–25. DOI: 10.1017/aju.2019.83

Merckaert, G. (2023a, January 20). 85+ Restaurant Industry Statistics That Every F&B Operator & Manager Needs to Know in 2023. *Apicbase*. Retrieved March 31, 2023, from https://get.apicbase.com/restaurant-industry-statistics-2023/

Merckaert, G. (2023b, March 17). Trends and Tactics: How to Navigate the Changing Landscape of the European Restaurant Industry. *Apicbase*. Retrieved March 30, 2023, from https://get.apicbase.com/restaurant-industry-europe/

Merga, M. K. (2021). How can Booktok on TikTok inform readers' advisory services for young people? *Library & Information Science Research*, 43(2), 101091. DOI: 10.1016/j.lisr.2021.101091

Merriam, S., & Bierema, L. (2005). *Adult Learning: Linking Theory and Practice*. Willey.

Meulen, N. V. D., & Wiesel, J. (2017). Plating Food On the Pictorial Arrangement of Cuisine on the Plate. In *Culinary Turn* (pp. 235–250).

Michel, C., Velasco, C., Fraemohs, P., & Spence, C. (2015). Studying the impact of plating on ratings of the food served in a naturalistic dining context. *Appetite*, 90, 45–50. DOI: 10.1016/j.appet.2015.02.030 PMID: 25728885

Michel, C., Velasco, C., Gatti, E., & Spence, C. (2014). A taste of Kandinsky: Assessing the influence of the artistic visual presentation of food on the dining experience. *Flavour (London)*, 3(1), 1–11. DOI: 10.1186/2044-7248-3-7

Miles, S. (1996). *Consumerism: As a way of life*. Sage Publications.

Millimen, S. K. (2015). *All made-up: the hyperfeminization of fat women*. The University of Toledo.

Minghini, M., Delucchi, L., Sarretta, A., Lupia, F., Napolitano, M., & Palmas, A. (2017). Collaborative mapping response to disasters through openstreetmap: The case of the 2016 Italian earthquake. In *Geoingegneria Ambientale e Mineraria* (Vol. 151, Issue 2).

Ministry of Electronic and information Technology, Government of India. Email Service. Retrieved from https://www.meity.gov.in/secure-email-service-government-india

Ministry of Health and Family Welfare, Government of India. Retrieved from https://www.mohfw.gov.in/

Mintel. (2022). Restaurant Breakfast and Brunch Trends Market Report – US – 2022. In https://store.mintel.com/report/us-restaurant-breakfast-and-brunch-trends-market-report

Miranda, J., Fidalgo, J., & Martins, P. (2021). Jornalistas em tempo de pandemia: Novas rotinas profissionais, novos desafios éticos [Journalists in times of pandemic: New professional routines, new ethical challenges]. *Comunicação e Sociedade*, 39, 287–307. DOI: 10.17231/comsoc.39(2021).3176

Mirzaei, A., Wilkie, D. C., & Siuki, H. (2022). Woke brand activism authenticity or the lack of it. *Journal of Business Research*, 139, 1–12. DOI: 10.1016/j.jbusres.2021.09.044

Mishrif, A., & Khan, A. (2023). Technology adoption as survival strategy for small and medium enterprises during COVID-19. *Journal of Innovation and Entrepreneurship*, 12(1), 53. DOI: 10.1186/s13731-023-00317-9

Misra, P., Chopra, G., & Bhaskar, P. (2023). Continuous usage intention for digital library systems among students at higher learning institutions: Moderating role of academic involvement. *Journal of Applied Research in Higher Education*, 15(5), 1752–1766. DOI: 10.1108/JARHE-06-2022-0185

Mittal, S., Khan, M. A., Purohit, J. K., Menon, K., Romero, D., & Wuest, T. (2020). A smart manufacturing adoption framework for SMEs. *International Journal of Production Research*, 58(5), 1555–1573. DOI: 10.1080/00207543.2019.1661540

Mittal, S., Khan, M. A., Romero, D., & Wuest, T. (2018). A critical review of smart manufacturing & Industry 4.0 maturity models: Implications for small and medium-sized enterprises (SMEs). *Journal of Manufacturing Systems*, 49, 194–214. DOI: 10.1016/j.jmsy.2018.10.005

Mkamboi, A., Ndavula, J., & Mwangi, S. W. (2022). Utilizing Social Media Platforms for Online Crisis Management: Focus on Always Pads Crisis of 2019. *African Multidisciplinary Journal of Research*, 7(1), 83–94.

Moglia, A., Ferrari, V., Morelli, L., Ferrari, M., Mosca, F., & Cuschieri, A. (2016). A Systematic Review of Virtual Reality Simulators for Robot-assisted Surgery. *European Urology*, 69(6), 1065–1080. DOI: 10.1016/j.eururo.2015.09.021 PMID: 26433570

Mokhtar, S. S. M., Javeed, A., & Khan, M. Y. (2021). *Halal Marketing: Concept and Strategies (UUM Press)*. UUM Press.

Moore, K. S. (2021). The Impact of COVID-19 on the Hispanic and Latino Population: A Scoping Literature Review. *Public Health Nursing*. https://www.ncbi.nlm.nih.gov/pmc/articles/PMC8251024/

Moore, M. (1989). Three types of interaction. *American Journal of Distance Education*, 3(2), 1–7. DOI: 10.1080/08923648909526659

Morales, D. X., & Paat, Y. F. (2024). Hesitancy or resistance? Differential changes in COVID-19 vaccination intention between black and white Americans. *Journal of Racial and Ethnic Health Disparities*, 11(1), 23–35. DOI: 10.1007/s40615-022-01494-1 PMID: 36547772

Morales-Vives, F., Dueñas, J. M., Ferrando, P. J., Vigil-Colet, A., & Varea, M. D. (2022). Compliance with pandemic Commands Scale (COCOS): The relationship between compliance with COVID-19 measures and sociodemographic and attitudinal variables. *PLoS One*, 17(1), e0262698. https://journals.plos.org/plosone/article?id=10.1371/journal.pone.0262698. DOI: 10.1371/journal.pone.0262698 PMID: 35045104

Moreira, A. C., Fortes, N., & Santiago, R. (2017). Influence of sensory stimuli on brand experience, brand equity and purchase intention. *Journal of Business Economics and Management*, 18(1), 68–83. DOI: 10.3846/16111699.2016.1252793

Moreira, T., Hernandez, D. C., Scott, C. W., Murillo, R., Vaughan, E. M., & Johnston, C. A. (2018). *Susto, Coraje, y Fatalismo*: Cultural-Bound Beliefs and the Treatment of Diabetes Among Socioeconomically Disadvantaged Hispanics. *American Journal of Lifestyle Medicine*, 12(1), 30–33. DOI: 10.1177/1559827617736506 PMID: 29541008

Mosley, R. (2007). Customer experience, organisational culture and the employer brand. *Journal of Brand Management*, 15(2), 123–134. DOI: 10.1057/palgrave.bm.2550124

Mostafa, R. B., & Kasamani, T. (2020). Brand experience and brand loyalty: Is it a matter of emotions? *Asia Pacific Journal of Marketing and Logistics*, 33(4), 1033–1051. DOI: 10.1108/APJML-11-2019-0669

Mott, M., Cutrell, E., Gonzalez Franco, M., Holz, C., Ofek, E., Stoakley, R., & Ringel Morris, M. (2019). Accessible by Design: An Opportunity for Virtual Reality. 2019 IEEE International Symposium on Mixed and Augmented Reality Adjunct (ISMAR-Adjunct). DOI: 10.1109/ISMAR-Adjunct.2019.00122

Moyce, S., Velazquez, M., Claudio, D., Thompson, S., Metcalf, M., Aghbashian, E., Vanderwood, K., & Sisson, N. (2021). Exploring a rural Latino community's perception of the COVID-19 pandemic. *Ethnicity & Health*, 26(1), 126–138. DOI: 10.1080/13557858.2020.1838456 PMID: 33126820

Mudička, Š., & Kapica, R. (2022). Digital Heritage, the Possibilities of Information Visualisation through Extended Reality Tools. *Heritage*, 6(1), 112–131. DOI: 10.3390/heritage6010006

Munodawafa, D., Ford, C., Oni, T., & Agyemang, C. (2020). Culture Matters in Communicating the Global Response to COVID-19. *PCD Staff, 17*, E60. https://www.cdc.gov/pcd/collections/pdf/US-Public-Health-Response-to-COVID-19.pdf#page=30

Muñoz-Saavedra, L., Miró-Amarante, L., & Domínguez-Morales, M. (2020). Augmented and Virtual Reality Evolution and Future Tendency. *Applied Sciences (Basel, Switzerland)*, 10(1), 322. DOI: 10.3390/app10010322

Murdoch, M., Simon, A. B., Polusny, M. A., Bangerter, A. K., Grill, J. P., Noorbaloochi, S., & Partin, M. R. (2014). Impact of different privacy conditions and incentives on survey response rate, participant representativeness, and disclosure of sensitive information: A randomized controlled trial. *BMC Medical Research Methodology*, 14(1), 1–11. https://bmcmedresmethodol.biomedcentral.com/articles/10.1186/1471-2288-14-90he. DOI: 10.1186/1471-2288-14-90 PMID: 25027174

Myers, C. (2021). *The Effects of Pedestrian Gestures on Motorist Yielding at Two-Lane Crosswalks* (Doctoral dissertation, University of Kansas).

Mystakidis, S., Fragkaki, M., & Filippousis, G. (2021). Ready teacher one: Virtual and augmented reality online professional development for K-12 school teachers. *Computers*, 10(10), 134. DOI: 10.3390/computers10100134

Nabavi, R. T. (2012). Bandura's social learning theory & social cognitive learning theory. *Theory of Developmental Psychology*, 1(1), 1–24.

Namboodiri, A. M., & Jain, A. K. (2007). Document structure and layout analysis. In Chaudhuri, B. B. (Ed.), *Digital document processing: Advances in pattern recognition* (pp. 29–48). Springer. DOI: 10.1007/978-1-84628-726-8_2

National Digital Communication Policy. Preamble (2018). 'National Digital Communication Policy 2018'. Retrieved from https://www.meity.gov.in/writereaddata/files/National_Digital_Communications_Policy%E2%80%932018.pdf

National Restaurant Association. (2022). *2022 State of Restaurant Industry*. Retrieved April 3, 2023, from https://xchange.avixa.org/documents/pdf-report-2022-state-of-restaurant-industry

National Restaurant Organization. (2019). *Restaurant Industry 2030*. Retrieved March 29, 2023, from https://restaurant.org/research-and-media/research/research-reports/restaurant-industry-2030/

Naveneethan, S. Madhan E.S., Kathirvel A. (2022) Identifying and Eliminating the Misbehavior Nodes in the Wireless Sensor Network. In: Reddy V.S., Prasad V.K., Wang J., Reddy K. (eds) Soft Computing and Signal Processing. ICSCSP 2021. Advances in Intelligent Systems and Computing, vol 1413. Springer International Publishing, Springer, Singapore. https://doi.org/DOI: 10.1007/978-981-16-7088-6_36

Newman, N. (2023). Journalism, media, and technology trends and predictions 2023. In ora.ox.ac.uk. Reuters Insitute for the Study of Journalism. https://ora.ox.ac.uk/objects/uuid:38c0393f-3634-459a-a505-c53f2eb637a3

News, A. B. C. (2024). Virtual Reality. ABC News. https://abcnews.go.com/VR

Nguyen, L., & Mikolajczyk, K. (2021, July 21). *A Beginner-Friendly Guide to Using Instagram Reels for Business*. Hootsuite. https://blog.hootsuite.com/instagram-reels/

Nguyen, V. T., & Dang, T. (2017). Setting up virtual reality and augmented reality learning environment in unity. In In 2017 IEEE international symposium on Mixed and augmented reality (ISMAR-Adjunct) (pp. 315–320). IEEE. DOI: 10.1109/ISMAR-Adjunct.2017.97

Nguyen, V. T., Jung, K., & Dang, T. (2019). Creating virtual reality and augmented reality development in classroom: Is it a hype? In In 2019 IEEE international conference on artificial Intelligence and virtual reality (AIVR) (pp. 212–2125). IEEE.

Nguyen, Q. C., Kath, S., Meng, H. W., Li, D., Smith, K. R., VanDerslice, J. A., Wen, M., & Li, F. (2016). Leveraging geotagged Twitter data to examine neighborhood happiness, diet, and physical activity. *Applied Geography (Sevenoaks, England)*, 73, 77–88. Advance online publication. DOI: 10.1016/j.apgeog.2016.06.003 PMID: 28533568

NHP INDIA -National Health Portal. Retrieved from https://www.nhp.gov.in/

Noe-Bustamante, L., Krogstad, J. M., & Lopez, M. H. (2021, July 16). *For U.S. Latinos, COVID-19 Has Taken a Personal and Financial Toll*. Race & Ethnicity. https://www.pewresearch.org/race-ethnicity/2021/07/15/for-u-s-latinos-COVID-19-has-taken-a-personal-and-financial-toll/

Noland, A. (2019). "Like, Share, Retweet: Testing Competing Models of the Theory of Planned Behavior to Predict Slacktivism Engagement". Journal of Nonprofit & Public Sector Marketing, 1 - 22, DOI: .DOI: 10.1080/10495142.2019.1589626

Nur, F., Akmaliah, N., Chairul, R., & Safira, S. (2021). Green purchase intention: The power of success in green marketing promotion. *Management Science Letters*, 11(5), 1607–1620.

Nysveen, H., Pedersen, P. B., & Skard, S. (2012). Brand experiences in service organizations: Exploring the individual effects of brand experience dimensions. *Journal of Brand Management*, 20(5), 404–423. DOI: 10.1057/bm.2012.31

O'Brien, M., & Shea, J. (2011). Disparities in patient satisfaction among Hispanics: The role of language preference. *Journal of Immigrant and Minority Health*, 13(2), 408–412. DOI: 10.1007/s10903-009-9275-2 PMID: 19626440

O'Neil, A. (2021, December 10). Latin America- Statistics & Facts. Statista. Retrieved from https://www.statista.com/topics/3287/latin-america/

Oberdorfer, ̈. S., Birnstiel, S., Latoschik, M. E., & Grafe, S. (2021). Mutual benefits: Interdisciplinary education of pre-service teachers and hci students in vr/ar learning environment design. [Frontiers Media SA]. *Frontiers in Education*, 6, 693012. DOI: 10.3389/feduc.2021.693012

Oetzel, J., De Vargas, F., Ginossar, T., & Sanchez, C. (2007). Hispanic women's preferences for breast health information: Subjective cultural influences on source, message, and channel. *Health Communication*, 21(3), 223–233. DOI: 10.1080/10410230701307550 PMID: 17567254

Ogunwole, S., Rabe, M., Roberts, A., & Caplan, Z. (2021). Population Under Age 18 Declined

Okano-Heijmans, M., & Vosse, W. (2021). Promoting open and inclusive connectivity: The case for digital development cooperation. *Research in Globalization*, 3, 100061. DOI: 10.1016/j.resglo.2021.100061

Okorafor, K., Arioie, I. A., & Anyalebechi, L. I. (2017). Awareness and Use of Reference and Information Services amongst Library Users in University Libraries in two States in Nigeria. *Journal of Applied Information Science and Technology*, 10(3), 109–116. https://www.jaistonline.org/10vol3/14.pdf

Olagunju, A. (2021). The Impact of Race and History on COVID Vaccination Acceptance.

Olbina, S., & Glick, S. (2022). Using integrated hands-on and virtual reality (VR) or augmented reality (AR) approaches in construction management education. *International Journal of Construction Education and Research*, , 1–20.

Olsson, T., Kärkkäinen, T., Lagerstam, E., & Ventä-Olkkonen, L. (2012). User evaluation of mobile augmented reality scenarios. *Journal of Ambient Intelligence and Smart Environments*, 4(1), 29–47. DOI: 10.3233/AIS-2011-0127

Omar, B., & Dequan, W. (2020). Watch, share or create: The influence of personality traits and user motivation on TikTok mobile video usage. *International Journal of Interactive Mobile Technologies*, 14(4), 121–137. DOI: 10.3991/ijim.v14i04.12429

Ong, C. H., Wei, L. H., & Ramayah, T. (2018). Impact of brand experience on loyalty. *Journal of Hospitality Marketing & Management*, 27(7), 1–20. DOI: 10.1080/19368623.2018.1445055

Online Education. (n.d.). *Artificial intelligence innovations in online learning*. https://www.onlineeducation.com/features/ai-in-distance-learning

Onwuegbuzie, I. U. (2022). 5G: next generation mobile wireless technology for a fast pacing world. [JPAS]. *Journal for Pure and Applied Sciences*, 1(1), 1–9. DOI: 10.56180/jpas.vol1.iss1.57

Ooi, C. S. (2022). Sustainable tourism and the moral limits of the market: can Asia offer better alternatives. In *Asian tourism sustainability* (pp. 177–197). Springer Nature Singapore. DOI: 10.1007/978-981-16-5264-6_10

Ortega-Martínez, E., Pacheco-Mendoza, J., García Meléndez, H. E., Ortiz-Díaz, E. M., & Saavedra-Alamillas, C. (2021). Digital services adapted by libraries in Mexico to COVID-19 pandemic: A critical review. *Digital Library Perspectives*, 37(1), 3–17. DOI: 10.1108/DLP-07-2020-0063

Ortega, P., Shin, T. M., Pérez-Cordón, C., & Martínez, G. A. (2020). Virtual Medical Spanish Education at the *Corazón* of Hispanic/Latinx Health During COVID-19. *Medical Science Educator*, 30(4), 1661–1666. DOI: 10.1007/s40670-020-01058-0 PMID: 32839674

Owner. (n.d.). The State of The Restaurant Industry In 2023 (Free Report). *Owner*. Retrieved March 26, 2023, from https://www.owner.com/blog/restaurant-industry

Paakki, M., Aaltojärvi, I., Sandell, M., & Hopia, A. (2019). The importance of the visual aesthetics of colours in food at a workday lunch. *International Journal of Gastronomy and Food Science*, 16(100131), 1–9. DOI: 10.1016/j.ijgfs.2018.12.001

Padilla, M. E., Frietze, G., Shenberger-Trujillo, J. M., Carrillo, M., & Loya, A. M. (2020). Influenza and intentions to vaccinate in an underserved Hispanic population: The role of theoretically derived constructs. *Journal of Pharmacy Practice*, 33(3), 326–333. DOI: 10.1177/0897190018810595 PMID: 30453817

Paintball, V. R. (2024). Fun in Prague. Retrieved July 20, 2024, from https://funinprague.eu/en/vr-paintball

Palfray, T., Hebert, D., Nicolas, S., Tranouez, P., & Paquet, T. (2012). Logical segmentation for article extraction in digitized old newspapers. In *Proceedings of the 2012 ACM symposium on Document engineering (DocEng '12)* (pp. 129–132). USA. DOI: 10.1145/2361354.2361383

Palmas, F., Reinelt, R., Cichor, J. E., Plecher, D. A., & Klinker, G. (2021, March 1). Virtual Reality Public Speaking Training: Experimental Evaluation of Direct Feedback Technology Acceptance. IEEE Xplore. .DOI: 10.1109/VR50410.2021.00070

Pánek, J., Marek, L., Pászto, V., & Valůch, J. (2017). The Crisis Map of the Czech Republic: The nationwide deployment of an Ushahidi application for disasters. *Disasters*, 41(4), 649–671. Advance online publication. DOI: 10.1111/disa.12221 PMID: 28059447

Pan, X., Zheng, M., Xu, X., & Campbell, A. G. (2021). Knowing Your Student: Targeted Teaching Decision Support Through Asymmetric Mixed Reality Collaborative Learning. *IEEE Access : Practical Innovations, Open Solutions*, 9, 164742–164751. DOI: 10.1109/ACCESS.2021.3134589

Pan, Z., Sun, Y., Yao, Z. W., & Li, M. (2021). Application of virtual reality in English teaching in 2021. *3rd World Symposium on Artificial Intelligence (WSAI)*, IEEE, pp. 64–71. DOI: 10.1109/WSAI51899.2021.9486322

Parepa, L. A. (2020). The Belt and Road Initiative as continuity in Chinese foreign policy. *Journal of Contemporary East Asia Studies*, 9(2), 175–201. DOI: 10.1080/24761028.2020.1848370

Park, J., & Bennis, M. (2018). URLLC-eMBB Slicing to Support VR Multimodal Perceptions over Wireless Cellular Systems. *IEEE Conference on Global Communications (GLOBECOM)*. https://doi.org/DOI: 10.1109/GLOCOM.2018.8647208

Patel, S., Panchotiya, B., & Ribadiya, S. A. (2020). Survey: Virtual, augmented and mixed reality in education. *IJERT*, 9, 1067–1072.

Paul, P. K., ... Cloud Computing and its Impact in Education, Teaching and Research-A Scientific Review, https://www.researchgate.net/publication/371834367_Cloud_Computing_and_its_Impact_in_Education_Teaching_and_Research-A_Scientific_Review https://www.researchgate.net/publication/255749774_Cloud_Computing_Applications

Pausch, R., Crea, T., & Conway, M. (1992). A Literature Survey for Virtual Environments: Military Flight Simulator Visual Systems and Simulator Sickness. *Presence (Cambridge, Mass.)*, 1(3), 344–363. DOI: 10.1162/pres.1992.1.3.344

Pechmann, J., & de los Reyes, C. (2023, March 17). *Using OSM Data in the Turkey and Syria Earthquake Response*. https://www.hotosm.org/updates/using-osm-data-for-the-turkey-and-syria-earthquake-response/

Pérez, E., HyunJeong Lee, J., Oaxaca Carrasco, A. L., Matthews, C., & Ritsema, M. (2023). Unexpected, but consistent and pre-registered: Experimental evidence on interview language and Latino views of COVID-19. *Research & Politics*, 10(2), 20531680231168736. DOI: 10.1177/20531680231168736

Pérez, M., Carou, D., Rubio, E. M., & Teti, R. (2020). Current advances in additive manufacturing. *Procedia*, 88, 439–444.

Pérez-Seijo, S., & Vicente, P. N. (2022). After the hype: How hi-tech is reshaping journalism. In Vázquez-Herrero, J., Silva-Rodríguez, A., Negreira-Rey, M. C., Toural-Bran, C., & López-García, X. (Eds.), *Total Journalism. Studies in Big Data* (pp. 41–52). Springer. DOI: 10.1007/978-3-030-88028-6_4

Permanasari, A. E., Nugroho, H. E., Ardhanarko, H. B. S., Fauziati, S., Utami, I. K., & Sholihin, M. (2022). Development of VR and AR learning media for enterprise business ethic scenarios. In In2022 international conference on information technology Systems and innovation (ICITSI) (pp. 225–230). IEEE. DOI: 10.1109/ICITSI56531.2022.9970892

Pestana, O. (2021). O acesso aos jornais históricos: Considerações sobre o desenvolvimento de coleções digitalizadas [Access to historical newspapers: Considerations for developing digitized collections]. *Media & Jornalismo*, 21(39), 161–174. DOI: 10.14195/2183-5462_39_8

Peteet, B., Belliard, J. C., Abdul-Mutakabbir, J., Casey, S., & Simmons, K. (2021). Community-academic partnerships to reduce COVID-19 vaccine hesitancy in minoritized communities. *EClinicalMedicine*, 34, 34. DOI: 10.1016/j.eclinm.2021.100834 PMID: 33880441

Peters-Desteract, L. (2020). *Innovative business plan for a brunch restaurant in Toulouse, France* [MA dissertation]. ISCTE Business School.

Pfanzelter, E., Oberbichler, S., Marjanen, J., Langlais, P., & Hechl, S. (2021). Digital interfaces of historical newspapers: Opportunities, restrictions and recommendations. *Journal of Data Mining & Digital Humanities*, 1-26.

Phakamach, P., Senarith, P., & Wachirawongpaisarn, S. (2022). The Metaverse in education: The future of immersive teaching & learning. *RICE Journal of Creative Entrepreneurship and Management*, 3(2), 75–88.

Pilcher, J. M. (2012). *The Oxford Handbook of Food History*. Oxford University Press. DOI: 10.1093/oxfordhb/9780199729937.001.0001

Pina, R., & Dias, Á. (2021). The influence of brand experiences on consumer-based brand equity. *Journal of Brand Management*, 28(2), 99–115. DOI: 10.1057/s41262-020-00215-5

Pine, B. J., & Gilmore, J. H. (1998). Welcome to the experience economy. *Harvard Business Review*, 76(4), 97–105. PMID: 10181589

Pittman, M., & Sheehan, K. (2021). Brand authenticity and strategic response to crises: Symbolic effects of donation type on purchase intent and digital engagement. *Journal of Current Issues and Research in Advertising*, 42(3), 215–235. DOI: 10.1080/10641734.2020.1734503

Plohl, N., & Musil, B. (2021). Modeling compliance with COVID-19 prevention guidelines: The critical role of trust in science. *Psychology Health and Medicine*, 26(1), 1–12. DOI: 10.1080/13548506.2020.1772988 PMID: 32479113

Poláková, P., & Klímová, B.Pol'akov´a. (2019). Mobile technology and Generation Z in the English language classroom—A preliminary study. *Education Sciences*, 9(3), 203. DOI: 10.3390/educsci9030203

Powe, M. (2020) New Survey Results: How Small Businesses are Managing the Pandemic and Social Unrest. *Main Street America*. Retrieved from https://www.mainstreet.org/blogs/national-main-street-center/2020/09/28/new-survey-results-how-small-businesses-are-managi

Powell, R. A., & Single, H. M. (1996). Focus groups. *International Journal for Quality in Health Care : Journal of the International Society for Quality in Health Care*, 8(5), 499–504. DOI: 10.1093/intqhc/8.5.499 PMID: 9117204

Powell, T., & Paynter, G. (2009). Going Grey? Comparing the OCR Accuracy Levels of Bitonal and Greyscale Images. *D-Lib Magazine : the Magazine of the Digital Library Forum*, 15(3/4).

Preece, J., Nonnecke, B., & Andrews, D. (2004). The top five reasons for lurking: Improving community experiences for everyone. *Computers in Human Behavior*, 20(2), 201–223. DOI: 10.1016/j.chb.2003.10.015

Private cloud http://http.networkcomputing.com/private-cloud-tech-center

Public cloud, http://www.networkcomputing.com/public-cloud-tech-center

Purnomo, H. (2021). Software trends in improving local tourism product marketing in Indonesia: A systematic review. *International Journal of Business. Economics and Management*, 4(1), 77–84.

Purwaningsih, S., & Pernamasari, R. (2024). Financial Performance Analysis in Predicting Corporate Bankruptcy: Before and During the COVID-19 Pandemic. *Journal of Business and Management*, 2(2), 36–48.

Qian, J. (2022). Research on artificial intelligence technology of virtual reality teaching method in digital media art creation. *Journal of Internet Technology*, 23(1), 125–132.

Qin, T., Wang, L., Zhou, Y., Guo, L., Jiang, G., & Zhang, L. (2022). Digital technology-and-services-driven sustainable transformation of agriculture: Cases of China and the EU. *Agriculture*, 12(2), 297. DOI: 10.3390/agriculture12020297

Qi, Q., & Tao, F. (2019). A Smart Manufacturing Service System Based on Edge Computing, Fog Computing, and Cloud Computing. *IEEE Access : Practical Innovations, Open Solutions*, 7, 86769–86777. DOI: 10.1109/ACCESS.2019.2923610

Quercus (s.d.). "Quem Somos – Apresentação". Available: https://quercus.pt/apresentacao/ (last access 14/06/2023).

Quisenberry, P. N. (2015). Texting and driving: Can it be explained by the general theory of crime? *American Journal of Criminal Justice*, 40(2), 303–316. DOI: 10.1007/s12103-014-9249-3

Radianti, J., Majchrzak, T. A., Fromm, J., & Wohlgenannt, I. (2020). A systematic review of immersive virtual reality applications for higher education: Design elements, lessons learned, and research agenda. *Computers & Education*, 147(7), 103–178. DOI: 10.1016/j.compedu.2019.103778

Raja, M., & Lakshmi Priya, G. G. (2022). Using virtual reality and augmented reality with ICT tools for enhancing quality in the changing academic environment in COVID-19 pandemic: An empirical study. In *InTechnologies, artificial Intelligence and the Future of learning post-COVID-19* (pp. 467–482). Springer. DOI: 10.1007/978-3-030-93921-2_26

Rajkumar, R. P. (2021). The relationship between measures of individualism and collectivism and the impact of COVID-19 across nations. *Public Health in Practice (Oxford, England)*, 2, 100143. https://www.sciencedirect.com/science/article/pii/S2666535221000689. DOI: 10.1016/j.puhip.2021.100143 PMID: 34494009

Ramya, R., & Ramamoorthy, S. (2022). Analysis of machine learning algorithms for efficient cloud and edge computing in the loT, Challenges and Risks Involved in Deploying 6G and NextGen Networks, pp. 72–90.

Ramya, R., & Ramamoorthy, S. "Analysis of machine learning algorithms for efficient cloud and edge computing in the loT", Challenges and Risks Involved in Deploying 6G and NextGen Networks, pp. 72–90, 2022.

Ramya, R. (2024). *Analysis and Applications Finding of Wireless Sensors and IoT Devices With Artificial Intelligence/Machine Learning. AIoT and Smart Sensing Technologies for Smart Devices*. IGI Global., DOI: 10.4018/979-8-3693-0786-1.ch005

Ramya, R., & Ramamoorthy, S. (2022). Survey on Edge Intelligence in IoT-Based Computing Platform. *Survey on Edge Intelligence in IoT-Based Computing Platform, Lecture Notes in Networks and Systems, Springer*, 356, 549–556. DOI: 10.1007/978-981-16-7952-0_52

Ratcliffe, J., Soave, F., Bryan-Kinns, N., Tokarchuk, L., & Farkhatdinov, I. (2021). Extended Reality (XR) Remote Research: a Survey of Drawbacks and Opportunities. *Proceedings of the 2021 CHI Conference on Human Factors in Computing Systems*. DOI: 10.1145/3411764.3445170

Rau, P. L. P., Zheng, J., Guo, Z., & Li, J. (2018). Speed reading on virtual reality and augmented reality. *Computers & Education*, 125, 240–245. DOI: 10.1016/j.compedu.2018.06.016

Rauschnabel, P. A., Babin, B. J., tom Dieck, M. C., Krey, N., & Jung, T. (2022). What is augmented reality marketing? Its definition, complexity, and future. *Journal of Business Research*, 142(1), 1140–1150. DOI: 10.1016/j.jbusres.2021.12.084

Rawson, K., & Shore, E. (2019). *Dining Out: A Global History of Restaurants*. Reaktion Books.

Rede Ex Aequo (s.d.). "A Rede Ex Aequo". Available: https://www.rea.pt/quemsomos/ (last access 11/06/2023).

Reeves, M., Chandra, D., & Kim, K. (2022). Cause Marketing in a Time of Polarization. *Advertising & Society Quarterly, 23*(1).

Remolar, I., Rebollo, C., & Fern'andez-Moyano, J. A. (2021). Learning history using virtual and augmented reality. *Computers*, 10(11), 146. DOI: 10.3390/computers10110146

Ren, X., Zeng, G., & Gozgor, G. (2023). How does digital finance affect industrial structure upgrading? Evidence from Chinese prefecture-level cities. *Journal of Environmental Management*, 330, 117125. DOI: 10.1016/j.jenvman.2022.117125 PMID: 36603250

Rheingold, H. (1992).. . *Virtual Reality (Waltham Cross)*.

Rhodes, N., Ralston, R., & Bigsby, E. (2016). Teens' Reactance to Anti-Smoking Public Service Announcements: How Norms Set the Stage. *Journal of Health Communication*, 21(5), 575–582. DOI: 10.1080/10810730.2015.1114052 PMID: 27116415

Rhyno, A. (2017, April). *Historical newspaper digitization on a shoestring* [Paper presentation]. 2017 IFLA International News Media Conference. Landsbókasafn Íslands – Háskólabókasafn, Reykjavík, Iceland.

Ribeiro, L., & Rodrigues, S. (16 de novembro de 2018). "Animais maltratados: Conheça o IRA, o grupo que faz justiça pelas próprias mãos". Available: https://visao.sapo.pt/atualidade/sociedade/2018-11-16-animais-maltratados-conheca-oira-o-grupo-que-faz-justica-pelas-proprias-maos/ (last access 30/05/2023).

Ringel, S. (2023). Digitizing the paper of record: Archiving digital newspapers at the New York Times. *Journalism, 24*(2), 245-261.

Roberts, H., Cowls, J., Morley, J., Taddeo, M., Wang, V., & Floridi, L. (2021). The Chinese approach to artificial intelligence: An analysis of policy, ethics, and regulation. *AI & Society*, 36(1), 59–77. DOI: 10.1007/s00146-020-00992-2

Robitaille, J. (2023, March 13). The Importance of Restaurant Branding. *Casa Media*. Retrieved June 10, 2023, from https://casamedia.com/branding/importance-restaurant-branding/

Rodrigues, R. A. (2016). *Femvertising: empowering women through the hashtag? A comparative analysis of consumers' reaction to feminist advertising on twitter* (Doctoral dissertation, Universidade de Lisboa (Portugal)).

Rojas, R. A., & Rauch, E. (2019). A conceptual framework of enablers for smart manufacturing control. *International Journal of Advanced Manufacturing Technology*, 104(1-4), 517–533. DOI: 10.1007/s00170-019-03854-4

Ronald, L. W. (2008). *Kleppner's advertising procedure*. Pearson Education India.

Røstvik, C. M. (2022). *Cash Flow: The businesses of menstruation*. UCL Press. DOI: 10.14324/111.9781787355385

Ryan, C. B. (2019). # BPCares: Sinking Oil, Spreadable Détournement. *ISLE: Interdisciplinary Studies in Literature and Environment*, 26(4), 882–900. DOI: 10.1093/isle/isy093

Salant, S. W. (2016). The Strategic Use of Early Bird Discounts for Dealers. *Quantitative Marketing and Economics*, 14(2), 97–127. DOI: 10.1007/s11129-016-9167-4

Sánchez Abril, P., Oliva Blázquez, F., & Martínez Evora, J. (2018). The Right of Withdrawal in Consumer Contracts: A comparative analysis of American and European law. *InDret*, 3, 18–13.

Sandoval-Almazan, R., & Gil-Garcia, J. (2014). "Towards cyberactivism 2.0? Understanding the use of social media and other information technologies for political activism and social movements". Government Information Quarterly, Vol. 31, n.º3, pp. 365-378. DOI:DOI: 10.1016/j.giq.2013.10.016

Santos, T. T., Piuvezam, G., Medeiros, G. C. B. S., & Mata, Á. N. de S. (2023). Extended reality as a health education strategy of adolescents at school: Protocol for systematic review and meta-analysis. *BMJ Open*, 13(7), e072438. DOI: 10.1136/bmjopen-2023-072438 PMID: 37407033

Sanusi, N. (2019). Maskulinitas dalam Iklan Gillette We Believe: What Best Man Can Be. *Media Komunika (Jurnal Komunikasi). Universitas Sangga Buana YPKP*, 4(2), 61–69.

Sathianarayanan, M., Hsu, P. H., & Chang, C. C. (2024). Extracting disaster location identification from social media images using deep learning. *International Journal of Disaster Risk Reduction*, 104, 104352. Advance online publication. DOI: 10.1016/j.ijdrr.2024.104352

Scavarelli, A., Arya, · Ali, Robert, ·, & Teather, J. (2021). Virtual reality and augmented reality in social learning spaces: a literature review. 25, 257–277. .DOI: 10.1007/s10055-020-00444-8

Scharp, K. M., & Sanders, M. L. (2018). What is a theme? Teaching thematic analysis in qualitative communication research methods. *Communication Teacher*, 33(2), 117–121. DOI: 10.1080/17404622.2018.1536794

Scherr, S., & Wang, K. (2021). Explaining the success of social media with gratification niches: Motivations behind daytime, nighttime, and active use of TikTok in China. *Computers in Human Behavior*, 124, 106893. DOI: 10.1016/j.chb.2021.106893

Schifferstein, H. N., Kudrowitz, B. M., & Breuer, C. (2020). Food Perception and Aesthetics - Linking Sensory Science to Culinary Practice. *Journal of Culinary Science &Amp. Technology*, 20(4), 293–335.

Schlukbier, G. (2008). Digitization of old newspapers: Software developments to enable successful conversion. In Walravens, H. (Ed.), *Newspapers collection management: Printed and digital challenges* (pp. 143–146). K. G. Saur. DOI: 10.1515/9783598441011.143

Schmeiser, S. (2018). Online Advertising Networks and Consumer Perceptions of Privacy. *Applied Economics Letters*, 25(11), 776–780. DOI: 10.1080/13504851.2017.1366634

Schmidt, S. (2016, November). When Service à la Française Met Service à la Russe. *Manuscript Cookbooks Survey*. Retrieved May 7, 2023, from https://www.manuscriptcookbookssurvey.org/when-service-a-la-francaise-met-service-a-la-russe/

Schmitt, B. H. (1999a). *Experiential marketing: how to get customers to sense, feel, think, act, and relate to your company and brands*. The Free Press New York.

Schmitt, B. H. (1999b). Experiential marketing. *Journal of Marketing Management*, 15(1–3), 53–67. DOI: 10.1362/026725799784870496

Schmitt, B. H., Brakus, J. J., & Zarantonello, L. (2014). The current state and future of brand experience. *Journal of Brand Management*, 21(9), 727–733. DOI: 10.1057/bm.2014.34

Schneid, R., Perez-Moreno, H., & Branson-Potts, H. (2022). 'I'm over it.' Many in L.A. shrug off COVID-19 wave despite super-infectious subvariants. *L.A. Times*. https://www.latimes.com/california/story/2022-07-21/la-me-covid-mood-mask-mandate

Schultze, C., Kerkfeld, N., Kuebart, K., Weber, P., Wolter, M., & Selgert, F. (2024). Reading yesterday's news. Layout recognition by segmentation of historical newspaper pages.

Schwab, K. (2017). *The fourth industrial revolution*. Crown Publishing Group.

Shamim, A., & Butt, M. A. (2013). A critical model of brand experience consequences. *Asia Pacific Journal of Marketing and Logistics*, 25(1), 102–117. DOI: 10.1108/13555851311290957

Sharpe, J. D., Kaufman, J. A., Goldman, Z. E., Wolkin, A., & Gribble, M. O. (2019). Determinants of Oil-Spill Cleanup Participation Following the Deepwater Horizon Oil Spill. *Environmental Research*, 170, 472–480. DOI: 10.1016/j.envres.2019.01.009 PMID: 30640081

Shelus, V. S., Frank, S. C., Lazard, A. J., Higgins, I. C., Pulido, M., Richter, A. P. C., Vandegrift, S. M., Vereen, R. N., Ribisl, K. M., & Hall, M. G. (2020). Motivations and barriers for the use of face coverings during the COVID-19 pandemic: Messaging insights from focus groups. *International Journal of Environmental Research and Public Health*, 17(24), 9298. https://www.mdpi.com/1660-4601/17/24/9298/pdf. DOI: 10.3390/ijerph17249298 PMID: 33322672

Shen, H. (2021). *Alibaba: Infrastructuring global china*. Routledge. DOI: 10.4324/9781003169055

Shim, J., Moon, J., Lee, W. S., & Chung, N. (2021). The impact of CSR on corporate value of restaurant businesses using triple bottom line theory. *Sustainability (Basel)*, 13(4), 2131. DOI: 10.3390/su13042131

Shorey, S., Ang, E., Yap, J., Ng, E. D., Lau, S. T., & Chui, C. K. (2019). A virtual counseling application using artificial intelligence for communication skills training in nursing education: Development study. *Journal of Medical Internet Research*, 21(10), e14658. DOI: 10.2196/14658 PMID: 31663857

Siawsh, N., Peszynski, K., Vo-Tran, H., & Young, L. (2023). Toward the creation of disaster-resilient communities: The Machizukuri initiative – The 2011 Tōhoku Great East Japan Earthquake and Tsunami. *International Journal of Disaster Risk Reduction*, 96, 103961. Advance online publication. DOI: 10.1016/j.ijdrr.2023.103961

Silin, Y., Kapustina, L., Trevisan, I., & Drevalev, A. (2018). The silk road economic belt: Balance of interests. *Economic and Political Studies*, 6(3), 293–318. DOI: 10.1080/20954816.2018.1499072

Silva, D. S., & Granado, A. (2021). *Cobertura jornalística dos números da Covid-19: Casos de inovação em Portugal* [Journalistic coverage of Covid-19 numbers: Cases of innovation in Portugal]. Obi.Media.

Silva, P. (2018). A Ação Coletiva: O Desafio da Mobilização. *Revista Movimentos Sociais e Dinâmicas Espaciais*, 7(2), 62–87. DOI: 10.51359/2238-8052.2018.238618

Silvia. (2022, January 14). *The Evolution & Future of Short Film Video*. Bbgg. https://bbggadv.com/insight-short-form-video/

Simon, L. (2021, March 18). *#LibraryTikTok Takes You Behind the Shelves*. I Love Libraries. https://ilovelibraries.org/article/librarytiktok-takes-you-behind-shelves/

Singh, S. K., & Sanjeev, R. (2022). Antecedents of Green Marketing Initiatives: A Hierarchical Regression Model. [IJSESD]. *International Journal of Social Ecology and Sustainable Development*, 13(1), 1–15. DOI: 10.4018/IJSESD.315310

Singireddy, M. (2020). Mcdonald's: Global Marketing. *International Journal of Health and Economic Development*, 6(2), 16–27.

Sitto, K., & Lubinga, E. (2020). A Disease of Privilege? Social Representations in Online Media about COVID-19 among South Africans during Lockdown. *Papers on Social Representations*, 29(2), 6–1.

Skelbye, M. B., & Dannells, D. (2021). OCR Processing of Swedish Historical Newspapers Using Deep Hybrid CNN–LSTM Networks. In *Proceedings of Recent Advances in Natural Language Processing* (pp. 190–198). Online.

Skilton, M., & Hovsepian, F. (2018). *The 4th Industrial Revolution: Responding to the impact of artificial intelligence on business*. Palgrave Macmillan. DOI: 10.1007/978-3-319-62479-2

Sloan, D., & Leith, P. (2012). *Culinary Taste*. Routledge. DOI: 10.4324/9780080476957

Smith, A., & Nagele-Piazza, L. "Employers React to Workers Who Refuse a COVID-19 Vaccination." *SHRM*, SHRM, (3 Feb. 2022), https://www.shrm.org/resourcesandtools/legal-and-compliance/employment-law/pages/if-workers-refuse-a-covid-19-vaccination.aspx

Smith, M. (2022, February 12). *20 Best Short Video Apps*. Marketing Scoop. https://www.marketingscoop.com/short-video-app/?utm_source=rss&utm_medium=rss&utm_campaign=short-video-app

Snibbe, S. S., & Raffle, H. S. (2009). Social immersive media: Pursuing best practices for multi-user interactive camera/projector exhibits. Conference on Human Factors in Computing Systems - Proceedings, 1447–1456. DOI: 10.1145/1518701.1518920

Sobreira, L. T. P., Mepe, G., & Nardoto, G. B. (2018). Anthropology of Food: An Essay on Food Transition and Transformations in Brazil. *Journal of Food Nutrition and Population Health*, 02(01), 1–8.

Sousa, J. P. (2021). *Portugal. Pequena história de um grande jornalismo I. Da manufatura à indústria* [Portugal. Small story of great journalism I. From manufacturing to industry]. Livros ICNOVA.

Sousa, J. P. (2018). Eduardo Coelho e a fundação do Diário de Notícias. [Eduardo Coelho and the founding of Diário de Notícias] In Sousa, J. P. (Ed.), *Notícias em Portugal: estudos sobre a imprensa informativa (século XVI-XX)* [News in Portugal: studies on the informative press (16th-20th century)]. (pp. 163–192). Colecção Livros ICNOVA.

Speaking, V. R. LLC (2022). Ovationvr.com. https://www.ovationvr.com/

Spence, C. (2012). Auditory contributions to flavour perception and feeding behaviour. *Physiology & Behavior*, 107(4), 505–515. DOI: 10.1016/j.physbeh.2012.04.022 PMID: 22579696

Spence, C. (2015). Just how much of what we taste derives from the sense of smell? *Flavour (London)*, 4(1), 30. DOI: 10.1186/s13411-015-0040-2

Spence, C. (2017). *Gastrophysics: The new science of eating*. Penguin UK.

Spence, C. (2021). What's the Story With Blue Steak? On the Unexpected Popularity of Blue Foods. *Frontiers in Psychology*, 12, 1–17. DOI: 10.3389/fpsyg.2021.638703 PMID: 33737898

Spence, C., Motoki, K., & Petit, O. (2022). Factors influencing the visual deliciousness / eye-appeal of food. *Food Quality and Preference*, 102(104672), 1–16. DOI: 10.1016/j.foodqual.2022.104672

Spence, C., Okajima, K., Cheok, A. D., Petit, O., & Michel, C. (2016). Eating with our eyes: From visual hunger to digital satiation. *Brain and Cognition*, 110, 53–63. DOI: 10.1016/j.bandc.2015.08.006 PMID: 26432045

Spence, C., & Piqueras-Fiszman, B. (2014). *The Perfect Meal: The Multisensory Science of Food and Dining*. Wiley-Blackwell. DOI: 10.1002/9781118491003

Spence, C., Piqueras-Fiszman, B., Michel, C., & Deroy, O. (2014). Plating manifesto (II): The art and science of plating. *Flavour (London)*, 3(1), 1–12. DOI: 10.1186/2044-7248-3-4

Spence, C., & Van Doorn, G. (2022). Visual communication via the design of food and beverage packaging. *Cognitive Research: Principles and Implications*, 7(1), 42. DOI: 10.1186/s41235-022-00391-9 PMID: 35551542

Spence, C., Youssef, J., Michel, C., & Woods, A. T. (2019). Assessing the aesthetic oblique effect in painting and plating. *International Journal of Gastronomy and Food Science*, 17(100168), 1–29. DOI: 10.1016/j.ijgfs.2019.100168

Spina, S. (2023). Artificial Intelligence in archival and historical scholarship workflow: HTS and ChatGPT. *Umanistica Digitale*, 16, 125–140.

Statista - The Statistics Portal. (n.d.). Statista. Retrieved July 20, 2024, from https://www.statista.com/search/?q=VR+headsets+sold&p=1

Statista (2022) Share of small businesses that have closed due to the COVID-19 pandemic in the United States from October 2020 to January 2022. Statista.com. https://www.statista.com/statistics/1222202/us-COVID-19-closings-small-businesses/

Statista. (2022, January). *Instagram: users by country*. https://www.statista.com/statistics/578364/countries-with-most-instagram-users/

Steed, A., Pan, Y., Watson, Z., & Slater, M. (2018). "We Wait"—The Impact of Character Responsiveness and Self Embodiment on Presence and Interest in an Immersive News Experience. *Frontiers in Robotics and AI*, 5, 112. Advance online publication. DOI: 10.3389/frobt.2018.00112 PMID: 33500991

Stephenson, J. (2021, January). While Waiting for Vaccination, Most US Adults Will Wear Masks, Follow Social Distancing Guidelines, Poll Finds. In *JAMA Health Forum* (Vol. 2, No. 1, pp. e210024-e210024). American Medical Association.https://jamanetwork.com/journals/jama-health-forum/fullarticle/2775257

Sterling, C. H. (2009). *Encyclopedia of Journalism*. Sage Publications. DOI: 10.4135/9781412972048

Sternadori, M., & Abitbol, A. (2022). How Male Consumers Respond to "Enlightened Manvertising" Campaigns: Gender Schema, Hostile Sexism, And Political Orientation Feed Attitudes. *Journal of Advertising Research*, 62(1), 87–101. DOI: 10.2501/JAR-2022-006

Steuer, J. (1992). Defining Virtual Reality: Dimensions Determining Telepresence. *Journal of Communication*, 42(4), 73–93. DOI: 10.1111/j.1460-2466.1992.tb00812.x

Subramaniam, M., Kathirvel, A., Sabitha, E., & Anwar Basha, H. (2021b) "Modified Firefly Algorithm and Fuzzy C-Mean Clustering Based Semantic Information Retrieval", Journal of Web Engineering, Vol. 20 1, 33–52,. Publisher: © River Publishers, 15 Feb 2021, ISSN: 1540-9589 (Print Version). ISSN: 1544-5976 (Online Version), https://doi.org/DOI: 10.13052/jwe1540-9589.2012

Subramanian, S. (2020). Bahujan girls' anti-caste activism on TikTok. *Feminist Media Studies*, 21(1), 154–156. DOI: 10.1080/14680777.2021.1864875

Sudha, D., & Kathirvel, A. (2022a).An Intrusion Detection System to Detect and Mitigating Attacks Using Hidden Markov Model (HMM) Energy Monitoring Technique. Stochastic Modeling an Applications, 26(3) (January – June, Special Issue 2022 Part – 9), 467-476, ISSN: 0972-3641.

Sudha, D., & Kathirvel, A. (2023a) The performance enhancement of Aodv protocol using GETUS. International Journal of Early Childhood Special Education (INT-JECSE), Vol. 15, Issue 2, Pp. 115-125. ISSN: 1308-5581. DOI:DOI: 10.48047/INTJECSE/V15I2.11

Sudha, D., & Kathirvel, A. (2023b)The effect of ETUS in various generic attacks in mobile ad hoc networks to improve the performance of Aodv protocol. International Journal of humanities, law, and social sciences, Kanpur philosophers, Vol. 9 Issue 1, N0.1, 467-476, 2022. ISSN: 2348-8301.

Sukarjono, S., Nasrulloh, A., & Fakhrurozi, Z. A. (2023). Effect of Literature Availability in Sports and Quality of Service of Library towards the Satisfaction and Learning Motivation of the Students Major of Sports. *INTERNATIONAL JOURNAL OF MULTIDISCIPLINARY RESEARCH AND ANALYSIS*, 06(11). Advance online publication. DOI: 10.47191/ijmra/v6-i11-46

Sundar, S. S., & Limperos, A. M. (2013). Uses and Grats 2.0: New Gratifications for New Media. *Journal of Broadcasting & Electronic Media*, 57(4), 504–525. DOI: 10.1080/08838151.2013.845827

Superlibrarymarketing. (2020). *The 2022 Guide to Social Media for Libraries: Everything You Need to Know About Instagram – Super Library Marketing: Practical Tips and Ideas for Library Promotion*. https://superlibrarymarketing.com/2021/11/22/instagramforlibraries2022/

Supernatural. (n.d.). Supernatural. https://www.getsupernatural.com/

Survey Monkey. (2021). Making Responses Anonymous. (2021). *Survey Monkey*. Retrieved from https://help.surveymonkey.com/articles/en_US/kb/How-do-I-make-surveys-anonymous

Survey Monkey. (2022). Sample Size Calculator. *Survey Monkey*. Retrieved from https://www.surveymonkey.com/mp/sample-size-calculator/

Survey Monkey. (n.d.) Non-Probability Sampling. *Survey Monkey*. https://www.surveymonkey.com/mp/non-probability-sampling

Sylaiou, S., Mania, K., Karoulis, A., & White, M. (2010). Exploring the relationship between presence and enjoyment in a virtual museum. *International Journal of Human-Computer Studies*, 68(5), 243–253. DOI: 10.1016/j.ijhcs.2009.11.002

Tai, D. B. G., Shah, A., Doubeni, C. A., Sia, I. G., & Wieland, M. L. (2020). The disproportionate impact of COVID-19 on racial and ethnic minorities in the United States. *Clinical Infectious Diseases*, 72(4), 703–706. DOI: 10.1093/cid/ciaa815 PMID: 32562416

Takeuchi, H., & Nonoka, I. (2008). *Gestão do conhecimento* [Knowledge management] (A Thorell, Transl.). Bookman. (First published in 2004).

Tang, Q., Yu, F. R., Xie, R., Boukerche, A., Huang, T., & Liu, Y. (2022). Internet of Intelligence: A survey on the enabling technologies, applications, and challenges. *IEEE Communications Surveys and Tutorials*, 24(3), 1394–1434. DOI: 10.1109/COMST.2022.3175453

Tao, F., Qi, Q., Liu, A., & Kusiak, A. (2018). Data-driven smart manufacturing. *Journal of Manufacturing Systems*, 48, 157–169. DOI: 10.1016/j.jmsy.2018.01.006

Tarnavsky, G. A., & Vorozhtsov, E. V. (2011). Cloud Computing in Science and Engineering and the "SciShop. ru" Computer Simulation Center. *Engineering, Technology &. Applied Scientific Research*, 1(6), 133–138.

Tasmin, M., Ahmed, N., & Motahar, T. (2019, December). Gender Disparity in Computer Science Education in Bangladesh: A Study of Women's Participation in Computer Science. In *2019 IEEE International Conference on Engineering, Technology and Education (TALE)* (pp. 1-7). IEEE. DOI: 10.1109/TALE48000.2019.9225981

Teel, Z. (2024). Artificial Intelligence's Role in Digitally Preserving Historic Archives. *Preservation. Digital Technology & Culture*, 53(1), 29–33.

Telo, J. (2023). Smart City Security Threats and Countermeasures in the Context of Emerging Technologies. *International Journal of Intelligent Automation and Computing*, 6(1), 31–45.

Ternikar, F. (2014). *Brunch: A History*. Rowman & Littlefield. DOI: 10.5771/9781442229433

Theofanidis, D., & Fountouki, A. (2018). Limitations and delimitations in the research process. Perioperative nursing, 7(3), 155-163. https://www.spnj.gr/articlefiles/volume7_issue3/pn_sep_73_155_162b.pdf

Thomas, A., Gaizauskas, R., & Lu, H. (2024, May). *Leveraging LLMs for Post-OCR Correction of Historical Newspapers* [Paper presentation]. Lingotto Conference Centre, Torino, Italy.

Tian, T., Zhang, Y., & Mei, Y. (2022). Intelligent analysis of precision marketing of green agricultural products based on big data and GIS. *Earth Science Informatics*, 15(3), 1–12. DOI: 10.1007/s12145-022-00789-5

Tilly, C., Castañeda, E., & Wood, L. J. (2020). *Social Movements, 1768 - 2018* (4.ª ed.). Routledge. UNICEF (s.d.). "A UNICEF". Available: https://www.unicef.pt/unicef/a-unicef/ (last access 12/12/2022).

Times, T. N. Y. (2016, May 19). Seeking Pluto's Frigid Heart. The New York Times. https://www.nytimes.com/interactive/2016/05/19/science/space/seeking-plutos-frigid-heart-nytvr.html

Tirupathi, R., Muradova, V., Shekhar, R., Salim, S. A., Al-Tawfiq, J. A., & Palabindala, V. (2020). COVID-19 disparity among racial and ethnic minorities in the US: A cross sectional analysis. *Travel Medicine and Infectious Disease*, 38, 101904. https://www.ncbi.nlm.nih.gov/pmc/articles/PMC7603979/. DOI: 10.1016/j.tmaid.2020.101904 PMID: 33137491

Tolentino, J. (2019, September 23). *How TikTok Holds Our Attention*. The New Yorker. https://www.newyorker.com/magazine/2019/09/30/how-tiktok-holds-our-attention

TouchBistro. (2023). *2023 Dining Trends Report*. Retrieved April 1, 2023, from https://www.touchbistro.com/blog/diner-trends-report/

Training, R. (2016)... *University Transportation Center to Support DOTD Research Needs*, 31(1), 1–5.

Training, R. (2016)... *University Transportation Center to Support DOTD Research Needs.*, 31(1), 1–5.

Trihas, N., Perakakis, E., Venitourakis, M., Mastorakis, G., & Kopanakis, I. (2013). Social media as a marketing tool for tourism destinations: The case of Greek municipalities. *Journal of Marketing Vistas*, 3(2), 38–48.

Turnbull, P. R., Khanal, S., & Dakin, S. C. (2021). The effect of cellphone position on driving and gaze behaviour. *Scientific Reports*, 11(1), 1–10. DOI: 10.1038/s41598-021-87120-2 PMID: 33833300

Udem, O. K., & Ogungbeni, J. I. (2023). Use of Mobile Technology-Based Library Services and Information Needs Satisfaction of Undergraduates in Universities in South-West Nigeria. *Folia Toruniensia*, 23, 15–33. DOI: 10.12775/FT.2023.001

Uhlemann, T. H.-J., Lehmann, C., & Steinhilper, R. (2017). The Digital Twin: Realizing the Cyber-Physical Production System for Industry 4.0. *Procedia CIRP*, 61, 335–340. DOI: 10.1016/j.procir.2016.11.152

Ulrich, A., Thompson, B., Livaudais, J. C., Espinoza, N., Cordova, A., & Coronado, G. D. (2013). Issues in biomedical research: What do Hispanics think? *American Journal of Health Behavior*, 37(1), 80–85. DOI: 10.5993/AJHB.37.1.9 PMID: 22943104

Um, N. (2021). Antecedents and consequences of attitude toward femvertising. *The Journal of the Korea Contents Association*, 21(1), 66–74.

Uriel, E. (2013). Hypothesis testing in the multiple regression model. Introduction to Econometrics, 1-49. https://d1wqtxts1xzle7.cloudfront.net/34695537/4_Hypothesis_testing_in_the_multiple_regression_model-with-cover-page-v2.pdf?Expires=1643629129&Signature=RExpdIJ0dE2MAqguFrUwoxWAw9WXvfG-3~PGfl-ddw~91sbjF5tprG6STN0pNCNTIpD7uO5qvHR184rs6LTkp~xB5886sMY98GsielVy7L88cA4iiDfj94aPfpX9Gil3Q5Mp9xqiMcOHdPZOYguWZI3CYt0FjQ8otQRlsShII4t8VNlWZnaAAweuKlwnTlRhg3bYiZ1TIaI9Y3Kg9eh4jJLyAmmZMy1393B1kbKTeSJD56yiUI1LEHxV9iYIomA1Q2G85YTFbmGo3bUtj2iiEXfimT1IJRc8V0qsWnVvwgcGfA5aebHjrpQIzqNHIlw66q7qCwZyFUQB0aHZ7Ao1vA__&Key-Pair-Id=APKAJLOHF5GGSLRBV4ZA

USCensusBureau. (2020). Quick Facts United States._https://www.census.gov/quickfacts/fact/table/US/RHI725219

Valente, J., António, J., Mora, C., & Jardim, S. (2023). Developments in image processing using deep learning and reinforcement learning. *Journal of Imaging*, 10(9), 1–22. DOI: 10.3390/jimaging9100207 PMID: 37888314

Van Dijck, J. (2013). Engineering Sociality in a Culture of Connectivity. In *J. Van Dijck, The culture of connectivity: a critical history of social media* (pp. 3–23). Oxford University Press. DOI: 10.1093/acprof:oso/9780199970773.003.0001

Van Wees, J. D., van der Kuip, M., Osinga, S., Keijser, B., van Westerloo, D., Hanegraaf, M., & van Furth, M. T. (2021). SIR model for assessing the impact of the advent of Omicron and mitigating measures on infection pressure and hospitalization needs. *MedRxIv*. DOI: 10.1101/2021.12.25.21268394

Vázquez-Herrero, J., Negreira-Rey, M. C., & López-García, X. (2020). Let's dance the news! How the news media are adapting to the logic of TikTok. *Journalism*. Advance online publication. DOI: 10.1177/1464884920969092

Vegh, S. (2003). Classifying Forms of Inline Activism: The Case of Cyberprotests against the World Bank. In McCaughey, M., & Ayers, M. D. (Eds.), *Cyberactivism - online activism in theory and practice* (pp. 71–95). Routledge.

Viana, L., & Sarmento, R. (2021). "Ativismo Digital em Âmbito Regional: Um Estudo Sobre a Presença Online de Movimentos Sociais no Interior de Minas Gerais#. Revista Teoria & Pesquisa, Vol.30, n.°1, 2-26. DOI: DOI: 10.31068/tp.30101

Vicente, P. N., & Pérez-Seijo, S. (2022). Spatial audio and immersive journalism: Production, narrative design, and sense of presence. *El Profesional de la Información*, 31(5), 1–14. DOI: 10.3145/epi.2022.sep.19

VirtualSpeech - Soft Skills Training with VR. (2019). Virtualspeech.com. https://virtualspeech.com/

Voicu, L. (2023, March 27). 10 Restaurant Sustainability Practices You Can Implement Immediately. *ORACLE GloriaFood*. Retrieved May 4, 2023, from https://www.gloriafood.com/restaurant-sustainability-practices

Vredenburg, J., Kapitan, S., Spry, A., & Kemper, J. A. (2020). Brands taking a stand: Authentic brand activism or woke washing? *Journal of Public Policy & Marketing*, 39(4), 444–460. DOI: 10.1177/0743915620947359

Vygotsky, L. S. (1987). Thinking and speech. In R.W. Rieber & A.S. Carton (Eds.), *The collected works of L.S. Vygotsky, Volume 1: Problems of general psychology (pp. 39–285)*. Plenum Press. (Original work published 1934.)

Vygotsky, L. S. (1962). *Thought and language*. MIT Press. DOI: 10.1037/11193-000

Vygotsky, L. S. (1978). *Mind in society: The development of higher psychological processes*. Harvard University Press.

Wachyuni, S. S., & Yusuf, L. (2021). Camera eat first: Tourist motivation in sharing food photograph on Instagram. *International Journal of Tourism & Hospitality Reviews*, 8(1), 62–70. DOI: 10.18510/ijthr.2021.815

Wagner, D. A. (2014). *The Marketing of Global Warming: A Repeated Measures Examination of the Effects of Cognitive Dissonance, Endorsement, and Information on Beliefs in a Social Cause* (Doctoral dissertation, Trident University International).

Walker, J. R. (2017). *Restaurant concepts, management and operations*. Wiley Global Education.

Walker, J. R. (2021). *The Restaurant: From Concept to Operation*. John Wiley & Sons.

Wang, W., & Gang, J. (2018). Application of Convolutional Neural Network in Natural Language Processing. In *2018 International Conference on Information Systems and Computer Aided Education (ICISCAE)* (pp. 64-70). Jilin Finance Building Hotel. DOI: 10.1109/ICISCAE.2018.8666928

Wang, D., Zhou, T., & Wang, M. (2021). Information and communication technology (ICT), digital divide and urbanization: Evidence from Chinese cities. *Technology in Society*, 64, 101516. DOI: 10.1016/j.techsoc.2020.101516

Wang, J., Ma, Y., Zhang, L., Gao, R. X., & Wu, D. (2018). Deep learning for smart manufacturing: Methods and applications. *Journal of Manufacturing Systems*, 48, 144–156. DOI: 10.1016/j.jmsy.2018.01.003

Wang, L., Wu, Y., Huang, Z., & Wang, Y. (2022). How big data drives green economic development: Evidence from China. *Frontiers in Environmental Science*, 10, 1055162. DOI: 10.3389/fenvs.2022.1055162

Wan, J.. (2016). Software-Defined Industrial Internet of Things in the Context of Industry 4.0. *Journal of Sensors*, 16(20), pp7373–pp7380.

Wasserstrom. (2019, August 30). Dietary Restrictions for Restaurants: Understanding Fads, Trends & Lifestyle. *The Official Wasserstrom Blog*. Retrieved May 4, 2023, from https://www.wasserstrom.com/blog/2019/08/08/dietary-restrictions-for-restaurants-understanding-fads-trends-lifestyle/

We Are Collider. (2022, November 22). Millennials fuelled the experience economy; will Gen Z kill it? *We Are Collider*. Retrieved August 10, 2023, from https://wearecollider.com/insights/millennials-fuelled-the-experience-economy-will-gen-z-kill-it/

Wease, G., Boateng, K., Yu, C. J., Chan, L., & Barham, H. (2018). Technology assessment: cloud service adoption decision. *Infrastructure and technology management: Contributions from the energy, healthcare and transportation sectors*, 447-471.

Webster, G. D., Howell, J. L., Losee, J. E., Mahar, E. A., & Wongsomboon, V. (2021). Culture, COVID-19, and collectivism: A paradox of American exceptionalism? *Personality and Individual Differences*, 178, 110853. DOI: 10.1016/j.paid.2021.110853 PMID: 36540785

Wellman, M. L. (2022). #Black Squares for Black Lives? Performative Allyship as Credibility Maintenance for Social Media Influencers on Instagram#. *Social Media + Society*, 8(1), 1–10. DOI: 10.1177/20563051221080473

Wenger, E. (1998). *Communities of practice: Learning, meaning, and identity*. Cambridge University Press., DOI: 10.1017/CBO9780511803932

Wenger-Trayner, E., & Wenger-Trayner, B. (2015). Communities of practice: A brief introduction. Wenger-Taylor. https://wenger-trayner.com/introduction-to-communities-of-practice/

Wenger-Trayner, E., Fenton-O'Creevy, M., Hutchison, S., Kubiak, C., & Wenger-Trayner, B. (2014). *Learning in Landscapes of Practice: Boundaries, Identity, and Knowledgeability in Practice-Based Learning*. Routledge., DOI: 10.4324/9781315777122

Wenger-Trayner, E., & Wenger-Trayner, B. (2020). *Learning to Make a Difference: Value Creation in Social Learning Spaces*. Cambridge University Press., DOI: 10.1017/9781108677431

Wertsch, J. V., & Sohmer, R. (1995). Vygotsky on learning and development. *Human Development*, 38(6), 332–337. DOI: 10.1159/000278339

Whembolua, G. L., & Tshiswaka, D. I. (2020). Public trust in the time of the Coronavirus Disease 2019 (COVID-19): the case of the DR Congo. *The Pan African Medical Journal, 35*(Suppl 2).

Willan, M. (2021). *The essentials of contemporary marketing*. Bloomsbury Publishing.

Wolfsdorf, A., Scott, A., & Herzog, S. (2019). What happened on July 21st: An investigation of trauma and trigger warnings in the English classroom. *Changing English*, 26(2), 198–215. DOI: 10.1080/1358684X.2018.1552514

Wood, N., & Reinert, C. (2019). NONPROFITS AND FOR PROFITS. WHO BENEFITS? *Interdisciplinary Insights: The Journal of Saint Leo University's College of Education and Social Services*, 1(2), 14–33.

Wood, S., & Schulman, K. (2021). Beyond politics—Promoting COVID-19 vaccination in the United States. *The New England Journal of Medicine*, 384(7), e23. https://www.nejm.org/doi/full/10.1056/nejmms2033790. DOI: 10.1056/NEJMms2033790 PMID: 33406324

Workneh, T. W. (2016). Chinese multinationals in the Ethiopian telecommunications sector. *Communication, Culture & Critique*, 9(1), 126–147. DOI: 10.1111/cccr.12129

Wright, C. (2017). Are beauty bloggers more influential than traditional industry experts? *Journal of Promotional Communications*, 5(3), 303–322. https://promotionalcommunications.org/index.php/pc/article/view/113

Wright, C. (2021). China's Digital Colonialism: Espionage and Repression Along the Digital Silk Road. *SAIS Review (Paul H. Nitze School of Advanced International Studies)*, 41(2), 89–113. DOI: 10.1353/sais.2021.0020

Wu, M. (2006). Hofstede's cultural dimensions 30 years later: A study of Taiwan and the United States. *Intercultural Communication Studies*, 15(1), 33. http://www.homecont.ro/pitagora/Hofstede-04-Ming-Yi-Wu.pdf

Wu, X., Huang, Z., Peng, X., Chen, Y., & Liu, Y. (2018). Building a spatially-embedded network of tourism hotspots from geotagged social media data. *IEEE Access : Practical Innovations, Open Solutions*, 6, 21945–21955. Advance online publication. DOI: 10.1109/ACCESS.2018.2828032

Xiaohong, W., & Yanzheng, W. (2021). The application of artificial intelligence in modern foreign language learning. *4th International Conference on Big Data and Education*, pp. 34–37. DOI: 10.1145/3451400.3451406

Xiaoxia, L. (2022). Practice of Artificial Intelligence and Virtual Reality Technology in College English Dialogue Scene Simulation. Wireless Communications and Mobile Computing, vol. 2022, Article ID 4922675, 9 pages. https://doi.org/DOI: 10.1155/2022/4922675

Xiao, Y., Wang, L., & Wang, P. (2019). Research on the Influence of Content Features of Short Video Marketing on Consumer purchase intentions. *Proceedings of the 2019 4th International Conference on Modern Management, Education Technology and Social Science (MMETSS 2019)*, 415–422. DOI: 10.2991/mmetss-19.2019.82

Xu, H., Yu, W., Griffith, D., & Golmie, N. (2018). A Survey on Industrial Internet of Things: A Cyber-Physical Systems Perspective. *IEEE Access : Practical Innovations, Open Solutions*, 6, pp78238–pp78259. DOI: 10.1109/ACCESS.2018.2884906 PMID: 35531371

Xu, W., Zhao, C., & Wang, X. (2021). The Role of the Internet Technology in the Employment Structural Transformation under Background of "Internet Plus" in China. *Theoretical Economics Letters*, 11(5), 1020–1037. DOI: 10.4236/tel.2021.115065

Yang, B., Kim, Y., & Yoo, C. (2013). The integrated mobile advertising model: The effects of technology-and emotion-based evaluations. *Journal of Business Research*, 66(9), 1345–1352. DOI: 10.1016/j.jbusres.2012.02.035

Yao, X., Zhou, J., Lin, Y., Li, Y., Yu, H., & Liu, Y. (2019). Smart manufacturing based on cyber-physical systems and beyond. *Journal of Intelligent Manufacturing*, 30(8), pp2805–pp2817. DOI: 10.1007/s10845-017-1384-5

Yin, S., Chen, S., & Ge, Y. (2024). Dynamic Associations Between Centers for Disease Control and Prevention Social Media Contents and Epidemic Measures During COVID-19: Infoveillance Study. *JMIR Infodemiology*, 4(1), e49756. Advance online publication. DOI: 10.2196/49756 PMID: 38261367

Yong, J. Y., Tong, E. M. W., & Liu, J. C. J. (2020). When the camera eats first: Exploring how meal-time cell phone photography affects eating behaviours. *Appetite*, 154(104787), 1–7. DOI: 10.1016/j.appet.2020.104787 PMID: 32579971

Yuan, S., & Chu, H. (2021). Vaccine for yourself, your community, or your country? Examining audiences' response to distance framing of COVID-19 vaccine messages. *Patient Education and Counseling*. https://www.ncbi.nlm.nih.gov/pmc/articles/PMC8384529/

Yuan, Y., & Medel, M. (2016). Characterizing international travel behavior from geotagged photos: A case study of Flickr. *PLoS One*, 11(5), e0154885. Advance online publication. DOI: 10.1371/journal.pone.0154885 PMID: 27159195

Yun-chao, M., Ren, M., & Li, X. (2023). Application of virtual simulation technology in sports decision training: A systematic review. *Frontiers in Psychology*, 14, 1164117. Advance online publication. DOI: 10.3389/fpsyg.2023.1164117 PMID: 37275736

Zampini, M., & Spence, C. (2004). The role of auditory cues in modulating the perceived crispness and staleness of potato chips. *Journal of Sensory Studies*, 19(5), 347–363. DOI: 10.1111/j.1745-459x.2004.080403.x

Zander, K. K., Garnett, S. T., Ogie, R., Alazab, M., & Nguyen, D. (2023). Trends in bushfire related tweets during the Australian 'Black Summer' of 2019/20. *Forest Ecology and Management*, 545, 121274. Advance online publication. DOI: 10.1016/j.foreco.2023.121274

Zeithaml, V. A., Verleye, K., Hatak, I., Koller, M., & Zauner, A. (2020). Three decades of customer value research: Paradigmatic roots and future research avenues. *Journal of Service Research*, 23(4), 409–432. DOI: 10.1177/1094670520948134

Zhang, Y., & Wang, J. (2021). "Through the Solar System" - XR science education system based on multiple monitors. 2021 IEEE Conference on Virtual Reality and 3D User Interfaces Abstracts and Workshops (VRW). DOI: 10.1109/VRW52623.2021.00096

Zhang, H. (2020, February). The application of cloud computing in government management. [). IOP Publishing.]. *IOP Conference Series. Materials Science and Engineering*, 750(1), 012166.

Zhang, L., Qu, J., Jie, J., & Liang, N. (2022). Big Data in the Internet Industry and Other Emerging Industries. In *China's Publishing Industry in the Era of Big Data* (pp. 27–45). Springer. DOI: 10.1007/978-981-19-0428-8_2

Zhang, M. L., & Chen, M. S. (2019). *China's digital economy: Opportunities and risks*. International Monetary Fund.

Zhang, X., Xu, X., Liu, A., Lu, Q., Xu, L., & Tao, F. (2019). Blockchain-Based Trust Mechanism for IoT-Based Smart Manufacturing System. *IEEE Transactions on Computational Social Systems*, 6(6), pp1386–pp1394. DOI: 10.1109/TCSS.2019.2918467

Zhou, B., Zou, L., Mostafavi, A., Lin, B., Yang, M., Gharaibeh, N., Cai, H., Abedin, J., & Mandal, D. (2022). VictimFinder: Harvesting rescue requests in disaster response from social media with BERT. *Computers, Environment and Urban Systems*, 95, 101824. Advance online publication. DOI: 10.1016/j.compenvurbsys.2022.101824

Zhu, C., Rodrigues, J. J. P. C., Leung, V. C. M., Shu, L., & Yang, L. T. (2018). Trust-Based Communication for the Industrial Internet of Things. *IEEE Communications Magazine*, 56(2), 16–22. DOI: 10.1109/MCOM.2018.1700592

Zou, L., Liao, D., Lam, N. S. N., Meyer, M. A., Gharaibeh, N. G., Cai, H., Zhou, B., & Li, D. (2023). Social media for emergency rescue: An analysis of rescue requests on Twitter during Hurricane Harvey. *International Journal of Disaster Risk Reduction*, 85, 103513. Advance online publication. DOI: 10.1016/j.ijdrr.2022.103513

Zreik, M. (2021, September 1). The Potential of a Sino-Lebanese Partnership through the Belt and Road Initiative (BRI). *Contemporary Arab Affairs*, 14(3), 125–145. DOI: 10.1525/caa.2021.14.3.125

Zreik, M. (2022). The Chinese presence in the Arab region: Lebanon at the heart of the Belt and Road Initiative. *International Journal of Business and Systems Research*, 16(5-6), 644–662. DOI: 10.1504/IJBSR.2022.125477

Zreik, M. (2023). Analytical study on Foreign Direct Investment Divestment Inflows and Outflows in Developing Economies: Evidence of China. *Chinese Economy*, 56(6), 415–430. DOI: 10.1080/10971475.2023.2193118

About the Contributors

Albérico Travassos Rosário, Ph.D. Marketing and Strategy, Universities of Aveiro (UA), Minho (UM) and Beira Interior (UBI). Post-doctorate - Economics, Management and Industrial Engineering and Tourism, University of Aveiro, with affiliation to the research center GOVCOPP of the University of Aveiro. Master in Marketing and Degree in Marketing, Advertising and Public Relations, degree from ISLA Lisbon Campus. He has the title of Specialist in Marketing and teaches as Assistant Professor with Aggregation at IADE-Faculty of Design, Technology and Communication of the European University and as Guest Assistant Professor with Aggregation at the School of Management and Technology of Santarém (ESGTS) of the Polytechnic Institute of Santarém. He taught at IPAM-Escola de Marketing, ISLA-Instituto Superior de Gestão e Administração de Santarém (ISLA-Santarém), was Director of the Commercial Management Course, Director of the Technical Professional Course (TeSP) of Sales and Commercial Management, Chairman of the Pedagogical Council and Member of the Technical Council and Scientific Researcher of ISLA-Santarém. Consultant on marketing and strategy for SMEs. His main research interests are: Competitive Strategy, Competitive Dynamics, Strategic Groups and Marketing.

Anna Carolina Boechat completed her PhD in Strategic Communication with the highest grade at University NOVA of Lisbon, and has worked as an Assistant Professor at prestigious Portuguese institutions such as IADE, University NOVA of Lisbon, Lusíada University of Lisbon, and the Portuguese Catholic University. She is also Marketing & Communication Scientific and Pedagogical Coordinator at IADE, Academic Supervisor, and participates in national and international scientific events as a speaker/conferencist. She develops Business Consulting projects in the areas of Marketing and Strategic Communication and is specialized in the following topics: Marketing, Strategic Marketing, Sports Marketing, Advertising, Communication, Strategic Communication, Project Management, Consumer Psychology, Market

Research, Brand Management, and Corporate Social Responsibility. As an Integrated Researcher at UNIDCOM/IADE, Anna Carolina Boechat has been developing projects and publications in the areas of Strategic Communication, Marketing, Brands, Consumer Behavior, Corporate Social Responsibility, Sustainability, and Metaverse. She has over 10 years of exclusive corporate experience in Marketing

***

Popoola Adeolu Gabriel teaches Language and Communication Education (General English) with twenty years of experience nurturing teachers-in-training.

Adebowale Adetayo is an academic staff of Adeleke University. His research interests are Library Science, Information Science, Knowledge Management, and Business Information Management. He has published many articles in reputable journals and currently working on projects relating to pandemics, vaccines and virtual learning. He is a graduate of Babcock University.

Munir Ahmad is a seasoned professional in the realm of Spatial Data Infrastructure (SDI), Geo-Information Productions, Information Systems, and Information Governance, boasting over 25 years of dedicated experience in the field. With a PhD in Computer Science, Dr. Ahmad's expertise spans Spatial Data Production, Management, Processing, Analysis, Visualization, and Quality Control. Throughout his career, Dr. Ahmad has been deeply involved in the development and deployment of SDI systems specially in the context of Pakistan, leveraging his proficiency in Spatial Database Design, Web, Mobile & Desktop GIS, and Geo Web Services Architecture. His contributions to Volunteered Geographic Information (VGI) and Open Source Geoportal & Metadata Portal have significantly enriched the geospatial community. As a trainer and researcher, Dr. Ahmad has authored over 50 publications, advancing the industry's knowledge base and fostering innovation in Geo-Tech, Data Governance, and Information Infrastructure, and Emerging Technologies. His commitment to Research and Development (R&D) is evident in his role as a dedicated educator and mentor in the field.

Sharon Andrews is Program Director and Professor of Software Engineering at the University of Houston-Clear Lake. She earned her Doctor of Philosophy in Computer Science from the Center for Advanced Computer Studies of the University of Louisiana in Lafayette, Louisiana. Dr. Andrews's areas of research surround the issues of software architecture, software reuse, architectural erosion and drift, domain and knowledge engineering, innovation, and creativity in engineering.

Kathirvel Ayyaswamy, acquired, B.E.(CSE), M.E. (CSE) from Crescent Engineering College affiliated to University of Madras and Ph. D (CSE.) from Anna University. He is currently working as Professor, Dept of Computer Science and Engineering, Karunya Institute of Technology and Sciences, Coimbatore. He is a studious researcher by himself, completed 18 sponsored research projects worth of Rs 103 lakhs and published more than 110 articles in journals and conferences. 4 research scholars have completed Ph. D and 3 under progress under his guidance. He is working as scientific and editorial board member of many journals. He has reviewed dozens of papers in many journals. He has author of 13 books. His research interests are protocol development for wireless ad hoc networks, security in ad hoc network, data communication and networks, mobile computing, wireless networks, WSN and DTN. He is a Life member of the ISTE (India), IACSIT (Singapore), Life Member IAENG (Hong Kong), Member ICST (Europe), IAES, etc. He has given a number of guest lecturers/expert talks and seminars, workshops and symposiums.

Miguel Boavida has a degree in Computer Engineering and a Master's in Computer Science from the Faculty of Science and Technology at Universidade Nova de Lisboa. He has been a lecturer and researcher at IADE - Universidade Europeia since 2016. In 2008, he co-founded SportTools - Technology for Sport, a pioneering company specializing in training support systems for various sports. His professional and academic work focuses on designing innovative information systems for sport and health, with particular expertise in neuropsychological rehabilitation and elite sports performance. His involvement in creating solutions for major sporting events sparked a profound interest in communication and the transformative potential of technology in this area. Outside academia and his professional career, Miguel pursues his passion for music and explores the use of emerging technologies to enhance interaction in art events.

Caroline M. Crawford, Ed.D., is a Professor of Instructional Design and Technology at the University of Houston-Clear Lake in Houston, Texas, United States of America. She earned her doctoral degree from the University of Houston in Houston, Texas, United States of America, in 1998, with specialization areas in Instructional Technology and Curriculum Theory, and began her tenure at the University of Houston-Clear Lake (UHCL) the same year. At this point in Dr. Crawford's professional career, her main areas of interest focus upon communities of learning and the appropriate and successful integration of technologies into the learning environment; the learning environment may be envisioned as face-to-face, blended and online (virtual or text-driven) environments, as well as microlearning and micro-assist deliverables.

João A. Dias holds a degree, a master's, and a Ph.D. in Computer Science from the University of Beira Interior. He is an associate professor at Universidade Europeia, where he acts as the scientific coordinator for the IADE technological area, the coordinator of the doctoral program in digital games development, and the bachelors in games development. His previous roles included coordinating the degree in Creative Technologies and the master's in Creative Computing and Artificial Intelligence. João is also a Member of the Pedagogic and Scientific Councils of the Universidade Europeia. He is a senior researcher at UNIDCOM and the author and co-author of about five dozen scientific articles published in international conferences and journals, some of which have been awarded prizes of scientific merit. Outside the academic context, he is the Chairman of the Board of Directors of inKlusion Entertainment, SA, of which he has been one of the co-founders and programmers since 2013.

Jacinto Estima is an assistant professor of information systems at the Department of Informatics Engineering at the University of Coimbra, Portugal. He is also a senior researcher at the Centre for Informatics and Systems of the University of Coimbra (CISUC). Jacinto holds a Ph.D. in information management with specialties in geographic information systems and an M.Sc. in science and geographic information systems from NOVA University of Lisbon. His research interests are related to the sociotechnical aspects of information systems, particularly the impact of emerging technologies such as artificial intelligence and machine learning on people and society. He also works on problems related to the general area of geographical information science, particularly the use of geographic data obtained through crowdsourcing initiatives to extract relevant information using machine learning automatically.

Muhammad Farrukh is an esteemed academician with a passion for mass communication, public relations, and advertising. He holds a Ph.D. degree in Mass Communication, which he successfully completed in 2022. With over 13 years of teaching experience, Dr. Farrukh has become a respected figure in the field of communication studies. Since 2016, Dr. Farrukh has been serving at The University of Lahore, where he has made significant contributions to the Department of Mass Communication. In addition to his regular teaching responsibilities, He also leads the Ph.D. course on Cross Cultural and International Communication, where he imparts his knowledge and expertise to doctoral students. His deep understanding of online identities, new media, public relations, and mass communication enables him to provide valuable insights to his students. Dr. Farrukh's research interests lie in exploring the impact of online identities on real life relarions, impact of social media, changing communication trends and practices, analyzing new media trends, and examining the evolving field of public relations. He has published several research papers in renowned academic journals and has presented his work at national and international journals and conferences. .

Bruno Frutuoso Costa is a PhD student in Communication Sciences and a postgraduate student in Data Analysis in Social Sciences at the University Institute of Lisbon (ISCTE-IUL). Bruno is a PhD research fellow of the Portuguese Foundation for Science and Technology (FCT - Fundação para a Ciência e a Tecnologia) at CIES-Iscte, where he is developing his funded project "Aversion2agony: A cross-national comparative analysis of journalistic mediation of euthanasia in Portugal and the United Kingdom" (Reference 2023.04877.BD). In 2021, Bruno conducted a pioneering study in Portugal about the phenomenon of digital and face-to-face violence against women journalists, which was published as a book by Media XXI. His main areas of research include journalism, media, and gender studies, focusing on topics such as public space and news production, participation and the media, digital platforms and sociability, gender-based violence and hate speech, as well as the mediation of death and mourning rituals.

Caroline Howard (PhD) is an award winning author, editor, consultant, and educator. Prior to joining Touro University Worldwide, she was on the faculty of Emory University's Goizueta Business School and the University of Colorado at Colorado Springs. Caroline holds an MBA from The Wharton School, University of Pennsylvania and a PhD in management information systems from the University of California - Irvine where she received honors for her teaching and research. She has published a number of articles on technology and learning. Her books include the first and 2nd editions of the Encyclopedia of Distance Learning (2005 and 2009), Winning the Net Game: Becoming Profitable Now that the Web Rules have Changed (Entrepreneur Press, 2002), The Design and Management of Effective Distance Learning Programs (2002), and Distance Learning and University Effectiveness: Changing Educational Paradigms for Online Learning (2005).

Ana Inácio is an Assistant Professor at ISG. Business & Economics School and Guest A.P. at Catholic University of Lisbon. Coordinator of the Degree of Tourism Management at ISG and an integrated researcher at the Centre of Geographical Studies of the University of Lisbon, since 2008. PhD in Geography from the Faculty of Arts of the University of Lisbon since 2009. She has a Master's degree in Communication, Culture and Information Technology from ISCTE and a Degree in Social and Cultural Communication - Tourism Variant from the Universidade Católica Portuguesa (1997). She also has a Bachelor degree in Tour Operator Management from ESHTE. She has been an Assistant Professor since

2009, teaching in higher education in the areas of Tourism, Communication and Heritage for over 25 years, in institutions where she held various Coordination, Direction and Management positions. Collaborates on Master's and Doctorate courses, having supervised both national and foreign theses. Participated in several funded research projects and other consultancy projects. Author of several scientific articles in national and international research journals, books, book chapters either individually or in co-authorship in the areas of Tourism, Wine Tourism, Tourism Communication and Marketing, Ethics and Social Responsibility and Health and Well-being Tourism.

Pratap Chandra Mandal is an Assistant Professor (Marketing) at Indian Institute of Management, Shillong, India. He has completed graduate degree from the reputed Indian Institute of Technology, Kharagpur (IIT Kharagpur), India (Bachelor of Technology in Mechanical Engineering), post-graduate degree from Vinod Gupta School of Management, IIT Kharagpur (Masters in Business Adminstration), PhD (Marketing) from Vinod Gupta School of Management, IIT Kharagpur. His research concerns customer relationship management, customer satisfaction, services marketing, marketing intelligence, and qualitative methods in management. He is the editor-in-chief of two international journals and is on the editorial board of journals like Journal of Global Marketing. Pratap has won several prestigious scholarships and awards throughout his academic career.

Janice Moore Newsum received a bachelor's degree from Rhodes College in Memphis, Tennessee. She earned a master's degree in early childhood education from the University of St. Thomas, Houston, Texas. Janice has an MLIS and her PhD in Library and Information Studies from Florida State University. Dr. Newsum served the students of the Houston Independent School District (HISD) in the Department of Library Services as a school librarian, a Library Services Specialist, and as Manager of the Department of Library Services. She is currently employed as an Associate Professor at the University of Houston Clear Lake. Dr. Newsum has served as the Representative at Large on the Texas Library Association (TLA) Executive Board, as Co-chair of the TLA Diversity and Inclusion Committee, as a member of the AASL/CAEP Coordinating Committee, and as a Councilor for the TLA Library Instruction Round Table. She is the Chair of the Coretta Scott King Book Awards Round Table, and a member of the Florida State University iSchool iadvocacy Board. Her research interests include school librarian leadership and administration, library advocacy, diversity in literature, and technology applications.

Marcel Odhiambo obtained a BSc degree in Electrical Engineering from university of Nairobi in 1980, MSc in Microprocessor Technology and Applications from Brighton Polytechnic, UK in 1990 and PhD degree in Parallel and Distributed Computer Architecture from University of Surrey, UK in 1998. He joined the department of Civil Aviation, Ministry of Transport & Communication as Trainee Engineer in 1980, arising through the ranks to become Planning Engineer in-charge of acquisition, installation and commissioning of Navigational Aids systems (NAVAIDS) in Kenya. He joined Moi University as a Lecturer in the department of Electrical and Communications Engineering in 1992, rising to the post of Senior Lecturer in 2001. He joined Tshwane University of Technology, South Africa in 2004 as a Postdoctoral Fellow. In 2006, he joined the department of Electrical & Mining Engineering, University of South Africa (UNISA), South Africa as Senior Lecturer. He was appointed head of head of the department of Electrical & Mining Engineering 2009 - 2011. He was appointed Associate Professor in the same department in January 2012. He was appointed an Associate Professor and Head: Department of Process Control & Computer Systems, Vaal University of Technology, South Africa since 1st December 2014. He retired from this post on 31st March 200. Prof. Odhiambo is currently Research Professor, Faculty Engineering, Mangosuthu University of Technology, South Africa and a Dean: Faculty of Engineering, Chuka University, Kenya wef 1st November 2023. Prof. Odhiambo is a Fellow of the Institution of Engineers and Technologists (IET), UK, Senior member of South African Institute of Electrical Engineers (SAIEE) and a member of the Institution of Engineers of Kenya (IEK). He is a Registered Engineer (REng): Engineers Registration Board of Kenya (ERBK), a Chartered Engineer (CEng): Engineering Council (UK) and a Professional Engineer (PrEng): Engineering Council of South Africa (ECSA). Prof. Odhiambo research interest is in Computer Systems Architecture, Intelligent Agents, Communication Systems Architecture, Electronic Systems and Devices, Process Control/Instrumentation, Internet of Things (IoT), Artificial Intelligence. Prof. Odhiambo has supervised 19 postgraduate students to completion and 75 Book/Book chapter/journal/conference publications.

Anne Pesavento is a driven educator and accomplished marketing professional with a deep commitment to fostering student engagement and achieving academic excellence. With extensive experience in research, communication, and educational support, Anne brings a meticulous and proactive approach to all professional endeavors, Anne holds a Doctor of Business Administration (DBA) degree from Touro University Worldwide, graduating in September 2022 with the highest honors, Summa Cum Laude. She previously earned her Master of Business Administration in Marketing (2020) and Bachelor of Business Administration in Marketing (2018) from the same institution, both Summa Cum Laude. Her foundational studies began with an Associate of Arts in Management from MiraCosta College in Oceanside,

CA. Anne is known for her expertise in learning management, research, scheduling, and task management. Her background in marketing, paired with her academic achievements and experience in client-centered consulting, underscores her dedication to meticulous planning, strategic thinking, and impactful communication.

Sabyasachi Pramanik is a professional IEEE member. He obtained a PhD in Computer Science and Engineering from Sri Satya Sai University of Technology and Medical Sciences, Bhopal, India. Presently, he is an Associate Professor, Department of Computer Science and Engineering, Haldia Institute of Technology, India. He has many publications in various reputed international conferences, journals, and book chapters (Indexed by SCIE, Scopus, ESCI, etc). He is doing research in the fields of Artificial Intelligence, Data Privacy, Cybersecurity, Network Security, and Machine Learning. He also serves on the editorial boards of several international journals. He is a reviewer of journal articles from IEEE, Springer, Elsevier, Inderscience, IET and IGI Global. He has reviewed many conference papers, has been a keynote speaker, session chair, and technical program committee member at many international conferences. He has authored a book on Wireless Sensor Network. He has edited 8 books from IGI Global, CRC Press, Springer and Wiley Publications.

Ramya R. Completed her B.E and M.E degree in Computer science and engineering. She had 14 years of teaching experience in various Engineering colleges. She completed her PhD in computer science and engineering and published many articles and book chapters in her domain. She specialises in edge computing, cloud computing, Machine learning, artificial intelligence, Quality of services and resource management.

S. Ramamoorthy is currently working as an Associate Professor at the SRM Institute of Science and Technology and completed his Ph.D. degree in big data and cloud computing with the SCSVM University at Kancheepuram, India. He earned his Ph.D. scholarship position as a talented student. Having 18 Years of Overall Teaching Experience and 10 Years of Research Experience in Computer Science and Engineering discipline. He has authored or co-authored over 40 publications in high- ranked journals like IEEE, Springer, Wiley, Taylor, and Francis and international conferences. He has more than 10 patents published and granted. His research interests are security and privacy of the Internet of Things, artificial intelligence, software-defined networking, cybersecurity, and blockchain. Received the Travel Grant from TNSCST to present a paper at an International Conference held in Malaysia. He acted as an Editorial board member and reviewer for many international research forums.

Rachel A. Sale, Ph. D., is the Dean of Digital Learning at South Texas College. Her experience in faculty professional development, enhancing digital and course design, and facilitation has earned her Quality Matters recognition as a Master Reviewer, Certified Online Instructor, and Peer Reviewer Course facilitator. With over 25 years of instructional design leadership, she guides faculty to use exceptional communication and course design as a foundation of student success. Rachel began her career teaching music and art. After 16 years in the K-12 world, including creating professional development for instructors to a new tool called the Internet and curriculum for early-use students in the early 2000s, she moved to the higher ed learning environment. She supervised instructional design teams since 2008,and has developed courses and programs for several institutions across the United States.

Subir Sinha is an Indian scholar and a Lecturer (State Aided College Teacher) in the Department of Journalism and Mass communication of Dum Dum Motijheel College (Affiliated under West Bengal State University). He was formerly attached with the Department of Journalism of Surendranath College (Affiliated under University of Calcutta) and Vijaygarh Jyotish Ray College (Affiliated under University of Calcutta)as a Guest Faculty . He was also remained attached with Indra Gandhi National Open University (IGNOU) under convergence scheme with Dum Dum Motijheel College as an Academic Counsellor of PGJMC for a short period.

Patricia Tavares has a PhD in Management from ISEG - Lisbon School of Economics and Management (University of Lisbon), she is an Assistant Professor at the Faculty of Human Sciences of the Universidade Católica Portuguesa. She teaches courses in the Bachelor's degree in Social and Cultural Communication, including Marketing, Marketing and Advertising (in English), Digital Marketing, and Integrated Marketing Communication (in English). In the Master's in Communication Sciences, she teaches Research Methods (in Portuguese and English), and in the Master's in Asian Studies, she teaches Strategies of Communication and Consumers' Behaviour in Asian Markets (in English). She has served as a Guest Assistant Professor in the Department of Management at ISEG - Lisbon School of Economics and Management, University of Lisbon, in the scientific area of Strategy and Marketing. From 2010 to 2019, she co-coordinated the Postgraduate Program in Marketing Management, Communication, and Multimedia. Concurrently, she has worked in communication consultancy, focusing on crisis communication and organizational reputation management. Her professional experience began in 1995 as a trainee journalist at the newspaper Semanário, where she worked until 1997. From 1997 to 2002, she was a Director at Burson-Marsteller, a multinational communication consultancy. Her experience in national and international projects across various sectors includes developing strategic communication plans, media relations, media training, product communication, event management, crisis training, corporate social responsibility, institutional communication, and reputation management. From 2004 to 2010, she was an Assistant at UAL – Universidade Autónoma de Lisboa, where she taught Communication Psychosociology, Public and Political Communication, Organizational Consulting, and Organizational Change and Development in the Psychology degree. She holds a Bachelor's degree in Communication Sciences from Universidade Nova de Lisboa, a postgraduate degree in Marketing Management from ISEG - Lisbon School of Economics and Management (University of Lisbon), and a Master's in Marketing, also from ISEG. Her research interests include Integrated Marketing Communication, Organizational Reputation, Stakeholder Theory, Individual and Organizational Values and Reputation, Marketing Ethics, Internal Communication, Sustainability Communication, and Responsible and Sustainable Marketing.

Balraj Verma works as an Assistant Professor at Chitkara Business School-Doctoral Research Centre with Chitkara University, Rajpura, Punjab. He had his PhD from Jaypee University of Information Technology (JUIT), Waknaghat and holds a master's degree in Business Administration and has 16 years of academic and corporate experience. He has qualified National eligibility test (NET) for teaching in the management discipline. He had been teaching courses like Marketing Management, Strategic Management, Business Statistics and Research Methodology. He believes in quality research and has published many papers in ABDC listed/ Unpaid Scopus Index journals. He has authored/edited books and many book chapters so far published by reputed major presses. He has been actively involved at different levels in organizing Workshops/ Conferences by the Department/ University.

H M Atif Wafik is a multifaceted professional and educator with a global perspective. Being a proud Bangladeshi citizen, Atif holds degrees from the University of Asia Pacific and Pittsburgh State University. He is a Senior Assistant Professor and Addl. Director of Brand & Communication at the University of Scholars. Atif wrote “Impress Your Audience” as an accomplished author to enhance his public speaking skills. As a TV host, his shows like “Live with a Purpose” inspire viewers to rethink their approach to life. His diverse experience spans advertising and the food industry in Bangladesh and the U.S. A passionate researcher, Atif focuses on Education, Innovation in Learning, and GenAI. He is deeply committed to his family and driven by his favourite quote, “Teach to Learn and Learn to Teach,” reflecting his belief in the transformative power of education.

Mohamad Zreik, a Postdoctoral Fellow at Sun Yat-sen University, is a recognized scholar in International Relations. His recent work in soft power diplomacy compares China's methods in the Middle East and East Asia. His extensive knowledge spans Middle Eastern Studies, China-Arab relations, East Asian and Asian Affairs, Eurasian geopolitics, and Political Economy, providing him a unique viewpoint in his field. Dr. Zreik is a proud recipient of a PhD from Central China Normal University (Wuhan). He's written numerous acclaimed papers, many focusing on China's Belt and Road Initiative and its Arab-region impact. His groundbreaking research has established him as a leading expert in his field. Presently, he furthers his research on China's soft power diplomacy tactics at Sun Yat-sen University. His significant contributions make him a crucial figure in understanding contemporary international relations.

Index

Symbols

A

B

C

D

E

F

G

H

I

www.ingramcontent.com/pod-product-compliance
Lightning Source LLC
LaVergne TN
LVHW080614300926
849162LV00005B/1155